FUNDAMENTALS OF ENHANCED OIL AND GAS RECOVERY FROM CONVENTIONAL AND UNCONVENTIONAL RESERVOIRS

FUNDAMENTALS OF ENHANCED OIL AND GAS RECOVERY FROM CONVENTIONAL AND UNCONVENTIONAL RESERVOIRS

ALIREZA BAHADORI PhD, CEng
School of Environment, Science and Engineering, Southern Cross University, Lismore, NSW, Australia
Australian Oil and Gas Services, Pty. Ltd, Lismore, NSW, Australia

Gulf Professional Publishing is an imprint of Elsevier
50 Hampshire Street, 5th Floor, Cambridge, MA 02139, United States
The Boulevard, Langford Lane, Kidlington, Oxford, OX5 1GB, United Kingdom

British Library Cataloguing-in-Publication Data
A catalogue record for this book is available from the British Library

Library of Congress Cataloging-in-Publication Data
A catalog record for this book is available from the Library of Congress

ISBN: 978-0-12-813027-8

For Information on all Gulf Professional Publishing publications
visit our website at https://www.elsevier.com/books-and-journals

Publishing Director: Joe Hayton
Senior Acquisition Editor: Katie Hammon
Editorial Project Manager: Serena Castelnovo/Joshua Mearns
Production Project Manager: Sruthi Satheesh
Cover Designer: Miles Hitchen

Typeset by MPS Limited, Chennai, India

CONTENTS

LIST OF CONTRIBUTORS

Mohammad Ali Ahmadi
Department of Chemical and Petroleum Engineering, University of Calgary, Calgary, AB, Canada

Hamed Akhondzadeh
School of Engineering, Edith Cowan University, Perth, WA, Australia

Amirhossein Mohammadi Alamooti
Department of Chemical and Petroleum Engineering, Sharif University of Technology, Tehran, Iran

Ali Alashkar
School of Chemical Engineering, Iran University of Science and Technology, Tehran, Iran

Forough Ameli
School of Chemical Engineering, Iran University of Science and Technology, Tehran, Iran

Pouria Behnoudfar
Department of Petroleum Engineering, Amirkabir University of Technology, Tehran, Iran

Abdolhossein Hemmati-Sarapardeh
Department of Petroleum Engineering, Shahid Bahonar University of Kerman, Kerman, Iran

Alireza Keshavarz
School of Engineering, Edith Cowan University, Perth, WA, Australia

Ehsan Mahdavi
Department of Chemical and Petroleum Engineering, Sharif University of Technology, Tehran, Iran

Farzan Karimi Malekabadi
Department of Chemical and Petroleum Engineering, Sharif University of Technology, Tehran, Iran

Ramin Moghadasi
Department of Petroleum Engineering, Petroleum University of Technology (PUT), Ahwaz, Iran

Alireza Rostami
Department of Petroleum Engineering, Petroleum University of Technology (PUT), Ahwaz, Iran

Mohammad Sayyafzadeh
Australian School of Petroleum, The University of Adelaide, Adelaide, SA, Australia

Afshin Tatar
Islamic Azad University, Tehran, Iran

Masoumeh Zargar
School of Engineering, Edith Cowan University, Perth, WA, Australia

Fatemeh Sadat Zebarjad
Mork Family Department of Chemical Engineering and Material Science, University of Southern California, Los Angeles, CA, United States

BIOGRAPHY

Alireza Bahadori, PhD, CEng, MIChemE, CPEng, MIEAust, RPEQ, NER is a research staff member in the School of Environment, Science and Engineering at Southern Cross University, Australia, and managing director and CEO of Australian Oil and Gas Services, Pty. Ltd. He received his PhD from Curtin University, Perth, Western Australia. During the past 20 years, Dr. Bahadori has held various processes and petroleum engineering positions and was involved in many large-scale oil and gas projects. His multiple books have been published by multiple major publishers, including Elsevier. He is Chartered Engineer (CEng) and Chartered Member of Institution of Chemical Engineers (MIChemE), London, United Kingdom. He is Chartered Professional Engineer (CPEng) and Chartered Member of Institution of Engineers Australia, Australia, Registered Professional Engineer of Queensland (RPEQ), Australia, Registered Chartered Engineer of Engineering Council of United Kingdom, United Kingdom, and Engineers Australia's National Engineering Register (NER), Australia.

CHAPTER ONE

An Introduction to Enhanced Oil Recovery

Amirhossein Mohammadi Alamooti and Farzan Karimi Malekabadi
Department of Chemical and Petroleum Engineering, Sharif University of Technology, Tehran, Iran

1.1 OVERVIEW

In the early stages of oil field development, reservoirs are mainly planned to produce oil naturally by intrinsic energy. The recoverable oil by the natural forces, including all various mechanisms (gas cap drive, water drive, solution gas drive, rock and fluid expansion, and gravity drainage), can be extracted up to 50% of original oil in place (OIP) (averagely 19%), and most of the oil will remain untouched in the reservoir. For extracting more oil, other methods are chronologically utilized after the first natural flow mechanisms. Thus the first and second actions for enhancing oil recovery (EOR) after primary recovery are called secondary and tertiary recovery, respectively.

In the secondary recovery period most focuses are on the reservoir energy maintenance. This aim is performed by waterflooding or gas injection. In gas injection, gas is injected to the gas cap to prepare the required energy of oil drive. The process of gas injection to the gas cap is not as effective as waterflooding. This fact and the vast usage of waterflooding as the most common reservoir energy-saving method have made many references consider waterflooding equivalent to the secondary recovery method.

Tertiary recovery processes include all methods conducted to extract irrecoverable oil by the two first production stages. Also it should be noted that many reservoirs, such as high viscous oil reservoirs or very tight reservoirs, are not capable of producing oil without the tertiary action. Thus in many cases the chronological-based classification of EOR methods fails, especially when the oil is not producible by natural forces or energizing methods. Therefore considering the tertiary actions as exclusive EOR methods is not out of mind. Almost all procedures classified in this category can be categorized to thermal, chemical, microbial, miscible, and immiscible gas injection actions. The mechanism of increasing the oil recovery varies along these methods.

Fundamentals of Enhanced Oil and Gas Recovery from Conventional and Unconventional Reservoirs.
DOI: https://doi.org/10.1016/B978-0-12-813027-8.00001-1

Surface tension reduction, oil swelling, relative permeability improvement, wettability alteration, etc. are different mechanisms that can work together simultaneously or separately.

Today, using progressive EOR processes is unavoidable, particularly in reservoirs that are in their second half of life or cannot produce oil naturally. Achievement and prosperity of EOR methods definitely depend on comprehensive studying of rock and fluid properties and reservoir condition. These features describing reservoirs are the main controlling parameters on which an appropriate EOR method is selected.

1.2 RESERVOIR ROCK PROPERTIES

A reservoir rock is a rock providing a condition to trap oil in porous media. The reservoir rock contains pores and throats, creating flow path and an accumulating system for hydrocarbon and also consist of a sealing mechanism for prohibiting hydrocarbon penetration to surface layers. The reservoir rock appears in different forms, from loose sands to dense and tight rocks. Reservoir rocks are totally classified as conventional and unconventional rocks. In the case of conventional type, the rock consists of grains bound together by a bunch of material such silica, calcite, and clay. These rocks provide appropriate storativity and conductivity for accumulating and flowing hydrocarbon. To evaluate and understand reservoir behavior and also improvement of reservoir performance, studying reservoir rock properties is vital.

Most reservoir rock properties are determined by lab-based works. In order to perform experimental tests, the reservoir rock should be sampled. The special sample of reservoir rocks is called the core. The lengths of cores are varied, from a few inches in core plugs to several meters in whole cores. For the following experimental tests these cores are maintained under reservoir condition (temperature, pressure), otherwise the cores are aged to reservoir condition.

Rock properties analysis is mainly subcategorized to advanced core analysis or special core analysis (SCAL) and routine core analysis (RCAL). In SCAL all saturation-dependent or multiphase flow properties including relative permeability, capillary pressure, compressibility, and wettability are determined, and other parameters such as porosity, permeability, saturation and lithology are characterized by RCAL. The abovementioned properties significantly influence hydrocarbon distribution along reservoirs, thus a comprehensive analysis of reservoir rock properties is definitely essential, especially in the case of EOR methods selection.

1.3 POROSITY

The grains and particles forming rocks make irregular shapes in the internal structure of rock. This leads to the creation of void space in rocks, called pore space. The porosity is the fraction of pore volume over the bulk volume occupied by fluid. In other words, porosity is the ability of rock to storage and hold fluid. Mathematically porosity is defined as follows:

$$\varnothing = \frac{pore\ volume}{bulk\ volume}$$

where $\varnothing$ is porosity.

During sedimentation some pores are isolated from the interconnected pore network. The oil trapped in isolated pores is inaccessible and remains in rock during the flowing period. Thus an effective porosity is defined as the fraction of interconnected pore volumes over the total bulk volume.

Porosity is determined by laboratory measurements and well logs like sonic, neutron, and density logs. The laboratory measurements are done by both liquid and gas phases separately. Methods of calculation are based on basic physical laws including buoyancy law or Boyle's law.

1.4 SATURATION

All pores in porous media are filled with different fluids. The portion of each fluid volume over the total pore volume is called saturation. Therefore the sum of all saturation is 100%.

The fluids existing in reservoirs are under equilibrium condition. These fluids are distributed along reservoirs according to present forces including gravity, capillary, and viscous forces. Gas is placed on the top, oil is in the middle, and water is on the bottom of the reservoir. The water trapped in reservoirs after oil migration is called connate water. The portion of water remaining in rock that cannot be reduced by oil flooding is called irreducible water saturation.

For the oil phase there are some similar expressions. The saturation of oil in which the oil becomes moveable is called critical oil saturation. Another term is residual oil saturation, which is the oil saturation after wet-phase flooding. The residual oil saturation is more than critical oil saturation and can be reduced by side works such as EOR.

1.5 PERMEABILITY

The rock's potential to conduct single phase fluid is called permeability, measured in Darcy. Ranges of permeability vary from 0.1 in tight limestone to more than 1000 md in loose sands. Higher permeability values allow fluids to flow fast through porous media. The term is independent of fluid type. This concept was founded by Henry Darcy. The following equation is recognized as Darcy's equation, showing the momentum equation in porous media:

$$q = \frac{kA\Delta P}{\mu \Delta x}$$

where q is volumetric rate, k is permeability, A is surface area, ΔP is differential pressure along the media, μ is fluid viscosity, and Δx is the length of media. In SI unit, unit of permeability is m^2 equaled to 1.013E + 12 Darcy. The mentioned equation is applicable under linear, laminar, steady state condition, and exclusively for incompressible fluid and homogenous media. For multiphase flow the relative permeability function is applicable. Also for turbulent flow, a parameter is considered for adjusting Darcy equation. This parameter should be considered in gas wells with high velocity.

1.6 WETTABILITY

When two immiscible fluids are in contact with solid surface, a contact angle is created between them, showing the tendency of fluids to spread on solid surface. The tendency of fluids to adhere to the solid surface is called wettability, and the fluid tending to have maximum contact surface with solid is called wetting fluid. Wettability is one of the main forces in reservoirs determining the fluid distribution in porous media. The contact angle between two immiscible liquids is an index for the degree of wettability. A zero contact angle shows the completely wetting phase, and a degree of 180 illustrates the completely nonwetting phase. As the wetting fluids tend to spread on solid surface, the small pores in porous media are occupied by wetting fluids, and the large pore throats are filled by nonwetting fluids. Therefore in water-wet porous media, water adheres to small pores and oil flows in open channels. This distribution occurs because of attractive forces between wetting phase and solid surface and repulsive forces between nonwetting phase and solid surface.

The forces at the boundary of oil—water can be drawn as oil—solid interfacial energy, water—solid interfacial energy, and oil—water interfacial tension, which are shown in Fig. 1.1.

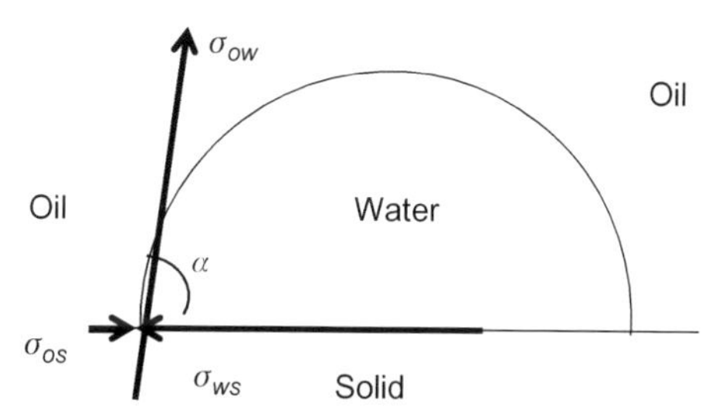

Figure 1.1 Interfacial energy distribution.

where σ_{os} is the interfacial energy between oil and solid, σ_{ws} is interfacial energy between water and solid, and σ_{ow} is interfacial tension between oil and water. For this system under static equilibrium condition, it can be written:

$$\sigma_{ws} - \sigma_{os} = \sigma_{ow}\cos\alpha$$

The right side of the equation for water-wet system is positive, for oil-wet system it is negative, and for neutral-wet it is zero.

Depending on reservoir condition different degrees of wettability are desirable. Therefore wettability alteration can be drawn as EOR method to improve oil movement in reservoir. This aim is done by chemical treatment. Such methods such as surfactant flooding, alkaline flooding, low salinity water injection, and mixed methods like alkaline-surfactant-polymer (ASP) flooding can be used for chemical treatment.

1.7 CAPILLARY PRESSURE

The differential pressure between two immiscible fluids that are in contact is called capillary pressure. In other words, the differential pressure between the nonwet phase and wet phase pressures is as follows:

$$P_c = P_{nw} - P_w$$

where P_c is capillary pressure, P_{nw} is nonwet phase pressure, and P_w is wet phase pressure.

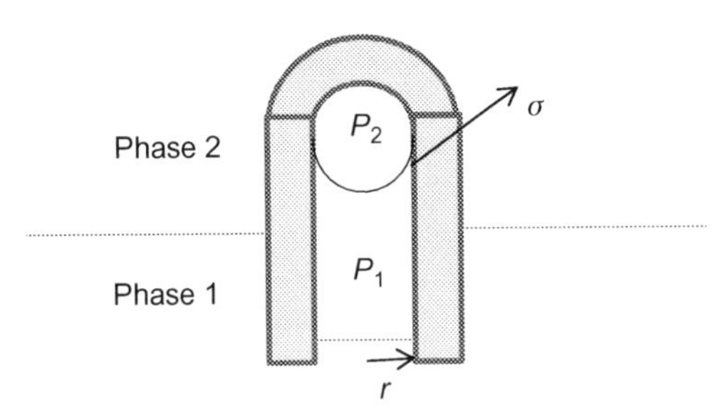

Figure 1.2 Capillary tube force balance.

For a two-phase system when a capillary tube is set below the interface, a force balance can be written as follows (Fig. 1.2):

$$2\pi r\sigma\cos\theta = \pi r^2(P_2 - P_1)$$

$$P_c = P_2 - P_1 = \frac{2\sigma\cos\theta}{r}$$

According to wettability of capillary tube, liquid rises up or depression occurs. If the liquid below the interface is wet, liquid comes, concave is upward, and the degree between the two fluids is less than 90; otherwise, liquid comes down below the interface and has downward concave.

When the saturation of the wetting phase increases in porous media, the mobility of wetting phase increases, like water flooding in a water-wet reservoir. This process is called imbibition. The reverse process "increasing the saturation of nonwetting phase" is called drainage.

During drainage process the nonwet phase firstly invades the largest pores and then occupies the smaller ones. The required pressure to enter the largest pore is called threshold pressure. By inserting more pressure, the saturation of the nonwet phase increases until the saturation of the wetting phase cannot be reduced. The schematic curve of this process is illustrated in Fig. 1.3.

The capillary pressure curve for imbibition is not the same as drainage and passes from a path below drainage curve. This effect is called hysteresis and should be considered during different drive mechanisms in reservoirs.

1.8 RELATIVE PERMEABILITY

The concept of permeability is defined for a single-phase system and is only a function of rock properties. This permeability is known as absolute permeability.

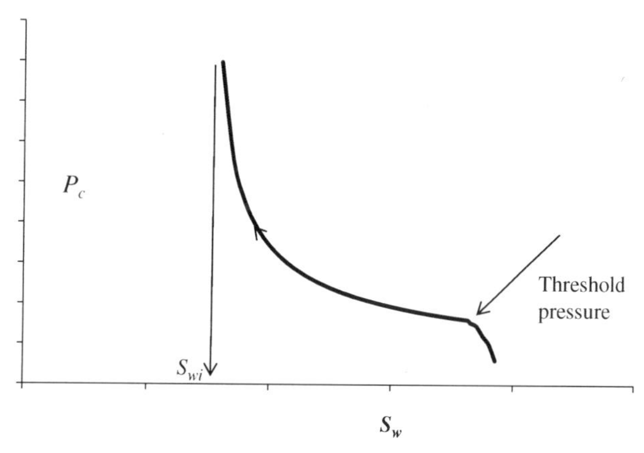

Figure 1.3 Capillary pressure curve for imbibition process.

Commonly, reservoirs contain two or three fluids. Consequently, this concept is modified to an improved saturation-based function, "effective permeability." Effective permeability is the potential of a porous medium to be saturated with other fluids. Also relative permeability is defined as the ratio of effective permeability to absolute permeability. Relative permeability shows the ability of a system for flowing fluid in the presence of other fluids. The range of this dimensionless parameter is between 0 and 1. The value and curvature of the relative permeability definitely depends on wettability of the rock. Maximum water relative permeability for a strongly water-wet system will not exceed more than 0.2. Also the cross point of oil and water relative permeability curves occurs in water saturation more than 0.5. Another important parameter influencing relative permeability is saturation history. Relative permeability functions are strongly sensitive to hysteresis effect. In other words, like capillary pressure function, relative permeability is different for both drainage and imbibition processes.

1.9 RESERVOIR FLUID PROPERTIES

Hydrocarbons accumulated in reservoirs treating multiphase consist of complex mixtures. The range of pressure and temperature varies largely in the petroleum industry. The differences in mixture composition, pressure, and temperature lead to the formation of different reservoir types. To predict phase behavior, different experiments are carried out on reservoir fluids. Consequently, for modeling these experiments and fluid phase behavior many equations of states are developed. Deep cognition of phase behavior of reservoir hydrocarbon is vital during the first production period and consequently in enhanced production design.

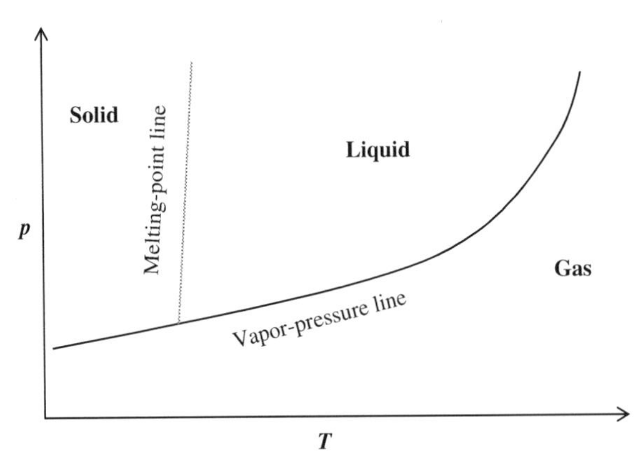

Figure 1.4 Phase diagram schematic.

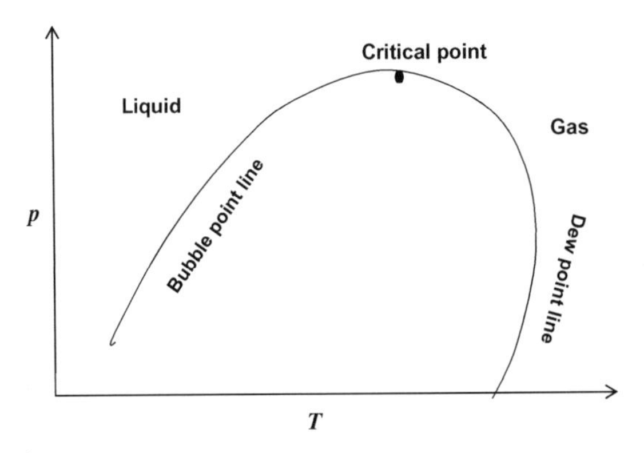

Figure 1.5 Phase envelope.

1.9.1 Hydrocarbon Phase Behavior

A phase diagram for a single component system is a pressure–temperature graph on which different states of the pure substance is shown. Fig. 1.4 illustrates the phase diagram schematic for a pure substance.

The vapor-pressure line separates gas and liquid phases, so the pressure–temperature condition below this line substance is gaseous and the pressure–temperature condition above the vapor-pressure line is liquid. In the case of a multicomponent system, the phase diagram is more complicated. In other words, it is represented by a broad region instead of a single line. In the region called "phase envelope," two phases exist simultaneously.

Fig. 1.5 illustrates the phase envelope for a multicomponent system. The curve is formed by connection of bubble point and dew point lines. The intersection of these two lines is critical point.

1.9.2 Classification of Reservoir Based on Reservoir Fluid

Generally, reservoirs are classified into oil and gas reservoirs from the reservoir fluid point of view. Based on the phase diagram plot and the point representing reservoir condition (reservoir pressure–temperature), the following subdivisions are drawn for oil and gas reservoirs:

Oil reservoirs:

- Ordinary black oil
- Volatile crude oil
- Near-critical crude oil

Gas reservoirs:

- Retrograde gas reservoirs
- Wet gas reservoirs
- Dry gas reservoirs

1.9.3 Natural Gas Properties

Gas is defined as a light fluid with low viscosity and density having significant compressibility. Natural gas is composed of hydrocarbon and nonhydrocarbon compounds. Hydrocarbons composing gas fluid are chiefly the lightest, including methane, ethane, propane, butane, pentane, and a small amount of heavier components. Nitrogen, carbon dioxide, and hydrogen sulfide are the most nonhydrocarbon compounds found in natural gas. In order to simulate reservoirs and investigate different scenarios, modeling reservoir fluid behavior is essential. In other words, pressure–volume–temperature study and physical properties for reservoir fluids should be carried out. The required physical properties in gas reservoirs for solving problems are: apparent molecular weight, density, specific gravity, compressibility factor, gas formation volume factor, and gas viscosity.

1.9.3.1 Apparent Molecular Weight

Apparent molecular weight for gases is mathematically defined as follows:

$$MW_a = \sum y_i MW_i$$

where MW_a is apparent molecular weight of mixture, y_i is mole fraction of component I, and MW_i is molecular weight of component i.

1.9.3.2 Density

In order to analyze fluid phase behavior, studying the relationship between pressure, volume, and temperature is necessary. For ideal gases the mathematical relationship is expressed as:

$$PV = nRT$$

where P is pressure, V is volume, R is universal gas constant, and T is temperature. The term n is numbers of moles of gas. On the other hand, molecular weight is defined as the weight of the gas divided by numbers of moles:

$$MW = \frac{m}{n}$$

So the pressure–volume–temperature relationship for ideal gases can be rewritten as:

$$PV = \frac{m}{MW} RT$$

Density is defined as the ratio of mass to the corresponding volume:

$$\rho = \frac{m}{V}$$

So,

$$\rho = \frac{P \cdot MW}{RT}$$

where ρ is density, m is mass, V is volume, R is universal gas constant, and T is temperature.

1.9.3.3 Specific Gravity

The term specific gravity is expressed as the gas density to the air density under the same condition (pressure and temperature).

$$\gamma = \frac{\rho_{gas}}{\rho_{air}}$$

where γ is specific gravity, ρ_{gas} is gas density, and ρ_{air} is air density.

In the case of standard conditions, the behavior of gases is close to ideal gases; therefore the specific gravity can be rewritten as:

$$\gamma = \frac{(P_{sc} \cdot MW_{gas})/(RT_{sc})}{(P_{sc} \cdot MW_{air})/(RT_{sc})}$$

$$\gamma = \frac{MW_{gas}}{MW_{air}} = \frac{MW_{gas}}{28.96}$$

1.9.4 Compressibility Factor

The ideal gas relationship can be used for real gases at very low pressure. Evaluation of the relationship proves it works under low pressure conditions with error less than 2–3%. In spite of that, for high pressure, especially the pressure usual in the petroleum industry, this relationship cannot be used anymore. Volumes of molecules, the repulsive and attractive forces between molecules, are the parameters ignored in an ideal gas relationship. Therefore a modification factor should be considered in the basic equation to improve the prediction of fluid behavior. Gas compressibility factor, or z-factor, is the term introduced to the basic ideal gas relationship:

$$PV = ZnRT$$

where the gas compressibility factor z is defined as the actual volume of n-moles of gas at T, and p divided by the ideal volume of the same number of moles at the same condition.

1.9.5 Gas Formation Volume Factor

The volume of gas in reservoir condition is definitely less than the volume in surface condition. The term gas formation volume factor is used to relate the volumes in reservoir and standard condition. This term is defined as the ratio of volume of certain amount of gas in reservoir condition to the same amount of gas in standard condition, as follows:

$$B_g = \frac{V_{g_{reservoir}}}{V_{g_{S.C}}}$$

where

B_g is gas formation volume factor
$V_{g_{reservoir}}$ is volume of gas in reservoir condition
$V_{g_{S.C}}$ is volume of gas in standard condition

1.9.6 Gas Viscosity

The viscosity of a fluid represents the resistance to flow. In the case of low viscosity, by inserting a certain shearing force, the fluid starts to flow and a sharp velocity gradient is created between fluid layers. Viscosity is defined as the ratio of shear stress to the velocity gradient.

Experimental methods are not common for gas viscosity measurement. Many empirical correlations are developed for gas viscosity measurements, which are mostly the function of pressure, temperature, and gas composition.

1.9.7 Crude Oil Properties

A complex mixture of hydrocarbonic and nonhydrocarbonic components that are liquid under reservoir condition is called crude oil. The physical and chemical properties of crude oil are mainly obtained by experimental approaches. These properties are strongly dependent on the reservoir conditions and fluid composition. In the absence of experimental methods, empirical correlations are used to estimate the crude oil properties. Crude oil gravity, specific gravity of the oil, solution gas, gas solubility, bubble-point pressure, oil formation volume factor, crude oil viscosity, and surface tension are the main crude oil properties.

1.9.8 Crude Oil-Specific Gravity

Crude oil-specific gravity is defined as the ratio of the oil density to the water density at ambient pressure and temperature.

$$\gamma_o = \frac{\rho_o}{\rho_w}$$

One of the most practical parameters in the oil industry is API gravity. This parameter, which is measured in the field and laboratory, is calculated as follows:

$$API = \frac{141.5}{\gamma_o} - 131.5$$

1.9.9 Solution Gas Ratio

The amount of standard cubic feet of gas dissolved in one stock tank barrel of oil at a certain pressure and temperature is called solution gas ratio, R_s. This parameter depends on gas gravity, API gravity, temperature, and especially pressure. For certain undersaturated oil reservoirs, by reducing the pressure, the solution gas ratio remains constant. When the pressure drops below the saturation pressure, the gas solubility decreases by pressure decline. Fig. 1.6 shows a typical curve for the solution gas ratio as a function of pressure.

1.9.10 Bubble Point Pressure

Bubble point pressure is defined as the pressure at which the first gas bubble appears in crude oil. In other words, above the bubble point pressure the crude oil is

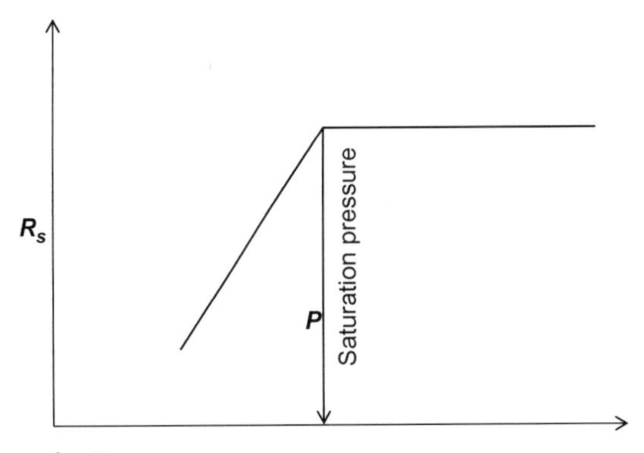

Figure 1.6 Solution gas oil ratio.

a single-phase liquid, and below the bubble point oil and gas are in equilibrium. Although this parameter is mainly measured by experimental methods, many empirical correlations are developed to be used in the absence of experimental instruments.

1.9.11 Oil Formation Volume Factor

The oil formation factor is equal to the volume of the oil under reservoir condition divided by the volume of the oil at standard condition.

$$B_o = \frac{V_{o_{reservoir}}}{V_{o_{S.C}}}$$

where

B_o is oil formation volume factor

$V_{o_{reservoir}}$ is volume of oil in reservoir condition

$V_{o_{S.C}}$ is volume of oil in standard condition

Above the bubble point pressure the oil formation volume factor increases by pressure drop. This is due to oil expansion in reservoir by pressure drop. On the contrary, below the bubble point pressure oil formation volume factor decreases by pressure decline. The reason is that below the bubble point pressure when the pressure drops, gas is liberated from oil, and consequently, the volume of oil in reservoir decreases.

1.9.12 Crude Oil Viscosity

Crude oil viscosity is one of the effective properties in porous media controlling the fluid flow. The viscosity definition for crude oil is similar to gas viscosity, resistance to flow. This parameter depends on oil composition, gas solubility, temperature, and pressure. Fig. 1.7 shows typical behavior of oil viscosity along pressure drop.

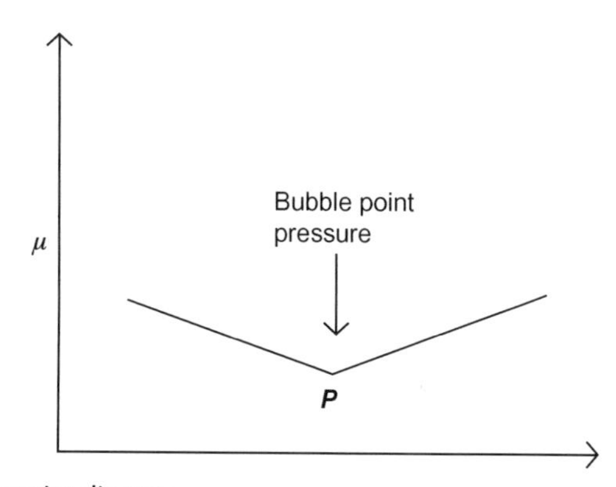

Figure 1.7 Crude oil viscosity diagram.

1.9.13 Surface Tension

Differential molecular forces make an imbalance in the boundary layer of liquid phase and vapor phase. The force exerted on the interface of oil and gas is called surface tension. For a pure substance the surface tension decreases by increasing temperature.

For hydrocarbon liquid and gas at equilibrium the surface tension is calculated as follows:

$$\sigma = \left[\sum_i p_i \left(x_i \frac{\rho_l}{M_l} - y_i \frac{\rho_g}{M_g}\right)\right]^4$$

where σ is surface tension; p is parachor, which is constant for a pure substance; x is molar percentage of liquid components; ρ_l is liquid density; M_l is apparent liquid molecular weight; y is molar percentage of gas components; ρ_g is gas density; and M_g is apparent gas molecular weight.

When the surface tension approaches zero, gas tends to be solved in oil. In other words, chance of miscibility increases by surface tension reduction.

1.10 RESERVOIR DRIVE MECHANISMS

A lot of forces, including capillary, viscous, and gravity forces, influence reservoir fluid in the porous media. Along the reservoir formation time these forces are equalized through the reservoir. The balance of these forces determines how the fluids flow toward the well. In other words, these forces as well as overburden pressure and reservoir temperature are forming the reservoir energy, and whenever the energy is not exhausted, the reservoir can produce oil naturally. The production period in which reservoir energy is enough to produce oil naturally is called primary recovery. During the primary period oil can be produced naturally by the primary reservoir energy, which can be categorized in different drive mechanisms as follows: rock and liquid expansion drive, solution gas drive, gas cap drive, water drive, gravity drainage drive, and combination drive.

1.10.1 Rock and Liquid Expansion

This mechanism is dominant when the reservoir is undersaturated. When the pressure decreases above the bubble point pressure, the rock and fluid expand. By decreasing the liquid pressure formation, compaction occurs and the pore volume is willing to be reduced. On the other hand, the liquid and individual rock grains tend to expand. Consequently, the crude oil will come out of pores and conduct to the well.

According to the low compressibility of oil and rock reservoir, the pressure drop is comparatively rapid, and this leads to the minimum efficiency in this mechanism in comparison to the rest of drive mechanisms.

1.10.2 Solution Gas Drive

This mechanism is active in a saturated oil reservoir, where the reservoir pressure is below bubble point pressure. As the pressure decreases, the gas bubbles are liberated from the oil and provide the needed force for oil production. The liberated gas supports reservoir pressure by gas bubble expansion and may help oil movement by oil viscosity reduction. Lack of external fluid drive such as gas cap or water influx leads to a relatively high reservoir pressure drop in this drive mechanism. As the reservoir pressure drops, the liberated gas may move vertically and create a secondary gas cap. This phenomenon definitely depends on the vertical permeability. Forming the secondary gas cap may significantly reduce reservoir pressure. The ultimate oil recovery for solution gas drive varies from 5% to 30%.

1.10.3 Gas Cap Drive

In the saturated oil reservoir with a primary gas cap, the dominant drive mechanism is gas cap drive. As the oil pressure decreases, gas expands and fills the extracted pore volume. By gas expansion, stored energy in gas is evolved and the gas oil contact comes down; therefore to avoid gas production from the cap, most wells are drilled in oil zones. High compressibility of gas causes a slow pressure drop in the reservoir. The level of pressure maintenance is relatively higher than the two abovementioned mechanisms. Ultimate oil recovery of the gas cap drive varies from 20% to 40% of original OIP.

1.10.4 Water Drive

The required energy for the water drive mechanism is provided by a bounded aquifer. By oil production the water oil contact comes up, and water replaces the oil withdrawals. Size of the aquifers varies from so small "which have negligible effects on the reservoir performance" to so large "which act as an infinite source in comparison to the reservoir size." According to the shape and structure of water sources, the water influxes into the reservoir are divided into edge water and bottom water. For a typical water drive reservoir the pressure decline is generally gradual. Many reservoirs exist all over the world that have one psi pressure drop per million barrel of oil. The ultimate oil recovery for water drive systems ranges from 30% to 70% of original OIP.

1.10.5 Gravity Drainage Drive

Gravity drive mechanism is active in both gas cap and water drives, where the differences in densities act as source of energy. The energy is provided by upward liberated gas movement to the primary or secondary gas cap, and also downward movement of water to the aquifer. The main gravity drainage drive mechanisms are found in saturated reservoirs, where the upward movements of gas bubbles push down the oil toward the well. In order to have optimized recovery of the gravity drainage drive, wells should be completed near the lowest possible part of oil layer. Numerous factors affecting the efficiency of this mechanism include vertical permeability, dip of the reservoir, relative permeabilities, and production rates.

In many reservoirs where both water and primary gas cap are available, these drive mechanisms exist simultaneously. In other words, the dominant drive mechanism is a combination of drive mechanisms.

1.11 MECHANISMS OF OIL TRAPPING AND MOBILIZATION

1.11.1 EOR: What, Why, and How?

Oil recovery through the reservoir life is classified in three main steps: primary, secondary, and tertiary, called EOR. During the reservoir recovery period, primary oil production is the early stage in which oil comes up to the surface by natural energy or some artificial lift tools, including gas lift or pumps. Consequently, the secondary recovery is followed to maintain the reservoir pressure by water or gas injection to aquifer or gas cap, respectively. In many reservoirs the secondary recovery is launched immediately after primary production. The vast and techniques implemented to extract residual oil are categorized in tertiary recovery.

The reservoirs that cannot produce oil by natural depletion are candidates for the tertiary recovery implementation. Numerous effective parameters in the oil reservoir shift the production plan to the tertiary recovery. Wettability adversity, reservoir depth, and hydrocarbon viscosity are some parameters making the EOR usage inevitable. Also some part of the residual oil immobilized in primary or secondary production can be displaced by some advanced EOR methods.

Depending on reservoir characteristics, various EOR methods can be carried out. The EOR processes are categorized as the thermal method, which incorporates heat transfer to bring up the viscous crude oil; gas injection, which uses nitrogen and carbon dioxide in both miscible and immiscible approaches; and chemical techniques, which are used not only to improve the Waterflood sweep efficiency, but also to reduce the oil surface tension.

1.11.2 Different EOR Processes

EOR methods are mainly categorized into four major techniques: gas injection, thermal injection, chemical injection, and other methods like microbial EOR. Gas injection is the most popular method in the world, then thermal injection is second.

1.11.3 Gas Injection

This technique includes miscible injection of different gases such as carbon dioxide, nitrogen, flue, and natural gas. The objective of miscible injection is to improve oil displacement and reservoir pressure maintenance by forming a single phase between the injected gas and oil. The reservoir conditions including temperature, pressure, and composition of the oil significantly influence oil displacement during miscible gas injection. Based on reservoir condition and phase behavior of the crude oil, the miscible processes are divided into two majors: first-contact miscible and multiple-contact miscible gas injection. In the first-contact miscible process, the injected gas is solved immediately in reservoir crude oil and a single-phase fluid is formed under reservoir condition. When the fluid is injected to the reservoir, the interference between the slug of injected fluid and reservoir oil is dropped because of the miscibility of two fluids. This process improves oil mobility toward the production well.

In the multiple-contact miscible process the injected fluid cannot be solved in reservoir oil at first contact. In this process the composition of the injected fluid strongly influences the final efficiency. A modified composition of injected fluid can lead to better mass transfer between injected fluid and reservoir oil through multiple contacts between them. The miscibility between the injected fluid and reservoir oil is formed, and subsequently, oil displacement is improved.

1.11.4 Thermal Injection

The thermal methods imply on the processes in which the oil displacement is improved by heat transfer through the reservoir. Thermal process can be categorized into two majors: steam drive and in fire flooding.

Steam drive is a means for heat transfer to reservoir oil by injecting steam from the surface into the reservoir. This process subdivides into two methods: cyclic steam injection and continuous steam injection. Cyclic steam injection, or the huff and puff process, is a method in which three stages of injection, soaking, and production are followed in a single well. Steam is injected for a determined amount of time, then the well is closed for a certain period, which is called the soak time. In this stage the well is allowed to be closed for days to allow heat transfer from the steam to viscose/heavy oil. Then the well is opened for a while and hot oil is produced. Again this process is repeated. Different mechanisms are active in this process, including viscosity reduction, oil swelling, and steam stripping. The second process in this classification is steamflood, in which steam is injected into the injection well, moves toward the production well, and oil is produced from production well. By heat loss near the steam front, steam condenses to hot water. Consequently, oil swells, viscosity is decreased, and oil displacement is improved.

In fire flooding method heat is supplied by in situ combustion in the reservoir. Air or a mixture of light hydrocarbon and oxygen is continuously injected into the reservoir to maintain the combustion near the well. The fire front moves through the production well, and heat is transferred to the reservoir fluid, including oil and water. The reservoir water evaporates to steam, light hydrocarbons are vaporized, and oil viscosity drops. Different mechanisms including steam drive, hot water, and light hydrocarbon solvent help oil movement.

1.11.5 Chemical Injection

Chemical injection includes a vast range of chemicals to help oil movement with different mechanisms. Three major mechanisms can be considered for chemical injection surface tension reduction, water shut-off, and wettability alteration. Although many chemicals are developed to EOR, the classifications can be limited to ASP and Polymer flooding. The objective of ASP injection is interfacial tension reduction between oil and water to improve movement of trapped oil after waterflooding. Alkaline chemicals react with reservoir oil and create in situ surfactant. This surfactant is relatively cheaper than commercial surfactant. Also synthetic surfactant is injected with the alkaline. Another component of ASP is polymer, which is used to increase the viscosity of injected slug. This chemical controls the mobility of the ASP to increase efficiency. The polymer flooding process is used in high permeable reservoirs with high watercut. Polymers that are soluble in water are injected to the water sources in reservoirs to control the mobility of the water by viscosity thickening. Polymer injection is usually used in the first stages of the waterflooding to postpone the water breakthrough.

1.11.6 Screening Criteria

The applicability of different EOR processes depends on reservoir condition, rock, and fluid properties. Many technical screening criteria are suggested based on the reservoir properties. The ranges proposed in these criteria are not absolute. Today, some artificial intelligence (AI) methods are developed to describe the applicability of different EOR process more realistically. Taber has gathered data on EOR projects all over the world and suggested a table in which the applicability of different EOR methods is investigated. Table 1.1 shows the Taber screening criteria.

As it can be seen, there are restrictions on the feasibility of processes. For example, because of heat losses in the wellbore, thermal methods have depth limitation. On the

Table 1.1 Screening Criteria for EOR

Summary of Screening Criteria for EOR Methods

No.	EOR Method	Oil Properties Reservoir Characteristics			Oil Properties Reservoir Characteristics					
		Gravity (*API*)	Viscosity (cp)	Composition	Oil Saturation (%PV)	Formation Type	Net Thickness (ft)	Average Permeability (md)	Depth (ft)	Temperature (F)
Gas Injection Methods (Miscible)										
1	Nitrogen and flue gas	>35 ↗48 ↗	<0.4 ↘0.2 ↘	High percent of C_1-C_7	>40 ↗75 ↗	Sandstone or carbonate	Thin unless dipping	NC	>6000	NC
2	Hydrocarbon	>23 ↗41 ↗	<3 ↘0.5 ↘	High percent of C_2-C_7	>30 ↗80 ↗	Sandstone or carbonate	Thin unless dipping	NC	>4000	NC
3	CO_2	>22 ↗36 ↗	<10 ↘1.5 ↘	High percent of C_5-C_{12}	>20 ↗55 ↗	Sandstone or carbonate	Wide range	NC	>2500	NC
1−3	Immiscible gases	>12	<600	NC	>35 ↗70 ↗	NC	NC if dipping and/or good vertical permeability	NC	>1800	NC
Chemical Injection										
4	Micellar/ ASP, and alkaline	>20 ↗35 ↗	<35 ↗13 ↗	Light, intermediate, some organic acid for alkaline floods	>35 ↗53	Sandstone preferred	NC	>10 ↗450 ↗	>9000 ↘3250	> 200 ↘80
5	Polymer flooding	>15	<150, >10	NC	>50 ↗80 ↗	Sandstone preferred	NC	>10 ↗800 ↗[a]	<9000	>200 ↘140
Thermal/Mechanical										
6	Fire flooding	>10 ↗16 ↗?	<5000 ↓ 1200	Some asphaltic components	>50 ↗72 ↗	High-porosity sand/ sandstone	>10	>50[b]	<11,500 ↘3500	> 100 ↗135
7	Steam	>8−13.5 ↗?	<200,000 ↓ 4700	NC	>40 ↗66 ↗	High-porosity sand/ sandstone	>20	>200 ↗2, 540 ↗[c]	<4500 ↘1500	NC
8	Surface mining	7−11	Zero cold flow	NC	>8 wt% sand	Mineable tar sand	>10	NC	>3:1 overburden to sand ratio	NC

NC, not critical.

[a] > 3 md from some carbonate reservoir.

[b] Transmissibility >20 md-ft/cp.

[c] Transmissibility >50 md-ft/cp.

other hand, most gas injection processes are applicable in light oil reservoirs. Temperature is a restriction for chemical injection processes, making the design of a stable chemical injectant difficult.

1.12 VISCOUS, CAPILLARY, AND GRAVITY FORCES

The success of any EOR method depends on microscopic displacement efficiency to extract oil from pores in porous media. All of the EOR processes are conjugated with slug injection through the porous media. Capillary and viscous forces in reservoirs influence the efficiency of displacement of the injected slug. In other words, these forces with gravity forces determine distribution of all phases in the reservoir and also cause trapping or mobilization of phases in a multiphase system. Understanding and studying these forces in reservoirs is significantly necessary in the design of any EOR process.

Differential pressures across the porous media reflect opposing or driving forces to displace fluids. Wherever two immiscible fluids coexist in a pore or tube, a curved surface is formed between the fluids. The pressure at the interface of two phases is not the same, and a differential pressure is created, which is called capillary pressure. Typically, the not-wet phase pressure more than the wetting phase pressure determines the curvature of the interface. The capillary pressure can work as a resisting force or driving force in different conditions. These forces come from the interfacial tension, contact angle of the fluids, and radius of the pore. Many equations are suggested to describe this force. The Young–Laplace equation is one of the most famous equations.

$$P_c = \frac{2\gamma\cos\theta}{r_p}$$

where

γ is the interfacial tension

r is the pore radius

θ is the contact angle of two phases

Viscous forces in reservoirs are the result of fluid flow through porous media. These forces can be shown by differential pressure parameters in Darcy's equation for porous media. If the porous media is considered as a bundle of tubes, the concept of viscous force can be formulated as follows:

$$DP = \frac{\mu L v}{r^2}$$

where

DP is differential pressure across the capillary tube
μ is viscosity of flowing fluid
L is tube length
V is average velocity in tube

1.13 PORE SCALE TRAPPING, MOBILIZATION OF TRAPPED OIL

During primary and secondary production some parts of oil remain in the reservoir. From a microscopic point of view, this oil is trapped in the pores surrounded with a second fluid. The trapping mechanism depends on the pore geometry, rock wettability, and interfacial tension. These parameters govern fluid trapping and mobilization through porous media. Here, trapping mechanisms are investigated in some simplified models.

In some cases a high driving force is required to push a globe of oil trapped in a tube. Fig. 1.8 shows the schematic of an oil drop in a tube surrounded with water.

By assuming a constant pressure profile in the oil drop, it can be written that:

$$P_B - P_A = \left(\frac{2\sigma_{ow}\cos\theta}{r}\right)_A - \left(\frac{2\sigma_{ow}\cos\theta}{r}\right)_B$$

Although the oil pressure would be more than the water pressure, the net pressure along the tube is zero.

In the case of difference in contact angles at points A and B, receding and advancing contact angles respectively, the pressure required to mobilize oil can be calculated as (Fig. 1.9):

$$P_B - P_A = \frac{2\sigma_{ow}}{r}(\cos\theta_A - \cos\theta_B)$$

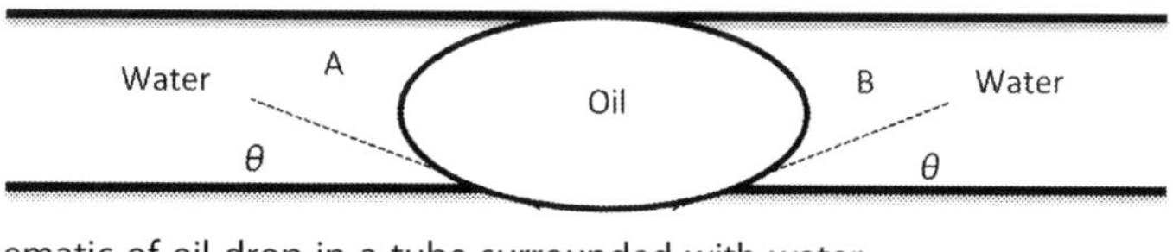

Figure 1.8 Schematic of oil drop in a tube surrounded with water.

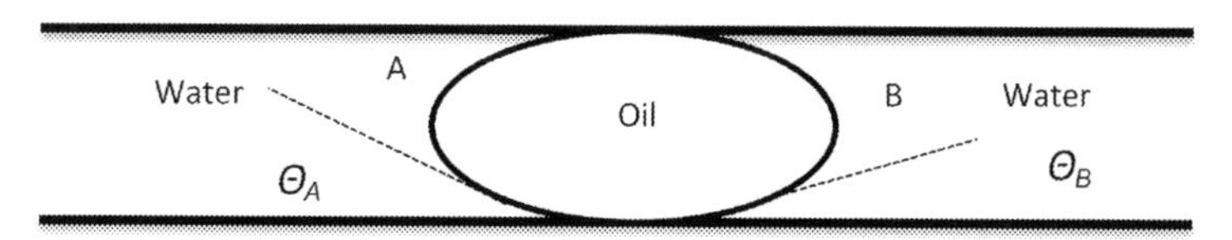

Figure 1.9 Difference in contact angles in oil drop in a tube surrounded with water.

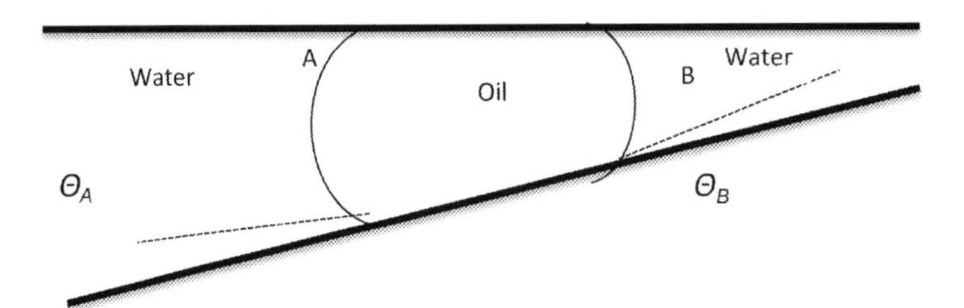

Figure 1.10 Difference in radius in oil drop in a tube surrounded with water.

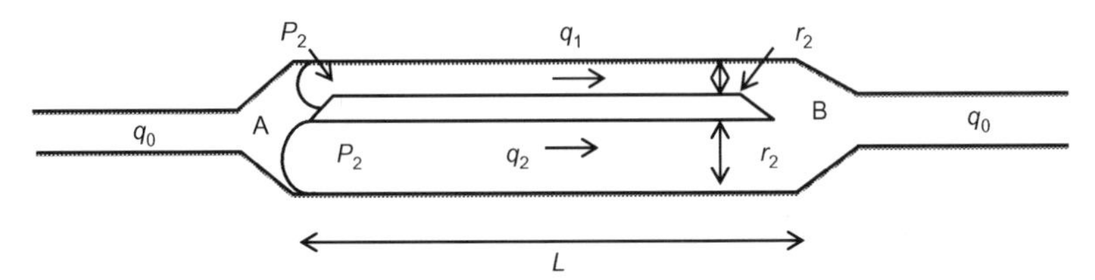

Figure 1.11 Pore doublet model.

The advancing contact angle is more than receding contact angle; therefore the pressure at point A is more than the pressure at point B, and the direction of flow is from A to B.

If the contact angle across the oil drop is the same and the radius of the capillary tube is narrowed, the pressure required for mobilizing oil is determined as follows (Fig. 1.10):

$$P_B - P_A = 2\sigma_{ow}\cos\theta\left(\frac{1}{r_A} - \frac{1}{r_B}\right)$$

The driving force required to push oil into the narrower part of the tube should be increased until the oil drop comes out.

Another typical model for pore structure is the pore doublet model. This model is more complicated, consisting of two connected parallel tubes. Fig. 1.11 shows the schematic of this model.

Assumptions in this mode are water-wet connected tubes with different radiuses, one of which is smaller than the other, and fluid flowing from point A to point B.

In the case of faster flow in one of the tubes and also inadequate driving force to push oil from the tube with a slower flow rate, oil will be trapped.

The pressure difference between point A and point B can be written as:

$$P_A - P_B = P_A - P_{wi} + P_{wi} - P_{oi} + P_{oi} - P_B$$

For downstream minus upstream pressure

$$P_A - P_B = \frac{8\mu_w L_w v_1}{r_1^2} - \frac{2\sigma\cos\theta}{r_1} + \frac{8\mu_o L_o v_1}{r_1^2}$$

$$\mu_o + \mu_w = \mu \text{ and } L_w + L_o = L, \text{so}$$

$$\Delta P_{AB} = \frac{8\mu L v_1}{r_1^2} - \Delta P_{c1}$$

For displacement of trapped oil fluid velocity in both pores must be positive; therefore ΔP_{AB} should be more than ΔP_{c1} and also more than ΔP_{c2}. Because of the size structure of the pores, ΔP_{c1} is more than ΔP_{c2}.

$$\Delta P_{AB} > \Delta P_{c1}$$

$$\frac{8\mu L v_1}{r_1^2} - \Delta P_{c1} > \Delta P_{c2}$$

By applying capillary pressure equations, the required velocity to push oil in a bigger pore (for having positive velocity in a bigger pore) can be calculated as:

$$v_1 > \frac{\sigma\cos\theta r_1^2}{\mu L}\left(\frac{1}{r_1} - \frac{1}{r_2}\right)$$

1.14 MICROSCOPIC DISPLACEMENT OF FLUIDS IN THE RESERVOIR (ED)

An essential part of any EOR process consists of the ability of the injected fluids to displace oil in the pore space at a microscopic scale. The microscopic displacement efficiency, E_D, has a significant impact on the success or failure of a project. For crude oil, microscopic efficiency depends on the magnitude of residual oil (S_{or}) at the end of the process where the displacing fluid is in contact with the displaced fluid. Nonetheless, since the EOR processes are usually associated with the injection of many slugs, the efficiency of each of the fluids is different in the porous environment of the reservoir. Moreover, low efficiency leads to early fingering phenomenon and consequently to poor performance in the injection process [1].

Capillary and viscous forces and viscosity of fluids and their mobility in porous media are among the important parameters affecting the microscopic displacement that are worth investigating as a research topic.

1.14.1 Macroscopic Displacement Efficiency

The oil efficiency in each displacement depends on the volume of the reservoir in contact with the fluid injected. The Volumetric Sweep Efficiency is a measure that accounts for this contact area. The sweep or volumetric displacement is also a macroscopic displacement, which includes the portion of the volume of the pore space that is occupied by the injected fluid or is affected by it. Total displacement efficiency, E_c, in a displacement process is defined by multiplying the microscopic displacement, E_D, in macroscopic displacement, E_v.

In general, the level of reservoir fluid contact during the displacement process is controlled by four factors:

1. Characteristics of the injected fluid
2. Properties of displaced fluid
3. Petrophysical properties of reservoir rock
4. Injection and production wells placement

1.14.2 Macroscopic Displacement Mechanism

Macroscopic displacement refers to a displacement where an injector fluid can replace the fluid inside the pores of the reservoir rock and cause the fluid to move from the reservoir and thus be positioned around the rock.

The macroscopic efficiency is one of the solutions used to determine displacement efficiency when flooding takes place in a specific volume of reservoir. The process of oil displacement almost always changes with time; hence the macroscopic displacement efficiency will also vary accordingly. There are a number of factors affecting this efficiency, some of which are:

- Mobility of the displacing fluid in comparison with that of the displaced fluid
- Homogeneity or heterogeneity of the reservoir
- Arrangement of injection and production wells
- Reservoir rock matrix

1. Mobility of the displacing fluid in comparison with that of the displaced fluid

   The mobility ratio of the displacing and displaced fluids is a relative phenomenon that occurs in the porous medium. If the displacing fluid moves more than the displaced fluid, the latter will move forward, and thus viscosity fingering will occur and areal sweep efficiency will be highly dependent on the mobility of the two fluids.

2. Homogeneity or heterogeneity of the reservoir

   If heterogeneity in the hydrocarbon layer of the reservoir is caused by factors such as porosity, permeability, and cementitious properties of the reservoir, it will prevent fluid movement across the reservoir from being homogeneous, causing a significant effect on macroscopic displacement efficiency. In parentheses, homogeneous limestones and sand formations generally have wide fluctuations in terms of porosity and permeability. Also many formations have small and large fractures that result in heterogeneity in the reservoir. When a fracture occurs in the reservoir, the fluid will try to move through the fracture because fractures usually have high permeability and can transfer hydrocarbons quickly. Many oil-producing regions have a permeability that varies in two directions (i.e., vertical and horizontal) and reduces vertical and areal sweep efficiencies. Areas and layers with low or high permeability create a transverse continuity in all or part of the reservoir. Where such layering exists in permeability, water will flow faster through the permeable layers.

3. Arrangement of injection and production wells

   The shape of the flow in the reservoir (depending on the arrangement of the injection and production wells) will have an impact on areal sweep efficiency, so that the higher the level of contact between the displacing and displaced fluids is, the greater the sweep efficiency will be. Therefore it can be said that when a complete contact surface appears between two fluids, a linear displacement will follow, and sweep efficiency can become 100%.

4. Reservoir rock matrix

   The rock has been used with respect to its chemical composition, and its water-wettability or oil-wettability has a decisive role here. Consequently, the better the reservoir material is, the easier it will be for the fluid to flow through it. In contrast, the weak quality of the reservoir rock will stop the flow of fluid in the reservoir. In the case of rocks such as inderite and limestone, hydrocarbon cannot flow in there at all [1].

1.14.3 Volumetric Displacement Efficiency and Material Balance

The volumetric (sweeping) displacement often uses the principles of material balance to calculate the rate of efficiency. For example, the displacement process in which the percentage of initial oil saturation is reduced to the percentage of residual oil saturation occurs in its contact point with the displaced fluid. If this process is assumed to be piston-like, the displaced oil can be expressed using the following formula:

$$N_P = V_P\left(\frac{S_{o1}}{B_{o1}} - \frac{S_{o2}}{B_{o2}}\right) V_P E_V$$

where N_P refers to displaced oil, S_{o1} represents the percentage of oil saturation at the beginning of the displacement process, S_{o2} denotes the percentage of oil saturation remaining at the end of the process in the area where the oil is in contact with the fluid, B_{o1} signifies the volumetric coefficient of oil formation at the beginning of the displacement process, B_{o2} points to volumetric coefficient of oil at the end of displacement, and V_P is the volume of the pore space of the reservoir. In the above equation, the division of the two sides of the equation into the amount of OIP at the beginning of the process (N_1) expresses the amount of fractional recovery, microscopic displacement efficiency (E_v), and macroscopic displacement efficiency (E_D) as follows:

If displacement performance data are available, one can estimate the volumetric (sweep) efficiency using the above equation. For example, if water injection efficiency data are at hand, the equation can be changed as follows in order to solve volumetric efficiency:

$$E_V = \frac{N_P}{V_P((S_{o1}/B_{o1}) - (S_{o2}/B_{o2}))}$$

where N_P is the amount of oil produced in the injection operation.

1.14.4 Areal and Vertical Sweep Efficiency

Volumetric displacement efficiency is defined based on areal sweep efficiency (E_A) and vertical sweep efficiency (E_I). A reservoir has several layers, and each porosity layer has the same thickness and percentage of saturated hydrocarbon. Therefore the definition of volumetric displacement efficiency will be as follows:

$$E_v = E_A \times E_I$$

In the above equation, E_A represents areal displacement efficiency in an ideal reservoir and E_I is vertical displacement efficiency. In a region with high areal displacement efficiency, E_I is small and limited. A real reservoir is characterized by conditions such as porosity, thickness, saturation percentage, and regional hydrocarbon; hence E_V is expressed as pattern sweep efficiency:

$$E_v = E_P \times E_I$$

In the above relation, E_P signifies pattern sweep efficiency, which is defined as hydrocarbon of the pore space behind the injected-fluid front divided by the pore space of the region or project. Thus E_P is the areal displacement efficiency that is expressed based on changes in thickness, porosity, and saturation percentage. Total efficiency coefficient is expressed as:

$$E = E_P \times E_I \times E_D$$

One has to estimate and calculate E_P (or E_A) and E_I in order to use the above equations. It is difficult to estimate and compute these parameters because E_A and E_I are not independent from one another during three-dimensional (3D) displacement. In the absence of vertical displacement factors, areal sweep efficiency can be obtained through equations developed on the basis of a physical or mathematical model. Also the methods of using these models are subject to limitations. E_V is usually calculated on the basis of suitable functional relationships or by using mathematical models based on a three-dimensional system that does not depend on E_A and E_I [1].

1.14.5 Areal (Sweep) Displacement Efficiency

Volumetric displacement efficiency is defined in both areal and vertical terms. These definitions are useful and cause volumetric displacement efficiency to be expressed based on two different independent methods. Of course, the approximation that E_A is independent of E_I is not true in reality and in the reservoir.

1.14.5.1 Factors Affecting E_A

Areal efficiency and displacement are controlled by four factors:

1. Injection/production well patterns
2. Reservoir permeability homogeneity
3. Mobility ratio
4. Viscosity and gravity forces

Studies that are based on physical models are analyzed using parameters 1 and 2 mentioned above. These factors include parameters that can be controlled in the laboratory and have an important effect on areal displacement efficiency. Thin-shaped models minimize the gravity segregation phenomenon and allow areal displacement efficiency to be calculated independently from vertical efficiency. In other words, areal efficiency is 100%. Various models of injection/production wells are used in the process of repositioning the reservoir. Fig. 1.12 shows a number of these patterns. Spot model 5.4 is the most widely used method in recovery and injection methods. The rules and factors that apply to the five-spot state can also be used for other models. This applies even if special equations are obtained for this type of pattern under its direct influence. Permeation changes often have a special effect on areal efficiencies. This impact may vary from one reservoir to another. Therefore developing a model for studying and calculating this parameter is a difficult task [1].

1.14.6 Vertical Displacement Efficiency

Areal displacement efficiency is measured by calculating vertical efficiency to obtain total efficiencies. For this purpose, simple models that are based on linear displacement are used to show the factors affecting vertical efficiency.

1.14.6.1 Factors Affecting Vertical Displacement Efficiency

The factors influencing this displacement are:

1. Mobility ratio
2. Changes in the horizontal and vertical permeability of the layers
3. Capillary forces
4. Gravity segregation due to differences in density

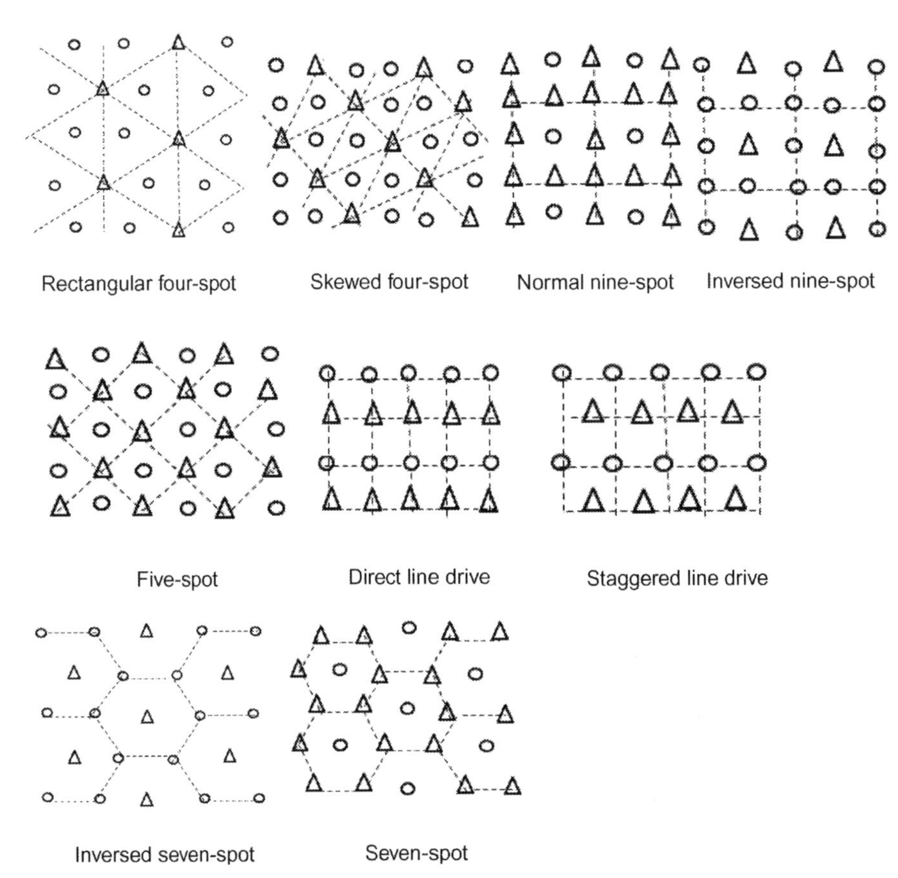

Figure 1.12 Injection patterns used in enhanced oil recovery methods.

The effects of these factors on efficiency are discussed in many studies. In these reports, physical and mathematical models are used for numerical simulation. These factors have been reviewed below.

1.14.6.2 Effect of Gravity Segregation on Vertical Displacement Efficiency

Gravity segregation occurs because of the difference in the density of the injector and displaced fluids. When the injector fluid has a lower density than the displaced fluid, gravity segregation happens and the displacing fluid moves higher than the fluid displaced, as shown in Fig. 1.13A.

This situation is observed in steam injection displacement, in situ combustion, carbon dioxide injection, and solvent injection processes. In the opposite case, when the displacing fluid is heavier than the displaced fluid, the heavier fluid tends to move from the bottom. This mode is presented in Fig. 1.13B. Here, gravity segregation results in early fingering and reduced sweep efficiency.

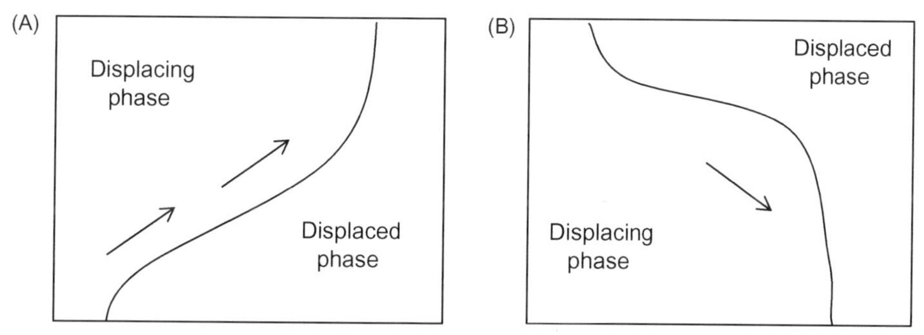

Figure 1.13 (A) Gravity segregation in the case of lighter injection fluid, (B) gravity segregation in the case of heavier injection fluid.

Craig et al. explain factors affecting gravity segregation as follows:

1. Gravity segregation intensifies with increasing vertical and horizontal permeability;
2. Gravity segregation intensifies with increasing density difference of the displacing fluid;
3. Gravity segregation intensifies with increasing mobility ratio;
4. Gravity segregation intensifies with decreasing flow rate due to fingering phenomenon;
5. Gravity segregation decreases with increasing viscosity ratio.

1.14.6.3 Gravity Segregation in Dipping Reservoirs

The difference in density between the displacing and displaced fluids has a significant effect on the displacement process in dipping reservoirs. When the reservoir is dipped, gravity force can improve displacement. For example, in the case of the oil being driven by a fluid that has a lower density, gravity force will lead to displacement stability. Therefore if the speed of displacement is sufficiently low, gravity prevents fingering from occurring between solvent/oil. Similarly, in the down-dip injection of water, gravity stabilizes the movement of the front and prevents fingering.

1.14.6.4 Effect of Vertical Heterogeneity and Mobility Ratio on Vertical Displacement Efficiency

Vertical permeability changes usually occur in reservoirs. Fig. 1.14 shows permeability and thickness variations in the vertical direction. In this figure, the reservoir is divided into several layers with different characteristics. Geological models have ideal conditions because they do not have different characteristics in the direction of vertical layers. In other words, real-life reservoirs are layered as depicted in the figure. Permeation changes in the vertical direction or, in other words, the layered nature of reservoirs, lead to a decrease in efficiency at the time of fingering [1].

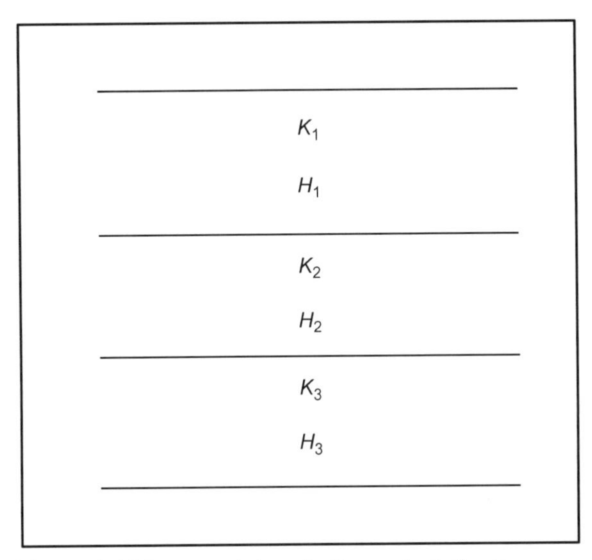

Figure 1.14 Permeability and thickness variations in the vertical direction.

1.14.6.5 Factors That Influence Displacement Efficiency

1. The size and shape of the reservoir

   Since the miscible thrust process includes injection operations, and since recovering the oil reservoir depends on the amount of reservoir surface coated by the displacing fluid, it is essential that the injected fluids have the maximum contact interface with pore spaces within the reservoir. This fact becomes more complicated when we know that the fluids are going to be injected into the reservoir from a certain spot. The coated surface in the project is controlled by the geometrical nature of the reservoir and the spots where the displacing fluids are injected into the reservoir. The basic shape of the reservoir is unchangeable.

2. Constructional slope

   Since the miscible thrust process is a process in which gas is the source of mobility and energy and causes the oil to displace, separation through density is a very important factor in providing maximum force to thrust the miscible mass and eventually displace the oil. Extreme steepness of the oil reservoirs allows the oil to be segregated from the gas and makes it move in front of gas, which leads to the formation of a complete and distinctive mass of solvent that will eventually separate gas and oil phases in motion.

3. The nature of the reservoir oil

   The properties of the oil in the reservoir are determined through laboratory techniques prior to launching the project. These properties determine the category of the practical procedure applied to recover oil from the reservoir. In general, oils with light or medium combinations or oils in which medium combinations have a greater portion than the other combinations are more suitable for miscible thrust operation.

4. Temperature and pressure properties of the reservoir

   A miscible thrust operation can be carried out successfully only if the pressure of the reservoir is adequately controlled, because pressure is a crucial element for the miscibility condition to remain stable between oil and the solvent and between the solvent and the injected gas.

5. Saturated fluids in the reservoir rock

   The saturated fluids within the reservoir affect the miscibility thrust in certain ways. Yet these fluids must be identified before launching the miscibility thrust project. In most cases, these fluids are produced within the reservoir before the project initiates, and therefore their essence can easily be discerned and their saturation can be obtained. In most reservoirs capillarity force makes the residual oil remain trapped in narrow and tight spaces.

6. Dispersion

   Dispersion coefficient affects the miscible displacement. The fingering phenomenon occurs when miscible thrust does not stimulate mobility. Whether longitudinal or latitudinal, dispersion coefficient affects the fingering phenomenon. Latitudinal dispersion has more severe effects on the fingers formed since it affects a vaster area and starts to move from the fingers, while longitudinal dispersion does not affect them so severely, as it covers a limited area.

1.15 MOBILITY RATIO CONTROL

Mobility ratio control is defined as any procedure that aims to reduce the mobility of the displacing fluid or the injector fluid within a reservoir. Mobility can improve the volumetric displacement in a process. It is usually analyzed in terms of mobility ratio. When mobility declines, volumetric displacement efficiency increases.

Some of mobility control techniques involve addition of some chemicals to the injector fluid. These chemicals increase the apparent viscosity in the injector fluid or reduce the efficient permeability of the injector fluid. The chemicals employed for this purpose include polymers when the injector fluid is water and foams when the injector fluid is gas. In some cases the mobility is controlled through WAG [1].

1.15.1 Mobility Ratio Control Processes

1.15.1.1 Polymers Along With Water Injection

Water and polymer solutions with a high molecular weight can partially increase the viscosity of water (Fig. 1.15). Two types of polymers can control water flooding: (1) hydrolyzed polyacrylamide and (2) xanthan polymers.

In polymer solutions containing polyacrylamide, increasing the viscosity and reducing the permeability of the rock that reacts to the chemicals causes the reduction of mobility. Xanthan polymers decrease the mobility of the solution while increasing the viscosity. In most cases, injecting polymers can improve macroscopic volumetric displacement.

In the process of polymer augment water flooding, polymer is persistently injected with its initial density for a specific time period. Polymer density drops regularly during the injection of the PVs. After polymer injection, it is mobilized within the reservoir via water.

1.15.1.2 Foam and Gas Injection

Employing foam is an efficient way to control mobility ratio of the miscible process and the injection of gas. When gas leads the oil forward, foam occupies the porous medium. Since foam has high viscosity, it compresses the gas into the water and keeps them separated through a thin film. So the mobility of gas decreases in this area as the outcome of reduction of permeability in the foamed area. A dry gas such as nitrogen or methane can be injected along with steam to further expand the foam in this process.

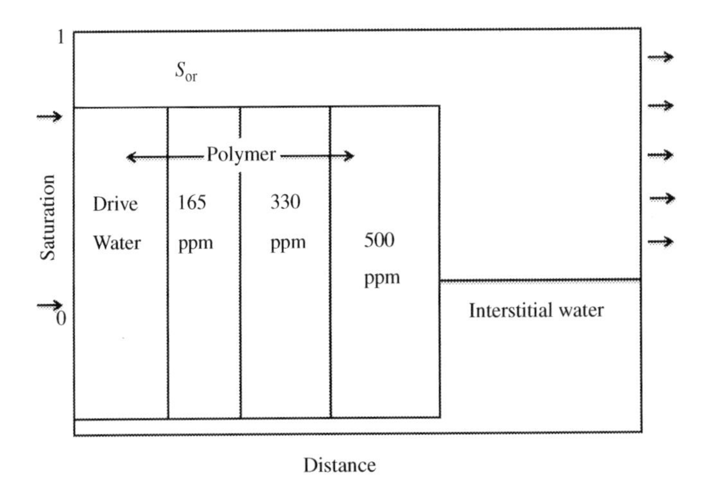

Figure 1.15 Polymer injection.

1.15.2 Mobility Ratio Control Through EOR Process

1.15.2.1 Chemical Injection

In this method surfactant injection or in situ surfactant production through injecting solution/crude oil causes oil to move. Because the polymer is expensive, a little bit of it, approximately 4%–5% PV, is injected and then displaced by water.

Mobility control in the chemical process is carried out in three stages:

- Preventing the fingering of the chemical material with oil
- Mobility control between the chemical solution and the minimized chemical slug
- Preventing the fingering of water injection front with the polymer containing chemical

1.15.2.2 Miscible Gas Injection

After one or more miscible contacts, since the viscosity of gas is much less than that of the water or oil, the mobility ratio would not be suitable. In addition to reducing sweep efficiency, this factor will also affect fingering and lead to the mobility of gas in a space with high permeability.

WAG injection is commonly carried out to control the mobility of this process (Fig. 1.16). Mobility ratio is improved by choosing the proper water/gas slug that minimizes gas recycling. Water and gas, which are injected as a slug, are mixed within the reservoir and cause increased injection efficiency.

1.15.3 Steam Flooding

When steam is injected, since density difference (gravity) starts to move towards the top of the layers and areal sweep efficiency is commonly high in this process, oil starts to move upwards from the lower layers when they become hot. This area, which

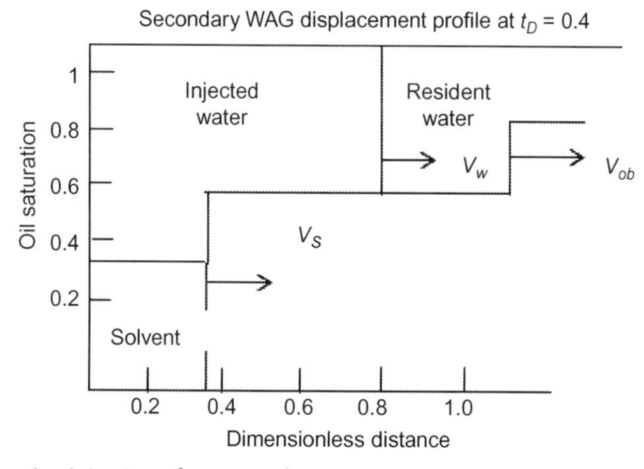

Figure 1.16 Alternative injection of water and gas.

comes into contact with oil, contains less residual oil (5%–10%). Steam mobility is high in this area, so mobility and steam presence decreases in production wells [1].

An oil reservoir naturally starts production using the potential energy saved within it. After a while, depending on the conditions of the reservoir, energy level declines and the reservoir will not have any production. Finally, it would reach a point where production is no longer economically worthwhile. Oil production through this natural mechanism is called initial exploitation, and EOR processes start at this point. EOR is a general term that differentiates oil recovery processes from the initial exploitation. Flooding and gas injection into the oil reservoirs are two commonplace gravity methods commonly known as secondary recovery. Gas or water is injected into the reservoir prior to tertiary recovery to save the internal pressure of the reservoir. Any other technique employed for further production after the secondary recovery is called tertiary recovery.

1.15.3.1 A Review on Enhanced Oil Recovery Methods From Reservoirs

The main purpose of the new EOR methods is to employ ways to recover more oil. Gas or water injection prevents the inner pressure of the reservoir from dropping quickly. Reservoir engineers have been looking for methods that would allow them to recover more or even all of the initial oil of a reservoir for years. Oil recovery methods fall into these three categories:

1. Natural deletion
2. Secondary recovery
3. Tertiary recovery

EOR methods include recovery techniques that can be classified under the second and third categories. EOR process is used as a supplementary mechanism for the natural thrust mechanism of the reservoir such as pressure fixation, wetness changes, and mobility coefficient control.

As shown in Figs. 1.17 and 1.18, the general categorization of enhanced recovery methods is as follows:

1. Natural production
   a. Water flooding
   b. Dissolved gas flooding

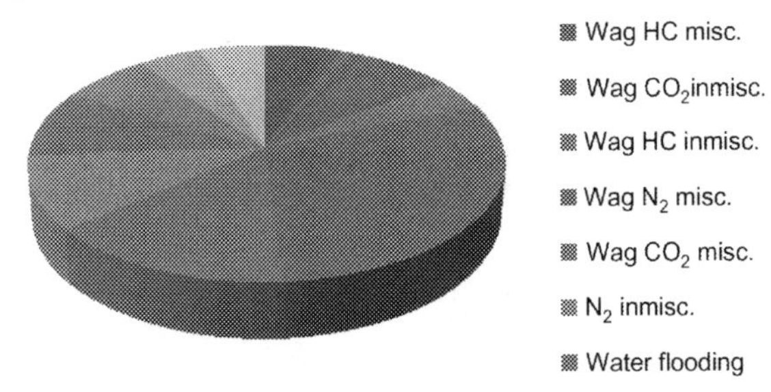

Figure 1.17 Distribution of methods employed in enhanced recovery.

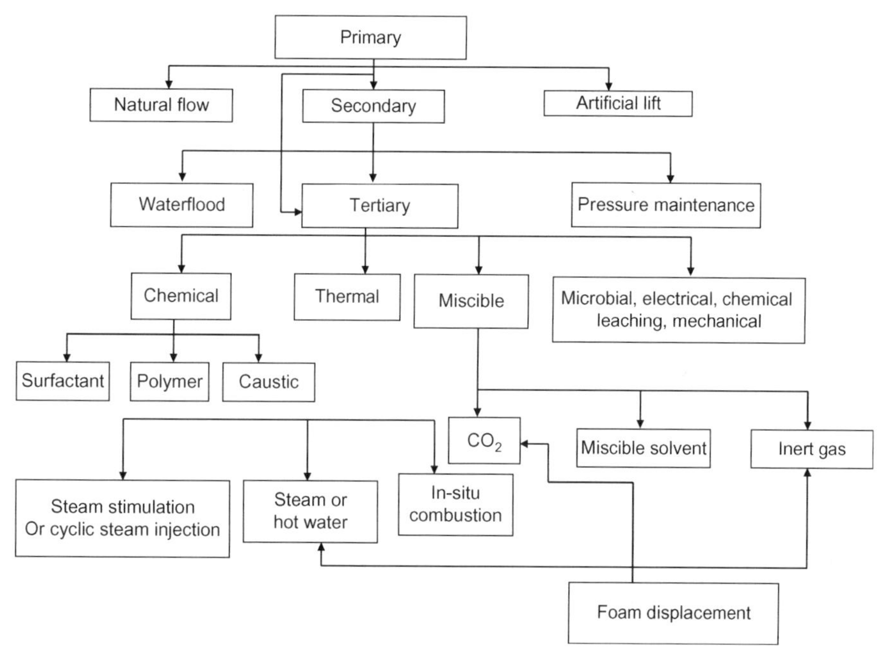

Figure 1.18 Categorization of enhanced recovery methods.

   c. Gas cap flooding
   d. Gravity segregation mechanism
   e. Rock and fluid expansion
2. Secondary recovery
   a. Water injection into the reservoir aquifer
   b. Gas injection into the gas cap
3. Tertiary recovery
   a. Thermal methods (steam injection, in-situ combustion)
   b. Miscible injection methods (injection of technical gases and carbon dioxide)
   c. Chemical injection (polymer surfactant and alkaline)
   Tertiary recovery has the following advantages:
   * Improved oil displacement efficiency
   * Improved oil sweep efficiency

1.15.4 Primary Recovery

Oil recovery by means of the natural energies of the reservoir is called primary recovery. Mechanisms employed to recover oil in primary recovery include: (1) the mechanism of solution gas drive energy, (2) gravity drainage mechanism, (3) gas cap expansion mechanism, (4) aquifer energy mechanism, and (5) rock and fluid density mechanism.

Two or more of the natural recovery mechanisms are frequently active at the time of natural reservoir production, one of which is the main and dominant one. But this main mechanism might change as natural production progresses. The nature of the main mechanism has great influence on the amount of oil recovered from the reservoir.

1.15.4.1 Dissolved Gas Mechanism

In most reservoirs, dissolved gas drive mechanism plays an important role in the oil recovery. This mechanism is especially effective in fractured reservoirs. When the inner reservoir pressure reaches bubble-point pressure, it becomes a saturated reservoir. Bubbles within the reservoir expand as pressure drops. Among especially influential elements to improve recovery coefficient is high API degree of the oil or low oil viscosity, high dissolved gas to oil ratio, and homogenizing structures. With the exception of high pressure reservoirs, which are under-saturated, and reservoirs with strong aquifer, all oil reservoirs are controlled through dissolved gas energy mechanism during their first years of life.

1.15.4.2 Gravity Drainage Mechanism

The gravity drainage process can occur in the reservoirs in two forms: free gravity drainage process and forced gravity drainage. Free gravity drainage occurs in reservoirs with high permeability and proper oil layer thickness that have reached low pressure levels, while forced gravity drainage occurs in reservoirs with dual porosity. In these reservoirs, gas proceeds to the highly permeable area (the fracture), and oil is left behind in the low permeable areas (the matrix). Pressure difference between matrix and fracture fluids provides the required force to drive gas from the fracture to the matrix and to displace and produce oil. Forced gravity drainage is demonstrated in Fig. 1.19. If the height of the matrix is affected by factors such as permeable layers or fracture and is thus limited, a high amount of unrecoverable oil is expectable in such reservoirs. While free gravity drainage in highly permeable structures with thick layers results in little unrecoverable residual oil in the reservoirs, this applies where oil recovery in a particular reservoir is considerably less than the critical level. In the case of gravity drainage processes, two basic points are noteworthy: (1). When the fluid moves under the influence of gravity drainage with a uniform pressure in the homogeneous reservoir, the flow is usually in the vertical direction. In other words, if we consider a homogeneous porous block that is in equilibrium with the displacing phase around the block and is applied in the gravity drainage, the production rate of the displacing phase would be independent of the opening or closure of the vertical margins of the block. (2). In forced gravity drainage, if the oil and gas contact area in the fracture exceeds the block or is in direct contact with the horizontal plane of the block, it would have little effect on the production of oil with time.

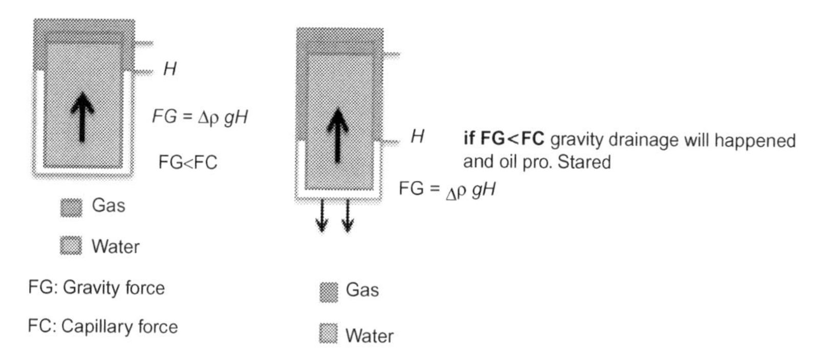

Figure 1.19 Gravity drainage mechanisms.

1.15.4.3 Gas Cap Expansion Mechanism

A gas cap might be formed under primary reservoir conditions, while a secondary cap can be shaped by free dissolved gas from oil due to reduced pressure resulting from oil recovery. In this case, it is necessary that the pressure of the reservoir be reduced to a lower pressure than the bubble point pressure. Reservoirs designed to produce gas cap by expansion mechanism usually have a lower pressure drop compared to reservoirs that are under the influence of a dissolved gas energy mechanism.

1.15.4.4 Water Flooding Mechanism

In reservoirs with aquifer, the energy required to drive the oil is either produced by the expansion of a large amount of water that is condensed in the adjacent (lower) aquifer pores, or it is provided by the hydrostatic pressure of an aquifer extending up to the surface. Oil production by aquifer energy is usually good for low-viscosity oils, provided that it is produced properly from the reservoir, but this method is not efficient for reservoirs with high viscosity oil.

1.15.4.5 Rock and Fluid Density Mechanism

The production of fluid from a reservoir increases the pressure difference between the upper layers of the reservoir and the porosity pressure and, as a result, reduces the volume of reservoir porosity and leads to increased oil production. If the compressibility coefficient of the structure is precisely determined, the reduction of porosity volume can be calculated. The production of oil by this mechanism would be high, provided that the density of the formation is high. Most of the reservoirs with high compressibility coefficients are shallow reservoirs that are not too dense.

1.15.5 Secondary Recovery

1.15.5.1 Miscible Gas Injection

Gas injection is one of the oldest methods of oil production and is known as an important method in EOR industry. As a technique that yields the highest rate of oil recovery, steam injection has also a special place in oil recovery industry, yet it comes second after the most important technique of oil recovery, i.e., gas injection. EOR using carbon dioxide not only reduces the production cost of oil and its price fluctuations, but it is the only technique that has developed steadily during recent years (13). After steam injection, injecting hydrocarbon has rendered the highest rate of oil recovery, and since there have been many efforts to reduce gas combustion, gas injection is considered as one of the most important techniques in oil recovery industry, both now and for the future. After many years of enhanced oil recovering from different reservoirs, gas recovery techniques are well understood now, and the recovery parameters are discussed with more certainty (13). Comprehensive and extensive studies have been conducted on carbon dioxide (CO_2) and nitrogen (N_2). But the range of gravity, pressure, and depth differs for each of these methods. Therefore these techniques are preferable for deep reservoirs, and the final decision often depends on local circumstances as well as the existence, amount, price, and availability of the given gas.

1.15.5.2 Hydrocarbon Injection

This technique is the oldest technique in EOR industry. Before the discovery and definition of concepts of "minimum miscibility pressure" and "minimum miscibility percentage," the injecting hydrocarbon technique has been practiced for years. It occupies a position between nitrogen, which needs a high-level pressure, and carbon dioxide, which requires an average level of pressure. This is true about methane as well. Nevertheless, in a reservoir with low depth that requires less pressure, one can do this process by increasing C_2-C_4. Various techniques that are used in hydrocarbon gas injection include:

1. Enriched gas drive
2. Lean gas drive
3. High pressure gas injection
4. LPG injection

1.15.5.3 Nitrogen and Generated Gases

Compressed air, nitrogen, and generated gases are the cheapest gases available, and this is considered a great advantage. These gases can be injected simultaneously, and because their minimum miscibility pressure is close to each other, they can be used alternatively for recovering oil.

Mechanisms: [2]

This technique includes the following mechanisms:

1. Evaporation of the lighter components of crude oil and generating miscibility in ample and high pressure.
2. Providing gravity and drift in a location where a huge portion of reservoir volume is filled with gas.
3. Increasing gravity drift mechanism which happens in dipping reservoirs.

1.15.6 Required Condition for Miscible Injection

The most important condition for a successful miscible injection is either attaining gravity equilibrium or stable displacement. These conditions are only found in a reservoir that has highly dipping formations with its slopes having a general high permeability. Although the permeability of fractures increase the permeability of the whole reservoir, these fractures reduce miscible drift efficiency. The main reason for the aforementioned problem is the gradual movement of the displacing gas alongside those fractures without having complete contact with matrix oil. Unless the layers are very thin, the stable miscible displacement cannot be sustained in vertical layers. The results of laboratory and field studies indicate that gravity, in this case, is dominant, and fingering can lead to low efficiency of productivity. On the other hand, the heterogeneity of reservoir can intensify this issue. Other optimum conditions for miscible injection are:

- Homogeneous formation
- High permeability of the matrix (in fractured reservoirs)
- Thick oil column
- Lack of gas cap
- Primary high pressure in reservoir (preferably more than miscible pressure)
- Appropriate properties of reservoir fluid (light oil, rich gas)

1.15.7 Immiscible Gas Injection

Immiscible gas injection is performed in two ways:

1.15.7.1 Gas Cap Injection

If a reservoir has a gas cap or when the gas cap is created by the movement of gas molecules emitted from oil toward the upper part of reservoir, during the initial production process gas will be injected to sustain and maintain pressure. Increasing the pressure of gas caps will drive oil towards production wells. This process resembles the rise of water and oil level during water injection to aquifer. Injecting gas to a gas cap is applicable when the reservoir has high permeability, the vertical thickness of oil layer is high enough, and there is a low oil viscosity. In such cases the recovery of oil will be high.

Gas is usually injected in a point below the reservoir crest. Gravity causes the gas to move upward and drives the oil towards the production well. If vertical permeability is low, the process of gravity separation will not be completed. If the intensity of oil production is more than the intensity of gravity separation process, the final recovery of oil will decrease.

1.15.7.2 Water Injection

Water and oil are two fluids that are immiscible. When these two fluids are situated side by side in the pores of a rock, they separate from each other according to their wettability. In other words, water usually sticks to the rock and separates and moves the oil. Given the existent forces, oil moves in a massive form until it is trapped in a location where its permeability is almost zero. Then over time, the phases of water and oil are separated completely, and due to the difference between the density of water and oil, water moves below the oil.

Flooding efficiency is dependent on several variables, the most important of which include the degree of oil saturation during the initiation of flooding, the saturation degree of the residual oil, water saturation degree, the saturation degree of free gas at the start of injection, the volume of floodable pores, oil and water viscosity, effective permeability of oil in nonmobile residual saturation, "and relative permeability of water and oil.

REFERENCES

[1] D.W. Green, G.P. Willhite, Enhanced Oil Recovery, SPE Textbook Series, vol. 6, 1998
[2] T.B. Jensen, K.J. Harpole, A. Osthus, EOR Screening for Ekofisk, SPE 65124, Presented at the 2000 SPE European Petroleum Conference Held in Paris, France, 24–25 October, 2000.

CHAPTER TWO

Screening Criteria of Enhanced Oil Recovery Methods

Ehsan Mahdavi[1] **and Fatemeh Sadat Zebarjad**[2]
[1]Department of Chemical and Petroleum Engineering, Sharif University of Technology, Tehran, Iran
[2]Mork Family Department of Chemical Engineering and Material Science, University of Southern California, Los Angeles, CA, United States

2.1 INTRODUCTION

Nowadays, oil is the dominant source of energy used worldwide, and no significant alternative is expected to affect the ascending trend of oil demand at least in next decade. On the other hand, most of producing oil fields are becoming mature; hence, they are close to the decline part of their production lifetime. Moreover, new oil field discoveries are not sufficient to maintain total world oil production stable. Therefore, application of new methods for increasing oil recovery from mature oil fields is mandatory. Enhanced oil recovery (EOR) methods involve application of external forces and materials to lead oil production mechanisms in a way to increase oil recovery. EOR methods are classified to three main categories: gas methods, chemical methods, and thermal methods. EOR methods are suitable for reservoirs with some specific characterizations; therefore, screening of reservoirs for EOR methods are performed prior to field design and implementation of the EOR methods. Based on world successful experiences of applying EOR methods in oil fields with different fluid and rock properties, some screening criteria are defined for each EOR method. Therefore, for any candidate oil field, screening studies must be carried out for selection of the most suitable EOR method leading to maximum oil recovery. In this chapter, EOR methods are explained precisely, and screening criteria for the methods are presented.

2.2 GAS METHODS

Gas flooding is known as a widely used EOR method for increasing recovery of light to moderate oil reservoirs. Injected gas can be a mixture of hydrocarbon (HC) gas or nonhydrocarbon gases. In the former, a mixture of hydrocarbons such as

Fundamentals of Enhanced Oil and Gas Recovery from Conventional and Unconventional Reservoirs.
DOI: https://doi.org/10.1016/B978-0-12-813027-8.00002-3

methane, ethane, and propane is injected into the reservoir in order to achieve a miscible or immiscible gas–oil system in the reservoir. In the latter method, nonhydrocarbon gases such as carbon dioxide (CO_2), nitrogen (N_2), and also some exotic gases are used as displacing fluid in the reservoir. It is noted that gas is also used as a secondary recovery method in oil reservoirs; in this method, HC gas is injected into the gas cap to compensate the reservoir's pressure decline. Based on some parameters such as operating condition and oil composition, gas flooding can be carried out in miscible or immiscible conditions. The primary mechanism of oil recovery in a gas flooding process is mass transfer between oil and gas phases. Under miscible condition, mass transfer increases and forms a miscible slug in front of gas phase. Moreover, swelling and viscosity reduction of oil phase are activated during gas flooding due to condensing of intermediate components of gas into the oil phase. Regarding gas flooding process, minimum miscibility pressure (MMP) is defined as the minimum operating pressure in which gas can reach miscibility with oil at reservoir conditions. Therefore, the efficiency of oil displacement by gas can be explained by the concept of MMP. During immiscible gas flooding, gas is injected below MMP; the gas increases reservoir pressure, and as a result the macroscopic displacement, efficiency is enhanced and cause oil swelling slightly. It is noted that mass transfer occurs not only in miscible gas flooding but also during immiscible process. Actually, gas always extracts some components of oil phase. Therefore, to be more precise, solubility of oil–gas system in immiscible gas flood is not zero, but it is very small and negligible. Higher efficiency of miscible gas flooding compared to immiscible process is attributed to the higher mass transfer between oil and gas phases causing oil viscosity reduction and oil swelling; thereby, greater viscous forces are obtained leading to better macroscopic and microscopic efficiencies.

As long as a gas flooding operation is implemented above MMP, high oil recovery factor is achieved. However, MMP is not the same for different gases, and usually it is much greater for N_2 than for CO_2; so for a N_2–oil system, miscibility condition is harder to be achieved. Usually, CO_2 injection is more efficient than the injection of other nonhydrocarbon gases, and the number of potential target reservoirs for CO_2 EOR is much greater than for other gas methods.

Gas methods can be implemented based on the reservoir rock and fluid parameters by different strategies as follows:

- Continuous gas injection
- Water alternating gas (WAG) injection
- Simultaneous water alternating gas (SWAG) injection
- Tapered WAG injection

The alternative methods were proposed and implemented in the recent decades to overcome the problem of an undesirable mobility ratio and low displacement efficiency of gas flooding. In WAG injection, gas and water slugs are injected alternately.

On the other hand, in SWAG, gas and water is injected simultaneously with a bubble structure using multistage pump systems; however, this method requires precise monitoring for injection of the two-phase (water–gas) fluid. Also, sometimes, gas is injected in a tapered mode; in this method, gas injection is changed to water injection once injection gas is detected in a production well (once breakthrough occurs).

2.2.1 CO_2 Injection

Since the beginning of the industrial revolution, human activities have produced about a 40% increase in the atmospheric concentration of carbon dioxide, which contributes to global warming [1]. Because of this, CO_2 capture and sequestration were proposed as a solution for environmental problems. On the other hand, not only the worldwide increase in demand for oil but also low recovery factor of oil reservoirs in primary production stage showed the necessity of implementation of EOR methods. Based on this fact, in recent decades, CO_2 injection has been used as an EOR method in oil reservoirs. The idea of using CO_2 in EOR is becoming more popular because of its aim to reduce greenhouse emission, its relatively lower cost, and high displacement efficiency; particularly in the United States, CO_2 injection is so cost effective because of available large pipeline of natural CO_2. Based on this, number of CO_2 EOR projects shows an increasing trend especially in the United States; Cranfield field, Lazy Creek field in Mississippi, and Sussex field in Wyoming are some recent CO_2 EOR projects in sandstone reservoirs [2]. CO_2 injection can be implemented in oil reservoirs in both miscible and immiscible conditions, and generally, miscible CO_2 injection is more efficient than the immiscible one. The dominant occurring mechanism for oil recovery by gas flooding is through mass transfer of components between oil and gas phases. Actually, gas is to be contacted with oil as much as the system reaches miscibility through condensing (condensing of heavy components of gas into oil phase) and vaporizing (vaporizing of light components of oil into gas phase) mechanisms. It should be noted that, although CO_2 is in gas form at atmospheric pressure and temperature, it may convert to supercritical fluid at some reservoir conditions [3]. In addition, the density decreases with temperature; thereby, the solubility of CO_2 in oil decreases, and as a result the required MMP increases. Therefore, the deeper the reservoir is, the higher the MMP of the gas–oil system (since the reservoir temperature usually increases with depth). Therefore, accurate determination of MMP, which can be carried out by experimental methods or correlations, is a must for a precise prediction of the efficiency of a CO_2 EOR process. In the case of asphaltenic crude oils, miscibility condition (MMP) of CO_2–oil system is adversely affected by asphaltene precipitation which may occur during CO_2 flooding [4,5].

Based on suitable fluid, rock, and reservoir properties, some criteria are defined for screening of CO_2 EOR method. Suitable reservoirs for CO_2 EOR are selected based on some factors such as reservoir geology, reservoir depth, MMP, API of reservoir oil, and oil viscosity. As discussed before, efficiency of miscible CO_2 EOR method is much higher than that of immiscible injection. Accordingly, screening of oil reservoirs plays a crucial role in identification of the most likely candidate reservoir for miscible CO_2 which directly affects efficiency of CO_2 EOR process. Aforementioned, MMP is the determinative factor in miscibility condition of a CO_2 EOR process. As a rule of thumb, oil reservoirs with a minimum mid-point reservoir depth of 3000 ft or deeper can be selected for CO_2 EOR due to appropriate reservoir pressure and temperature which facilitate the achievement of miscibility in CO_2–oil systems. Generally, MMP of CO_2–oil system increases as the viscosity of oil increases. Regarding CO_2 EOR, there is another rule of thumb stating that oils with bubble-point viscosities less than 10 cp and API of 25 or greater can become miscible with CO_2 at reservoir pressure greater than 1000 psia [3]. It is noted that reservoirs not having these criteria is not rejected for CO_2 EOR, because the above criteria are not strict and depends on reservoir size and potential of oil recovery.

The role of temperature is significant due to its effect on miscibility of CO_2–oil system; MMP for reservoirs having lower temperature are lower. A good example for this condition is the Permian Basin reservoirs having low geothermal gradient, leading to a lower required pressure to reach miscibility condition. Generally, if MMP and sufficient residual oil saturation conditions be satisfying, CO_2 EOR is not adversely affected by reservoir geological complexity; therefore, carbonate or sandstone reservoirs can be selected for CO_2 EOR. Typically, reservoirs with successful implementation of waterflooding can be suitable candidates for CO_2 EOR. Regarding fluid properties, most of CO_2 EOR projects have been performed on reservoirs with medium to light oils. Among all 123 CO_2 projects in the United States until 2012, 114 projects [6,7] were reported as miscible flooding in reservoirs with light to ultra-light oils and viscosity of less than 3 cp except for two reservoirs. Other nine immiscible projects were performed on reservoirs with heavy to light oils (11–35°API). Table 2.1 presents screening criteria of CO_2 EOR.

In addition to the above-mentioned projects in the United States, Joffre and Pembina fields in Canada, Buracica and Rio Pojuca fields in Brazil, Budafa and Lovvaszi in Hungary are some sandstone reservoirs in which CO_2 EOR projects have been carried out [16,17]. In the case of carbonate reservoirs, Permian Basin in the United States and Weyburn in Canada are two large projects that have made a major contribution to the total world oil production by CO_2 EOR. Judy Creek and Swan Hills in Canada, Bati Raman in Turkey, and Ghawar in Saudi Arabia are other examples of CO_2 EOR projects in Carbonate formations [17].

Currently, there are some ongoing CO_2 EOR projects worldwide which contributes in oil production of around 300,000 bbl/day. Most of the projects have been implemented in North America (United States and Canada). Actually, large part of

Table 2.1 Screening Criteria for EOR Methods [6,8–15]

		Oil Properties			Reservoir Properties						
		Gravity (°API)	Viscosity (cp)	Composition	Temperature (°F)	Porosity (%)	Permeability (mD)	Oil Saturation (% PV)	Net Thickness (ft)	Formation Type	Depth (ft)
Gas Methods											
HC	Range	23–57	0.04–18,000	High % C_2–C_7	85–329	4–45	0.1–5000	30–98	Thin unless dipping	Sandstone/carbonate	4040–15,900
	Average	38	286		202	14.5	726	71			8344
CO_2	Range	22–45	0–35	High % C_5–C_{12}	85–257	3–37	1.5–4500	15–89	Wide range	Sandstone/carbonate	1500–13,365
	Average	37	2		138	15	210	46			6230
N_2	Range	38–54	0–0.2	NC	190–325	7.5–14	0.2–35	0.76–0.8	Thin unless dipping	Sandstone/carbonate	10,000–18,500
	Average	48	0.07		267	11	15	0.78			14,633
Chemical Methods											
Polymer	Range	13–42.5	0.4–4000	NC	74–237.2	10.4–33	1.8–5500	34–82	NC	Sandstone	700–9460
	Average	26.5	123		167	22.5	834	64			4222
ASP	Range	23–34	11–6500	NC	118–158	26–32	596–1520	68–74.8	NC	Sandstone	2723–3900
	Average	32.6	875.8		121.6	26.6	–	73.7			2985
Surfactant + polymer/alkaline	Range	22–39	2.6–15.6	NC	122–155	14–16.8	50–60	43.5–53	NC	Sandstone	625–5300
	Average	31.8	7		126.3	15.6	56.67	49			3406
Thermal Methods											
SF	Range	8–33	3–5000,000	NC	10–350	12–65	1–15,001	35–90	> 20	Sandstone	200–9000
	Average	14.6	32,595		106	32	2670	66	–		1647
CSS	Range	8–35	50–350,000	NC	–	> 18	> 50	> 40	> 20	–	< 5000
	Average	14.4	5247		–	32	1736	–	79		1700
ISC	Range	10–38	1.5–2770	Some asphaltic components	64–230	14–35	10–15,000	50–94	> 10	Sandstone/carbonate (preferably carbonate)	400–11,300
	Average	24	505		176	23	1982	67	–		5570

ASP, alkaline–surfactant–polymer; SF, steam flooding; CSS, cyclic steam stimulation; ISC, in situ combustion.

the oil produced by CO_2 EOR is from two oil fields: Permian Basin in the United States and Weyburn in Canada. However, some oil fields in other countries are producing oil as a result of CO_2 EOR as well, such as Bati Raman in Turkey, Ghawar in Saudi Arabia, and some oil fields in Trinidad.

2.2.2 Hydrocarbon Gas Injection

Hydrocarbon gas flooding is known as the oldest EOR method [8]. In this method, a surplus of light associated or free hydrocarbon gases are injected to increase oil recovery. Usually, this method is used where large natural gas resources are available, but there is no transportation system in markets such as North Slope of Alaska (United States) [2]. It is worthwhile to note that HC gas injection can be implemented by first contact miscibility (FCM) or multiple contact miscibility (MCM) process. In FCM condition, the injected fluid forms a single phase with oil upon first contact. Actually, when liquefied high molecular weight HC gases (called liquefied petroleum gas) is injected, oil is displaced through a FCM process. However, when light gases such as methane are injected into oil reservoirs, they are not miscible in first contact; hence, miscibility is reached through multiple contacts and mass transfer between gas and oil phases. On the other hand, the required pressure to achieve miscible condition for HC method is greater than that one for CO_2 injection, while the highest miscibility pressure is required for N_2 injection process [18]. However, particularly for shallow reservoirs with low pressure, under a reasonable economic condition, light HC gas injection can be enriched by adding heavier HC gases (ethane, propane, and butane) to make miscibility easier to achieve [19]. There are some performed HC gas floods in sandstone reservoirs; in addition to miscible and immiscible HC gas flooding in Alpine [20], Kuparuk [21], and Prudhoe Bay [22,23] oil fields (all in the North Slope of Alaska, United States), a miscible HC gas flooding project has been implemented in Brassey field in Canada [17]. In the case of carbonate reservoirs, miscible HC gas flooding was carried out in South Swan Hills field in Canada [24]; furthermore, this EOR method has been reported in some carbonate reservoirs of the Middle East [25,26].

2.2.3 N_2-Flue Gas Injection

Availability and cost of the injected fluid are main restrictions for implementing EOR processes. Nitrogen and flue gas are the cheapest gases with widespread availability that can be used for improving oil recovery [8]. Although N_2 and flue gas are suitable for application as MCM displacement fluids like CO_2, they usually require much higher pressures to reach miscibility condition. On the other hand, this characteristic makes N_2 and flue gas more suitable for deep reservoirs with light oil where high-pressure conditions can be achieved without any concern about fracturing of reservoir rock. Although N_2 is known as a low-cost EOR method that can be implemented in miscible conditions for light oil reservoirs, no new N_2 injection project was reported in sandstone and carbonate reservoirs during the last few years; however, several N_2 flooding projects have been carried out particularly in the United States over the past decades [2]. In contrast, recent successes and field projects (in Montana, North and South Dakota) of high pressure air injection shows high potential of this method as a new option with lower cost than miscible N_2 injection [27,28].

In addition to all above-mentioned advantages and screening criteria for the gas EOR methods, the effect of gravity leading to gas override should be considered; during the immiscible gas injection, the injected gas can move upward through the reservoir due to its lower density compared to oil; this allows gas to bypass the oil phase. Under these circumstances, it is recommended to perforate the bottom part of the pay zone.

Table 2.1 presents screening criteria for the gas methods. As it was mentioned earlier, based on concept of miscibility, depth of reservoir and oil composition are the most significant factors that must be considered in screening process of the gas EOR methods.

2.3 CHEMICAL METHODS

Chemical EOR methods consist of injecting chemicals such as polymer, alkaline, surfactant, and their combinations to increase oil recovery by improving macroscopic and microscopic sweep efficiencies. Generally, only around 1% of the overall EOR projects have been allocated to chemical EOR which is directly affected by oil price. Although, high-performance chemicals were introduced to the industry in the last decade, number of chemical EOR projects compared to other EOR methods has decreased significantly due to the oil price crisis since 2014. An introduction of the chemicals and their fundamental mechanisms along with their screening criteria is provided in this chapter.

2.3.1 Polymer Flooding

Compared to oil phase, water movement is faster in reservoir; therefore, to avoid viscous fingering, polymers are added to water (displacing phase) in order to increase the viscosity and reduce the mobility of the water and finally increase the sweep efficiency.

Xanthan gum and partially hydrolyzed polyacrylamide (PHPA) are the most common types of polymer used in polymer flooding. Polysaccharide structure of xanthan gum is a great resource used by bacteria. Even adding the combination of both biocide and xanthan polymer to an oil well with bacteria issues will not be successful due to molecular weight difference of xanthan and biocide. Moreover, xanthan biopolymer has a lower molecular weight and higher price compared to PHPA which is less economical for large-scale projects. Therefore, PHPA is the most used polymer in field studies. On the other hand, high molecular weight PHPA increases water viscosity, while in low permeability formations, it causes polymer retention on rock surface.

Bailey [29] and Taber [8] have adopted reservoir screening criteria for polymer flooding. The reservoir properties that should be considered are reservoir type, permeability, oil viscosity, reservoir temperature, and formation water salinity. Table 2.1 shows the range and average value of some screening criteria for polymer flooding. Based on the literature, formation water salinity must be lower than 10,000 ppm for a successful polymer flooding.

Sandstone reservoirs are mostly the preferred type of reservoirs for polymer flooding projects and Daqing (1996–2010) as a large-scale project is an example of successful project with 10%–12% average recovery [30]. On the other hand, recent studies also show that polymer flooding is an option for unconventional reservoirs as well [31], while permeability of the reservoir is an essential factor in polymer solution propagation as mentioned earlier. The average permeability reported for 40 successful treatments was 563 mD compared to 112 mD for three discouraging projects [32]. Another sensible approximation is the pore throat radius that should at least be five times greater than the root mean square radius of gyration of the polymer [33].

Polymer concentration is another important factor that should be considered for successful polymer flooding, while low polymer concentration (213 ppm) [34] causes viscosity reduction; moreover, inappropriate mixing mechanism results in a similar effect [35]. In this regard, it is noted that special mixing equipment is required to mix the polymer with the injected water to avoid forming fish-eye. Also, oxygen jeopardizes the polymer stability while degrading the polymer structure; therefore, oxygen scavenger is used as a solution for this problem. Likewise, high salinity formation water has high negative effect on the PHPA structure; divalent and trivalent cations of salt interact with the PHPA structure and precipitate. A well-known solution for this problem is injecting a low salinity water preflush prior to polymer slug. It is worthwhile to note that Xanthan biopolymer can tolerate water salinity more than PHPA.

The maximum reservoir temperature reported for polymer flooding is reported to be 237.2°F, but in general, 62% of the projects were implemented in 108–158°F [14]; therefore, chemical EOR is not recommended for high-temperature wells. The mobility ratio of water and oil phase depends on the viscosity of the oil. It has been shown that incremental oil recovery increases with increasing oil viscosity lower than 30 cp, while at greater oil viscosities, the incremental oil recovery decreases [36]; moreover, based on data of 70 chemical projects (mainly polymer flooding), most of the projects have been carried out in reservoirs with oil viscosity over the range of 9–75 cp [14].

In addition to Daqing project and many polymer flooding projects in China, some pilot and large-scale polymer flooding projects have been reported worldwide. North Burbank in the United States, Pelican Lake in Canada, El Tordillo field in Argentina, Jhalora field in India, Buracica and Canto do Amaro fields in Brazil, and Marmul field in Oman are some examples of polymer flooding in Sandstone reservoirs [17,37,38].

2.3.2 Surfactant Flooding

Surfactants are amphiphilic organic molecules that possess hydrophilic and hydrophobic regions [39]. They have a long hydrocarbon tail and an ionic or polar head group. The surfactant molecules form an interface between two immiscible liquids, and larger quantities of surfactant lead to more interfacial area between two liquids until eventually they are considered miscible. Also, oil and water emulsion produced by surfactant flooding increases the displacement efficiency of the process. The main mechanisms for enhancing the displacement efficiency are interfacial tension (IFT) reduction, wettability alteration, and as a result reducing capillary force in porous media.

There are four types of surfactant categorized based on the ionic type of the head group as anionic, cationic, nonionic, and zwitterionic. The most used types in the chemical EOR are anionics while they do not adsorb on the negative charged clays of sandstone reservoirs (surface of the rock). In contrast, cationic surfactants are more expensive than anionics, and they are only used in carbonate reservoirs to change rock wettability, while ability of nonionic surfactants to reduce the IFT is less than anionic surfactants; therefore, they are mostly used as cosurfactant in chemical flooding.

Sometimes, surfactant flooding in sandstone reservoirs is combined with polymer, alkali, or even the both chemicals. Although surfactant flooding is more popular in sandstone reservoirs, recently few field studies were carried out in carbonate reservoirs [40,41]. Detailed screening criteria for the existing combinations are discussed further in details.

2.3.3 Alkaline Flooding

Alkaline flooding is the cheapest chemical EOR method, and the main alkali used in oil field is sodium hydroxide. Alkali ($NaOH$) interacts with the pseudo acid component (HA) of crude oil and creates the sodium salt of the organic acid (NaA) on the interface between the oleic and aqueous phases. In other words, in situ anionic surfactant is produced to decrease the IFT of the system (Fig. 2.1). This mechanism requires high pH condition; therefore, no promising result was noticed in waterflooding projects. Moreover, high viscosity crude oils are recommended for alkaline flooding since they contain high organic acid content.

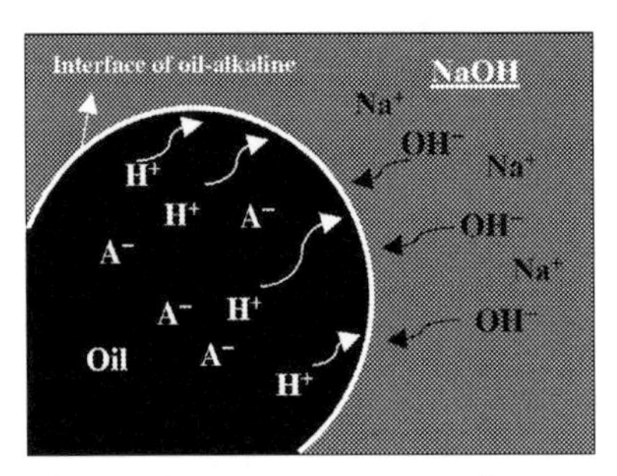

Figure 2.1 Mechanism of IFT reduction by in situ anionic surfactant.

Nowadays, alkali is less popular due to the complication it causes in field [42,43]. Divalent cations existing in clays and water interact with alkali and precipitates (scale problems) and causes formation damage. It is worthwhile to note that large amount of alkali is consumed during this reaction. Moreover, the emulsion mechanism that is supposed to assist the alkali to increase oil recovery usually forms stable emulsion that intensively increases operating and capital cost of project due to the required treatment on the produced fluids using surface facilities. Therefore, alkali is mostly combined with polymer and surfactant to obtain greater oil recovery factor.

2.3.4 Combination of Chemical Methods

2.3.4.1 Alkaline–Polymer Flooding and Alkaline–Surfactant Flooding

Mobility control is achieved due to the presence of polymer in the solution and alkaline generates in situ surfactants in alkaline–polymer (AP) flooding. On the other hand, the interaction between both chemicals has known to reduce the viscosity of polymer solution. However, polymer reduces the amount of alkaline consumption in the well. AP flooding was implemented on few pilot trail and field cases with low viscosity oil [44–46], and some were not considered as an economical treatment and scale problems were observed. David Pool in Canada and Xing Long Tai in China are examples of AP flooding that were reported in the literature [47,48].

Less pilot studies were conducted with alkaline–surfactant (AS) flooding since the mobility control is a crucial fact that is missing in this treatment. Moreover, salinity of the system increases as the alkaline is added to the surfactant solution and changes the optimum-salinity of the surfactant-alone sample. When the salinity of the system exceeds the optimum value, the IFT will not be at its lowest values, and when the salinity is lower than the optimum, the system requires salt addition to achieve the best results. It is noted that AS flooding is only used in sandstone reservoirs.

Generally, AP and AS are not as popular as the other chemical combinations; therefore, there is no specific screening criteria developed or reported for them, and the limitations mentioned earlier for AP and AS methods should be considered in these methods.

2.3.4.2 Alkaline–Surfactant–Polymer Flooding

Surfactant and alkaline reduce the IFT, and polymer assists in the sweep efficiency. In the recent years, it is generally believed that alkaline–surfactant–polymer (ASP) flooding is the most complex chemical method in which the cons are more than the pros because of the scale and precipitation formed due to the alkaline. Large cases were only implemented in China and Canada after 2005 [49]. Even with the recovery factor of 25% original oil in place (OOIP), still the complex interaction between the alkaline and reservoir rock and treatment of produced fluids remains as a great difficulty.

The screening criteria have been changed in past decade for ASP flooding. Limit of oil viscosity was updated as 1000 cp, while it was around 200 cp for ASP flooding for a long period of time [50,51]. Polymer maintains the temperature limit in ASP flooding, while new polymers can tolerate temperatures up to 100°C [52]. Recently, some new surfactants have been proposed for high-temperature wells up to 200°C [53]. Lithology is another screening criteria which changed over time. ASP flooding were mostly used in sandstone reservoirs, but in Saudi Arabia and West Texas (United States), this method was used in carbonate reservoirs [40,41]. Other criteria such as permeability, acid content, and low salinity water (low concentration of divalent cations) are still the same as before. West Kiehl, Sho-Vel-Tum, and Tanner fields in the United States [54–56], Daqing, Gudong, and Karamay in China [13,57,58], and Viraj field in India [59] are some ASP projects that have been performed in the recent two decades in sandstones.

2.3.4.3 Surfactant–Polymer Flooding

Usually, in real field scale projects of surfactant flooding, the surfactant will finger to the oil bank and decrease the sweep efficiency. The fundamental of surfactant and polymer flooding were explained earlier in this chapter. One of the highlights of this combination is excluding the alkaline and its complications in the treatment. Therefore, if the oil price increases to a level at which chemical flooding is considered economical, surfactant–polymer (SP) flooding is recommended prior other methods.

In the SP flooding, the main screening criteria are temperature and salinity. The formation water divalent ions should be less than 500 ppm, and the temperature limit is less than 100°C, the same as all chemical flooding methods which contain polymer. The limitation on oil viscosity is less than 35 cp as proposed by Taber [8]. However, other criteria like permeability are believed to be higher than 50 mD (compared to Taber value that is 10 mD) [60]. It is noted that these values are not universally agreed. Detailed screening criteria based on chemical EOR method is presented in Table 2.1. It is worthwhile to note that the parameters reported in the table have been determined based on four field scale projects.

2.4 THERMAL METHODS

Thermal EOR methods consist of injecting steam or hot water or creating combustion in the reservoir which all of them increase the temperature of the production zone that reduces the oil viscosity leading to greater oil recovery factor. The mechanism of thermal EOR is to enhance the microscopic displacement efficiency by decreasing interfacial tension and also the macroscopic displacement efficiency by decreasing viscous forces. Other mechanism may include gas drive, emulsification formed by oil/water and thermal swelling. A fundamental parameter in thermal EOR is oil–steam ratio (OSR) that is described as barrels of oil produced by injecting one barrel of steam. The minimum amount of OSR is 0.15, but normally the successful treatments that are considered economical have a higher OSR value [61].

2.4.1 Steam Flooding

One of the most famous methods in thermal EOR is steam flooding (SF) that steam is generated on the surface then injected to the well. SF application started in the early 1960s; it was used to extract viscous oils such as bitumen (20,000 cp). For oils with viscosities lower than 20 cp, waterflooding is a better option compared to SF. Steam is injected through the target zone and pushes the oil to the production well. Screening criteria for this type of thermal EOR is summarized by Green [62] and Taber [8]. Table 2.1 presents range and average of some reservoir parameters which have been determined based on field data of SF projects.

Permeability of the reservoir requires to be higher than 200 mD, since the steam should move fast enough through the porous media to avoid heat lost as much as possible. Reservoir thickness is another parameter that should be noticed due to the heat loss; a minimum of 20 ft is required to avoid heat lost. The average thickness is reported to be 70 ft. Moreover, the pressure over the well is an important aspect in heat lost; therefore, the depth and spacing between injection and production well should also be considered. As the target zone is deeper, the well spacing should increase [61]. SF treatments are most often reported with a recovery of 50% OOIP and OSR of 0.195 [63]. Some SF projects have been implemented in sandstones in the past four decades such as Yorba Linda and Kern River fields in the United States [64] and Mene Grande field in Venezuela [65]. In addition, steam injection projects have been reported in carbonates such as Garland field in the United States [66].

Sandstone is the mostly reported formation in which SF is implemented; except few cases, that formation was carbonate [67,68]. Type of the clays existing in the sandstone should be determined prior to the treatment, while some types of the clays swell as they get in contact with formation water. There are different injection pattern and well spacing used for SF. The most famous one is the inverted five-spot model, but

Chinese favorite pattern is inverted nine-spot. The maximum pressure that we can inject into the steam is formation fracture pressure while low injection pressures are not desirable, since low injection rate will end up in more heat lost. Zhang [69] proposed an injection strength around 1.3 bbl/(d ac ft) for a successful treatment.

Soaking process is recommended for more viscous oil the duration of which is around 4 years. Large amount of water is required in SF projects; the ratio of water required for oil production is 4–5. Therefore, water treatment plant is required to maintain the water quality for the boilers. Actually, water treatment plant removes the oxygen level (<0.05 ppm) and lowered the water hardness to 0.1 ppm. Moreover, sodium particles and any suspended particles are also removed in water treatment plants.

2.4.2 Cyclic Steam Stimulation

Cyclic steam stimulation (CSS) also known as huff-n-puff is another thermal EOR method which only requires one well, and it consists of three levels. First, high-pressure steam is injected through the target zone for several weeks to reduce the oil viscosity; then, in the next step, a soaking period is given to the steam to diffuse through the reservoir. Finally, oil is produced from the same well. In multilayer reservoirs, the treatment starts from the bottom layer and moves up to the top layer.

Screening criteria for steam treatments was reported by Taber [8] and Green [62] without considering the soaking time. It is noted that Taber and Green classified SF and CSS into one category. Sheng [15] presented modified screening criteria including the soaking effect and pointed out that CSS treatment can be performed in wider ranges of screening criteria, Accordingly, a soaking time of 1–4 days was proposed in each cycle of CCS process. Summarized general screening criteria are illustrated in Table 2.1.

The CSS process is generally carried out with a combination of SF after the CSS process when the oil viscosity range is 10,000–50,000 cp. In lower viscosities (average viscosity of 100 cp), waterflooding is more favorable before the SF. On the other hand, gas cap is not desirable in the CSS treatments since it increases the gravity override of the steam. Also, a bottom aquifer acts as a competitor for crude oil in receiving the heating energy of steam; therefore, CSS method is not recommended for this type of reservoir.

The soaking period is another crucial screening parameter. Optimum amount of time is required to allow the heat distribution through the reservoir to produce the maximum level of oil; also, fining optimum soaking time is mandatory to avoid heat loss with long soaking periods or heat accumulation in short period of times. For the production period of half a year, an average field data reported for soaking period is 6.25 days [70]. Liu [61] also noticed that 2–3 days of soaking is enough. High steam quality but low amount of steam should be injected into the well for the first cycle since the damage and plugging near the wellbore will be pushed far down in the reservoir and the flow back process to eliminate the plugging will get harder to perform. The production rate in the second and third injections are known be higher than the first cycle.

The number of the steam stimulation cycles is required to be economical (6–7 times) and not more than 10 times [61], while the maximum production rate is observed in the second and third cycles. It is also recommended that when oil production rate reaches one-third of initial value at start of the cycle, the cycle should be ended and the next cycle should be initiated; actually this is highly proposed to maintain the performance of the cycles high enough.

Cold Lake oil field in Canada, Midway-Sunset oil field in the United States, and Gudao field in China are some CCS projects worldwide [15].

2.4.3 Steam-Assisted Gravity Drainage

Steam-assisted gravity drainage (SAGD) is one of the in situ thermal recovery methods which was recently discovered to extract extraheavy oils and bitumen from Alberta's reservoirs [71]. 1.7 trillion barrels of bitumen is laying in Alberta making it second large hydrocarbon resource on earth [72]. High recovery rates have been reported up to 70% of OOIP, and most of the cases are in Canada. Most of the commercial SAGD projects have been carried out in Canada, where there are many heavy and extraheavy oil reservoirs; almost all successful commercial SAGD projects have been reported in the reservoirs of Athabasca region (McMurray formation), for instance, Hanginstone, Foster Creek, Christina Lake, and Firebag reservoirs [73,74]. In the case of fractured carbonate reservoirs, no promising economically efficient result is expected in a fractured reservoir, due to the existence of fracture networks leading to early breakthrough and low oil recovery factor.

Commonly in SAGD method, two horizontal wells are drilled in the target zone with the distance of 4–6 m from each other. The steam is injected through the upper well which is called "steam chamber," and the heat diffusing in the formation mobilizes the heavy crude oil and increases its viscosity; thus, oil flows to the lower well due to gravity and then is produced from the lower well.

The energy produced from the steam is roughly divided into three portions, one-third of which is lost in the formation rock. The second part remains in the chamber and last part is produced [75]. Therefore, large amount of natural gas is required as fuel to produce the steam for SAGD treatment that results in considerable amount of greenhouse emission, also high treatment cost. To overcome the aforementioned issues, a method recognized as solvent SAGD treatment adds a chemical to reduce the energy consumption [76,77]. It is worthwhile to note that an average cumulative steam-to-oil ratio of 3 t/m^3 is suggested for a SAGD process in order to be efficient from an economical point of view.

Resource quality is an essential parameter in SAGD treatment performance that includes

1. High thickness of the pay zone in order to be able to drill two vertically aligned horizontal wells; also a thick pay zone means less energy loss. A reasonable range for thickness is reported to be 10–15 m.
2. High vertical permeability (an average of 2700 mD) is necessary since low permeability limits the steam rise and from the chamber well and considerably reduces the drainage head. Therefore, geological studies before drilling are highly recommended to avoid any shale layers in sandstone formations located between the injection and production well.
3. High oil concentration (mostly high content of bitumen) while it is directly related to thermal efficiency; higher the oil content, more oil is produced from the formation with the same thermal energy and lower steam-to-oil ratio. The operations are economically viable if 10% bitumen content is present in the formation.

2.4.4 In Situ Combustion

In situ combustion (ISC) method was first presented in 1923, a recovery treatment by burning of oil originally existed in the reservoir [78], which oil ignition acquires spontaneously or artificially. ISC front produced from the ignition moves through the reservoir and pushes the heavy unburned oil out of the reservoir and to the production well. Continuous injection of air keeps the ISC front moving. The ISC treatment is an exothermic process that assists the improvement of oil recovery by reducing the viscosity of the oil by the generated heat from the burned oil.

ISC treatment generates less greenhouse gas emission since the compressed air is injected to the reservoir instead of steam. Moreover, less energy is consumed in ISC compared to other thermal recoveries. Although ISC is known as the second thermal EOR recovery [79], there are several drawbacks such as very low process control resulting in poor sweep efficiency and completions adversely getting effect by the ignition. Also, greater number of experienced and knowledgeable personnel is required compared to other thermal methods due to the complexity of the process.

ISC method is generally used for very light and very heavy crude oil, since low-pressure profile (due to shallower reservoirs) and low corrosion rate in heavy crude oil formations assist the ISC procedure. Also, in very light crude oils, integral oxygen consumption and eliminating the ignition process (due to deep reservoirs) are great impetuses for ISC treatment. ISC is not commercialized in viscosity range between 2 and 60 cp [80]. Formations with high permeability, shallow and homogenous sandstones are more favorable in ISC process. Crude oils with viscosities higher than 1500 cp are recommended for a preheating treatment (e.g., by CSS method). ISC is not commercially approved to be used as tertiary recovery after waterflooding and is known as a potentially hazardous treatment when it is used on a wrong formation. Detailed screening criteria are illustrated in Table 2.1.

REFERENCES

[1] T.F. Stocker, et al., 1535 pp Climate Change 2013: The Physical Science Basis. Contribution of Working Group I to the Fifth Assessment Report of the Intergovernmental Panel on Climate Change, Cambridge University Press, Cambridge, UK, and New York, 2013.

[2] V. Alvarado, E. Manrique, Enhanced oil recovery: an update review, Energies 3 (9) (2010) 1529–1575.

[3] R.T., Johns, B. Dindoruk,, Gas flooding, Enhanced Oil Recovery Field Case Studies, Gulf Professional Publishing, Elsevier, 2013, pp. 1–22.

[4] E. Mahdavi, et al., Experimental investigation on the effect of asphaltene types on the interfacial tension of CO_2–hydrocarbon systems, Energy Fuels 29 (12) (2015) 7941–7947.

[5] E. Mahdavi, et al., Effects of paraffinic group on interfacial tension behavior of CO_2–asphaltenic crude oil systems, J. Chem. Eng. Data 59 (8) (2014) 2563–2569.

[6] L. Koottungal, Worldwide EOR survey, Oil Gas J. 110 (2012) 57–69.

[7] V. Kuuskraa, QC updates carbon dioxide projects in OGJ's enhanced oil recovery survey, Oil Gas J. 110 (7) (2012) 72.

[8] J.J. Taber, F. Martin, R. Seright, EOR screening criteria revisited-Part 1: introduction to screening criteria and enhanced recovery field projects, SPE Reservoir Eng. 12 (03) (1997) 189–198.

[9] G. Moritis, EOR continues to unlock oil resources, Oil Gas J. 102 (14) (2004) 49–52.

[10] Awan, A.R., Teigland, R., Kleppe, J. A Survey of North Sea Enhanced-Oil-Recovery Projects Initiated During the Years 1975 to 2005, *SPE Reservoir Evaluation & Engineering* 11 (03), 2008, 497–512.

[11] Demin, W., et al., 1999. Summary of ASP pilots in Daqing oil field. SPE Asia Pacific Improved Oil Recovery Conference. Society of Petroleum Engineers.

[12] C. Cadelle, et al., Heavy-oil recovery by in-situ combustion-two field cases in Rumania, J. Pet. Technol. 33 (11) (1981) 2. 057-2,066.

[13] Li, H., et al., 2003. Alkaline/surfactant/polymer (ASP) commercial flooding test in central xing2 area of Daqing oilfield. SPE International Improved Oil Recovery Conference in Asia Pacific. Society of Petroleum Engineers.

[14] A. Al Adasani, B. Bai, Analysis of EOR projects and updated screening criteria, J. Pet. Sci. Eng. 79 (1) (2011) 10–24.

[15] J.J. Sheng, Cyclic steam stimulation, Enhanced Oil Recovery Field Case Studies, Gulf Professional Publishing, Elsevier, 2013, pp. 389–412.

[16] Doleschall, S., Szittar, A., Udvardi, G., 1992. Review of the 30 years' experience of the CO_2 imported oil recovery projects in Hungary. International Meeting on Petroleum Engineering. Society of Petroleum Engineers.

[17] G. Moritis, Worldwide EOR survey, Oil Gas J. 106 (2008) 41–42. 44–59.

[18] J. Taber, F. Martin, R. Seright, EOR screening criteria revisited—part 2: applications and impact of oil prices, SPE Reservoir Eng. 12 (03) (1997) 199–206.

[19] L. Sibbald, Z. Novosad, T. Costain, Methodology for the specification of solvent blends for miscible enriched gas drives (includes associated papers 23836, 24319, 24471 and 24548), SPE Reservoir Eng. 6 (03) (1991) 373–378.

[20] Redman, R.S., 2002. Horizontal miscible water alternating gas development of the Alpine Field, Alaska. SPE Western Regional/AAPG Pacific Section Joint Meeting. Society of Petroleum Engineers.

[21] Shi, W., et al., 2008. Kuparuk MWAG project after 20 years. SPE Symposium on Improved Oil Recovery. Society of Petroleum Engineers.
[22] Rathmann, M.P., McGuire, P.L., Carlson, B.H., 2006. Unconventional EOR program increases recovery in mature WAG patterns at Prudhoe Bay. SPE/DOE Symposium on Improved Oil Recovery. Society of Petroleum Engineers.
[23] M. Panda, et al., Optimized EOR design for the Eileen west end area, Greater Prudhoe Bay, SPE Reservoir Eval. Eng. 12 (01) (2009) 25–32.
[24] K. Edwards, B. Anderson, B. Reavie, Horizontal injectors rejuvenate mature miscible flood-south swan hills field, SPE Reservoir Eval. Eng. 5 (02) (2002) 174–182.
[25] Al-Bahar, M.A., et al., 2004. Evaluation of IOR potential within Kuwait. Abu Dhabi International Conference and Exhibition. Society of Petroleum Engineers.
[26] Schneider, C.E., Shi, W., 2005. A miscible WAG project using horizontal wells in a mature offshore carbonate middle east reservoir. SPE Middle East Oil and Gas Show and Conference. Society of Petroleum Engineers.
[27] Mungan, N., Enhanced Oil Recovery with High Pressure Nitrogen Injection. Journal of Petroleum Technology 53 (3), 2001, 81.
[28] Linderman, J.T., et al., 2008. Feasibility study of substituting nitrogen for hydrocarbon in a gas recycle condensate reservoir. Abu Dhabi International Petroleum Exhibition and Conference. Society of Petroleum Engineers.
[29] R. Bailey, Enhanced oil recovery, NPC, Industry Advisory Committee to the US Secretary of Energy, Washington, DC, USA, 1984.
[30] D. Wang, Polymer flooding practice in Daqing, Enhanced Oil Recovery Field Case Studies, Gulf Professional Publishing, Elsevier, 2013, pp. 83–116.
[31] R. Seright, Potential for polymer flooding reservoirs with viscous oils, SPE Reservoir Eval. Eng. 13 (04) (2010) 730–740.
[32] D.C. Standnes, I. Skjevrak, Literature review of implemented polymer field projects, J. Pet. Sci. Eng. 122 (2014) 761–775.
[33] J. Chen, D. Wang, J. Wu, Optimum on molecular weight of polymer for oil displacement, Acta Petrolei Sin. 21 (1) (2001) 103–106.
[34] H. Krebs, Wilmington field, California, polymer flood a case history, J. Pet. Technol. 28 (12) (1976) 1–473.
[35] H. Groeneveld, R. George, J. Melrose, Pembina field polymer pilot flood, J. Pet. Technol. 29 (05) (1977) 561–570.
[36] J.C. Zhang, Tertiary Recovery, Petroleum Industry Press, China, 1995, pp. 23–24.
[37] Moffitt, P., Mitchell, J., 1983. North Burbank Unit commercial scale polymer flood project-Osage County, Oklahoma. SPE Production Operations Symposium. Society of Petroleum Engineers.
[38] Shecaira, F.S., et al., 2002. IOR: the Brazilian perspective. SPE/DOE Improved Oil Recovery Symposium. Society of Petroleum Engineers.
[39] P. Renouf, et al., Dimeric surfactants: first synthesis of an asymmetrical gemini compound, Tetrahedron Lett. 39 (11) (1998) 1357–1360.
[40] Al-Hashim, H., et al., 1996. Alkaline surfactant polymer formulation for Saudi Arabian carbonate reservoirs. SPE/DOE Improved Oil Recovery Symposium. Society of Petroleum Engineers.
[41] Levitt, D., et al., 2011. Design of an ASP flood in a high-temperature, high-salinity, low-permeability carbonate. International Petroleum Technology Conference. International Petroleum Technology Conference.
[42] E. Mayer, et al., Alkaline injection for enhanced oil recovery—a status report, J. Pet. Technol. 35 (01) (1983) 209–221.
[43] Weinbrandt, R., 1979. Improved oil recovery by alkaline flooding in the Huntington Beach field. In: Proceedings of the 5th Annual DOE Symposium on Improved Oil Recovery, August 1979.
[44] Bala, G., et al., 1992. A flexible low-cost approach to improving oil recovery from a (very) small Minnelusa Sand reservoir in Crook County, Wyoming. SPE/DOE Enhanced Oil Recovery Symposium. Society of Petroleum Engineers.

[45] Yang, D.H., et al., 2010. Case study of alkali–polymer flooding with treated produced water. SPE EOR Conference at Oil & Gas West Asia. Society of Petroleum Engineers.
[46] J. Sheng, Modern Chemical Enhance Oil Recovery: Theory and Practice, Gulf Professional, London, Oxford, 2011.
[47] Pitts, M., Wyatt, K., Surkalo, H., 2004. Alkaline–polymer flooding of the David Pool, Lloydminster Alberta. SPE/DOE Symposium on Improved Oil Recovery. Society of Petroleum Engineers.
[48] Zhang, J., et al., 1999. Ultimate evaluation of the alkali/polymer combination flooding pilot test in XingLongTai oil field. SPE Asia Pacific Improved Oil Recovery Conference. Society of Petroleum Engineers.
[49] J.J. Sheng, ASP fundamentals and field cases outside China, Enhanced Oil Recovery Field Case Studies, Gulf Professional Publishing, Elsevier, 2013, pp. 189–201.
[50] D.L. Walker, Experimental Investigation of the Effect of Increasing the Temperature on ASP Flooding, (Doctoral dissertation), The University of Texas at Austin, 2011.
[51] Kumar, R., Mohanty, K.K., 2010. ASP flooding of viscous oils. SPE Annual Technical Conference and Exhibition. Society of Petroleum Engineers.
[52] Levitt, D., Pope, G.A., 2008. Selection and screening of polymers for enhanced-oil recovery. SPE Symposium on Improved Oil Recovery. Society of Petroleum Engineers.
[53] Zebarjad, F.S., Nasr-El-Din, H.A., Badraoui, D., 2017. Effect of Fe III and chelating agents on performance of new VES-based acid solution in high-temperature wells. SPE International Conference on Oilfield Chemistry. Society of Petroleum Engineers.
[54] Meyers, J., Pitts, M.J., Wyatt, K., 1992. Alkaline–surfactant–polymer flood of the West Kiehl, Minnelusa Unit. SPE/DOE Enhanced Oil Recovery Symposium. Society of Petroleum Engineers.
[55] T. French, Evaluation of the Sho-Vel-Tum Alkali–Surfactant–Polymer (ASP) Oil Recovery Project-Stephens County, OK, National Petroleum Technology Office (NPTO), Tulsa, OK, 1999.
[56] Pitts, M.J., et al., 2006. Alkaline–surfactant–polymer flood of the Tanner Field. SPE/DOE Symposium on Improved Oil Recovery. Society of Petroleum Engineers.
[57] Zhijian, Q., et al., 1998. A successful ASP flooding pilot in Gudong oil field. SPE/DOE Improved Oil Recovery Symposium. Society of Petroleum Engineers.
[58] Qi, Q., et al., 2000. The pilot test of ASP combination flooding in Karamay oil field. International Oil and Gas Conference and Exhibition in China. Society of Petroleum Engineers.
[59] Pratap, M., Gauma, M., 2004. Field implementation of alkaline–surfactant–polymer (ASP) flooding: a maiden effort in India. SPE Asia Pacific Oil and Gas Conference and Exhibition. Society of Petroleum Engineers.
[60] J.J. Sheng, Surfactant–polymer flooding. Enhanced Oil Recovery Field Case Studies, 2013, p. 117.
[61] W.-Z. Liu, Steam Injection Technology to Produce Heavy Oils, Petroleum Industry Press, China, 1997.
[62] D.W. Green, G.P. Willhite,, Enhanced oil recovery, in: L. Henry (Ed.), Doherty Memorial Fund of AIME. Vol. 6, Society of Petroleum Engineers, Richardson, TX, 1998.
[63] J.J. Sheng, Steam flooding, Enhanced Oil Recovery–Field Case Studies, Elsevier, Waltham, MA, USA, 2013.
[64] Hanzlik, E.J., Mims, D.S. Forty Years of Steam Injection in California – The Evolution of Heat Management. SPE International Improved Oil Recovery Conference in Asia, Society of Petroleum Engineers, 2003.
[65] J. Ernandez, EOR Projects in Venezuela: Past and Future, ACI Optimising EOR Strategy, 2009, pp. 11–12.
[66] Dehghani, K., Ehrlich, R., 1998. Evaluation of steam injection process in light oil reservoirs. SPE Annual Technical Conference and Exhibition. Society of Petroleum Engineers.
[67] Olsen, D., et al., 1993. Case history of steam injection operations at naval petroleum reserve no. 3, teapot dome field, Wyoming: a shallow heterogeneous light-oil reservoir. SPE International Thermal Operations Symposium. Society of Petroleum Engineers.
[68] B.C. Sahuquet, J.J. Ferrier, Steam-drive pilot in a fractured carbonated reservoir: Lacq Superieur field, J. Pet. Technol. 34 (04) (1982) 873–880.

[69] Y.T. Zhang, Thermal recovery, Technological Developments in Enhanced Oil Recovery, Petroleum Industry Press, Beijing, 2006, pp. 189–234.
[70] S. Ali, Current status of steam injection as a heavy oil recovery method, J. Can. Pet. Technol. 13 (01) (1974).
[71] Butler, R.M., Method for continuously producing viscous hydrocarbons by gravity drainage while injecting heated fluids. 1982, Google Patents.
[72] C. Shen, SAGD for heavy oil recovery, Enhanced Oil Recovery Field Case Studies, Gulf Professional/Elsevier, Oxford, 2013, pp. 413–445.
[73] Rottenfusser, B., Ranger, M., 2004. Geological comparison of six projects in the Athabasca oil sands. In: Proceedings of CSPG-Canadian Heavy Oil Association-CWLS Joint Conf. (ICE2004), Calgary, AB, Canada.
[74] Jimenez, J., 2008. The field performance of SAGD projects in Canada. International Petroleum Technology Conference. International Petroleum Technology Conference.
[75] C.-T. Yee, A. Stroich, Flue gas injection into a mature SAGD steam chamber at the Dover Project (Formerly UTF), J. Can. Pet. Technol. 43 (01) (2004).
[76] T. Nasr, et al., Novel expanding solvent-SAGD process "ES-SAGD", J. Can. Pet. Technol. 42 (01) (2003).
[77] Govind, P.A., et al., 2008. Expanding solvent SAGD in heavy oil reservoirs. International Thermal Operations and Heavy Oil Symposium. Society of Petroleum Engineers.
[78] C. Cheih, State-of-the-art review of fireflood field projects (includes associated papers 10901 and 10918), J. Pet. Technol. 34 (01) (1982) 19–36.
[79] A. Turta, et al., Current status of commercial in situ combustion projects worldwide, J. Can. Pet. Technol. 46 (11) (2007).
[80] A. Turta, In Situ Combustion. Enhanced Oil Recovery Field Case Studies, Gulf Professional Publishing, Boston, MA, 2013, pp. 447–541.

CHAPTER THREE

Enhanced Oil Recovery Using CO_2

Ramin Moghadasi[1], Alireza Rostami[1] and Abdolhossein Hemmati-Sarapardeh[2]
[1]Department of Petroleum Engineering, Petroleum University of Technology (PUT), Ahwaz, Iran
[2]Department of Petroleum Engineering, Shahid Bahonar University of Kerman, Kerman, Iran

3.1 INTRODUCTION

Naturally, oil reservoirs can produce up to 20% of total original oil in place (OOIP), which is termly considered as primary production. Any further increase in production will only be assisted through an implementation of the enhanced oil recovery (EOR) method. Generally, up to 20% of the remained oil in place could be produced by secondary EOR strategies, and the rest up to 30%, practically, is accessible to production by tertiary EOR techniques [1–4]. Among different types of tertiary methods, gas injection, or more specifically and commonly, CO_2 injection, has been much in practice. On average, an incremental oil recovery factor of 7%–23% of the remained oil in place has been reported for the CO_2 injection method. Although CO_2 injection would result in an improved oil recovery, the amount of oil recovered by this method is dependent on reservoir rock and fluid characteristic in whole, the specific properties of the injected CO_2, and operation conditions (e.g., injection rate, pattern) [1,2].

Carbon dioxide is found plentifully in our planet, and it is mostly sourced from power plants, petrochemical companies, etc. It is considered a greenhouse gas, and its harmful impacts to our environment have been recognized and addressed well. Not surprisingly, the use of CO_2 for oil recovery goes back to the early days of oil reserves production. However, it was after World War II that a great deal of progress was made in the development of CO_2-assisted oil recovery methods. The foundation of such developments has been laid by the works of Whorton et al. [3], Saxon Jr. et al. [4], Beeson and Ortloff [5], Holm [6], and Martin [7] during the 1950s [8]. Such advances have led to the first field-wide application of CO_2–oil recovery, which took place in 1972 at the SACROC (Scurry Area Canyon Reef Operators Committee) Unit in the Permian Basin. Currently, there are more than 70 major CO_2–EOR projects worldwide, most of which are in the United States.

Historically, there is an extreme interest toward the practical application of CO_2 sequestration, which takes the advantages of both improvements in oil recovery factor and reduction in CO_2 emission simultaneously. Undoubtedly, such a task will help us

Fundamentals of Enhanced Oil and Gas Recovery from Conventional and Unconventional Reservoirs.
DOI: https://doi.org/10.1016/B978-0-12-813027-8.00003-5

to positively impact global warming and assist in providing to the world's energy demand. Hopefully, there are number of ongoing field practices to CO_2 sequestration and EOR. Research has resulted a good understanding of associated mechanisms, effective screening parameters, and also operational conditions for an optimized process. But yet there is an amount of uncertainty about how efficiently this process could be implemented [9–12].

This chapter demonstrates the fundamentals of CO_2 injection process in both miscible and immiscible modes, explains how CO_2–EOR process could be facilitated in practice, discusses laboratory tests, illustrates some examples of reservoir simulation during CO_2 injection, details applicability of CO_2 EOR for unconventional resources, and finally depicts the environmental aspects of CO_2 injection.

3.2 CO_2 INJECTION FUNDAMENTALS

When CO_2 is injected to the reservoir, it interacts physically and chemically with reservoir rock and the existing hydrocarbon fluid. Such interactions are the base mechanisms to explain why and how injected CO_2 recovers the remained oil in place [13]. Majorly, these mechanisms are categorized as follows [8,14–16]:

1. Oil volume swelling
2. Oil and water density reduction
3. Oil viscosity reduction
4. Reducing the interfacial tension (IFT) between the reservoir rock and oil, which has previously inhibited oil flow through the pores
5. Vaporization and extraction of the trapped of oil portions (mostly light components).

Carbon dioxide has high solubility in oil, causing the oil to swell and consequently reducing the oil viscosity and density. Additionally, there is almost always some water in the reservoir, which is left from previous water flood; thereby injecting CO_2 will result in reduced water density because it is soluble in water to some extent. Eventually, it causes water and oil densities to be mostly similar, resulting in reduction of gravity segregation effects, less override flow, and lower occurrence of the fingering phenomenon [17].

The importance of each mechanism depends on the pressure and temperature of the reservoir. The miscible process occurs at high temperatures and pressures, and the immiscible process at lower pressure and temperature conditions. This makes a clear distinction between these processes, which in turn leads to different performances considering the incremental oil recovery associated with each of them [18].

But what does it mean to be miscible? Or more accurately, what conditions make a CO_2 flood considered as miscible? Theoretically, when CO_2 is injected into an oil reservoir, there is a minimum pressure level—below that value CO_2 and oil are no longer miscible. Increasing the pressure leads to an increase in CO_2 density, which reduces the density difference between crude oil and CO_2. As a result, the IFT between crude oil and CO_2 vanishes, and they will reach mutual solubility in each other. This minimum pressure is named as minimum miscibility pressure (MMP) [19–21]. A large amount of research has been implemented to determine the MMP parameter, and one could find several correlations and experimental methods as well, which are mainly applied for MMP prediction/measurement. The main factors affecting CO_2–oil MMP are reservoir temperature, oil composition, and purity of injected gas. Generally, low temperature reservoirs containing light crude oils have smaller CO_2–oil MMP. However, the impacts of impurity are not general and depend on the type of components [22]. Adding H_2S results in MMP reduction, while addition of N_2 leads to an increase in CO_2–oil MMP value [23,24].

Basically, oil recovery is higher when CO_2 and oil are miscible. In other words, there is a great deal of interest toward reaching miscibility when injecting CO_2. To give an explanation, imagine some oil on a surface. Water will get a little of oil off, but solvent will remove every trace of oil. This is because solvent can get mixed with the oil, creating a homogenous solution. Here, it can be stated water is immiscible with oil and solvent is just miscible [25,26].

3.2.1 Miscible Flooding

As discussed before, the pressure at which miscibility occurs is defined as MMP. Providing this condition during injection process will lead to a miscible CO_2 EOR, in which the recovery would be as high as 90%, theoretically. Indeed, oil recovery increases rapidly as the pressure increases and then flattens out when MMP is achieved [10,15,27,28].

Dealing with CO_2 injection, there are two types of miscible flooding, known as follows [15]:

1. First-contact miscibility (FCM): In this process CO_2 and crude oil are mixed in all proportions upon first contact, making a single homogenous solution.
2. Multiple-contact miscibility (MCM): Generally, CO_2 and crude oil are not miscible on the first contact. Indeed, miscibility occurs dynamically upon multiple contacts within the reservoir. This type of miscibility is called MCM. During this process, the composition of solutions (injection and reservoir fluids) are changed through a mass transfer between CO_2 and crude oil. This mass transfer phenomenon drives miscibility in two ways [12]:

a. Vaporizing gas drive (VGD): Miscibility is achieved through in situ vaporization of the intermediate molecular weight hydrocarbons from the reservoir oil into the CO_2.

b. Condensing gas drive (CGD): Miscibility is developed by an in situ transfer of CO_2 into the reservoir oil. In fact, CO_2 will be diffused into the crude oil.

When CO_2 interacts with reservoir oil, a dynamic miscibility zone would be developed. Therefore a CO_2-enriched crude oil is produced from the producing wells.

3.2.1.1 First-Contact Miscibility

Normally, through the FCM process, a relatively small slug volume is first injected. It is then followed by an injection of a larger and less expensive slug. These slugs are called the primary and secondary slug, respectively. Economically, these slugs should be miscible. Otherwise, a residual saturation of primary slug will be trapped within the reservoir.

In order to determine the miscibility conditions, or in other words, the possibility of the FCM process, it is essential to accurately predict fluids phase behavior in contact. Phase behavior can be shown on a ternary or pseudo-ternary diagram. Fig. 3.1 shows a typical pseudo-ternary diagram. Each of the vertices represents the pure components, and side edges of this equilateral triangle are scaled to represent the binary composition of the three possible pairs. A fluid system consisting of all three components, such as the typical crude oil characterized in this figure, is represented by points interior to the triangle.

As shown in this figure, C_1 and C_{2-6} can make a single-phase mixture in all proportions. The same is true for all mixtures of C_{2-6} and C_{7+}. However, C_1 and C_{7+}

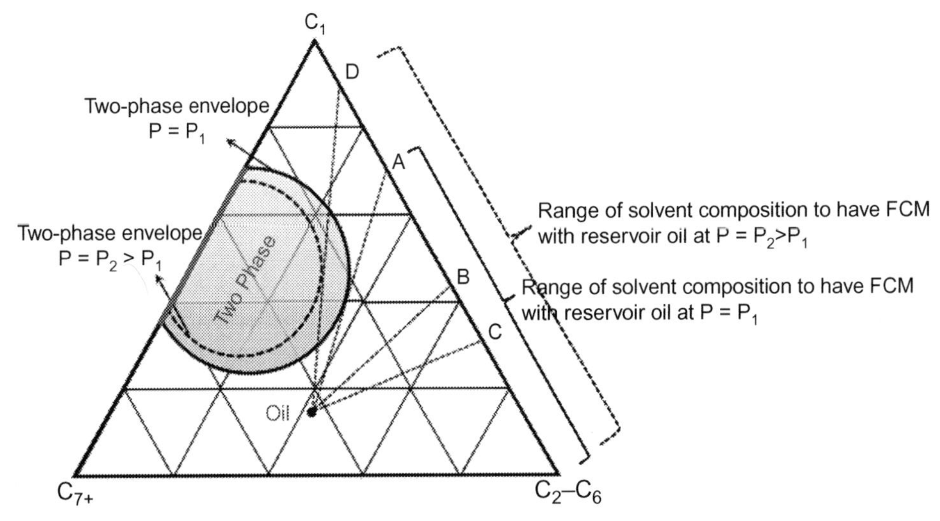

Figure 3.1 Phase behavior on ternary diagram for Methane and pseudo components [29].

could only make a one-phase mixture in a certain range of mixing ratios. The red (gray in print version) line on the edge side of C_1-C_{7+} represents the composition of mixed solutions, which will not make a single phase at a certain pressure of P_1. For pressures equal or more than P_1, the compositional ratio of the crude oil lies within the two-phase region.

Here in this typical representation, reservoir oil consists of all three components. In general, it includes intermediate to heavy compounds. As an option, we are planning to flood the reservoir with a mixture of $C_1 + C_{2-6}$. The question is: "What composition of this mixture will lead to a miscible process?" Considering that reservoir stays at pressure P_1, injection of pure C_1 will not make a miscible process. On the other hand, intermediate components are miscible with reservoir oil in all proportions. One could find this by drawing a line from reservoir oil to pure C_1 and C_{2-6}. For the case of pure C_1, the line that passes through the two-phase region, indicates an immiscible process. So the question still remains. In order to find the maximum mixture concentration of pure C_1 for the injection solution to be miscible with crude oil, a tangent line from reservoir oil to the two-phase curve is drawn. Its intersection with the side edge of C_1-C_{2-6} indicates the maximum concentration of C_1 to be added to the injection slug without altering the miscibility between crude oil and injection fluid. Any further addition of C_1 to the injection solution will lead to an immiscible process. As a result, a mixture of $C_1 + C_{2-6}$ is only a FCM process within a certain range of composition.

As shown, the pressure affects the two-phase region size. An increase in pressure will lead to a reduction in the two-phase region size. Therefore in this specific example, a higher concentration of C_1 could be used if higher pressure was set. Although pressure could modify the phase behavior toward a FCM process, it is not always possible to increase the pressure, as it may lead to formation fracture. CO_2-enriched mixtures have smaller two-phase regions compared with other gases at the fixed pressure. Thus CO_2 has been much in use for miscible injections.

As the reservoir pressure depletes, the two-phase region will be developed. Therefore CO_2 will not be totally miscible with reservoir oil anymore. A slug of C_4 enriched with CO_2 will possibly make a miscible solution with the crude oil at pressure of 1700 psi. Reservoir depletion occurs naturally, thus development of a two-phase region is inevitable. On the other hand, increasing the injection pressure is not always possible due to the operation costs and safety issues. As a result, CO_2 is not normally miscible with crude oil at first contact. Miscibility, however, could be assisted through multiple contacts.

3.2.1.2 Multiple-Contact Miscibility

Miscibility in multiple contacts occurs through two kind of mechanisms. Fig. 3.2 shows a ternary diagram, which depicts the process of VGD. Obviously, the injection solvent and reservoir oil are not miscible, as the line connecting them passes through

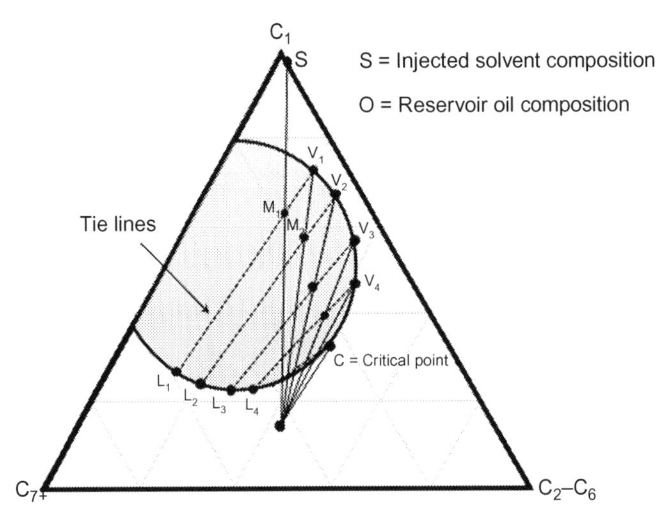

Figure 3.2 Representation of VGD process, development of miscibility [29].

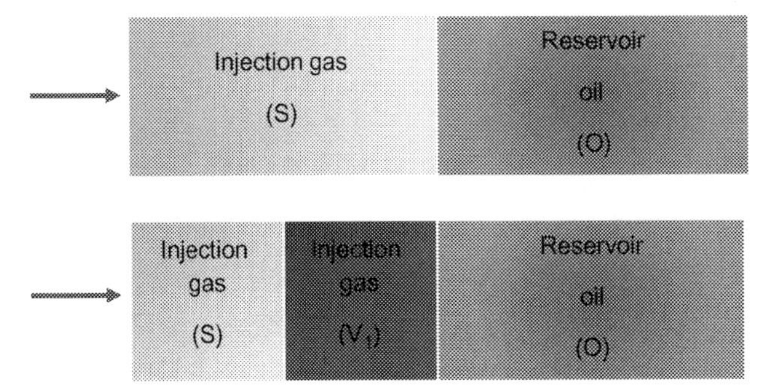

Figure 3.3 Frontal view of VGD process.

the two-phase region. However, as solvent invades through the porous media and contacts the reservoir oil, some components of oil will vaporize and transfer to the gas phase (solvent). As a result, gas and oil compositions change. Typically, the point M_1 represents the mixture composition. The new mixture consists of a gas phase (V_1) and a liquid phase (L_1). As shown in Fig. 3.3, the vapor phase V_1 moves ahead of liquid L_1 and contacts fresh oil (O). The resulting mixture will be along line V_1O, typically shown as point M_2. Mixture M_2 separates into gas phase V_2 and liquid phase L_2. The process continues till the vapor becomes miscible with oil, because the mixing line will lie entirely in the single-phase region.

Although the miscibility is developed through successive contacts within the reservoir, it will not be generated for all combinations of injection and reservoir fluid. Fig. 3.4 represents the process of vaporization, in which miscibility will not be

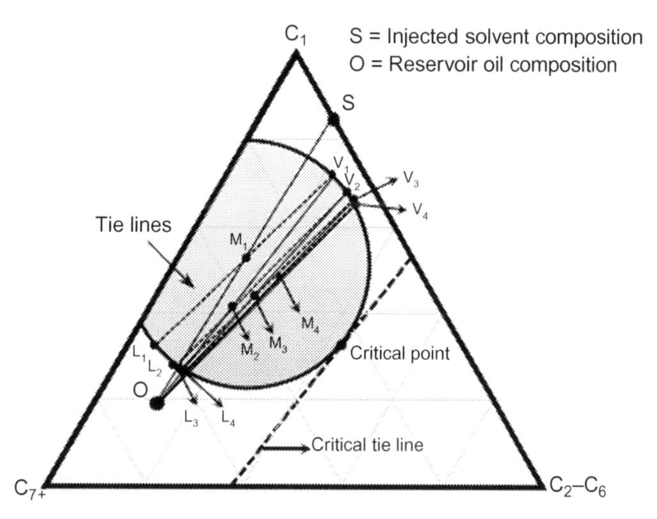

Figure 3.4 Representation of undeveloped miscibility during VGD process [29].

generated for the selected solvent. As shown in Fig. 3.4, the process of enrichment continues till the mixing line will be along a tie line. Enrichment then ceases, resulting in an immiscible process.

The limiting criteria for a multiple-contact process to be miscible through vaporization are determined by the so-called critical tie line. This line is tangent to the binodal curve at critical point. Conceptually, for a gas flooding process to be miscible through multiple contacts with VGD, the oil composition should lie on or to the right of critical line and injected fluid composition should lie to the left. This means that oil should be rich in intermediate components, while injection fluid can be a dry gas.

Consider Fig. 3.5—it represents the condensation gas drive process on a ternary diagram. Through this process, the solvent contacts the oil and some of gas components will condense and transfer to the oil phase. As a result, oil phase composition changes till it gets completely miscible with the gas phase. Similar to vaporization gas drive, miscibility will not occur for all combinations of solvent and reservoir oil. The limiting criteria is the critical tie line. Based on this criteria, the oil composition should lie to the left side of the critical tie line and injected fluid composition should lie to the right. In other words, oil should contain heavy components, whereas the injected fluid should contain a significant amount of intermediate components rather than being a dry gas. Fig. 3.6 shows a typical condition in which miscibility will not be generated.

3.2.1.2.1 Liquid (Vapor) Dropout

During vaporization process, there can be a liquid dropout behind the front. Once again, consider Fig. 3.3—the interface of new gas phase (V_1) and injection gas (S) is

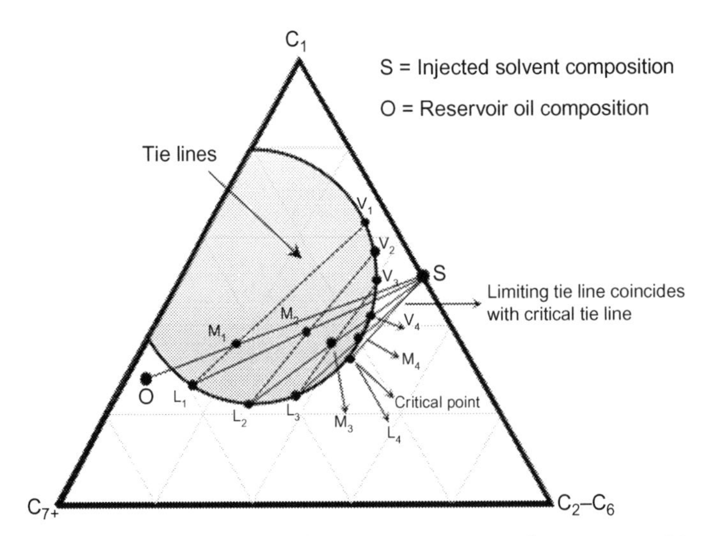

Figure 3.5 Representation of development of miscibility during condensation gas drive process [29].

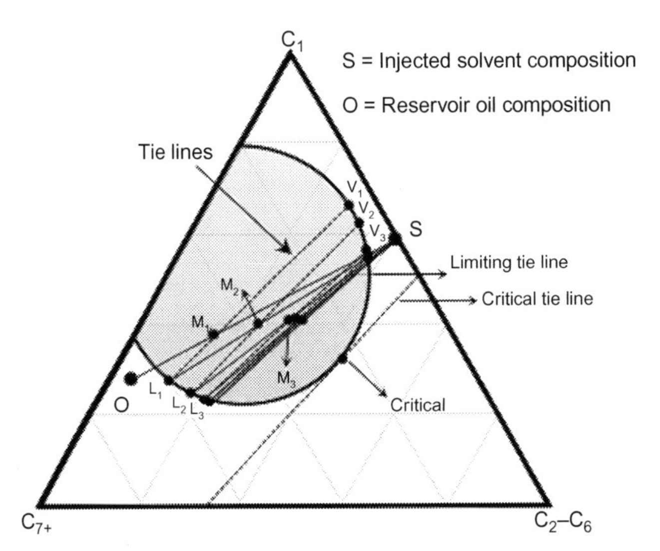

Figure 3.6 Representation of undeveloped miscibility during condensation gas drive process [29].

where liquid dropout may occur. To put this process into perspective, consider the mixing of two gas phases, which is discussed as follows:

Mixing of any two point of new gas phases is just slightly within the two-phase region. Therefore mixing leads to a relatively small amount of liquid generation. This amount of liquid could be vaporized again through multiple contacts. However, there is almost always a possibility that this small amount of liquid will remain in the reservoir. Similar to vaporization process, in condensation gas drive process a relatively small amount of vapor will be created. As the volume of dropouts is relatively small and could be followed by subsequent vaporization or condensation processes, it is of no significant concern in design process.

3.2.1.2.2 Vaporization/Condensation Gas Drive

Recent investigations on the process of MCM have suggested that an alternative mechanism, one that involves both vaporization and condensation, is often responsible for an efficient process of MCM. This is what has been shown by Zick [30] through a series of experiments on MCM. As explained in this work, reservoir oil was loaded into a PVT cell and a specific amount of gas was then injected to the cell. After equilibrium was reached, a small amount of liquid and gas phases were sampled and analyzed. Second contact was then proceeded with a fresh gas. This process continued for seven contacts.

Zick [30] noticed that this process was not only a CGD, because if it was, then the density of the liquid phase would decrease monotonically and that of the gas phase would increase monotonically. Zick [30] deduced that such a maximum–minimum-like behavior represents a combination of vaporization and condensation process, which is called vaporization/condensation gas drive.

Based on these observations, Zick [30] proposed a mechanism for the vaporization/condensation process as follows:

- First, it was assumed that oil/gas system is composed of four major groups of components as follows:
    - Lean components (C_1, N_2, and CO_2)
    - Light intermediate components, which are named as enriching components (C_1-C_4)
    - Middle intermediate components (ranging from C_4 through C_{10} on the low-molecular-weight side up to C_{30} on the high side that are generally not in the injected gas but in the reservoir oil and may be vaporized from the oil to the gas phase)
    - High molecular weight components that cannot be vaporized from the oil with significant amount.
- As enriched gas, which contains components of groups a and b contacts reservoir oil, light intermediates condense and transfer from gas phase to the oil. Eventually, oil gets lighter. The gas moves faster ahead, and fresh gas contacts the oil again. Thus oil density will further decrease. If this process continued until the oil became miscible with the injected gas, it would be the CGD process. But there is a counter effect explained in subsequent items.
- The middle intermediates are not originally in the gas phase; thus they are stripped from the oil into the gas. As lighter components are being stripped from oil, it tends to be enriched in very heavy fractions and thus becomes less similar to the

injection gas. This prevents the development of miscibility between injected gas and reservoir oil, as it was about to happen through CGD. However, if these all occurred, the process would not be very efficient. Fortunately, there is a positive mechanism as explained in subsequent lines.

- After some periods of injection, some of the oil will be rich in light intermediates. Thus as it contacts fresh gas, less condensation occurs. Nonetheless, the gas phase will strip some middle intermediates from the oil. As a result, the gas phase gets rich in both light and middle intermediates. This gas will contact the oil while stripping less middle intermediates, still losing light intermediates. Such a combination of vaporization/condensation continues until miscibility is developed in an efficient way.

3.2.1.2.3 Minimum Miscibility Enrichment

As discussed before, pressure would modify phase behavior and miscibility can be obtained for a combination of solutions that were not miscible previously. However, there is an alternative to changing pressure for miscibility to be reached. This alternative deals with injection fluid composition alternation. For instance, in a condensing gas process, injection fluid composition could be enriched to a minimum amount at which the critical tie line passes through its composition. This is to be done at a fixed pressure. This minimum enrichment is called the minimum miscibility enrichment (MME). As shown in Fig. 3.7, the miscibility at fixed pressure can only be obtained through enrichment process to MME. In this typical example, the injection fluid will be enriched with light intermediates, which decreases the density difference between reservoir oil and injection fluid. Indeed, this means more similarity between the oil and injection fluid.

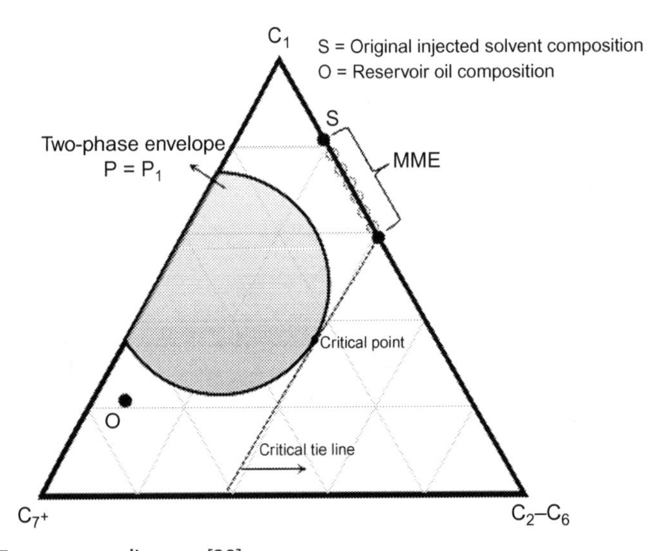

Figure 3.7 MME on ternary diagram [29].

As a general statement, heavy oil could only be assisted to flow by those solvents that contain more intermediates than lean components [12].

3.2.1.3 Screening Factors for Miscible Flooding

Theoretically, compared with other methods of EOR, miscible gas flooding results in more oil production. However, not all reservoirs can be candidates for a miscible gas flood. Generally, miscible flooding process require a deep depth for injection. At shallow depths miscibility pressure cannot be attained as it proceeds to formation fracture. There are other several screening criteria for evaluation of miscible flooding efficiency. As these criteria are met, a quick performance evaluation can depict how efficient a miscible process could be. As a key screening criteria, for instance, reservoirs with an API gravity of 30 and above are more appropriate for a successful miscible flooding. This is because high API crude oils are less viscous, and richer crude oils in intermediate components are required for miscibility to be reached through VGD or CGD process. Less viscosity also provides a more favorable mobility ratio. Moreover, it has been also reported that viscosity should be lower than 12 cP, residual oil saturation should be higher than 300 STB/acre-ft, and reservoir heterogeneity should be very low for CO_2 breakthrough to be delayed [31–34].

3.2.1.4 Miscible Flooding in Actual Fields

A large number of field applications of miscible CO_2 flooding have been implemented worldwide. Most of them had promising results. In 1989, Brock and Bryan [35] summarized their field tests with CO_2 flooding as an EOR candidate. They categorized projects into three classes: (1) field cases, (2) producing pilots, and (3) nonproducing pilots. Field cases are focused here.

In some of these projects, both continuous CO_2 flood and water alternative gas (WAG) were implemented. For instance, in Dollarhide field, CO_2 breakthrough was occurred only after 17 months. For a better mobility control process, a WAG process was then implemented.

3.2.2 Immiscible Flooding

Generally, immiscible flooding has been used for pressure maintenance. However, considerable numbers of immiscible gas injections have been applied directly as the EOR agents. Although in an immiscible process, injection gas does not mix with reservoir oil, but it partially dissolves in the oil phase, causing the crude oil to swell, and reduces its viscosity. The degree of swelling and viscosity reduction depends on the gas solubility; therefore it is essential to investigate the gas solubility into the reservoir oil. In literature, three other mechanisms have also been presented for immiscible flooding performance: (1) IFT reduction, (2) blowdown recovery, and (3) injectivity increase. These mechanisms are known as compositional effects, and they could increase oil recovery by improving oil mobility. The abovementioned mechanisms assist the oil in flowing easily; however, immiscible gas injection has a lower recovery factor compared with miscible flooding with the same operational conditions [25,36].

3.2.2.1 CO_2 Solubility in Oil

CO_2 solubility in crude oil is mainly under control of saturation pressure, temperature, and oil gravity. Generally, CO_2 solubility increases with pressure and API gravity and decreases with a rise in temperature. Oil composition and liquefaction pressure are two other factors that affect CO_2 solubility at temperatures less than CO_2 critical temperature. At this condition (i.e., subcritical CO_2 condition), CO_2 dissolves in oil as a gas rather than as a liquid. As CO_2 dissolves in oil, it affects oil viscosity, density, and IFT value, and it causes oil to swell. There are several correlations that can predict CO_2 solubility in crude oil.

3.2.2.1.1 Simon and Graue [37]

In 1965, they developed a graphical correlation for dead oils with temperatures ranging from 43.33°C to 121.1°C, pressures up to 15.86 MPa, and oil gravity from 12 to 33 °API. They presented solubility of CO_2 (mole fraction of CO_2 in a mixture of CO_2 + oil, x_{CO_2}) as a function of fugacity, saturation pressure, and temperature at Universal Oil Products Company (UOP) characterization factor (UOPK) equal to 11.7. For oils with different UOP characterization, they proposed a correction factor. Simon and Graue [37] have reported an average deviation of 2.3% between their predictions and experimental data.

3.2.2.1.2 Mulliken and Sandler [38]

In 1980, they argued about the inconvenience of the Simon and Graue [37] graphical method for reservoir simulation studies. They also stated that Simon's method is among the methods that are not applicable for impure CO_2 or mixed gases. Considering such shortcomings, they tried to develop a theoretical basis for predicting the CO_2 solubility in crude oils with a wide range of application. They applied Peng–Robinson's (PR) equation-of-state (EOS), which is as follows:

$$P = \frac{RT}{V-b} - \frac{a}{V(V+b)+b(V-b)} \tag{3.1}$$

where for mixtures:

$$a = \sum_i \sum_j x_i x_j (1-\delta_{ij})(a_i a_j)^{1/2} \tag{3.2}$$

$$b = \sum x_i b_i \tag{3.3}$$

where a and b are the constants of PR equation, x represents the mole fraction of each component in the mixture, δ is the binary interaction parameter, $a_{i,j}$ and b_i are also constants that are functions of critical pressure and temperature, and acentric factor.

Using PR EOS, Mulliken and Sadler [38] showed that it is possible to treat the oil as a single pseudo-component. Finally, they were able to characterize the system of CO_2−oil only when specific gravity, mean boiling point, or UOP factor is known. They used the same experimental data of Simon and Graue [37] for accuracy evaluation. Their model showed an average error of only 1.9% in the predicted CO_2 mole fraction in the oil; thus it shows better performance as compared with the Simon−Graue [37] correlation, which gives an average error of 2.3%.

3.2.2.1.3 Mehrotra and Svrcek [39]

They proposed a new correlation for prediction of CO_2 solubility, mainly in bitumen. However, their correlation has been applied for crude oil samples. They found a linear relationship between pressure and CO_2 solubility up to a pressure of about 6 MPa. Beyond this region predicted values deviate much from experimental data. This is indeed a limitation to the use of their correlation. Temperature range was also set to be from 23.89°C to 97.22°C. An average deviation of 6.3% is reported for their correlation. The correlation is as follows:

$$\text{Solubility } (\text{cm}^3\ \text{CO}_2/\text{cm}^3\ \text{mixture}) = b_1 + b_2 P + b_4 \frac{P}{T} + b_4 \left(\frac{P}{T}\right)^2 \tag{3.4}$$

with

$$b_1 = -0.0073508, b_2 = -14.794, b_3 = 6428.5, \text{ and } b_4 = 4971.39$$

where T is in K and P is in MPa.

3.2.2.1.4 Chung et al. [40]

Chung et al. [40] defined the solubility of CO_2 in a crude oil, R_s, as the volume (in scf) of CO_2 in the CO_2-saturated oil per barrel of dead-state oil at the temperature at which solubility is measured. They stated that CO_2 solubility mostly depends on temperature and pressure and only slightly on specific gravity. Referring to the proposed solubility diagram as a function of pressure by Chung et al. [40], they postulated the following discussions about CO_2 solubility.

"*As* demonstrated in the solubility diagram developed by Chung et al. [40], *solubility of CO_2 in heavy oil increases with pressure but decreases with temperature at pressures below 3000 psia. The line of the 75°F (24°C) isotherm shows that solubility of liquid CO_2 (at pressures greater than 1000 psia (6.9 MPa)) in oil is not strongly sensitive to pressure. The solubilities of gases in liquids are normally decreased with the increase of* *temperature at low pressures because light components (gas molecules) tend to vaporize at high temperature. As pressure increases, however, liquid becomes denser at lower temperatures—i.e., molecules in the liquid phase are packed more tightly and thus leave less room for gas molecules to enter. Therefore, at high pressures, the solubility of gas in liquid may increase with temperature because of the decrease in liquid density. This phenomenon was shown in the plot of CO_2 solubility. The isothermal line of 200°F (94°C) crosses the line of 140°F (60°C) at pressures above 3000 psia (20.7 MPa).*"

Based on these statements and experimental results, they proposed the following correlation:

$$R_s = 1/\left[a_1\gamma^{a_2}T^{a_7} + a_3T^{a_4}\ \exp\left(-a_5P + \frac{a_6}{P}\right)\right] \tag{3.5}$$

where R_s is solubility in scf/bbl, T is temperature in °F, P is pressure in psia, and γ is specific gravity. The empirical constants a_1 through a_7 are 0.4934×10^{-2}, 0.928, 0.571×10^{-6}, 1.6428, 0.6763×10^{-3}, 781.334, and -0.2499, respectively. They used the correlation for three different oil samples and have reported an average deviation of 5.9% for Cat Canyon oil, 7.6% for Wilmington oils, and 2% for Densmore oil.

Example 3.1: Consider the Chung et al. [40] correlation. Answer the following questions:

a. What would happen if the pressure is increased to a very high value?

b. What would happen if the temperature is increased to a very high value?

Solution:

a. As the pressure is increasing to a very high value, the term a_6/P would vanish. Therefore the term $\exp(-a_5P + a_6/P)$ is reduced to $\exp(-a_5P)$. Putting a high value in this remained term would also vanish the whole term $a_3T^{a_4}\exp(-a_5P)$. As a result, the R_s could be calculated by $1/[a_1\gamma^{a_2}T^{a_7}]$. This means that CO_2 solubility is only dependent on temperature and specific gravity of the crude oil at high pressures.

b. At very high temperatures, the term $a_1\gamma^{a_2}T^{a_7}$ is nearly equal to zero. What remains is the term $a_3T^{a_4}\exp(-a_5P + a_6/P)$, which would obviously result in $R_s = 0$ at very high temperatures.

3.2.2.1.5 Emera and Sarma [41]

In 2007, they used genetic algorithm (GA) and proposed a new set of correlations. Emera and Sarma [41] correlations are as follows:

- Dead oil:
  a. For temperatures above critical temperature of CO_2 at any pressure:

$$CO_2 \text{ solubility } \left(\frac{\text{mol}}{\text{mol}}\right) = 2.238 - 0.33y + 2.23y^{0.6474} - 4.8y^{0.25656} \quad (3.6)$$

where $y = \gamma\left(\frac{T^{0.8}}{P_s}\right)\exp\left(\frac{1}{\text{MW}}\right)$. For this correlation, the CO_2 solubility at P_b (bubble point pressure equal to 1 atm for the dead oil case) is taken to be equal to zero.

b. For temperatures below critical temperatures of CO_2 at pressures under liquefaction pressure of CO_2:

$$CO_2 \text{ solubility } \left(\frac{\text{mol}}{\text{mol}}\right) = 0.033 - 1.14y - 0.7716y^2 + 0.217y^3 - 0.2183y^4 \quad (3.7)$$

where $y = \gamma\left(\frac{P_s}{P_{\text{liq}}}\right)\exp\left(\frac{1}{\text{MW}}\right)$.

In Eqs. (3.6) and (3.7), γ is the oil-specific gravity (oil density at 15.6°C), T is the temperature (°F), P_s is the saturation pressure (psi), P_{liq} is the CO_2 liquefaction pressure at the specified temperature (psi), and MW is the oil molecular weight (g/mol).

For accuracy evaluation, they examined the model with those data of Simon and Graue [37], Mehrotra and Svrcek [39], and Chung et al. [40].

Emera and Sarma also conducted a sensitivity analysis for the factors that affect CO_2 solubility in dead oil. They found that the GA-based CO_2 solubility correlation for the dead oil depends, primarily, on the saturation pressure and temperature. Also, it depends to a lesser degree on the oil-specific gravity and oil molecular weight.

- Live oil:

a. In case of gaseous CO_2, for temperatures greater than CO_2 critical temperature ($T_{c,\text{CO2}}$) at all pressures, and for temperatures less than $T_{c,\text{CO2}}$ at pressures less than the CO_2 liquefaction pressure, CO_2 solubility can be calculated by the following correlation:

$$CO_2 \text{ solubility } \left(\frac{\text{mol}}{\text{mol}}\right) = 1.748 - 0.5632y + 3.273y^{0.704} - 4.3y^{0.4425} \quad (3.8)$$

where $y = \gamma\left(0.006897 \times \frac{(1.8T+32)^{1.125}}{P_s - P_b}\right)^{\exp\left(\frac{1}{\text{MW}}\right)}$. For this correlation, it is considered that the CO_2 solubility at P_b is equal to zero.

b. In case of liquid CO_2, for temperatures less than $T_{c,\text{CO2}}$ and pressures greater than CO_2 liquefaction pressure, they suggested the same correlation used for the solubility in the dead oil (as given in Eq. 3.7) can also be used for the live oil.

For accuracy evaluation, they examined the model with those data of Simon and Graue [37].

They also performed a sensitivity analysis of the factors affecting CO_2 solubility in live oil. As they compared the results of sensitivity analysis for the live and deal oil, they found that the saturation pressure effect on the CO_2 solubility in live oil is higher than that in the dead oil. The temperature effect, on the other hand, is lower in the live oil case.

3.2.2.1.6 Rostami et al. [42]

Recently, Rostami et al. [42] have proposed two new and accurate models for prediction of CO_2 solubility. In their research, they examined different correlations over a wide range of experimental data. In the second step, they developed two models using neural networks and gene expression programming (GEP). In a comparative study, they concluded that GEP is much more accurate than any other correlation both for live and dead crude oils. They have proposed this method as a feasible approach for CO_2 flooding simulations. Their proposed correlations are as follows:

- Dead oils:

  Based on the latest investigations, they considered that the key variables influencing CO_2 solubility in dead oil are oil molecular weight (MW), specific gravity of oil (γ), reservoir temperature (T), and saturation pressure (P_s). The CO_2 solubility in dead oils can be calculated by the following equation:

$$R_s = \frac{P_s T(5.6444 + 0.008756\text{MW})}{8.9318P_s^2 + 0.010819\text{MW}P_s T + T^2 + 41.105^{\gamma} T} \tag{3.9}$$

- Live oil:

  For the case of live oils, they considered that CO_2 solubility is primarily dependent on oil molecular weight (MW), specific gravity of oil (γ), reservoir temperature (T), saturation pressure (P_s), and bubble point pressure (P_b). Eq. (3.10) has been developed by Rostami et al. [42] for predicting CO_2 solubility in live oils as follows:

$$R_s = \frac{7.3695P_b - 7.3713P_s + 0.48618}{0.021262\text{MW} + 4.6233P_b - 5.0337P_s - \gamma T - A} \tag{3.10}$$

  In which the parameter A is a conditional function, which is defined by the following relationship:

$$A = \begin{cases} 0 & \text{if } \gamma \leq 0.849 \\ 0.042756\text{MW} & \text{if } \gamma > 0.849 \end{cases} \tag{3.11}$$

  For both cases, the units for MW, T, P_s, P_b, and R_s are g/mole, °C, MPa, MPa, and mole fraction, respectively.

3.2.2.2 Swelling Effects

This effect is the most obvious effect that gas injection could have on oil recovery during an immiscible process. When a reservoir oil is not saturated with gas or reservoir pressure increases due to gas injection so that more gas can be dissolved, the volume of gas dissolved in oil will increase until the oil is saturated at that pressure. As this phenomenon occurs, oil formation volume factor (FVF) will increase. This phenomenon is called oil swelling, which can significantly increase oil recovery. Swelling effects are less significant for those reservoirs with a gas cap; however, for the reservoir oils that do not have an associated gas cap or have low bubble point pressure, this mechanism has high effectiveness. Indeed, the swelling phenomenon is important for two reasons. Firstly, the residual oil saturation decreases as it is inversely proportional to swelling factor. The residual oil saturation is an important point in relative permeability curves and determines ultimate recovery. Secondly, as the crude oil swells, it drives trapped oil droplets out of pores, leading to a drainage process. It also increases oil saturation, which actually increases oil relative permeability. All of these result in increased oil recovery. In comparison to all common nonhydrocarbon gases used for immiscible gas injection, CO_2 promotes oil relative permeability to higher degrees [14].

Most common correlations used for the swelling factor of CO_2-saturated oil mixtures are as follows:

3.2.2.2.1 Welker [43]

In their model, the swelling factor is a linear function of CO_2 solubility. Although their correlation is simple, it is mainly applicable to crude oils with API gravities from 20 to 40 °API. The formula was also developed for dead oils at temperature equal to 80°F. Eq. (3.12) represents the Welker [43] correlation for the swelling factor as given as follows:

$$SF = 1.0 + \frac{1.96525 \times \mathrm{Sol}(\mathrm{m}^3/\mathrm{m}^3)}{1000} \tag{3.12}$$

Welker [43] examined their correlation over 13 crude oils and reported an average deviation of 0.01. Chung et al. [40] have also reported a good fit of their data on the Welker [43] correlation (Fig. 3.8).

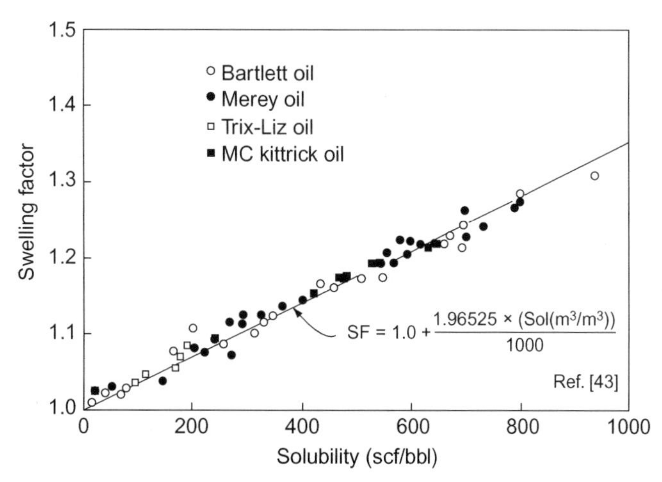

Figure 3.8 Correlation for swelling factor [43].

3.2.2.2.2 Simon and Graue [37]

In 1965, Simon and Graue [37] postulated that the swelling factor is not only a function of the amount of dissolved CO_2, but also of the size of the oil molecules (M/ρ cc/g mol). They presented a graphical tool for predicting the swelling factor, which is based on the relationship between swelling factor, mole fraction of CO_2 dissolved, and molecular size. Obviously, the swelling effect for heavy oil is not as drastic as for light oil.

Simon and Graue [37] have reported an average deviation of 0.5% from experimental data.

3.2.2.2.3 Mulliken and Sadler [38]

Mulliken and Sadler have tried to find a theoretical correlation for swelling factor prediction. They applied the PR [44] EOS once to the CO_2-saturated crude oil at saturation conditions, and then again to the crude oil at 1 atm and the same temperature. The swelling factor was defined as the ratio of the volume of the oil$-CO_2$ mixture at saturation pressure and temperature to the oil volume at the saturation temperature and atmospheric pressure.

3.2.2.2.4 Emera and Sarma [41]

Based on Emera and Sarma [41] correlation, oil swelling factor can be predicted as follows:

a. For oils with MW $\geq$ 300:

$$SF = 1 + 0.3302y - 0.8417y^2 + 1.5804y^3 - 1.074y^4 + 0.0318y^5 - 0.21755y^6 \quad (3.13)$$

b. For oils with MW $<$ 300:

$$SF = 1 + 0.48411y - 0.9928y^2 + 1.6019y^3 - 1.2773y^4 + 0.48267y^5 - 0.06671y^6 \quad (3.14)$$

where $y = 1000\left(\left(\left(\frac{\gamma}{MW}\right) \times \text{Sol}^2\right)^{\exp\left(\frac{\gamma}{MW}\right)}\right.$. γ is the oil-specific gravity (oil density at 15.6°C), SF is the oil swelling factor, Sol is the CO_2 solubility in oil in mole fraction, and MW is the oil molecular weight in g/mol.

The experimental data ranges used for developing and testing of the model for dead and live oil are given in the work of Emera and Sarma [41]. Emera and Sarma [41] also examined their model over that of Simon and Graue [37].

3.2.2.2.5 Viscosity Reduction

By definition, viscosity is the resistance of a fluid to flow on a solid surface. Therefore any flow equation accounts fluid viscosity for calculations. In a reservoir being flooded immiscibly with gas, there is system of two-phase fluid flow. It is therefore very important to accurately predict and model flow behavior of each fluid. Viscosity is a parameter that could assist flow modeling/prediction as it is involved in a dimensionless parameter called mobility ratio. Mobility relates fluid flow resistance to rock property of a porous medium [12]. Mobility ratio is defined as the mobility of displacing fluid (gas) divided by the mobility of displaced fluid (crude oil). Eq. (3.15) presents a two-phase mobility ratio as follows:

$$M = \frac{k_{rg}}{\mu_g} / \frac{k_{ro}}{\mu_o} \tag{3.15}$$

where k_{rg}, k_{ro} μ_g, and μ_o stand for gas relative permeability, oil relative permeability, gas and oil viscosity, respectively.

Fundamentally, for an injection process to be very efficient from both the microscopic and macroscopic point of view, it is essential to keep the mobility ratio low enough. For a typical gas injection process, M varies from 20 to 100. Such a high value for mobility ratio results in high instabilities and arises fingering potential. Consequently, it results in early breakthrough of injected gas, which in turn leads to a small incremental oil recovery [45,46]. Fig. 3.9 depicts the effects of mobility ratio on flow stability.

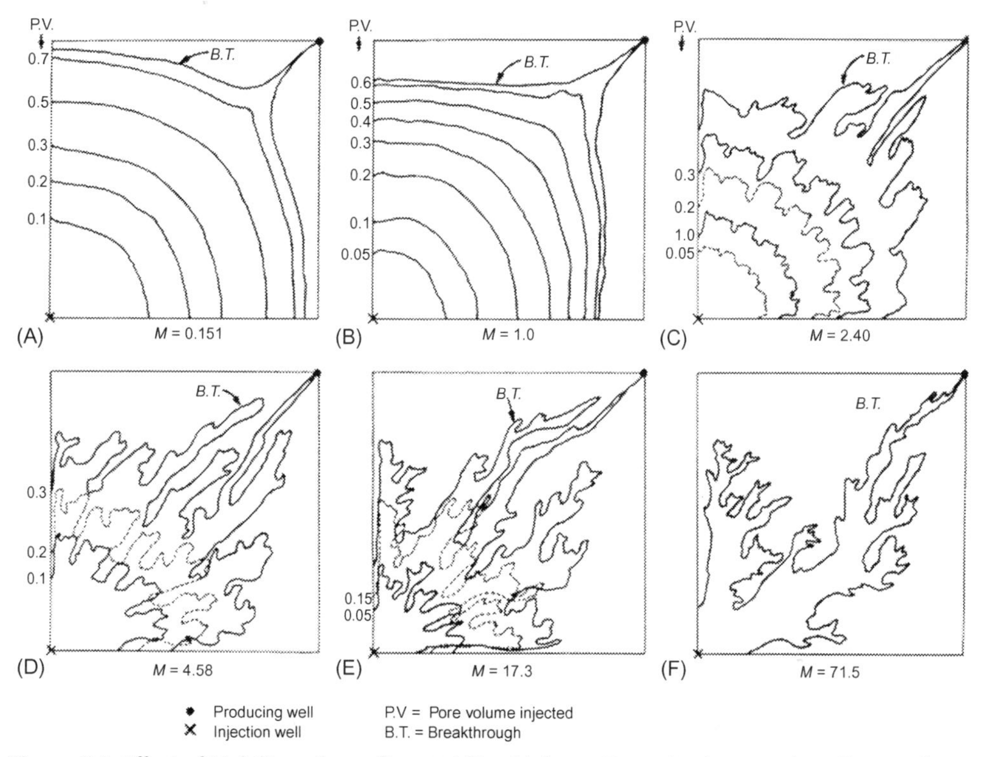

Figure 3.9 Effect of Mobility ratio on flow stability, high motility ratios (greater than 1) are unfavorable: (A) $M = 0.151$, (B) $M = 1.0$, (C) $M = 2.40$, (D) $M = 4.58$, (E) $M = 17.3$, and (F) $M = 71.5$ [47].

As it can be seen, the mobility ratio increases through A to F, and as a result, flow instability occurs. At high mobility ratios, gas will break through via narrow fingers at very low pore volume injected. This results in a low sweep efficiency, which means that low oil recovery would be achieved. Favorably, a stable displacement would occur as long as $M < 1$, and an unstable fingering displacement would occur for $M > 1$. The other plot for representing such instability is shown in Fig. 3.10 [12,48].

Based on the abovementioned discussions, a modification to M toward smaller values would stabilize the flow condition. When CO_2 is injected into the reservoir, it dissolves in oil and reduces the oil viscosity. Therefore mobility ratio is modified to a smaller value. However, CO_2 mobility is still high compared with that of the oil, and fingering potential could be very high, especially when permeability variations in a stratum are considerable.

Indeed, viscosity reduction is an effective mechanism accounting for increasing oil recovery during immiscible CO_2 injection; nonetheless, the degree of effectiveness is highly dependent on oil properties and rock characteristics. Viscosity effects are totally more profound on heavy oils than light ones. At low viscosity values, this is water flood, which is technically superior to immiscible CO_2 flood. This is due to the more favorable mobility control through the water flooding process. However, CO_2 immiscible flooding recovers significantly more oil than inert gas drive at low viscosities. This observation could be attributed to the better mobility ratio due to viscosity reduction and greater swelling of the oil, thus leaving less residual oil in place. At higher viscosities (i.e., 70–1000 mPa · s), CO_2 injection appears to be superior than the other methods. This is because of higher viscosity reduction and swelling associated with CO_2 injection. Typically, natural gas-saturated oils have viscosities in the range of 0.7 to 700 mPa · s, and carbonated oils have viscosities between 0.3 and 30 mPa.s. This serves less mobility ratio during CO_2 injection, which is more

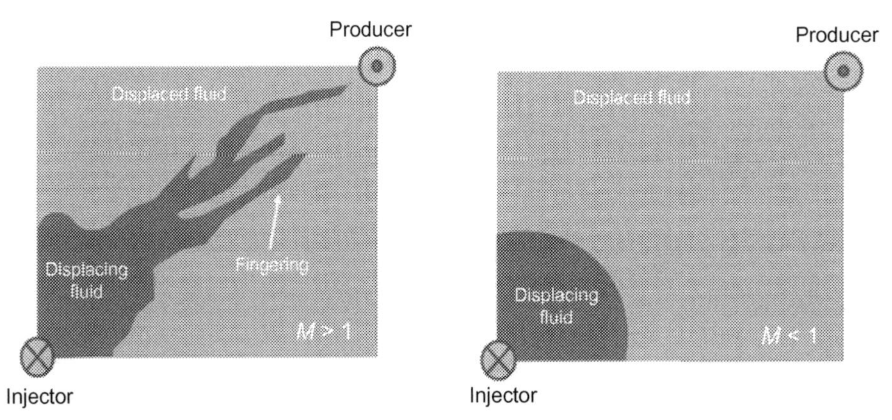

Figure 3.10 Effect of mobility ratio on flow stability for both values less than and more than unity [49].

favorable toward a higher sweep efficiency. In addition, it is interesting to note that CO_2 effects on viscosity reduction are minimal at temperatures above 150°C, because at this temperature solubility of CO_2 experiences large decreases.

A number of correlations can be found in literature, which are applicable for viscosity calculation of crude oils mixed with a gas (e.g., CO_2).

3.2.2.2.6 Welker and Dunlop [43]

They presented a graphical method for viscosity prediction of carbonated crude oils. They first designed an experimental setup, and based on the original Darcy's equation, developed their model of viscosity prediction. The setup basically consisted of a small-diameter steel tube, pressure gauges, flow controllers, and inlet–outlet accumulators. A detailed description of their setup has been given in their chapter. Having laminar stabilized flow rates, finally, they were able to measure the viscosity of different crude oil–CO_2 solutions. Fig. 3.11 shows the results of experiments conducted by Welker and Dunlop [43]. As it can be seen in this figure, they presented viscosity ratio values instead of viscosity itself. Indeed, such a presentation would give a good understanding of viscosity reduction due to CO_2 dissolution in crude oil. Based on this figure, they found that viscosity reduction is much more pronounced for those crude oils with a high dead oil viscosity. For this statement to be clear, consider a high viscosity oil (e.g., $\mu = 1000$ cP) and a low viscosity one (e.g., $\mu = 40$) both being kept at pressure of 200 psia. Based on Fig. 3.11, the viscosity has reduced to about 25% of dead oil viscosity for the high viscosity oil, while this reduction is read to be about 45% for low viscous one.

Welker [43] also stated that the viscosity reduction using dissolved CO_2 is greater than that for natural gas or pure methane (Fig. 3.12). This means that a higher

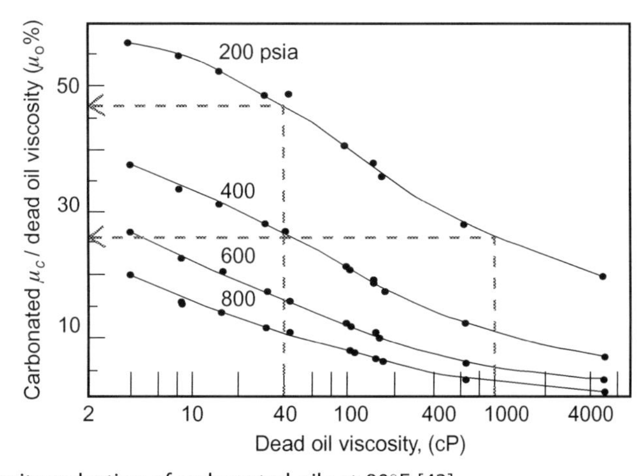

Figure 3.11 Viscosity reduction of carbonated oils at 80°F [43].

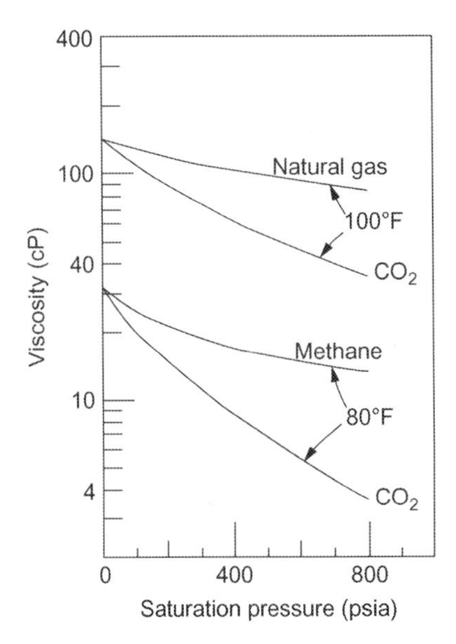

Figure 3.12 Comparison of viscosities of oils containing natural gas, methane, and CO_2 [43]

amount of oil could be recovered due to viscosity reduction during a conventional CO_2 flood compared with natural gas or pure methane flooding.

Finally, Welker [43] presented a quick graphical correlation for the prediction of carbonated crude oils as a function of dead oil viscosity and saturation pressure. The use of the Welker [43] correlation is limited to 80°F, saturation pressures up to 800 psia, and crude oils with viscosity in the range of 4–5000 cP.

3.2.2.2.7 Simon and Graue [37]

In their research, they gathered a set of experimental data measured in the temperature range of 110 to 250°F. These data were measured in two steps. First, the atmospheric viscosity was measured at a fixed temperature. Secondly, a mixture of CO_2–oil was prepared, and its viscosity and bubble point pressure were measured at the same temperature. A correlation relating the CO_2–oil viscosity (μ_m) to the mixture saturation pressure and to the original oil viscosity (μ_o) was then proposed by them. Basically, this correlation was prepared for 120°F.

Simon and Graue [37] have reported an average deviation of 9% for systems at 120°F. They have also reported an average and maximum deviation of 7 and 14, respectively, for systems at temperatures other than 120°F.

3.2.2.2.8 Beggs and Robinson [50]

Their model was basically developed by plotting $\log_{10}(T)$ versus $\log_{10}\log_{10}(\mu_{oD} + 1)$ on Cartesian coordinates. The plots revealed a series of straight lines of constant slope, each representing an oil of a particular °API. Beggs and Robinson [50] neglected the dependence of oil viscosity on the composition since oils of widely varying compositions have the same gravity. They also neglected the effects of pressure on viscosity. Formulations are as follows [50]:

- Dead oil:

$$\mu_{oD} = 10^X - 1 \tag{3.16}$$

where

$$X = yT^{-1.163}$$
$$y = 10^z$$
$$z = 3.0324 - 0.02023\gamma_0$$

μ_{oD} is dead oil viscosity in cP, T is temperature in °F, and γ_0 is oil-specific gravity in °API.

- Live oil:

$$\mu = A\mu_{oD}^B \tag{3.17}$$

where

$$A = 10.715(R_s + 100)^{-0.515}$$
$$B = 5.44(R_s + 150)^{-0.338}$$

In Eqs. (3.16) and (3.17), μ is live oil viscosity in cP, and R_s is dissolved gas ratio in SCF/STB.

3.2.2.2.9 Mehrotra and Svrcek [39]

They presented a double-logarithm type correlation as follows:

$$\log\,\log\mu = a_1 + a_2T + a_3P + a_4\frac{P}{(T + 273.16)} \tag{3.18}$$

where μ is the viscosity of the crude in mPa.s, T is the temperature in °C, P is the pressure in MPa, and $a_{1,2,3,4}$ are the correlation constants given as follows:

$$a_1 = 0.815991 \quad a_2 = -0.0044495 \quad a_3 = 0.076639 \quad a_4 = -34.5133$$

It should be mentioned that this correlation was basically developed for CO_2-saturated bitumen. However, it can be used for crude oils with an error. The use of their correlation is limited to a temperature range from 23.89°C to 97.22°C and pressures up to 6.38 MPa.

3.2.2.2.10 Chung et al. [40]

Chung et al. [40] developed a correlation for viscosity prediction of CO_2/heavy oil mixtures. In general, they stated: "*the viscosity of CO_2/heavy oil mixture is a function of composition. This composition-dependent function is extremely complex for CO_2/heavy oil mixtures.*" Due to this fact, they suggested to treat the mixture as a binary system with two components: pure CO_2 and heavy oil. Based on experimental results, they also concluded that the viscosity change of heavy oil is related to the quantity of CO_2 dissolved in the oil. Finally, they developed a model for predicting the viscosity of CO_2-satuated mixtures while the concentration of CO_2 in the oil and viscosities of both CO_2 and oil are known. Chung et al. [40] used the Lederer [51] equation as the basic formulation to their model. The formulations are as follows [40]:

$$\ln \mu_m = X_o \ln \mu_o + X_s \ln \mu_s \tag{3.19}$$

with

$$X_s = \frac{V_s}{\alpha V_o + V_s} \tag{3.20}$$

and

$$X_o = 1 - X_s \tag{3.21}$$

where V is volume fraction; μ is viscosity in mPa.s; and the subscripts m, o, and s stand for mixture, heavy oil, and CO_2, respectively. In Eq. (3.20), α is an empirical constant that can be determined by Eq. (3.22) as follows:

$$\alpha = 0.255\gamma^{-4.16} T_r^{1.85} \left[\frac{e^{7.36} - e^{7.36(1-p_r)}}{e^{7.36} - 1} \right] \tag{3.22}$$

where $T_r = T/547.57$ and $p_r = p/1071$ are reduced temperature and pressure, respectively; T is temperature in °R; and P is pressure in psia.

The volume fraction of CO_2 in the mixture can be obtained from the CO_2 solubility or swelling factor according to their definitions. Accordingly, X_s can be calculated by Eq. (3.23) as follows:

$$X_s = \frac{1}{\alpha F_{CO2}/(F_o R_s) + 1} \tag{3.23}$$

where F_{CO2} is the ratio of CO_2 gas volume at standard conditions to the volume at system temperature and pressure, and F_o is the ratio of oil volume at system temperature and 1 atm (0.101 MPa) to the volume at system temperature and pressure. It must be noted that for viscosity calculation, one needs to determine the CO_2 viscosity (μ_s) at the specified temperature and pressure conditions. Chung et al. [40] have reported an average deviation of 3.5% between measured and calculated viscosities for 429 data points.

3.2.2.2.11 Emera and Sarma [41]

Emera and Sarma [41] developed an empirical model based on the CO_2 solubility, initial oil viscosity, saturation pressure, temperature, and oil-specific gravity as follows:

$$\mu = y \times \mu_i + A\left(\frac{\text{Sol(mole fraction)}}{\mu_i}\right) \tag{3.24}$$

where

$$y = x^B$$

$$x = (C \times \mu_i(P_s/1.8T+32))D)^{(\gamma \times \text{Sol})}$$

$$A = -9.5, B = -0.732, C = 3.14129, D = 0.23$$

P_s is saturation pressure in psi, T is temperature in °F, Sol is CO_2 solubility in oil in mole fraction, and γ is oil-specific density.

3.2.2.2.12 IFT Reduction

IFT determines the mixing potential between two fluids. In an immiscible process, the IFT between CO_2 and oil is not very close to zero. However, when CO_2 dissolves in the oil, the IFT between CO_2 and the oil phase reduces. It is not only the IFT between CO_2 and the oil that decreases during an immiscible process, but also the IFT between the oil and water decreases. Such a drop in IFT value will result in an increase oil recovery when water is considered to be flooded alternatively/simultaneously with CO_2.

To make a clear picture of IFT importance and its effects on oil recovery, a dimensionless number is defined as the capillary number (N_{ca}). In essence, the capillary number is the ratio of viscous versus capillary forces. Commonly, capillary number is formulated as follows [12]:

$$N_{ca} = \frac{\text{Viscous force}}{\text{Capillary force}} = \frac{v \times \mu}{\sigma} \tag{3.25}$$

where v is the velocity, μ denotes the viscosity, and σ indicates the IFT.

In practice, it is preferred to have a viscous-dominated flow regime. In another words, a large capillary number leads to less residual trapped oil saturation. Looking at Eq. (3.25), it is assumed that any technique raising the product of velocity and viscosity does increase oil recovery, as it increases the capillary number. Owing to this fact that the prementioned product is proportional to the pressure drop; thereby, the injection pressure is limited to fracture pressure. Therefore it is of great interest to make a modification to the other parameter (i.e., σ) in order to raise the value of the capillary number. Indeed, CO_2 injection would practically decrease IFT value and consequently increase N_{ca}. Ultimately, it results in the mobilization of trapped oil droplets within the porous media. This means higher microscopic efficiency.

3.2.2.2.13 Blowdown Effects

After CO_2 injection terminates, pressure depletion into the reservoir will occur naturally. As a result, dissolved CO_2 will be liberated. This process acts just as the solution gas drive, which is responsible for oil recovery. This process is called the blowdown recovery effect. This is highly dependent to the compressibility of oil and CO_2 at reservoir condition as well as the amount of the CO_2 dissolved in oil [18].

3.2.2.2.14 Injectivity Increase

A mixture of water and CO_2 could create an acidic solution that can react with the carbonated portion of the rock matrix. Such reactions result in rock dissolution, thus increasing the permeability of the rock. As this acidic solution contacts the reservoir oil, it may lead to asphaltene precipitations, which is highly dependent on oil properties. Some oils have a high potential for asphaltene precipitation occurrence as they make contact with such acidic solutions [52].

3.2.2.3 Immiscible Flooding Field Cases

In most of the field case studies, miscible flooding has been actually implemented. However, as heavy oil resources are increasing in number, the use of immiscible flooding has also been attracting more global attention. The first field CO_2 immiscible project was carried out in 1949. Although at that time oil recovery was lower than that of measured in laboratory, it was proved that CO_2 immiscible flooding could improve oil recovery to an acceptable level. As examples, two field projects with detailed information are reviewed here.

3.2.2.3.1 Lick Creek Field, United States [53]

This field is located in southern Arkansas and was discovered in 1957. It is reported that production rate peaked at 1900 BPD and declined at 230 BPD in 1960 and 1976, respectively. Total oil recovery till the start of immiscible CO_2 flood in 1976 was 4.486 million bbl, equivalent to 28.3% OOIP.

The CO_2 was sourced from 65 miles away, transported through pipelines in supercritical state. As its production history showed, the field responded positively to the CO_2 project. Production increased from 8000 to 28000 bbl/month. Till 1990, CO_2 injection contributed to cumulative production of about 11% of OOIP. It has been reported that the relatively low tertiary recovery was due to the occurrence of CO_2 channeling in the high permeability sands [53].

3.2.2.3.2 Bati Raman Field, Turkey [54]

This field, which is the largest oil field in Turkey, was discovered in 1961 with an estimated OOIP of 1850 million STB. It was found that the formation is limestone with vertical and horizontal heterogeneity. At primary recovery, only 1.5% of OOIP was produced due to the high oil viscosity. Water flooding was carried out for 7 years (i.e., 1971–1978). Injecting 3.2 million bbl of water achieved only 5% of oil recovery.

A CO_2 immiscible flooding was planned with a gas containing 91% of CO_2. The injected CO_2 converted carbonate formation into water-soluble calcium bicarbonate. As a result, pore volume and permeability were increased. CO_2 injection was carried out in two phases. In the first phase, Huff and Puff was conducted from 1986 to 1988. Afterward, a WAG process was started in 1988.

3.2.2.3.3 Wilmington Field, United States [55]

The Wilmington field, located in Los Angeles, was discovered in 1936. This field is a layered formation, which is somewhat arbitrarily divided into seven reservoirs located at depths ranging from 2300 to 4800 ft. This formation contained a crude oil of 13–28 °API, and temperature was in the range of 123–226°F throughout the reservoirs. Such a low gravity crude oil contained in this multilayered formation along with many faults have created a significant technical and economic challenge to the profitable recovery of a significant fraction of Wilmington's original oil in place. Therefore different EOR strategies were studied. Water flooding started in 1961 and continued until the end of May 1980. Primary recovery and water flooding recovered only 30% OOIP. Five other EOR processes—polymer flooding, caustic flooding, micellar/polymer, CO_2, and steam flooding—were then piloted in this field. Among these process, CO_2 flooding showed a significant improvement.

The shallowest reservoir, Tar Zone, was chosen for pilot CO_2 injection. The injected gas was composed of 85% carbon dioxide and 15% nitrogen. CO_2 was injected in liquid state at the early stage of the project. However, it changed to gaseous state later. The WAG technique was also employed to slow down gas breakthrough. The CO_2 injection rate was kept in the range of 1000–1500 MSCF/D per well. The water injection rate was held at 1000 B/D per well to avoid formation fracturing.

3.2.2.3.4 Forest-Oropouche Reserves, Trinidad [56]

Between the years of 1973 and 1990, four major CO_2 immiscible flood projects (namely, EOR4, 26, 33, and 44) were implemented in Trinidad at Forest-Oropouche reserves. The detailed reservoir data are presented in the relevant literature [56].

In the EOR 4 project, the primary recovery was about 21.3% of OOIP. A secondary gas flood recovered about 20% of OOIP. CO_2 injection was then started in 1986. From 1992 to 1994, the average oil rate was around 60 BPD. From 1995 to 1998, the oil rate increased to 200 BPD. Overall, CO_2 injection resulted in 2.2% of OOIP recovery.

3.3 CO_2 INJECTION METHODS

3.3.1 Injection Location

Depending on the location of gas injection wells, there are two types of injection techniques, including crestal and pattern injections [57].

3.3.1.1 Crestal Injection

This type of injection is sometimes called external or gas cap injection. In this type, gas is usually injected into higher structural positions in primary or secondary gas caps. This method of injection is normally applied to reservoirs with thick oil columns and good vertical permeability. Indeed, recovery is assisted by the gravity drainage process. Crestal injection, if applicable, is superior to pattern injection due to its higher volumetric sweep efficiency [58].

3.3.1.2 Pattern Injection

Pattern injection is usually called the internal or dispersed injection method. Through this type of injection, a geometric arrangement of injection wells is designed for the purpose of uniform distribution of the injected gas throughout the oil-productive portions of the reservoir. This type of injection is normally used for reservoirs with a low vertical permeability and relatively homogeneous reservoirs with low permeabilities. Well spacing could be regular (e.g., five spot) or irregular. Injection and producing wells are located in an area, thus there is no great benefit of gravity drainage drive. Sweep efficiency decreases due to high potential of gas override and fingering. Few pattern gas injection projects have been implemented in recent years because this method is not very attractive economically [58].

3.3.2 Injection Mode

CO_2 injection could be implemented in different ways. Generally, CO_2 is injected either as continuous or injected alternatively with water. When CO_2 is injected continuously into the reservoir, it contacts with crude oil to recover the oil through different mechanisms, depending on the degree of miscibility. Although mixing CO_2 with reservoir oil can reduce oil viscosity, mobility ratio in a continuous CO_2 injection is unfavorable due to very low viscosity of CO_2. Such an unfavorable mobility ratio will promote fingering, and as a result, CO_2 will break through soon, leaving a high saturation of residual oil. This means a poor volumetric efficiency, which is not desirable [59].

In order to overcome this issue, it is recommended to lower CO_2 mobility by decreasing its saturation. This could be done by alternatively injecting CO_2 and water in cycles. This process is called WAG injection. A slug of CO_2 is injected and followed by water. As water enters into the reservoir, it fills some of the pore spaces, resulting in reduced mobility to CO_2.

Although WAG gives promising results, there are still some problems associated with this mode of injection. For instance, injected water may not be distributed uniformly in the reservoir, causing reduced microscopic efficiency. Indeed, nonuniform distribution can occur when there is a sufficient vertical permeability. This causes water to move downward while gas will tongue in an upward direction.

Another important problem associated with the WAG process is the so-called problem of "water-shielding." Injected water blocks any possible connections between CO_2 and trapped oil in the porous media that were swept by water injection, leaving a high residual oil saturation. Water blocking is much more significant for water-wet rocks and it is negligible for mixed-wet rocks.

The major design issues to be considered for an optimum WAG injection process are WAG ratio, ultimate CO_2 slug size, rock and fluid properties, reservoir characteristics and heterogeneity, injection rate, and injection pattern [63,64]. Recently, for an optimized WAG process, researchers are working on the brine composition of water being injected alternatively with CO_2. A new hybrid method has been proposed as low salinity water (LSW) alternative CO_2 injection. Results show that this new method enhances oil recovery by both LSW and CO_2 effects. Although the results of CO_2–LSW are promising, further research is needed to assess its efficiency [65].

3.4 CO_2 INJECTION LABORATORY TESTS

As discussed earlier in previous sections, when CO_2 is injected into a reservoir, a series of complex mechanisms will occur. These mechanisms are effective in improving oil recovery. Therefore it is of great importance to know the performance of each mechanism. The performance of a CO_2 flood could only be assessed by experimental investigations in simulated condition of the reservoir. Because reservoirs and CO_2 injection processes vary widely, it is not possible to list a set of laboratory experiments appropriate to all situations. Laboratory experiments performed most often fall into three general classes as follows [66]:

- *Standard PVT test*

  PVT experiments are commonly performed for determining the relationship between pressure–volume–temperature. Phase behavior and fluid properties that are obtained through a PVT test can directly provide information about oil–CO_2 mixtures. For instance, a ternary diagram could be provided, which helps understanding the occurrence of miscibility. In a typical PVT test, CO_2 and oil are injected into a high-pressure cell. The volume of the mixture is then changed. Such a change will also change the pressure of the system. This test resembles a standard constant–composition–expansion. At each pressure, the volume of any phases present are measured. In the case of one phase, the density of the mixture represents swollen oil density. Bubble point pressures for the CO_2–oil mixtures can also be measured accurately by plotting cell pressure versus cell volume [17].

- *Core flood experiments*

  Core flood experiments are useful for estimation of displacement efficiency at microscopic level. Normally, small cores, which are used for experimentation, are difficult to obtain from long cores. Without ignoring the usefulness of core flood experiments, their data are difficult to interpret because even in linear cores, displacement efficiency can be affected by viscous fingering, gravity segregation, channeling, or bypassing of oil due to core heterogeneities, and trapping or shielding of oil in contact with CO_2 by high mobility of water saturations, as well as by the complexities of CO_2–oil phase behavior. Core flood experiments could also be used to investigate whether unexpected problems can occur due to interactions of CO_2 with reservoir oil, brine, clay, and cementing materials. Any of these phenomena can result in increased or decreased permeability. For instance, asphaltene deposition due to CO_2–oil interactions leads to decreased permeability. On the other hand, rock dissolution due to CO_2–rock interactions in carbonates may lead to increased permeability. It should be mentioned that core flooding data cannot be readily extrapolated to field dimensions [9,67,68].

- *Slim-tube displacement*

  This type of test is mainly used for determination of MMP. Slim tube consists of a very slim coiled tube, which is filled with crushed core, sand, or glass bead materials. This tube is typically very long in order to allow development of dynamic miscibility. Displacements in slim tubes approach nearby the ideal displacements. Viscous fingering growth is inhibited by the walls of the tube. It is assumed that fluids are mixed well due to very small diameter of the tube and also nearly homogeneous porous media within the tube. For MMP measurements, the tube is first saturated with the oil while keeping the temperature at reservoir condition. The gas is then injected into the tube, and recovery is calculated as the amount of oil produced divided by the initial oil volume. Fixing the time, the pore volume recovered are plotted against pressure. Normally, the time is fixed at 1.2 hydrocarbon pore volumes injected. The same procedure is then carried out for higher pressures. The pressure at which a break or a sharp change occurs in the oil recovery at 1.2 pore volumes of injection, or the lowest pressure at which the recovery is about 90%–95%, is often used to define the minimum pressure [69].

Oil recovery from a slim tube depends not only on CO_2−oil phase behavior, but also on displacement rate and the level of dispersion, which in turn depends on displacement rate and the particle diameter of the packing material [20,70].

3.5 CO_2 INJECTION FACILITIES AND PROCESS DESIGN CONSIDERATIONS

3.5.1 Surface Facilities

When a reservoir becomes a candidate for a CO_2 flood, whether miscible or immiscible, it requires a special gathering of surface facilities. CO_2 is first supplied through pipelines by a high-pressure compressor, if sufficient pressure is not satisfied. Then it is directly injected into the reservoir through the injection well, which is located near the production well. Indeed, wells are spaced from each other depending on the injection pattern. CO_2 will then assist oil flow to the production wells. Some of this CO_2 might be stored within the reservoir, but the remaining ones will be produced as it can be dissolved in oil or as it breaks through the production path. Commonly, water will also be produced. This amount of water could be sourced from a previous water flood process, water alternative CO_2, or even formation water. When these products get into a high-pressure separator, CO_2 will be separated and then transmitted to a recycling compressor. Recycled CO_2 will be injected again for economic and environmental considerations. Produced oil and water will undergo a separation process in which water will be directed for disposal and oil for sales [13,71].

3.5.2 Process Design Considerations

After a reservoir is nominated for a CO_2 flood and a surface facilities configuration is designed, several further factors should be accounted for the project to be continued in an economic fashion. First of all, the CO_2 availability is of high concern. Majorly, CO_2 is sourced either from the atmosphere (anthropogenic CO_2) or from natural gas decomposition (flue gas). Availability is also affected greatly by the costs of transportation. Based on availability, an optimum slug size of injection CO_2 can be determined. Secondly, the corrosion potential should be evaluated carefully. CO_2−water mixture can be very corrosive, resulting in serious damages to flow lines and facilities. In order to prevent loss of cost due to corrosion, a dehydration unit is customarily installed in CO_2-transporting pipelines to release CO_2 from water. In addition, flow lines are normally coated with some especial materials that are resistant to CO_2 corrosion. Thirdly, the potential of asphaltene deposition occurrence needs to be predicted. As CO_2 interacts with the oil, asphaltenes instability arises, which may lead to formation damage because of permeability impairment [73]. It has also the potential to damage flow lines as asphaltene deposition occurs through the transporting lines. Finally, a CO_2 flood should be conducted in a way that gas mobility is controlled. Injection of water alternative CO_2 or injection of CO_2 foam using surfactants are the common options for a mobility control process. All of these mentioned situations need careful evaluation to avoid loss of cost and time.

3.6 CO_2 INJECTION IN TIGHT RESERVOIR

Tight reservoirs are categorized as unconventional resources due to their poor reservoir characteristics. At the time of discovery, these type of reservoirs were believed to be uneconomical because of their low production rate. However, in recent years, as the energy demand of the world is increasing, the interest toward production from these reservoirs is also increasing. New technologies like multilateral drilling have been implemented on these reservoirs for the purpose of oil recovery. On the other hand, the use of the appropriate EOR method would further increase the oil recovery from tight reservoirs [74,75].

Commonly, water flooding would be the first potential EOR method to all kinds of reservoirs. However, for the case of tight reservoirs, water flooding would be very challenging and in some cases impossible to be conducted. This is because of very low permeability of these reservoirs, which leads to large growth of unattainable pressure during injection process. In other words, injectivity of water into tight reservoirs is very low. Such a problem would be assisted by using an injection fluid, which is less viscous than water. Among all types of fluids, gases have much lower viscosity than any other fluid. As a result, gas injection can serve a high injectivity into tight reservoirs compared with common water flooding [76].

Among all types of gases suitable for gas injection, CO_2 has received much more attention as it serves lower MMP, which means higher potential for miscibility development. In tight reservoirs, viscous force and gravity drainage are less important while molecular diffusion will be the dominating mechanism. It has also been reported that in tight reservoirs, implementation of Huff-n-Puff injection is much more efficient compared with a WAG or continuous CO_2 injection process. In the following, two field case studies of CO_2 injection into tight reservoirs are discussed [77–81].

- *Yu-Shu-Lin Field, Daqing* [81]

  The reservoir of this field, Fuyang, has an absolute permeability of 0.96 mD. The oil viscosity was 3.6 cP. The reservoir was water flooded before CO_2 injection. There were seven injection wells in a five-spot pattern along with 17 producing wells. The injection of CO_2 started in December 2007 with two injection wells. The five injector wells were also added to the process in July 2008. Totally,

CO_2 injectivity was 4 times greater than that of water. The estimated oil recovery was 21% compared with 12% from a water flooding well pattern in the same field.

- *Song-Fang-Dun Field, Daqing* [81]

  The injection was carried out in a layer with an average gas permeability of 0.79 mD. The reservoir oil had a viscosity of 6.6 cP. The project started in March 2003 with one injector and five producing wells. The first response was observed in August 2004. The CO_2 injectivity was estimated to be 6.3 times higher than that of water. Then WAG process was applied to this field in 2014. Overall, CO_2 injection performance was not good due to high heterogeneity of the formation.

3.7 CO_2 INJECTION FOR ENHANCED GAS RECOVERY

CO_2 injection into gas reservoirs for enhanced gas recovery (EGR) is a new topic of concern, and no field trial has been reported so far. However, recent researches have shown that CO_2 injection would profit those countries that have very large gas reservoirs but limited number of oil reservoirs. Gas reserves can also provide good storage capacity for CO_2. The main mechanisms responsible for EGR due to CO_2 injection are: displacement, which is analogous to water flooding process in oil reservoirs; reservoir pressure support, which prevents the reservoir from being depleted and also inhibits from subsequent subsidence and water intrusion [13,82].

Despite the promising results of CO_2 EGR for gas reservoirs, it has never been tested in the field. This can be attributed to two main reasons. Firstly, CO_2 costs are high as a commodity, and thus geological storage is not accepted worldwide. Secondly, concerns exist about the potential of CO_2 mixing with natural gas, which could downgrade the natural gas resource.

CO_2 and CH_4 are able to be mixed at any pressure, and CO_2-CH_4 mixture has some characteristics that promote the CO_2 EGR efficiency. In summary, these characteristics are as follows [13]:

1. An achievable gravity-stable displacement due to higher density of CO_2 compared to that of CH_4 (normally 2–4 times higher). This also could be attributed to lower mobility of CO_2, as it is more viscous than CH_4 at reservoir condition.
2. A delayed CO_2 breakthrough due to higher solubility of CO_2 in formation water compared to that of methane.
3. Higher injectivity of CO_2 by reason of its nearly gas-like viscosity.

In order to optimize a CO_2 EGR project, a sensitivity analysis should be conducted on CO_2 resources, mixing of injected gas with methane and project time. Other effective factors like injection gas rate should also be accounted. In conclusion, more research works are needed to assess CO_2 EGR viability and effectiveness [11,83,84].

3.8 ENVIRONMENTAL ASPECTS OF CO_2 INJECTION

CO_2 has been known as a byproduct of industrial and domestic operations, which is undesirable due to its harmful impacts on the world's climate. Since the Industrial Revolution, a rise of about 30% has been reported in released CO_2 content in the atmosphere [85]. Many research studies have been conducted to achieve proper ways for mitigating CO_2. Recently, and to a greater extent of innovation, geological sequestration has been known to be a promising option that may lead to significant reduction of CO_2. Beside this sequestering opportunity, underground CO_2 injection serves as an opportunity for increased oil recovery. Indeed, nowadays CO_2 EOR is being studied along with the opportunities of CO_2 sequestration. This is because oil reservoir has such characteristics that could bring the opportunity of CO_2 sequestration. Generally, there are three major mechanisms by which CO_2 could be sequestered within the oil reservoirs. The first is hydrodynamic trapping, in which CO_2 gas would be trapped beneath a cap rock. The next is related to CO_2 solubility in water and oil phases, which is known as solubility trapping. Lastly, CO_2 can react with reservoir rock and organic matter, directly or indirectly, and can be converted into the solid phase. However, each of these mechanisms can alter the injection and production efficiency toward reduction of injection and production efficiency. For instance, when CO_2 is converted into a solid phase, it may lead to permeability impairment. As a result, the injectivity is reduced [11,86].

To measure the storage capacity of a reservoir, specific capacity of the rock could be used. Indeed, this value could be used to differentiate sequestration potential among reservoirs. Specific storage capacity is expressed by Eq. (3.8) as follows [87]:

$$C = \rho(1 - S_{or} - S_{wir})\varnothing + S_{wir}\varnothing C_s \tag{3.26}$$

where ρ is CO_2 density, which is a function of pressure and temperature; S_{or} indicates the residual oil saturation; S_{wir} stands for the irreducible water saturation; $\varnothing$ shows the rock porosity; and C_s symbolizes the mass of CO_2 dissolved per unit volume of water. Based on this equation, a reservoir that is deep and has a sizeable porosity in which a large fraction of moveable fluids is contained would lead to maximum CO_2 sequestration potential.

Eq. (3.26) describes the total possible amount of the CO_2 that could be stored within a reservoir. In real conditions, this capacity depends on many factors, including reservoir heterogeneity, aquifer availability, reservoir boundaries, and geophysical aspects. Reservoirs with a high degree of heterogeneity have less potential for combined efficient storage and EOR purposes. This is because CO_2 breakthrough occurs at early stages when there is high permeability variation within the reservoir (i.e., high degree of heterogeneity). Aquifers are categorized as bottom water or edge water on the basis of the thickness and orientation of the aquifer. Regardless of the type of the aquifer, those reservoirs that are closed and disconnected from huge aquifers are the most appropriate candidates for combined EOR and CO_2 storage operations. This is because when CO_2 is injected into a reservoir, it needs to force out the invaded water from the aquifer. As a result, injectivity is reduced compared to those reservoirs with closed boundaries. Lastly, most probably an efficient storage would occur in those reservoirs with a sealing structure at formation tops with less open fault and fracture systems [88]. Selection criteria for an efficient CO_2 EOR and storage are given in the relevant literature [32].

Commonly, a co-optimization function is defined for the combination of incremental oil recovery and CO_2 storage to be maximized. Based on Kamali and Cinar [89], this co-optimization function can be expressed as follows:

$$f = w_1 \frac{N_p}{\mathrm{OIP}} + w_2 \left(1 - \frac{M^P_{\mathrm{CO2}}}{M^I_{\mathrm{CO2}}} \right) \tag{3.27}$$

where N_p is the net oil production, OIP stands for the oil in place at the start of CO_2 injection, M^P_{CO2} shows the amount of produced CO_2, and M^I_{CO2} indicates the amount of CO_2 injected. Obviously, the amount of stored CO_2 can be found by subtracting M^I_{CO2} from $\mathrm{M}^P_{\mathrm{CO2}}$. The constants w_1 and w_2 are weighting factors for oil recovery and CO_2 storage, respectively ($0 \leq w_1, w_2 \leq 1$ and $w_1 + w_2 = 1$). If the aim is to maximize oil recovery, then $w_1 = 1$, similarly for maximum CO_2 storage $w_2 = 1$. When both aims are equally important, then w_1, $w_2 = 0.5$.

REFERENCES

[1] S. Kokal, A. Al-Kaabi, Enhanced oil recovery: challenges & opportunities, World Petroleum Council: Offic. Publicat. 64 (2010) 64–69.

[2] E.J. Manrique, C.P. Thomas, R. Ravikiran, M. Izadi Kamouei, M. Lantz, J.L. Romero, et al., EOR: current status and opportunities, SPE Improved Oil Recovery Symposium, Society of Petroleum Engineers, Tulsa, OK, 2010.

[3] Whorton, L.P., E.R. Brownscombe, and A.B. Dyes: inventors; Atlantic Refining Co, assignee, Method for producing oil by means of carbon dioxide. 1952, Google Patents, U.S. Patent No. 2,623,596.

[4] J. Saxon Jr, J. Breston, R. Macfarlane, Laboratory tests with carbon dioxide and carbonated water as flooding mediums, Prod. Monthly 16 (1951).

[5] D. Beeson, G. Ortloff, Laboratory investigation of the water-driven carbon dioxide process for oil recovery, J. Petrol. Technol. 11 (1959) 63–66.

[6] L. Holm, Carbon dioxide solvent flooding for increased oil recovery, Trans. AIME 216 (1959) 225–231.

[7] J.W. Martin, Additional oil production through flooding with carbonated water, Producers monthly 15 (1951) 18–22.

[8] L.W. Holm, Miscibility and Miscible Displacement, Society of Petroleum Engineers, New York, 1986.

[9] M. Bayat, M. Lashkarbolooki, A.Z. Hezave, S. Ayatollahi, Investigation of gas injection flooding performance as enhanced oil recovery method, J. Nat. Gas Sci. Eng. 29 (2016) 37–45.

[10] S. Asgarpour, An overview of miscible flooding, J. Can. Pet. Technol. 33 (1994).
[11] Z. Dai, H. Viswanathan, R. Middleton, F. Pan, W. Ampomah, C. Yang, et al., CO_2 accounting and risk analysis for CO_2 sequestration at enhanced oil recovery sites, Environ. Sci. Technol. 50 (2016) 7546−7554.
[12] D.W. Green, G.P. Willhite, Enhanced Oil Recovery, Henry L. Doherty Memorial Fund of AIME, Society of Petroleum Engineers, Richardson, TX, 1998.
[13] S. Kalra, X. Wu, CO2 injection for enhanced gas recovery, in: SPE Western North American, and Rocky Mountain Joint Meeting, Society of Petroleum Engineers, 2013.
[14] H. Li, S. Zheng, D.T. Yang, Enhanced swelling effect and viscosity reduction of solvent (s)/CO_2/heavy-oil systems, SPE J. 18 (2013) 695−707.
[15] N.J. Clark, H.M. Shearin, W.P. Schultz, K. Garms, J.L. Moore, Miscible drive—its theory and application, J. Petrol. Technol. 10 (1958).
[16] R.T. Johns, F.M. Orr Jr., Miscible gas displacement of multicomponent oils, SPE J. 1 (1996) 39−50.
[17] N. Mungan, Carbon dioxide flooding—fundamentals, J. Petrol. Technol. (April) (1981) 396−400.
[18] M.A. Klins, Carbon Dioxide Flooding: Basic Mechanisms and Project Design, Springer, The Netherlands, 1984.
[19] K. Ahmadi, R.T. Johns, Multiple-Mixing-Cell Method for MMP Calculations, SPE J. 16 (2011) 733−742.
[20] J.M. Ekundayo, S.G. Ghedan, Minimum miscibility pressure measurement with slim tube apparatus—how unique is the value? SPE Reservoir Characterization and Simulation Conference and Exhibition, Society of Petroleum Engineers, Abu Dhabi, UAE, 2013.
[21] A. Hemmati-Sarapardeh, M.H. Ghazanfari, S. Ayatollahi, M. Masihi, Accurate determination of the CO_2−crude oil minimum miscibility pressure of pure and impure CO_2 streams: a robust modelling approach, Can. J. Chem. Eng. 94 (2016) 253−261.
[22] A. Kamari, M. Arabloo, A. Shokrollahi, F. Gharagheizi, A.H. Mohammadi, Rapid method to estimate the minimum miscibility pressure (MMP) in live reservoir oil systems during CO_2 flooding, Fuel 153 (2015) 310−319.
[23] S. Ayatollahi, A. Hemmati-Sarapardeh, M. Roham, S. Hajirezaie, A rigorous approach for determining interfacial tension and minimum miscibility pressure in paraffin−CO_2 systems: application to gas injection processes, J. Taiwan Inst. Chem. Eng. 63 (2016) 107−115.
[24] P.Y. Zhang, S. Huang, S. Sayegh, X.L. Zhou, Effect of CO_2 impurities on gas-injection EOR processes, Journal of Canadian Petroleum Technology 48 (2004) 30−36.
[25] A.S. Bagci, Immiscible CO_2 flooding through horizontal wells, Energy Sources A Recov. Util. Environ. Effects 29 (2007) 85−95.
[26] R.M. Brush, H.J. Davitt, O.B. Aimar, J. Arguello, J.M. Whiteside, Immiscible CO_2 flooding for increased oil recovery and reduced emissions, SPE/DOE Improved Oil Recovery Symposium, Society of Petroleum Engineers, Tulsa, OK, 2000.
[27] F.F. Craig Jr., W.W. Owens, Miscible slug flooding—a review, J. Pet. Technol. 12 (1960) 11−16.
[28] B. Dindoruk, F.M. Orr Jr., R.T. Johns, Theory of multicontact miscible displacement with nitrogen, SPE J. 2 (1997) 268−279.
[29] J.B. Apostolos Kantzas, S. Taheri, PERM, Inc., published materials, Online e-book Available from: http://perminc.com/resources/fundamentals-of-fluid-flow-in-porous-media.
[30] A.A. Zick, A combined condensing/vaporizing mechanism in the displacement of oil by enriched gases, SPE Snnual Technical Conference and Exhibition, Society of Petroleum Engineers, New Orleans, LA, 1986.
[31] A. Al Adasani, B. Bai, Analysis of EOR projects and updated screening criteria, J. Petrol.Sci. Eng. 79 (2011) 10−24.
[32] A.R. Kovscek, Screening criteria for CO_2 storage in oil reservoirs, Petrol. Sci. Technol. 20 (2002) 841−866.
[33] J. Shaw, S. Bachu, Screening, evaluation, and ranking of oil reservoirs suitable for CO_2-flood EOR and carbon dioxide sequestration, J. Can. Petrol. Technol. 41 (2002).
[34] G.F. Teletzke, P.D. Patel, A. Chen, Methodology for Miscible Gas Injection EOR Screening, Society of Petroleum Engineers, Kuala Lumpur, Malaysia, 2005.

[35] W.R. Brock, L.A. Bryan, Summary Results of CO_2 EOR Field Tests, 1972–1987, Society of Petroleum Engineers, Denver, CO, 1989.
[36] C. Gao, X. Li, L. Guo, F. Zhao, Heavy oil production by carbon dioxide injection, Greenhouse Gases Sci. Technol. 3 (2013) 185–195.
[37] R. Simon, D. Graue, Generalized correlations for predicting solubility, swelling and viscosity behavior of CO_2–crude oil systems, J. Petrol. Technol. 17 (1965) 102–106.
[38] C.A. Mulliken, S.I. Sandler, The prediction of CO_2 solubility and swelling factors for enhanced oil recovery, Indus. Eng. Chem. Process Design Develop. 19 (1980) 709–711.
[39] A.K. Mehrotra, W.Y. Svrcek, Correlations for properties of bitumen saturated with CO_2, CH_4 and N_2, and experiments with combustion gas mixtures, J. Can. Petrol. Technol. 21 (1982).
[40] F.T. Chung, R.A. Jones, H.T. Nguyen, Measurements and correlations of the physical properties of CO_2–heavy crude oil mixtures, SPE Reservoir Eng. 3 (1988) 822–828.
[41] M. Emera, H. Sarma, Prediction of CO_2 solubility in oil and the effects on the oil physical properties, Energy Sources A 29 (2007) 1233–1242.
[42] A. Rostami, M. Arabloo, A. Kamari, A.H. Mohammadi, Modeling of CO_2 solubility in crude oil during carbon dioxide enhanced oil recovery using gene expression programming, Fuel 210 (2017) 768–782.
[43] J.R. Welker, Physical properties of carbonated oils, J. Petrol. Tech. 15 (1963) 873–875.
[44] D.-Y. Peng, D.B. Robinson, A new two-constant equation of state, Indus. Eng. Chem. Fund. 15 (1976) 59–64.
[45] B. Habermann, The efficiency of miscible displacement as a function of mobility ratio, Trans. AIME 219 (1960) 264–272.
[46] B.L. O'Steen, E.T.S. Huang, Effect of solvent viscosity on miscible flooding, SPE J. 7 (1992) 213–218.
[47] B. Habermann, The efficiency of miscible displacement as a function of mobility ratio, Petrol, Trans. AIME 219 (1960) 264–272.
[48] A. Emadi, M. Jamiolahmady, M. Sohrabi, S. Irland, Visualization of Oil Recovery by CO_2-Foam Injection; Effect of Oil Viscosity and Gas Type, Society of Petroleum Engineers, Tulsa, OK, 2012.
[49] G. Glatz, A primer on enhanced oil recovery, Physics 240 (2013).
[50] H.D. Beggs, J. Robinson, Estimating the viscosity of crude oil systems, J. Petrol. Technol. 27 (1975) 1,140–141.
[51] E. Lederer, Viscosity of mixtures with and without diluents, Proc. World Pet. Cong. Lond. 2 (1933) 526–528.
[52] I.M. Mohamed, J. He, H.A. Nasr-El-Din, Permeability Change during CO2 Injection in Carbonate Aquifers: Experimental Study, Society of Petroleum Engineers, Houston, TX, 2011.
[53] T.B. Reid, H.J. Robinson, Lick creek meakin sand unit immiscible CO_2 waterflood project, J. Pet. Technol. 33 (1981) 1723–1729.
[54] S. Sahin, U. Kalfa, D. Celebioglu, Bati Raman field immiscible CO_2 application—status quo and future plans, SPE Reserv. Eval. Eng. 11 (4) (2008) 778–791.
[55] W. Saner, J. Patton, CO_2 recovery of heavy oil: Wilmington field test, J. Petrol. Technol. 38 (1986) 769–776.
[56] L.J. Mohammed-Singh, K. Ashok, Lessons from Trinidad's CO_2 immiscible pilot projects 1973–2003, in: IOR 2005-13th European Symposium on Improved Oil Recovery, 2005.
[57] E. Leissner, Five-spot vs. crestal waterflood patterns, comparison of results in thin reservoirs, J. Petrol. Technol. 12 (1960) 41–44.
[58] H. Warner Jr, E. Holstein, Immiscible gas injection in oil reservoirs, in: E.D. Holstein (Ed.), Reservoir Engineering and Petrophysics: Petroleum Engineering Handbook, Society of Petroleum Engineering, Richardson, TX, 2007, pp. 1103–1147.
[59] J. Casteel, N. Djabbarah, Sweep improvement in CO_2 flooding by use of foaming agents, SPE Reservoir Eng. 3 (1988). 1,186-181,192.
[60] R. Ehrlich, J.H. Tracht, S.E. Kaye, Laboratory and field study of the effect of mobile water on CO_2-flood residual oil saturation, J. Petrol. Technol. (October, 1984) 1797–1809.
[61] D.L. Tiffin, W.F. Yellig, Effects of mobile water on multiple-contact miscible gas displacements, SPE J. 23 (1983) 447–455.

[62] R. Hadlow, Update of industry experience with CO_2 injection, SPE Annual Technical Conference and Exhibition, Society of Petroleum Engineers, Washington, DC, 1992.
[63] M. Sohrabi, M. Jamiolahmady, A. Al Quraini, Heavy oil recovery by liquid CO2/water injection, in: EUROPEC/EAGE Conference and Exhibition, London, 2007.
[64] M. Sohrabi, M. Riazi, M. Jamiolahmady, S. Ireland, C. Brown, Enhanced oil recovery and CO_2 storage by carbonated water injection, in: International Petroleum Technology Conference, 2009.
[65] C. Dang, L. Nghiem, N. Nguyen, Z. Chen, Q. Nguyen, Evaluation of CO_2 low salinity water-alternating-gas for enhanced oil recovery, J. Nat. Gas Sci. Eng. 35 (2016) 237−258.
[66] S.G. Ghedan, Global Laboratory Experience of CO_2−EOR Flooding, Society of Petroleum Engineers, Abu Dhabi, UAE, 2009.
[67] S.M. Fatemi, M. Sohrabi, Experimental investigation of near-miscible water-alternating-gas injection performance in water-wet and mixed-wet systems, SPE J. 18 (2013) 114−123.
[68] R.K. Srivastava, S.S. Huang, M. Dong, Laboratory investigation of Weyburn CO miscible flooding, J. Can. Petrol. Technol. 39 (2000) 41−51.
[69] O. Glass, Generalized minimum miscibility pressure correlation (includes associated papers 15845 and 16287), Soc. Petrol. Eng. J. 25 (1985) 927−934.
[70] A.M. Elsharkawy, F.H. Poettmann, R.L. Christiansen, Measuring CO_2 minimum miscibility pressures: slim-tube or rising-bubble method?, Energy Fuels 10 (1996) 443−449.
[71] H.N.H. Saadawi, Surface facilities for a CO_2−EOR project in Abu Dhabi, in: SPE-127765. SPE EOR Conference at Oil and Gas West Asia, Muscat, 2010.
[72] A. Amarnath, E.P.R. Institute, Enhanced Oil Recovery Scoping Study, EPRI, Palo Alto, CA, 1999.
[73] R.K. Srivastava, S.S. Huang, Asphaltene deposition during CO_2 flooding: a laboratory assessment, in: Paper SPE 37468, Proceedings of the 1997 SPE Productions Operations Symposium, OK, 1997.
[74] A. Arshad, A.A. Al-Majed, H. Menouar, A.M. Muhammadain, B. Mtawaa, Carbon dioxide (CO2) miscible flooding in tight oil reservoirs: a case study, in: Proceedings of the Kuwait International Petroleum Conference and Exhibition, Kuwait City, Kuwait.
[75] A.Y. Zekri, R.A. Almehaideb, S.A. Shedid, Displacement efficiency of supercritical CO_2 flooding in tight carbonate rocks under immiscible conditions, in: SPE Europec/EAGE Annual Conference and Exhibition, Society of Petroleum Engineers, 2006.
[76] C. Song, D.T. Yang, Optimization of CO_2 flooding schemes for unlocking resources from tight oil formations, in: SPE Canadian Unconventional Resources Conference, Society of Petroleum Engineers, 2012.
[77] A. Habibi, M.R. Yassin, H. Dehghanpour, D. Bryan, CO2-oil interactions in tight rocks: an experimental study, SPE Unconventional Resources Conference, Society of Petroleum Engineers, Calgary, 2017, February.
[78] K. Joslin, S.G. Ghedan, A.M. Abraham, V. Pathak, EOR in tight reservoirs, technical and economical feasibility, SPE Unconventional Resources Conference, Society of Petroleum Engineers, Calgary, 2017, February.
[79] P. Luo, W. Luo, S. Li, Effectiveness of miscible and immiscible gas flooding in recovering tight oil from Bakken reservoirs in Saskatchewan, Canada, Fuel 208 (2017) 626−636.
[80] J. Ma, X. Wang, R. Gao, F. Zeng, C. Huang, P. Tontiwachwuthikul, et al., Enhanced light oil recovery from tight formations through CO_2 huff "n" puff processes, Fuel 154 (2015) 35−44.
[81] J.J. Sheng, B.L. Herd, Critical review of field EOR projects in shale and tight reservoirs, J. Pet. Sci. Eng. 7 (2017) 147−153.
[82] W. Yu, E.W. Al-Shalabi, K. Sepehrnoori, A sensitivity study of potential CO_2 injection for enhanced gas recovery in Barnett shale reservoirs, in: SPE Unconventional Resources Conference, Society of Petroleum Engineers, 2014.
[83] J. Narinesingh, D. Alexander, CO_2 enhanced gas recovery and geologic sequestration in condensate reservoir: a simulation study of the effects of injection pressure on condensate recovery from reservoir and CO_2 storage efficiency, Energy Procedia 63 (2014) 3107−3115.
[84] C.M. Oldenburg, K. Pruess, S.M. Benson, Process modeling of CO_2 injection into natural gas reservoirs for carbon sequestration and enhanced gas recovery, Energy Fuels 15 (2001) 293−298.

[85] A.E. Peksa, K.H.A.A. Wolf, M. Daskaroli, P.L.J. Zitha, The Effect of CO_2 Gas Flooding on Three Phase Trapping Mechanisms for Enhanced Oil Recovery and CO_2 Storage, Society of Petroleum Engineers, Madrid, 2015.
[86] F.M. Orr Jr., Storage of carbon dioxide in geologic formations, J. Petrol. Technol. 56 (2004) 90–97.
[87] IPCC, Carbon Dioxide Capture and Storage: Special Report of the Intergovernmental Panel on Climate Change, Cambridge University Press, Cambridge, 2005.
[88] M.H. Holtz, E.K. Nance, R.J. Finley, Reduction of greenhouse gas emissions through CO_2 EOR in Texas, Environ. Geosci. 8 (2001) 187–199.
[89] F. Kamali, F. Hussain, Field scale co-optimisation of CO_2 enhanced oil recovery and storage through swag injection using laboratory estimated relative permeabilities, in: SPE Asia Pacific Oil & Gas Conference and Exhibition, Society of Petroleum Engineers, 2016.
[90] W. Dodds, L. Stutzman, B. Sollami, Carbon dioxide solubility in water, Indus. Eng. Chem. Chem. Eng. Data Series 1 (1956) 92–95.

CHAPTER FOUR

Miscible Gas Injection Processes

Pouria Behnoudfar[1], Alireza Rostami[2] and Abdolhossein Hemmati-Sarapardeh[3]

[1]Department of Petroleum Engineering, Amirkabir University of Technology, Tehran, Iran
[2]Department of Petroleum Engineering, Petroleum University of Technology (PUT), Ahwaz, Iran
[3]Department of Petroleum Engineering, Shahid Bahonar University of Kerman, Kerman, Iran

4.1 ENHANCED OIL RECOVERY

Applying primary and secondary oil recovery approaches leads to approximate remaining of 67% original oil in place (OOIP). As an example, in the known oil fields of the United States, this remaining oil in place is approximately equal to 377 billion barrels. Therefore, the so-called enhanced oil recovery (EOR) methods are increasingly being used in the oil fields due to rapid increase of the world oil price and oil consumption [1]. After primary and secondary recovery, a high portion of original oil remains in place that is entrapped in the pore spaces as a result of capillary and viscous forces [2]. Hence, the EOR processes contribute to a greater recovery efficiency from depleted oil reservoirs.

In general, EOR methods are categorized as waterflooding, chemical flooding, gas injection, and thermal techniques such as hot water or steam injection. Previously developed EOR methods are chiefly proposed for heavy oils, even though the latter is generally employed in light oil recovery [3]. By three key mechanisms of nonthermal EOR techniques including waterflooding, chemical flooding, and gas injection, recovery factor can be promoted. These mechanisms are as follows [4]:

- Viscosity modification of displaced and/or displacing phase,
- Reduction of interfacial tension (IFT) between the displaced and displacing phases,
- Approaching or accomplishing miscibility by diluent extraction.

Some of nonthermal EOR techniques, such as alkaline flooding, alkaline–surfactant–polymer, and polymer flooding, are exposed to some operational limitations, such as reservoir permeability and formation temperature. Besides, they are costly to be conducted in field-scale operation [5].

In the midst of all EOR methods, gas injection processes have been identified as one of the most efficient techniques. On the other hand, gas injection can enhance oil recovery by IFT reduction due to mass transfer between the displaced and

Fundamentals of Enhanced Oil and Gas Recovery from Conventional and Unconventional Reservoirs.
DOI: https://doi.org/10.1016/B978-0-12-813027-8.00004-7

displacing phases during vaporizing/condensing gas drive, oil swelling, and oil viscosity reduction leading to reservoir repressurization and alleviation of capillary forces [6,7]. Various forms of injected gases consist of hydrocarbon gases such as natural gas, a liquefied petroleum slug driven by natural gas and enriched natural gas, and nonhydrocarbon gases, including nitrogen, carbon dioxide, and flue gas, are extensively utilized to reduce the residual oil saturation.

The thermal and chemical projects are frequently used in sandstone reservoirs rather than other lithologies (e.g., carbonates and turbiditic formations) [8].

Since the 1970s, CO_2 injection has been documented as an encouraging operation for EOR [9]. Selection of CO_2 as an injection gas is more desired as it could be employed for the goals of both CO_2 sequestration resulting in the greenhouse gas reduction and also as an EOR agent. CO_2 as an EOR agent is suitable for recovering light-to-medium oils from deep reservoirs [10]. Furthermore, CO_2 storage into underground hydrocarbon reservoirs can result in declining the greenhouse gas emissions. For CO_2 injection processes, various difficulties including large requisite per incremental barrel of CO_2, corrosion problem in surface facilities, precipitation and deposition of asphaltene leading to formation damage and wettability alteration, and separation of CO_2 from the valuable hydrocarbons have been reported in the literature [10–13].

N_2-contaminated lean hydrocarbon gases or simply nitrogen injection are also another applicable EOR agents for high pressure and deep reservoirs with light or volatile crude oils containing hydrocarbon components of C_2-C_5. The advantages of N_2 gas are the availability, low cost, and abundance. By cryogenic processes, nitrogen is produced from air for a long time [14]. Based on the injection pressure at oil composition and reservoir temperature, injection of nitrogen process could be implemented in both miscible and immiscible means which is discussed subsequently [5].

4.2 IMMISCIBLE AND MISCIBLE PROCESSES

In immiscible flooding, an interface between the trapped crude oil and the injected fluid exists, and consequently, a capillary pressure also will be established due to the prementioned interface. The immiscible flood benefits are principally as a result of reservoir pressure maintenance. Residual oil saturations can probably be higher than that of miscible flood since the two fluids are immiscible [15]. Therefore, the miscible flood attains higher recovery factors than the immiscible flood. The miscibility is theoretically referred to the cases in which there is no interface between the two phases involved (i.e., zero equilibrium IFT). In other words, the miscible condition is when two phases can be mixed with each other at any ratio [16].

Miscible CO_2 injection improves the oil recovery efficiency by means of the below mechanisms [17–19]:

- Reduction in oil viscosity,
- Reduction in oil density,
- Improvement in volumetric sweep efficiency.

Fulfillment of a miscible or immiscible flood is imposed by the magnitude of minimum miscibility pressure (MMP) and the injection pressure of the gas into the oil reservoir. The MMP is stated as the lowest operating pressure at which the injected gas and the crude oil become miscible after their dynamic multicontact process at reservoir temperature [20]. In other words, the MMP parameter is defined as the pressure at which the local displacement efficiency approaches 100%. At pressures lower than the MMP, miscibility can be achieved through condensation, vaporization, or their combined process [21].

In a miscible gas injection process, the IFT between the trapped crude oil and the injected fluid is lowered to zero; consequently, the capillary forces will be decreased to a minimum value leading to miscibility achievement. As a result of miscibility development in a gas injection process, the huge amounts of trapped oil will be remobilized leading to the increased oil recovery factor. In a miscible gas injection process, by selecting the required operational conditions, oil recovery can be maximized [22,23].

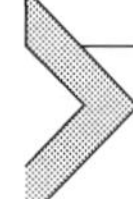

4.3 MINIMUM MISCIBILITY DETERMINATION

4.3.1 Minimum Miscibility Pressure and Interfacial Tension Measurement

Analytical and experimental methods have been introduced to approximate and measure the MMP value [24]. The most widely known experimental procedures to determine the gas–oil miscibility pressure under reservoir conditions are shown as follows [23]:

- Slim-tube displacement
- Rising bubble apparatus (RBA)
- Method of constructing pressure–composition (P–X) diagrams
- Vanishing interfacial tension (VIT) technique/axisymmetric drop shape analysis (ADSA)

The slim-tube test is considered as the most common practice and it has been widely accepted as a standard in petroleum industry to determine gas–oil miscibility. In spite of this fact, this approach suggests neither a standard design nor a standard operating procedure and criteria for the evaluation of miscibility conditions [25]. Furthermore, this technique has a time-consuming and costly experimentation, in which it takes more than a month to complete one miscibility measurement.

RBA is the other technique utilized for the estimation of the MMP parameter. Its normal usage is for a fast and approximate range of gas–oil miscibility. This technique is qualitative in nature, which recognizes miscibility from visual observations. Subjective interpretations of miscibility from visual observations are an advantage of this method, whereas lack of quantitative data to support the results is a main disadvantage [23].

P–X diagrams method is also time consuming, expensive, and requires a large amount of fluids, and it is also subjected to some experimental errors [23].

The VIT technique has been recently developed and employed to measure the miscibility conditions of various crude oils with different gases such as CO_2 [26–28]. The VIT technique is based on the measurement of equilibrium IFT between crude oil and CO_2 as the pressure increases for the equilibrium IFT between the two phases, which approach zero at miscible condition. Determination of MMP by using this approach takes 4–6 hours, while slim-tube method takes 4–6 weeks [25]. In VIT experiment, the equilibrium IFTs between an oil phase and a gas phase can be accurately measured at different equilibrium pressures and reservoir temperature by applying the ADSA technique for the pendant drop approach [29]. A schematic diagram of the experiment setup used for measuring the dynamic/equilibrium IFT between the dead/live crude oil and pure/impure CO_2 by applying the ADSA technique for the pendant drop case is reported in the relevant literature [30].

The major component of the above-mentioned experimental setup is a high-pressure IFT cell. The light crude oil and CO_2 are stored in two transfer cylinders. The temperature effects on the MMP and maximum injection pressure may be estimated by applying the VIT technique. The following correlations could be utilized for determination of MMP for cases which the test temperature T is considered as show below [29]:

$$\mathrm{MMP} = 0.116(T/\mathrm{K}) - 27.1 \quad \text{for dead oil and pure CO}_2\text{ system} \tag{4.1}$$

$$\mathrm{MMP} = 0.222(T/\mathrm{K}) - 51.0 \quad \text{for dead oil and impure CO}_2\text{ system} \tag{4.2}$$

$$\mathrm{MMP} = 0.168(T/\mathrm{K}) - 42.7 \quad \text{for live oil and pure CO}_2\text{ system} \tag{4.3}$$

$$\mathrm{MMP} = 0.194(T/\mathrm{K}) - 42.2 \quad \text{for live oil and impure CO}_2\text{ system} \tag{4.4}$$

where *MMP* and T stand for minimum miscibility pressure in MPa and reservoir temperature in K, respectively. In addition, the maximum injection pressure can be determined by the subsequent equations as follows [29]:

$$P_{\max} = 0.384(T/\mathrm{K}) - 102.8 \quad \text{for dead oil and pure CO}_2\text{ system} \tag{4.5}$$

$$P_{\max} = 0.281(T/\mathrm{K}) - 61.9 \quad \text{for dead oil and impure CO}_2\text{ system} \tag{4.6}$$

$$P_{\max} = 0.417(T/\mathrm{K}) - 113.5 \quad \text{for live oil and pure CO}_2\text{ system} \tag{4.7}$$

$$P_{\max} = 0.247(T/\mathrm{K}) - 50.8 \quad \text{for live oil and impure CO}_2\text{ system} \tag{4.8}$$

where P_{max} and T stand for maximum injection pressure in MPa and reservoir temperature in K, respectively. IFT is defined as the energy needed to build a unit surface area at the boundary of two immiscible phases. It is the consequence of the attraction of surface molecules to the interior molecules of liquid. Hence, it is very sensitive to temperature [31].

Example 4.1: Determine the MMP and maximum injection pressure for the following systems and then discuss about the effect of purity of carbon dioxide (assume the reservoir temperature is 80°C):
1. Dead oil and pure CO_2 system
2. Dead oil and impure CO_2 system

Solution: $T(\text{K}) = T(°\text{C}) + 273.15$
$T = 80 + 273.15 = 353.15\text{K}$
1. Using Eq. (4.1), MMP can be calculated as follows:

$$\text{MMP} = 0.116 \times 353.15 - 27.1 = 16.86 \text{ MPa}$$

and P_{max} can be calculated via Eq. (4.5) as follows:

$$P_{max} = 0.384 \times 353.15 - 102.8 = 32.80 \text{ MPa}$$

2. Using Eq. (4.2), MMP can be calculated as follows:

$$\text{MMP} = 0.222 \times 353.15 - 51.0 = 27.39 \text{ MPa}$$

and P_{max} can be calculated via Eq. (4.6) as follows:

$$P_{max} = 0.281 \times 353.15 - 61.9 = 37.33 \text{ MPa}$$

The surface energy alters due to the dissolution of gas into the liquid interface. Solubility is low at low pressures, and the effect of dissolved gas is negligible at low pressure conditions, whereas the increase of temperature can cause increase of solubility of CO_2 in the crude oil at low pressures and consequently, the IFT will decrease. The solubility of CO_2 in crude oil decreases with increasing temperature at higher pressures [32]. As a result, the IFT increases with increasing temperatures for the case of CO_2 gas injection. These effects can be studied by assembling an experimental setup similar to Fig. 4.1 [29].

The obtained results in the work of Hemmati-Sarapardeh et al. [29] are illustrated in Fig. 4.2. Also, the derived correlations, which are used to estimate IFT of crude oil/CO_2 system in different conditions, are inserted in Table 4.1.

Example 4.2: Calculate IFT for the following thermodynamic conditions:
1. $T = 60°\text{C}$, $P = 6$ MPa
2. $T = 80°\text{C}$, $P = 7.5$ MPa
3. $T = 100°\text{C}$, $P = 10$ MPa

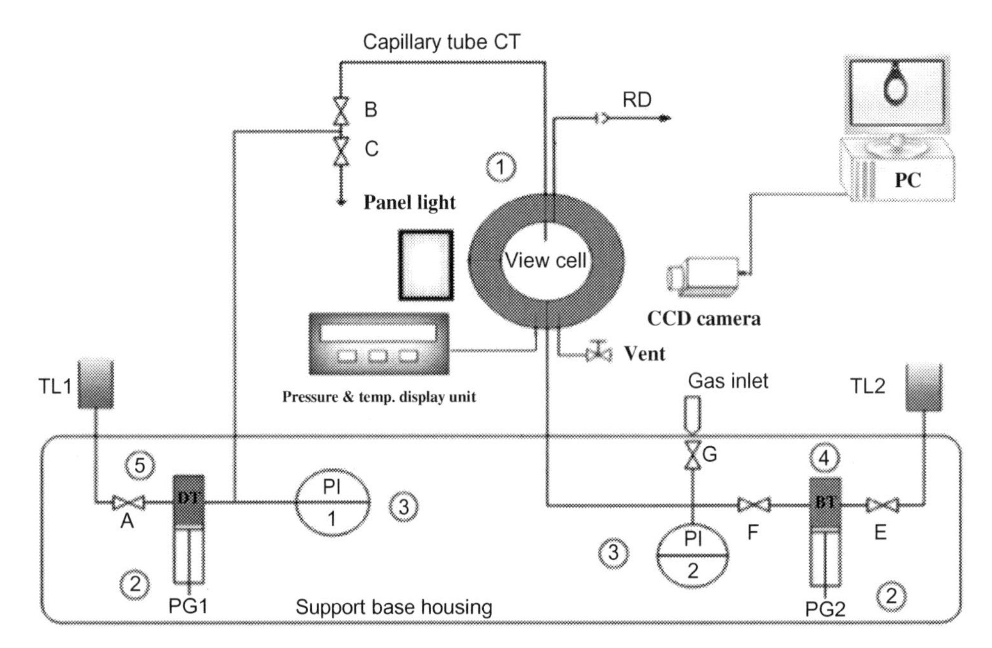

Figure 4.1 Schematic illustration of IFT determination in different temperatures [29]. *IFT*, interfacial tension.

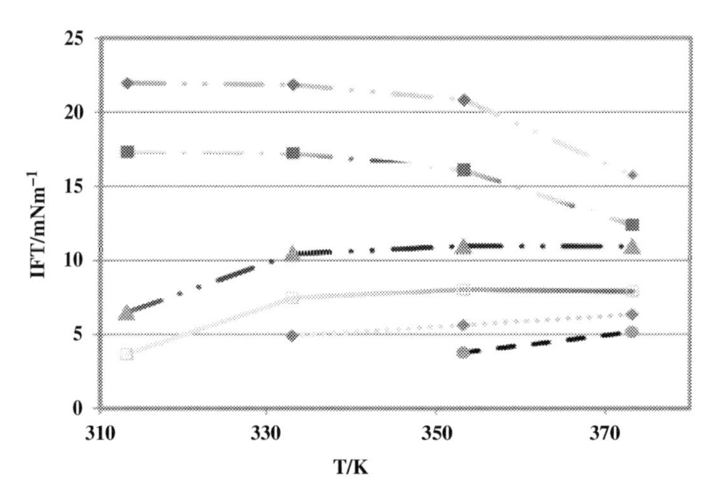

Figure 4.2 Measured equilibrium IFTs of crude oil/CO_2 system versus temperature at various pressures: -· ·♦, 0.68 MPa;-· ■, 3.44 MPa; -· · ▲, 6.89 MPa;, -□, 8.96 MPa;, · · ·♦, 11.72 MPa; ---, ● 15.16 MPa [29]. *IFT*, interfacial tension.

Table 4.1 The Governing Equations Derived for IFT Determination at Various Thermodynamic Condition [29]

Temperature (K)	Pressure (MPa)	IFT (mN/m)
313.15	$0.69 \leq P \leq 6.20$	$\text{IFT} = -2.3512P + 24.01$
313.15	$6.89 \leq P \leq 8.96$	$\text{IFT} = -1.3624P + 15.84$
333.15	$0.69 \leq P \leq 8.96$	$\text{IFT} = -1.7652P + 23.02$
333.15	$8.96 \leq P \leq 11.72$	$\text{IFT} = -0.8343P + 14.59$
353.15	$0.69 \leq P \leq 8.96$	$\text{IFT} = -1.5603P + 21.89$
353.15	$8.96 \leq P \leq 15.85$	$\text{IFT} = -0.6899P + 13.94$
373.15	$0.69 \leq P \leq 8.96$	$\text{IFT} = -0.9375P + 16.06$
373.15	$8.96 \leq P \leq 18.27$	$\text{IFT} = -0.4382P + 11.86$

IFT, interfacial tension.

Solution:

1. $\text{IFT} = -1.7652P + 23.02 = 12.40\ \text{mN/m}$
2. $\text{IFT} = -1.5603P + 21.89 = 10.18\ \text{mN/m}$
3. $\text{IFT} = -0.4382P + 11.86 = 7.47\ \text{mN/m}$

4.3.2 Minimum Miscibility Pressure Correlations

4.3.2.1 Cronquist [33]

In 1978, Cronquist [33] proposed a correlation in a mathematical form as function of reservoir temperature (T_R), molecular weight of C_{5+} (MW_{C5+}), and volatile components (Vol) for the first time in the history of petroleum industry. This correlation is as follows:

$$\text{MMP} = 0.11027 \times (1.8 \times T_R + 32)^Y \tag{4.9}$$

where

$$Y = 0.744206 + 0.0011038 \times MW_{C5+} + 0.0015279 \times \text{Vol} \tag{4.10}$$

where MMP is in MPa, MW_{C5+} is in g/mol, Vol is dimensionless, and T_R has the unit of °C. This correlation is valid for oil gravities from 23.7 to 44°API, reservoir temperature from 21.67 to 120°C, and MMP values from 7.4 to 34.5 MPa.

Example 4.3: Calculate the MMP for a typical CO_2 flood by using proposed correlation by Cronquist [33]. The required data are as follows:

$$T_R = 80°C,\quad MW_{C5+} = 240.7\ g/\text{mol},\quad \text{Vol} = 53.36\%,\quad \text{MMP}^{\text{exp.}} = 27.52\ \text{MPa}.$$

Solution: At first, exponent Y has to be calculated as follows:

$Y = 0.744206 + 0.0011038 \times MW_{C5+} + 0.0015279 \times \text{Vol} = 1.092$, from Eq. (4.10)

By substituting Y into Eq. (4.9), we have
$MMP = 0.11027 \times (1.8 \times T_R + 32)^Y = 31.217$ MPa, from Eq. (4.9)

4.3.2.2 Lee [34]

Lee [34] suggested a mathematical correlation as a function of reservoir temperature (T_R) which is as follows:

$$MMP = 7.3924 \times 10^b \tag{4.11}$$

where

$$b = 2.772 - \left(\frac{1519}{492 + 1.8 \times T_R}\right) \tag{4.12}$$

in which, T_R and MMP indicate the reservoir temperature in °C and minimum miscibility pressure in MPa, respectively. If MMP is less than bubble point pressure, bubble point pressure will be taken as MMP.

Example 4.4: Calculate the MMP by using Eq. (4.11) for the given data in *Example 4.3*.

Solution: At first, exponent b has to be calculated as follows:

$$b = 2.772 - \left(\frac{1519}{492 + 1.8 \times T_R}\right) = 0.38$$

from Eq. (4.12)

By substituting b into Eq. (4.11), we have

$$MMP = 7.3924 \times 10^b = 17.882 \text{ MPa}$$

from Eq. (4.11)

4.3.2.3 Yellig and Metcalfe [35]

Yellig and Metcalfe [35] developed a mathematical correlation as a function of reservoir temperature (T_R) which is as follows:

$$\begin{aligned} MMP = {} & 12.6472 + 0.015531 \times (1.8 \times T_R + 32) + 1.24192 \times 10^{-4} \\ & \times (1.8 \times T_R + 32)^2 - \frac{716.9427}{1.8 \times T_R + 32} \end{aligned} \tag{4.13}$$

in which, T_R and MMP indicate the reservoir temperature in °C and minimum miscibility pressure in MPa, respectively. This correlation is developed based on the reservoir temperature from 35 to 88.9°C. If MMP is less than bubble point pressure, bubble point pressure will be taken as MMP.

Example 4.5: Calculate the MMP by using the Eq. (4.13) for the given data in *Example 4.3.*

Solution: The MMP is calculated by means of Eq. (4.13) as follows:

$$\mathrm{MMP} = 12.6472 + 0.015531 \times (1.8 \times T_R + 32) + 1.24192 \times 10^{-4} \times (1.8 \times T_R + 32)^2 - \frac{716.9427}{1.8 \times T_R + 32} = 15.154\ \mathrm{MPa}$$

from Eq. (4.13)

4.3.2.4 Orr and Jensen [36]

In addition to Yellig and Metcalfe [35], Orr and Jensen [36] established a new MMP correlation as a function of reservoir temperature (T_R) by the following equation:

$$\mathrm{MMP} = 0.101386 \times \mathrm{EXP}\left(10.91 - \frac{2015}{255.372 + 0.5556 \times (1.8 \times T_R + 32)}\right) \tag{4.14}$$

in which, T_R and MMP indicate the reservoir temperature in °C and minimum miscibility pressure in MPa, respectively. This correlation is developed based on the reservoir temperature from 35 to 88.9°C. If MMP is less than bubble point pressure, bubble point pressure will be taken as MMP.

Example 4.6: Calculate the MMP by using the Eq. (4.14) for the given data in *Example 4.3.*

Solution: The MMP is calculated by means of Eq. (4.14) as follows:

$$\mathrm{MMP} = 0.101386 \times \mathrm{EXP}\left(10.91 - \frac{2015}{255.372 + 0.5556 \times (1.8 \times T_R + 32)}\right) = 18.458\ \mathrm{MPa}$$

from Eq. (4.14)

4.3.2.5 Alston et al. [37]

Alston et al. [37] created a new MMP correlation as a function of reservoir temperature (T_R), MW_{C5+}, and volatile to intermediate components ratio (Vol/Int) by the following relationships:

If $P_b \geq 0.345$ MPa,

$$\mathrm{MMP} = 6.056 \times 10^{-6} \times (1.8 \times T_R + 32)^{1.06} \times (MW_{C_{5+}})^{1.78} \times \left(\frac{\mathrm{Vol}}{\mathrm{Int}}\right)^{0.136} \tag{4.15}$$

$$\text{If } P_b < 0.345 \text{ MPa},$$
$$\text{MMP} = 6.056 \times 10^{-6} \times (1.8 \times T_R + 32)^{1.06} \times (\text{MW}_{C_{5+}})^{1.78} \tag{4.16}$$

where MMP is in MPa, MW_{C5+} is in g/mol, Vol/Int is dimensionless, and T_R has the unit of °C. If MMP is less than bubble point pressure, bubble point pressure will be taken as MMP.

Example 4.7: Calculate the MMP for a typical CO_2 flood proposed by Alston et al. [37]. The required data are as follows:

$$T_R = 34.4°\text{C}, \text{MW}_{C5+} = 212.56\ g/\text{mol}, \frac{\text{Vol}}{\text{Int}} = 1.56, \text{MMP}^{\text{exp.}} = 10\ \text{MPa}, P_b = 6.5\ \text{MPa}.$$

Solution: Bubble point pressure is greater than 0.345 MPa. Therefore, the MMP is calculated by means of Eq. (4.15) as follows:

$$\text{MMP} = 6.056 \times 10^{-6} \times (1.8 \times T_R + 32)^{1.06} \times (\text{MW}_{C_{5+}})^{1.78} \times \left(\frac{\text{Vol}}{\text{Int}}\right)^{0.136}$$
$$= 11.562\ \text{MPa}$$

from Eq. (4.16)

4.3.2.6 Impurity Correction Factor by Alston et al. [37]

Alston et al. [37] also applied a correction factor (F_{impure}) to his previous MMP correlation as a function of modified impurities critical temperature (T_{cm}), and their weight fraction (w_i) by the following relationships:

$$F_{\text{impure}} = \left(\frac{87.8}{1.8 \times T_{\text{cm}} + 32}\right)^{\left(\frac{(1.935 \times 87.8)}{(1.8 \times T_{\text{cm}} + 32)}\right)} \tag{4.17}$$

where

$$T_{\text{cm}} = \sum w_i + T_{ci} \tag{4.18}$$

where T_{ci}, T_{cm}, F_{impure}, and w_i indicate the critical temperature of each component in °C, modified impurities critical temperature in °C, impurity correction factor, and weight factor of each component in fraction, respectively. H_2S and CO_2 critical temperatures are modified to 51.678°C.

Example 4.8: In earlier example, calculate the MMP by Alston et al. [37] considering H_2S weight fraction of 0.1 as an impure component.

Solution: At the beginning, the modified critical temperature is computed by the followings:

$T_{cm} = \sum w_i + T_{ci} = 51.77°C$, from Eq. (4.18)

Then, the effect of H_2S impurity on MMP can be taken into account through Eq. (4.17) as follows:

$$F_{impure} = \left(\frac{87.8}{1.8 \times T_{cm} + 32}\right)^{\left(\frac{(1.935 \times 87.8)}{(1.8 \times T_{cm} + 32)}\right)} = 0.618,$$

from Eq. (4.17)

Multiplying this factor by the predicted value of Eq. (4.15), we have

$MMP = 0.618 \times 11.562 = 7.145$ MPa, from Eq. (4.15)

4.3.2.7 Impurity Correction Factor by Sebastian et al. [38]

In addition to Alston et al. [37], Sebastian et al. [38] also applied a correction factor (F_{impure}) to Alston et al. [37] MMP correlation, which is a function of modified impurities critical temperature (T_{cm}) and their mole fraction (x_i) by the following relationships:

$$F_{impure} = 1.0 - 2.13 \times 10^{-2}(T_{cM} - 304.2) + 2.51 \times 10^{-4}(T_{cM} - 304.2)^2 - 2.35 \times 10^{-7}(T_{cM} - 304.2)^3 \ldots \quad (4.19)$$

where

$$T_{cM} = \sum x_i \times T_{ci} \quad (4.20)$$

where T_{ci}, T_{cM}, F_{impure}, and x_i indicate the critical temperature of each component in K, modified impurities critical temperature in K, impurity correction factor, and mole fraction of each component, respectively. H_2S critical temperature is modified to 51.67°C.

Example 4.9: In Example 4.7, calculate the MMP by Sebastian et al. [38] considering H_2S mole fraction of 0.1 as an impure component.

Solution: At the beginning, the modified critical temperature is computed as follows:

$$T_{cM} = \sum x_i \times T_{ci} = 278.317 \text{ K}$$

from Eq. (4.20)

The effect of H_2S impurity on MMP can be taken into account through Eq. (4.19) as follows:

$$F_{impure} = 1.0 - 2.13 \times 10^{-2}(T_{cM} - 304.2) + 2.51 \times 10^{-4}(T_{cM} - 304.2)^2 - 2.35 \times 10^{-7}(T_{cM} - 304.2)^3 = 1.723$$

from Eq. (4.19)

Multiplying this factor by the predicted value of Eq. (4.15), we have

$$MMP = 1.723 \times 11.562 = 19.926 \text{ MPa}$$

from Eq. (4.15)

4.3.3 CO_2 Flooding Properties and Design

By far, the most common application of solvent methods is in a displacement mode, but injection and production through the same wells have been reported [39–41].

Carbon dioxide can be injected and dissolved in water in a distinctly immiscible fashion that recovers oil through swelling and viscosity reduction [12]. If the solvent is completely miscible with the oil (first contact), the process has a very high ultimate displacement efficiency since there can be no residual phases. If the solvent is only partially miscible with the crude, the total composition in the mixing zone between the solvent and the oil can change to generate or develop in situ miscibility. Regardless of whether the displacement is developed or first contact miscible (FCM), the solvent must immiscibly displace any mobile water present with the resident fluids. The economics of the process usually dictates that the solvent cannot be injected indefinitely. Therefore, a finite amount or a slug of solvent is usually followed by a chase fluid whose function is to drive the solvent toward the production wells. This chase fluid—N_2, air, water, and dry natural gas seem to be the most common choices—may not itself be a good solvent. But it is selected to be compatible with the solvent and because it is available in large quantities [4].

A phase behavior or pressure–temperature plot (P–T diagram) for different pure components and air is reported in the work of McCain [4,42]. For each curve, the line connecting the triple and critical points is the vapor pressure curve; the extension below the triple point is the sublimation curve. The P–T diagram for air is really similar to an envelope, although its molecular weight (MW) distribution is so narrow that it appears as a line. The critical pressures for most components fall within a relatively narrow range of 3.4–6.8 MPa (500–1000 psia), although critical temperatures vary over a much wider range. The critical temperatures of most components increase with increasing MW. Carbon dioxide (MW = 44) is an exception to this trend with a critical temperature of 304K (87.8°F), which is closer to the critical temperature of ethane (MW = 30) than to propane (MW = 44). Most reservoir applications would be in the temperature range of 294–394K (70–250°F) and at pressures greater than 6.8 MPa (1000 psia); hence, air, N_2, and dry natural gas will all be supercritical fluids at reservoir conditions. Solvents such as LPG, in the MW range of butane or heavier, will be liquids. Carbon dioxide will usually be a supercritical fluid since most reservoir temperatures are above the critical temperature. The proximity to its critical temperature gives CO_2 more liquid-like properties than the lighter solvents [4,42].

Two charts of compressibility factor for air and carbon dioxide is reported in the work of Lake [4]. So, the fluid density ρ_g can be calculated as follows:

$$\rho_g = \frac{PMW}{zRT} \tag{4.21}$$

where pressure, molecular weight, compressibility factor, temperature, and global gas constants are shown with P, MW, z, T, and R. The gas formation volume factor (B_g) at any temperature and pressure is given as follows:

$$B_g = z\frac{P_s}{P}\frac{T}{T_s} \tag{4.22}$$

where T_s and P_s are the standard temperature and pressure, respectively. All fluids become more liquid-like, at a fixed temperature and pressure, as the MW increases. The anomalous behavior of CO_2 is again manifested by comparing its density and formation volume factor to that of air.

4.3.4 CO_2 Field Case Study

Some field case studies are considered here which have lessons in gas injection EOR. As discussed before, the recovery in oilfield depends on both volumetric and displacement sweep efficiencies. Summary of gas floods performed can be found in the works of Manrique et al. [43] and Christensen et al. [44]. The followings are some examples of miscible gas injections.

4.3.4.1 Slaughter Estate Unit CO_2 Flood

The miscible flood in the West Texas San Andres dolomite is an example of a gas flood with very good oil recovery. The average permeability is low around 4 mD at a depth of about 5000 ft. A waterflood in the early 1970s prior to gas flooding led to a good recovery. A volume of 72 mol% CO_2 and 28 mol% H_2S were mixed for injection. The MMP of approximately 1000 psia with this gas and moderate API oil (32°API) is substantially less than the average reservoir pressure of 2000 psia. Therefore, this flood is multicontact miscible (MCM).

Water was injected alternately with the acid gas with a water-alternating-gas (WAG) injection ratio of about 1.0. A 25% hydrocarbon pore volume (HCPV) slug of acid gas was injected. The chase gas was also alternated with water, and eventually, the gas—water ratio was reduced to 0.7 to improve vertical sweep. The cycles are shown in figure.

Incremental tertiary recovery was 19.6% OOIP, which is largely the result of good WAG management and the use of H_2S in the gas [45]. H_2S, although very dangerous, is a very good miscible agent. When added to the primary and secondary recovery (waterflood) of about 50%, the total recovery in this pilot is expected to be around 70% OOIP, which is well above the average for most fields.

4.3.4.2 Immiscible Weeks Island Gravity Stable CO_2 Flood

This sand, which is up against a salt dome, is highly dipping (30 degree dip) and is very permeable both vertically and horizontally. The initial reservoir pressure for this sand was 5100 psia at a reservoir temperature of 225°F, but at the time of the pilot, the pressure was lower. The pilot lasted 6.7 years and consisted of one up-dip injector and two producers about 260 ft down-dip. Following a waterflood, gas was injected up-dip so that gravity would stabilize the front in a relatively horizontal interface. The main idea of this gravity stable flood is that the gas—oil contact (oil bank) will move down vertically recovering oil and displacing it to the down-dip production wells. This process can be highly efficient (good volumetric sweep) as long as there is good vertical permeability, and the gas interface is stable and moves vertically downward.

The gas injected at Weeks Island was a mixture of CO_2 and about 5% plant gas. The plant gas was used to lighten CO_2 so that the gas—oil interface is more stable. Injection of plant gas with CO_2, however, was found unnecessary to ensure a gravity stable flood at Weeks Island as CO_2 was effectively diluted by dissolved gas (methane) from the reservoir oil [46].

At the reservoir temperature of 225°F and pressure at gas injection, the flood was immiscible, not miscible. Nevertheless, a pressure core taken in zones where the gas traversed was nearly "white" with average oil saturations in the CO_2 swept zone of approximately 1.9%. This low oil saturation value is lower than miscible flood residuals, S_{orm}, that are typically observed due to oil-filled bypassed pores. The unexpectedly good recovery demonstrates that even immiscible floods when properly designed can achieve good extraction of oil components by gravity drainage. A subsequent commercial test of the gravity stable process was not as successful largely because of significant water influx down-dip of the production wells. Injection of CO_2 largely pressurized the gas cap but did not cause the gas—oil interface to move vertically downward. A gravity stable process like this would be very effective as long as water influx is relatively small. Perhaps, one solution could have been outrunning the aquifer with water production wells or trying to plug off water influx. One difficult problem also encountered was the production of the thin oil bank owing to both gas and water coning. The second producer was not planned but was drilled to measure saturations in the oil bank and to speed oil bank capture. Immiscible gas floods in general can achieve better displacement efficiency as a secondary recovery method if gravity override is controlled, than for waterfloods owing to decreased oil viscosity, oil swelling, IFT lowering, extraction of oil components, and the potential for gravity drainage as occurred at Weeks Island. Immiscible gas floods could also be a good alternative for reservoirs with injectivity issues when water is used. Two main disadvantages of immiscible gas flooding over waterflooding are the potential for poor sweep due to its adverse mobility ratio and gravity override due to higher contrast between oil and gas gravities [46].

4.3.4.3 Jay Little Escambia Creek Nitrogen Flood

The Jay field near the Alabama–Florida border is one of the few nitrogen floods ever conducted. The reservoir is in the Smackover carbonate at a depth of 15,000 ft. Nitrogen is a good miscible gas in this reservoir because of its very light sour crude (50°API) and high reservoir pressure around 7850 psia. The average formation permeability is about 20 mD. A significant advantage of nitrogen is that it is readily available via separation from the air, is relatively cheap, and does not cause corrosion unlike CO_2. Nitrogen was injected using a WAG ratio near 4.0, which is greater than typical. The overall recovery at Jay is expected to be near 60% OOIP. Incremental recovery beyond waterflood recovery from miscible nitrogen injection is forecast to be around 10% OOIP. The high primary and secondary recovery of around 50% OOIP is likely the result of low vertical permeability coupled with good horizontal permeability in the dolomite.

Miscible nitrogen injection is expected to give an incremental recovery of 10% OOIP over waterflooding alone [47]. Low vertical permeability caused by shale lenses or in this case, cemented zones associated with thin stylolites is an ideal candidate for both water and gas flooding as fluids are less able to segregate vertically so that gravity override is reduced. This is especially true in this flood since nitrogen has very low density compared to the resident fluids and would likely have gone to the top of the reservoir otherwise.

4.3.4.4 Overview of Field Experience

Gas flooding technology is well developed and has demonstrated good recoveries in the field [43]. Recoveries from both immiscible and miscible gas flooding vary from around 5% to 20% OOIP, with an average of around 10% incremental OOIP for miscible gas floods [44]. Tertiary immiscible gas flooding recoveries are less on average, around 6% OOIP. Although recovery by gas flooding is very economic at these levels, 55% OOIP still remains on average post miscible gas flooding assuming 65% OOIP prior to gas flooding. The significant amount of oil that remains is largely the result of gas channeling through the formation owing to large gas mobility, reservoir heterogeneity, dispersion (mixing), and gravity effects. Channeling also results in early breakthrough of the gas (gas), typically around the same time that oil is produced. This is in contrast to surfactant/polymer flooding, which almost always exhibits an oil bank prior to surfactant breakthrough. Poor volumetric sweep is not as much of a problem for surfactant–polymer floods or waterfloods, which have more favorable mobility ratios. Nevertheless, miscible flooding is generally very economic and less complex than chemical flooding, especially for deeper reservoirs that are more technically challenging for surfactant/polymer floods.

4.4 FIRST CONTACT MISCIBLE VERSUS MULTICONTACT MISCIBLE

In some cases, the injected gas cannot yield the FCM with the reservoir crude oil once it is injected. However, it may gradually develop dynamic miscibility with the residual oil through MCM under actual reservoir conditions [48,49].

The result of recovery governed by immiscible gas injection is limited basically by three factors [50]:

- Areal sweep efficiency
- Volumetric sweep efficiency
- Microscopic sweep efficiency

Because of viscous fingering, gravity segregation, permeability stratification, IFT, wettability, and pore structure, ultimate oil recovery is always much less than 100%. Hence, the reason of the interest in the miscible injection method can be explained as it is a more efficient recovery method [18,21,51,52]. In fact, it is usually not economical and in some cases, not technically feasible to inject a gas that create the FCM with the oil; therefore, the injected gas is designed to develop miscibility by the net transfer of components from the oil into the gas (a vaporizing gas drive) or from the gas to the oil (a condensing gas drive) [7].

Many phenomena are studied as they can limit the efficiency of the miscible flooding process. For instance, permeability heterogeneity is considered as a strong limitation on recovery because it can lead to flow channeling and poor sweep efficiency. Small-scale heterogeneities are particularly problematic for all secondary and tertiary recovery processes, because they can have a significant effect on recovery, which cannot be modeled explicitly in field-scale simulations [53].

Compositional simulation is usually utilized to estimate the performance of the recovery schemes on the basis of equation-of-state properties developed from the regression on data obtained from laboratory experiments. The accuracy of the measurements is highly dependent on the validity of the assumptions used in these simulations [54,55].

4.5 HEAVY OIL RECOVERY USING CO_2

Recently, recovery of heavy oil by CO_2 displacement methods has gained much popularity worldwide. This approach especially received attention in reservoirs where steam flooding is not applicable. First, efforts in large-scale field tests such as Lick Creek field in Arkansas and Wilmington field in California have proved the applicability of CO_2 immiscible displacement for heavy oil recovery [56,57]. The high oil volume remaining in reservoirs and also its economic benefits lead to production enhancement via CO_2 for more than 40 years [58].

Heavy crude oils have a wide range of viscosities from about several hundred to hundred thousand centipoises. Concerning mobility of oil under reservoir conditions and reducing its value is an important factor in oil recovery. Carbon dioxide can be used as an effective agent for reducing viscosity of the heavy oil which results in oil mobility reduction [59].

Various mechanisms are observed during injection of carbon dioxide such as oil viscosity reduction, oil swelling, and relative permeability hysteresis due to reduced water saturation, wettability alternation, depressurization, diffusion, and IFT reduction in the zone near the wellbore [12,42,60–63].

Some physical properties such as viscosity, density, and CO_2 solubility in heavy oils are required to design and simulate a heavy oil recovery process. Determination of the effects of CO_2 on the physical properties of heavy crude oils is the first step to design an effective displacement process. Hence, different studies have focused on predictive methods for properties of heavy oil/CO_2 mixtures. For instance, Simon and Graue [64] published data for mixtures of CO_2 and nine different oils ranging from 11.9 to 33.3°API. Experimental conditions covered a range of temperatures from 38 to 121°C and pressures up to 15.9 MPa. In addition, they presented correlations for predicting solubility, swelling, and viscosity behavior of CO_2/crude oil systems, which are the principal correlations currently used in reservoir engineering. Miller and Jones [65] investigated the properties of four dead heavy oils obtained from Cat Canyon (10°API) and Wilmington (15–17°API) oil fields in California, and Densmore (19.8°API) oil field in Kansas. Data were presented on solubility of CO_2 in the pre-mentioned heavy oils, oil swelling by CO_2, and effect of CO_2 on heavy oil viscosity. Sankur et al. [66] also presented some data on properties of CO_2/reservoir oil mixture.

The abovementioned methods used to characterize CO_2/reservoir oil mixture apply mole fraction instead of volume fraction for indicating composition. Therefore, determination of the MW of heavy crude is essential. For convenient application, correlations developed in the work of Chung et al. [67] used CO_2 volume fractions in the crude oil leading to minimum experimental work.

4.5.1 Vapor ExtractionsHeavy Oil

The recovery of heavy oil using thermal methods from reservoirs with low porosity, low thermal conductivity, and high fractures and/or fissures can be problematic. Moreover, the production from these reservoirs is not satisfactory from economic aspects. Hence, the vapor extractions (VAPEX) process can be proposed as an alternative approach [68].

The VAPEX process was introduced by Mokrys and Butler [69] and Butler and Mokrys [70]. In this process, a pure hydrocarbon vapor or a vaporized hydrocarbon mixture diffuse and dissolve in heavy oil leading to a reduction in oil viscosity and an increase in the oil mobility. This method is mostly used in cases in which solvent-assisted gravity drainage (SAGD) process is not applicable. In other words, VAPEX is utilized in cases where SAGD leads to excessive heat losses and extra costs. These reservoirs are as follows [68]:

- Thin reservoirs
- Low permeability carbonate reservoirs with high heat capacity per unit volume of contained oil
- Reservoirs underlain by aquifers and/or gas cap

Solvents used frequently include [71]

- Ethane
- Propane
- Butane

However, it should be noticed that pure solvents are applicable for only limited conditions. In the most reservoirs, pressure and temperature do not allow to inject a pure solvent in its saturated vapor (dew point) condition [72].

4.5.1.1 The Solvent Requirement for the Vapor Extractions Process

Amount of the used solvent is a critical parameter which affects the costs of operation. The injected solvent undergoes different phenomena including dissolution into heavy oil and then stripping from heavy oil in order to create a recycling process.

It is shown that one barrel of propane is required for recovery of one barrel of oil in a condition in which there is no stripper and recycle loop; however, a drastic decrease as much as 74% of the injected propane as solvent can be recovered when recycling is employed [73].

Various studies have been done to model solvent material balance in a reservoir stimulated by the VAPEX recovery process. The method used by Butler et al. [74], which was based on specific volumes of the solvent and heavy oil, failed as it was not able to approximate the specific volumes of the two solvents correctly. Afterward, Mokrys [72] proposed an equation based on Gibbs theory [75] for calculating the volume of solvent mixture required for recovery. The mathematical form of this model can be states as follows [75]:

$$\overline{M_i^{\text{ig}}}(T \cdot P) = M_i^{\text{ig}}(T \cdot p_i) \tag{4.23}$$

where

$$M = \sum_i x_i \overline{M}_i \tag{4.24}$$

Therefore, it can be concluded that

$$M_i^{ig}(T \cdot P) = \sum_i x_i M_i^{ig}(T \cdot p_i) \tag{4.25}$$

In Eqs. (4.23)–(4.25), T, P, p_i, M, and $\overline{M}$ are temperature, pressure, partial pressure, molar, and partial molar property, respectively. The superscript ig shows the ideal gas. The Gibbs theorem states that the parameter M can be any thermodynamic property in exception of the volume parameter. Application of Eq. (4.25) for estimation of volume may lead to inaccurate results. In approximating the amount of solvent for a specific reservoir conditions, thermodynamic calculations should be considered for characterizing the behavior of solvent–heavy oil system [59].

4.5.1.2 Diffusion Coefficient for Solvent–Heavy Oil System

Injection of miscible solvent in the reservoir is an example of mass transfer process that is governed by a diffusion coefficient. It is suggested that precise diffusion data for this process are essential in order to calculate the following parameters as follows [76]:

- The amount of gas flow rate for injection,
- The extent of heavy oil the its viscosity would be decreased,
- The needed time to reach to desired mobility,
- The oil production rate.

It is only possible when constant diffusion coefficient is assumed, which is in the need of the following conditions [76]:

- Having similarity between molecular diameter and shape of molecules,
- Negligibility of intermolecular forces within diffusion mixture,
- Nonreacting environment.

In most of the reservoir situations, the solvent–heavy oil system meets the third condition, except in place where asphaltene deposition can happen and consequently, none of the above conditions are fulfilled [77].

Chang and Chang [78] developed a finite volume approach for the inverse estimation of thermal conductivity in one-dimensional domain. The method can also be utilized to convert the differential equation governing the mass diffusion into a system of linear equations in matrix form. The unknown diffusion coefficients are obtained by solving the system directly. Hence, no prior information about the functional form of the diffusion coefficient is required and no iterations in the calculation process are needed. Thus, starting from Fick's second law, which is as follows [77]:

$$\frac{\partial c}{\partial t} = \frac{\partial}{\partial x}\left(D\frac{\partial c}{\partial x}\right) \tag{4.26}$$

The equation can be written in terms of density as follows [77]:

$$\frac{\partial \rho}{\partial t}\frac{\partial \rho}{\partial c} = \frac{\partial}{\partial x}\left(D\frac{\partial \rho}{\partial x}\frac{\partial \rho}{\partial c}\right) \tag{4.27}$$

Here, the dependency of density with respect to composition is assumed as it is unknown. Hence, a linear approach is proposed base on the small composition gradients expected for the diffusion process of gases in heavily oil and bitumen. Thus, Eq. (4.27) can be simplified as follows [77]:

$$\frac{\partial \rho}{\partial t} = \frac{\partial}{\partial x}\left(D\frac{\partial \rho}{\partial x}\right) \tag{4.28}$$

In the next step, the medium domain is discretized with mesh size Δx and time step Δt, which is illustrated in Fig. 4.3.

It can also be assumed that concentration in the boundary A is constant, and there is no density flux condition in the boundary B. The followings are obtained by discretization of equations:

Internal points,

$$-\left(\rho_i^n - \rho_{i-1}^n\right)D_{i-0.5}^n + \left(\rho_{i+1}^n - \rho_i^n\right)D_{i+0.5}^n = \frac{\Delta x^2}{\Delta t}\left(\rho_i^{n+1} - \rho_i^n\right) \tag{4.29}$$

Boundary A,

$$-2\left(\rho_i^n - \rho_{\mathrm{A}}^n\right)D_{\mathrm{A}}^n + \left(\rho_{i+1}^n - \rho_i^n\right)D_{i+0.5}^n = \frac{\Delta x^2}{\Delta t}\left(\rho_i^{n+1} - \rho_i^n\right) \tag{4.30}$$

Boundary B,

$$-\left(\rho_i^n - \rho_{i-1}^n\right)D_{i-0.5}^n + \left(\rho_{\mathrm{B}}^n - \rho_i^n\right)D_{i+0.5}^n = \frac{\Delta x^2}{\Delta t}\left(\rho_i^{n+1} - \rho_i^n\right) \tag{4.31}$$

Eqs. (4.29)–(4.31) can be arranged in a matrix form as $Ax = b$. Matrix A can be constructed from the discretization of governing equations. Vector b is made of the density measurement at specific grid locations along the medium and boundary conditions. x is also the unknown diffusion coefficients.

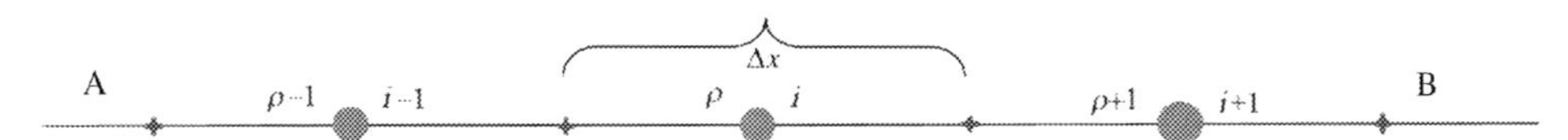

Figure 4.3 A medium domain with mesh size Δx and time step Δt.

4.6 HYDROCARBON: LPG, ENRICHED GAS, AND LEAN GAS

This process involves the continuous injection of high-pressure methane, ethane, nitrogen, or flue gas into the reservoir. The lean gas process, similar to enriched gas, involves multiple contacts between reservoir oil and lean gas before forming a miscible bank. But, there is a difference in the enriched gas process where light components condense out of the injected gas and into the oil, then intermediate hydrocarbon fractions (C_2-C_6) are stripped from the oil into the lean gas phase.

In a reservoir with initial reservoir pressure of 6425 psi, the swelling test was simulated by various proportions of injection gas mixed with original reservoir oil. The bubble point is calculated as 2302 psi. A comparison of the effects of CO_2 and lean gas injection on the saturation pressure change is reported in the relevant literature [79]. It is observed that the saturation pressure can be reduced by increasing the percentage of injected CO_2, whereas the saturation pressure increases as the percentage of injected lean gas increases. Swelling factor shows the same pattern in injection of CO_2 or lean gas.

The effect of injected gas mole fraction on the relative volume is reported in the relevant literature [79]. The relative volume goes up by increase in gas mole fraction. The reason lies in the fact that more gas can evolve from the oil when pressure goes below the bubble point pressure [80].

4.7 RESERVOIR SCREENING

It should be noted that CO_2 EOR and storage do not lead to efficient results in all oil reservoirs due to different technical and economic reasons. Hence, selecting a screening procedure seems to be vital. Shaw and Bachu [81] established a number of suggestions for doing basic evaluations on some selected oil reservoirs for conducting simultaneous CO_2 EOR and CO_2 storage before considering other economic criteria. These criteria are as follows [81]:

- Screening for EOR and storage suitability,
- Technical ranking of suitable reservoir,
- Improved oil recovery (IOR) and CO_2 storage capacity predictions.

A series of criteria were recommended by various authors for the technical screening of CO_2 EOR by miscible flood [8,83–85]. These criteria are based on the optimization of reservoir performance to have better IOR efficiency. These criteria allow a rapid screening and evaluation of oil reservoirs which are suitable for CO_2 EOR based on general reservoir characteristics and oil properties. Rivas et al. [82] studied the impact of different reservoir parameters on CO_2 EOR efficiency by carrying a comprehensive simulation out. They achieved a set of optimum values for reservoir and oil properties, which are best suited for CO_2 EOR operation. Based on this table, weighting factors indicate the relative importance of each parameter.

4.8 CORROSION

4.8.1 Facility and Corrosion

One of the main challenges to oil industry has always been CO_2-related problems on facilities. In addition, its great impact on economic aspects of projects is remarkable. For instance, CO_2 may cause severe corrosion on pipelines, well tubing, and pumping equipment [86]. Effect of CO_2 injection on the reservoir formation and reservoir fluid should also be taken into account. One example of these impacts is solid deposition caused by CO_2 mixing with reservoir fluids, such as scale formation and asphaltene precipitation. For offshore projects, possible problems are related to platforms, well completion, and pipelines to handle CO_2, such as extra weight on injection and production platforms, and hydrate formation [87].

By increasing CO_2 injection operations and also the fact that its stages are taken place in different locations, numerous technological and engineering advances made over the past 35 years in CO_2 injection were well designed, including [88,89]:

- Corrosion-resistant materials such as stainless and alloy steels [316 SS, nickel, Monel, corrosion resistant alloy (CRA), etc.] for piping and metal component trim,
- Swell-resistant elastomer materials such as Buna-N and Nitrile rubbers for downhole packers, and Teflon or Polytetrafluoroethylene (PTFE) and Nylon for seals,
- Fiberglass lined (GRE) and internally plastic coated (IPC) pipe (phenolics, epoxies, urethanes, and novolacs) tubing strings to retard corrosion,
- Acid-resistant cements containing latex, pozzolan, alumina, and other additives,
- Automatic control systems that not only regulate flows but also provide real-time monitoring which is capable of initiating well shutdowns in an unsafe condition.

Various innovations in tools and hardware are also manufactured which have had an important role in development of operational and safety practices, including [88]:

- Use of corrosion protection of the casing strings via impressed and passive currents and chemically inhibited fluid (oxygen, biocide, corrosion inhibitor) in the casing–tubing annulus,
- Use of special procedures for handling and installing the production tubing to provide gas tight seals between adjacent tubing joints and to avoid coating damage or liner damage,

- Use of tubing and casing leak detection methods and repairing techniques, using both resin and cement squeeze technologies as well as insertion of fiberglass and steel liners,
- Formulation and implementation of criteria unique to siting wells in or near populated areas incorporating: fencing, monitoring, and atmospheric dispersion monitoring elements to protect public safety.

4.8.2 Corrosion Control

Carbon steel casing is used for CO_2 EOR injection wells and as such, it is susceptible to corrosion. To mitigate corrosion, several techniques are typically used as follows [88]:

- *Correct cement placement.* To minimize contact between carbonic acid and the steel casing, great care is used to assure that the cement, used to bond casing to the formation, is adequately distributed along its entire axis. This requires some treating actions, including careful removal of residual drilling mud from the hole, use of centralizers to center the casing string in the bore hole, and full circulation of the cement returns to the surface. With a well-formed cement sheath in place, the rate of permeation of corrosive material is greatly reduced.
- *Placement of acid resistant cements in zones susceptible to cement carbonation.* As appropriate, operators will incorporate specialty cements or specialty slurry designs adjacent to and above the CO_2 injection zone. These cements are more resistant to CO_2 attack and hence dramatically reduce the rate of CO_2 degradation.
- *Cathodic protection of the casing string.* Operators employ both impressed and passive current techniques on the casing string to counteract naturally occurring galvanic action, which leads to corrosion. Both methods are used widely in many industrial applications.
- *Corrosion inhibitor.* After completing the well, a biocide/corrosion inhibitor-laden fluid is placed in the annular space between the casing and tubing string to further suppress any corrosive tendency.

4.9 DESIGN STANDARDS AND RECOMMENDED PRACTICES

Oil and gas wells have existed for almost 150 years, since the time of first efforts in Pennsylvania in 1859. Hence, well technology has developed over these years, and also professional organizations, such as American Petroleum Institute, American Society of Mechanical Engineers, the National Association of Corrosion Engineers, and others, have continued to evaluate the catalog and technical requirements and contributed in designing the operational practices based on formal engineering standards and recommended practices. For well technology and field piping, common standards are available in relevant literature [88].

There are two basic elements which should be considered in order to design a well, which are as follows:

- The wellbore, consisting of casing, cement, and wellhead,
- The mechanical completion equipment consisting of valves, tubulars, and packers.

4.9.1 Wellbore Design

The well designs are almost similar for different cases of CO_2 injection, consisting of surface casing and production casing. The reason of using multiple casings is isolating groundwater resources from potential sources of contamination and maintaining the stability of the wellbore. A typical CO_2 wellhead is available in relevant literature [90].

From the mechanical point of view, thickness and weight of a casing are selected based on maximum potential burst and collapse pressures plus safety factors. The safety factors are function of injection and production pressures, well trajectory, and reservoir conditions. Carbon steel casing is a common casing type used for wells of 10,000 ft or less in depth and usually, in these cases, grades of J-55 and K-55 are typical. In deep, high pressure and high temperature environments, higher strength grades should be used, and CRA are used in wells susceptible to H_2S and CO_2 leaks [90].

For new construction, almost all wells are cased-hole completions. In isolated cases, depending on reservoir conditions, open-hole completions are still used; however, they are rare [92,93]. Since cased-hole completions are amenable to a larger variety of profile management techniques (mechanical isolation, chemicals, squeeze cementing, etc.) than open-hole completions, they are the more common completion strategy [91].

It should also be considered that pore pressure variation, which is caused by injection (or production), changes the in-situ stresses. This fact may lead to some huge influences on wellbore designs. For instance, fault reactivation is possible and consequently casings will be sheared. In addition, wellbore instability is very common [94,95]. Besides, induced thermal stresses should be estimated during thermal recoveries, in wellbore trajectory design as well as completion design [96].

4.9.2 Cement Technology

Cementing is critical to the mechanical performance and integrity of a wellbore both in terms of its method of placement and cement formulation used. Chemically, the degradation of Portland-based cements by carbonic acid (H_2CO_3) is well known and documented [97]. The basic chemical mechanism is described below [98]:

$$CO_2 + H_2O \xrightarrow{\text{yields}} H_2CO_3$$

$$H_2CO_3 + C - S - H \xrightarrow{\text{yields}} \text{amorphous silica gel} + CaCO_3$$

$$H_2CO_3 + Ca(OH)_2 \xrightarrow{\text{yields}} CaCO_3 + 2H_2O$$

$$H_2CO_3 + CaCO_3 \xrightarrow{\text{yields}} Ca(HCO_3)_2$$

In the foregoing reactions, calcium–silica–hydrate, C–S–H, compounds are major components in Portland cements, whereas free lime, $Ca(OH)_2$, constitutes about 20% of the cement composition in set Portland cements. Because CO_2 corrosion of cement is thermodynamically favored and cannot be entirely prevented, various solutions have been developed to limit CO_2 attack on the cement sheath. Most of these approaches involve substituting materials such as fly ash, silica fume, or other nonaffected filler or other cementation materials for a portion of the Portland cement. The water ratio of the cement slurry is designed to be low to reduce the permeability of the set cement. The permeability of the set cement may be further lowered through the addition of materials such as latex (styrene butadiene) to the compound [99].

4.10 WATER-ALTERNATING-GAS PROCESS

WAG injection is a tertiary oil recovery process. First, it was introduced in the 1950s and its popularity has grown since then. WAG is a combination of the two secondary recovery processes of waterflooding and gas injection, and its original aim for the ideal system of oil recovery is to enhance the macroscopic and microscopic sweep efficiency simultaneously [100].

It is classified based on the types of fluid involved and the manner in which they are injected to different groups. In general, it can be divided into miscible and immiscible displacement processes [7,16,102]. In the miscible WAG process, the injected gas is miscible with the reservoir oil under the prevailing conditions. In the immiscible WAG process, the injected gas is not miscible with the reservoir oil and it displaces the oil while maintaining its gaseous phase, with a front between the two phases. Further classification of WAG process is given below, depending on the injection technique used in the process [103,104].

Further classification of WAG process depends on the injection method used in the process. A WAG process can be implemented as hybrid WAG injection, in which injection of a large volume of gas is done at the first step, and then small volumes of water and gas are injected maintaining a WAG ratio of 1:1. Moreover, simultaneous water and gas injection is also one of the popular approaches [103,105].

Mobility of gas can be controlled by water injection. The cyclic nature of the WAG process leads to an increase in water saturation during the water injection half cycle and a decrease of water saturation during the gas injection half cycle. The processes of imbibition and drainage taken by inducing cycles cause the residual oil saturation to be usually lower than that of waterflooding and similar to those of gas flooding [106]. The oil recovery factor can be described by two factors that are the macroscopic sweep efficiency and the microscopic sweep efficiency. Furthermore, the macroscopic sweep efficiency is defined by the horizontal and the vertical sweep efficiencies. This can be formulated as follows [106]:

$$R_f = E_v E_h E_m \tag{4.32}$$

$$M = \frac{k_{rg}/\mu_g}{k_{ro}/\mu_o} \tag{4.33}$$

$$R_{\frac{v}{g}} = \left(\frac{v\mu_o}{kg\Delta\rho}\right)\left(\frac{L}{h}\right) \tag{4.34}$$

where E_v, E_h, E_m, k_r, μ, $\Delta\rho$, g, L, and h stand for vertical sweep efficiency, horizontal sweep efficiency, microscopic sweep efficiency, relative permeability, viscosity of fluid, density difference, constant of gravity, the length of porous media, and net-pay thickness, respectively. The subscripts o and g are in correspondence with oil and gas.

Literature on the WAG process typically discusses two major management parameters that affect the economics of a WAG project. These operational aspects are the half-cycle slug sizes and the WAG ratio. The two major problems faced are early breakthrough and injectivity losses. It is therefore proposed that the third parameter to be studied is the operation of the smart wells. The two most common distinctions in the classification of the WAG process are miscible WAG injection and immiscible WAG injection. Miscible WAG injection occurs when the reservoir is above the MMP, and it is immiscible when injection pressure is below the MMP value. When the initial reservoir pressure is just above the MMP, it often moves in and out of miscibility condition in part of the reservoir or all of it [101].

4.10.1 Factors Influencing Water-Alternating-Gas

Considerable parameters in the design of the WAG process are as follows [7,16,107–109]:

Fluid properties and rock–fluid interaction. The fluid behavior within the reservoir is a key parameter which still requires more knowledge and development for better understanding. This phenomenon becomes even more complex when the prevailing conditions within the reservoir change as a result of undergoing processes. Variations in rock–fluid interaction with changing conditions in a reservoir result in wettability variations, which in turn affect flow parameters such as capillary pressure and relative permeability [110,111].

Availability and composition of injection gas. The availability of gas, in terms of quantity and composition, plays a vital role. Usually, the gas produced with oil from a reservoir is reinjected during the WAG process.

WAG ratio. The WAG ratio is highly significant in WAG process design [112]. A WAG ratio of 1:1 is normally used in field applications.

Heterogeneous permeability. The vast majority of reservoirs have nonuniform pore size distribution with varying degrees of interconnectivity, giving rise to heterogeneous permeability. Sometimes, the heterogeneity can be segregated in the form of layers, constituting homogeneous layers in the reservoir [113].

Injection pattern. Well spacing is critical in WAG process design [7,104]. The five-spot injection pattern is very popular, as it can provide better control on frontal displacement.

The other affecting parameters are capillary pressure, relative permeability, and wettability.

4.10.2 WAG Ratio Optimization

The WAG ratio has a key role in determining the shape of the oil production performance and the CO_2 utilization curves. As the WAG ratio increases, the peak oil production rate decreases, the time to reach the peak is delayed, and CO_2 utilization decreases [114]. For different conditions of WAG process, a reservoir simulation is implemented, in which its results are available in the work of Ettehadtavakkol et al. [115]. Time dependency of oil production rate and CO_2 utilization and also the trend of oil production rate which generally rises to a peak and then follows an exponential-like decline can be observed in this figure [115].

Christensen et al. [44] studied the use of WAG in different formations with varying injecting gases and drive mechanisms. In fact, several projects faced either channeling problems or reduced injectivity, whereas optimal flow allocation has the potential to vanish these two primary problems and increase the recovery through WAG project.

4.11 ESTIMATING RECOVERY

The theoretical carbon dioxide sequestration capacity may be calculated using the reservoir data on the basis of reservoir rock volume, porosity, and oil saturation [116,117]. It should be noted that in reservoirs flooded using water, the available volume is reduced by the volume of invading water. Therefore, the mass of required CO_2 in a reservoir, M_{CO_2}, is obtained by [58]

$$M_{CO_2} = \rho_{CO_2}\left[\mathrm{RF} \cdot A \cdot h \cdot \varnothing \cdot (1 - S_{wi}) - V_{iw} + V_{pw}\right] \tag{4.35}$$

where RF, A, h, ρ_{CO_2}, V_{iw}, V_{pw}, and S_{wi} are recovery factor, area, net pay thickness, CO_2 density at reservoir conditions, reservoir volume of invading water, reservoir volume of produced water, and initial water saturation, respectively. Most of this information can, generally, be found. The volume of injected or produced water can be calculated from production records.

The same methodology applies in the case of solvent- or gas-flooded reservoirs. If CO_2 EOR is done in a reservoir after a hydrocarbon flood, then immiscible gas is displaced along with oil by CO_2 and the vacant volume will be available for CO_2 sequestration [118].

In waterflooding, a common practice to obtain the ultimate recovery is to plot fractional recovery with respect to WOR on semilog paper and extrapolate the results. The same logic can be utilized to determine the recovery in gas floods [116].

The overall reservoir recovery for a fully developed waterflood reservoir can be estimated as follows:

$$E = mX + n \tag{4.36}$$

where

$$X = \ln\left[\frac{1}{f_w} - 1\right] - \frac{1}{f_w}, \tag{4.37}$$

$$m = \frac{1}{b}(1 - S_{wi}) \tag{4.38}$$

$$n = -\frac{1}{1 - S_{wi}}\left[S_{wi} + \frac{1}{b}\ln A\right] \tag{4.39}$$

$$A = a\left(\frac{\mu_w}{\mu_o}\right) \tag{4.40}$$

and a and b are obtained from $(k_o/k_w) = e^{bS_w}$.

According to this procedure, recovery of gas flooding can be estimated by substituting gas parameters instead of water ones into the relations. Hence, the final equations can be simplified as

$$\frac{5.615}{R} = \frac{1}{f_g} - 1 \tag{4.41}$$

Therefore, $X = -\ln R - 1 - \left(5.615/R\right) + \ln 5.615$. Taking the limit as R reaches to infinity lead to the followings:

$$X = -\ln R + 0.725 \tag{4.42}$$

Hence, $E_R = m^{'}(\ln R) + n^{'}$

For example, if oil recovery versus GOR is plotted on a semilog scale, a straight line should be obtained.

4.12 CO_2 PROPERTIES AND REQUIRED VOLUMES

4.12.1 Correlation of CO_2/Heavy Oil Properties

The following correlations are introduced to determine the solubility of CO_2, swelling factor, and viscosity of the CO_2/heavy oil mixture. For this purpose, temperature, pressure, specific gravity of oil, and oil viscosity at any temperature and pressure condition are required to be specified. The temperature dependence of heavy oil viscosity can be correlated as follows [67]:

$$\log\left(\frac{\mu_2}{\mu_1}\right)_{1\ \text{atm}} = 5707\left(\frac{1}{T_2} - \frac{1}{T_1}\right) \tag{4.43}$$

where μ_2 and μ_1 are the viscosities of heavy oil at temperatures of T_2 and T_1 (in °R), respectively. This equation is a modified version of proposed correlation by Reid et al. [119]. The pressure dependency of the heavy oil viscosity could be estimated as follows [67]:

$$\log\left(\frac{\mu_2}{\mu_1}\right) = A_T\left(\frac{p}{14.7} - 1\right) \tag{4.44}$$

in which, p shows pressure in psia and A_T is a function of temperature. It should be noticed that Eq. (4.44) is valid for pressures less than 3000 psia. It is notable that Eq. (4.44) is not applicable for a highly viscous oil. μ_2 and μ_1 are the oil viscosities at temperature T and pressures of p and 14.7 psi, respectively. The proportionality constant, A_T, can be correlated as a function of temperature and the specific gravity of oil as follows [67]:

$$A_T = \frac{13.877\exp(4.633\gamma)}{T^{2.17}} \tag{4.45}$$

In Eq. (4.45), T and γ denote the temperature (in °R) and oil specific gravity, respectively. The solubility of CO_2 in a crude oil is defined as the volume of CO_2 in the CO_2-saturated oil per barrel of dead oil at the temperature in which solubility is measured.

CO_2 solubility depends mostly on temperature, pressure, and specific gravity of the oil. A correlation for solubility of CO_2 in heavy oil is given by the followings [67]:

$$R_s = \frac{1}{a_1 \gamma^{a_2} T^{a_7} + a_3 T^{a_4} \exp\left(- a_5 p + \frac{a_6}{p} \right)} \tag{4.46}$$

where γ is the specific gravity of heavy oil, T is temperature in °F, p stands for pressure in psia, and R_s indicates the solubility of CO_2 in a crude oil in scf/bbl. The empirical constants a_1 through a_7 are $0.4934\text{E}-2$, 4.0928, $0.571\text{E}-6$, 1.6428, $0.6763\text{E}-3$, 781.334, and -0.2499, respectively. The above correlation can approximate the CO_2 solubility for pressures below 3000 psia.

The swelling factor (F_s) is defined as the ratio of the volume of CO_2-saturated oil at the temperature and pressure of the reservoir to the volume of the dead oil at reservoir temperature and atmospheric pressure [120]. The magnitude of swelling for heavy oil is not as drastic as for light oil, which can be swelled more than two times of its original volume. The following correlation for determining swelling factor could be proposed as follows [120]:

$$F_s = 1 + \frac{0.35 R_s}{1000} \tag{4.47}$$

In above equation, R_s and F_s denote, respectively, CO_2 solubility in crude oil (in scf/bbl) and swelling factor as ratio. In general, viscosity of CO_2/heavy-oil mixture is a function of composition. This composition-dependent function is extremely complex for the CO_2/heavy-oil mixtures due to the fact that it is not possible to reach the detailed composition of heavy oil and the contribution of each component to the viscosity of the mixture cannot accurately be determined. Hence, the CO_2/heavy oil mixture is treated as a binary system with two components [5]:

- Pure CO_2
- Heavy oil

Chung et al. [67] stated that the viscosity variation of heavy oil is related to the quantity of CO_2 dissolved in the oil. Therefore, if the concentration of CO_2 in the oil and both the viscosities of CO_2 and heavy oil is determined, a relationship can be obtained to estimate the viscosity of the CO_2/heavy-oil mixture.

The viscosity ratio between the two components of this system (i.e., heavy oil and CO_2) is in the range of 103–106. The following equation can be utilized in such high ratio of viscosity combinations [121].

$$\ln \mu_m = X_0 \ln(\mu_0) + X_s \ln(\mu_s) \tag{4.48}$$

where

$$X_s = \frac{V_s}{\alpha V_o + V_s} \tag{4.49}$$

$$X_o = 1 - X_s \tag{4.50}$$

where V is volume fraction and μ is viscosity in cp. The subscripts O, S, and m stand for heavy oil, CO_2, and CO_2/heavy-oil mixture, respectively. The empirical parameter α is determined by Eq. (4.51) as follows [121]:

$$\alpha = 0.255\gamma^{-4.16}T_r^{1.85}\left[\frac{e^{7.36} - e^{7.36(1-p_r)}}{e^{7.36} - 1}\right] \tag{4.51}$$

where

$$T_r = \frac{T}{547.57} \tag{4.52}$$

$$p_r = \frac{p}{1071} \tag{4.53}$$

where specific gravity of the heavy oil, temperature, and pressure are indicated by γ, T in R, and p in psi, respectively. The volume fraction of CO_2 in the mixture (X_s) can be estimated from the CO_2 solubility or swelling factor, which is as follows [67]:

$$X_s = \frac{1}{\left(\alpha F_{CO_2}/F_0 R_S\right) + 1} = \frac{F_0 F_s - 1}{\alpha + F_0 F_s - 1} \tag{4.54}$$

where F_{CO_2} is the ratio of CO_2 gas volume at standard conditions to the volume at system temperature and pressure, and F_o is the ratio of oil volume at system temperature and 14.7 psi to the volume at system temperature and pressure.

4.12.2 Required Volume

The CO_2 storage capacity of a reservoir can be defined as the CO_2 remained in the reservoir at the end of EOR operation and any extra CO_2 that can be injected after the EOR project. It is showed that about 40% of the originally injected CO_2 is being produced in the producer wells and can be reinjected [12,81]. Shaw and Bachu [81] introduced an approach to determine the CO_2 storage capacity in the reservoir during EOR process. At breakthrough time, the CO_2 storage capacity can be calculated as follows [81]:

$$M_{CO_2} = \rho_{CO_2 res} RF_{BT} \frac{\text{OOIP}}{S_h} \tag{4.55}$$

where the CO_2 storage capacity in million tone (Mt), density of CO_2 at reservoir condition in kg/m^3, the recovery factor in percent, and oil shrinkage in 1/oil formation volume factor are exhibited by the symbols M_{CO_2}, $\rho_{CO_2 res}$, RF_{BT}, OOIP, and S_h, respectively. At any HCPV injection, the generalized form of Eq. (4.55) can be proposed as follows [122]:

$$M_{CO_2} = \rho_{CO_2res}[RF_{BT} + 0.6(RF_{\%HCPV} - RF_{BT})]\frac{OOIP}{S_h} \quad (4.56)$$

where the CO_2 storage capacity in Mt, density of CO_2 at reservoir condition in kg/m^3, the recovery factor at breakthrough time, recovery factor at any HCPV injection, in percent, original oil in-place in percent, and oil shrinkage factor in 1/oil formation volume factor are exhibited by the symbols M_{CO_2}, ρ_{CO_2res}, RF_{BT}, $RF_{\%HCPV}$, OOIP, and S_h, respectively. The oil shrinkage factor is defined as the inverse of the oil formation volume factor.

ECL Technology Limited (United Kingdom) used a similar method to determine the net CO_2 retained in the reservoir for different EOR operations. For WAG injection, the net CO_2 retained in the reservoir is calculated as follows [123]:

$$\begin{aligned} \text{Net } CO_{2\text{retained}} &= WAG_{\text{IOR efficieny}} \times WAG_{\text{score efficiency}} \times OOIP \\ &\times WAG_{CO_2 \text{ factor alpha}} \times \frac{B_0}{B_g} \end{aligned} \quad (4.57)$$

where $WAG_{\text{IOR efficiency}}$, $WAG_{\text{score efficiency}}$, and $WAG_{CO_2 \text{ factor alpha}}$ are targeted incremental oil recovery factor, a factor between 0 and 1 (it is 1 for an efficiently and fully implemented WAG project). The WAG_{CO_2} factor alpha varies between 1 and 2 and is related to the net CO_2 utilization efficiency when expressed in reservoir volumes, indicating more gas may be stored in the reservoir than required for WAG operation, respectively. For gravity stable gas injection (GSGI), the net CO_2 retained in the reservoir is calculated as follows [122]:

$$\text{Net } CO_{2\text{retained}} = \left(GSGI_{CO_2\text{factor}}\right) \times \left(GSGI_{\text{score } CO_2\text{factor}}\right) \times OOIP \times 0.7\frac{B_0}{B_g} \quad (4.58)$$

where $GSGI_{CO_2\text{factor}}$, $GSGI_{\text{score } CO_2\text{factor}}$, B_o, and B_g illustrate targeted incremental oil recovery by GSGI operation, factor between 0 and 1, and gas volume factor, respectively. The $GSGI_{\text{score } CO_2\text{factor}}$ permits the user to reduce the injected CO_2 volume in comparison with the potential target volume. For a fully implemented project, $GSGI_{\text{score } CO_2\text{factor}}$ is equal to 1. The factor "0.7" is responsible for the fraction of OOIP left in the formation at the end of gas flood and a small amount of mobile water which is left in the swept region by the injected gas [124]. The GSGI process differs from the WAG operation. The amount of CO_2 retained in GSGI is proportional to the pore volume, rather than the recovery of IOR process. More CO_2 is needed in a GSGI process; thereby, this process is more favorable for CO_2 storage. Numerical reservoir simulations may also be used, which may take into account the impact of water invasion, gravity segregation, reservoir heterogeneity, and CO_2 dissolution in formation water [124].

REFERENCES

[1] J.J. Sheng, Enhanced oil recovery in shale reservoirs by gas injection, J. Nat. Gas Sci. Eng. 22 (2015) 252–259.
[2] D. Rao, Gas injection EOR—a new meaning in the new millennium, J. Can. Pet. Technol. 40 (2001).
[3] D.O. Shah, Improved Oil Recovery by Surfactant and Polymer Flooding, Elsevier, Amsterdam, 2012.
[4] Lake, L.W., 1989. Enhanced Oil Recovery, Prentice Hall, Englewood Cliffs, NJ. ISBN: 0132816016.
[5] V. Alvarado, E. Manrique, Enhanced oil recovery: an update review, Energies 3 (2010) 1529.
[6] A. Firoozabadi, A. Khalid, Analysis and correlation of nitrogen and lean-gas miscibility pressure (includes associated paper 16463), SPE Reservoir Eng. 1 (1986) 575–582.
[7] Christensen, J.R., Stenby, E.H., Skauge, A. Review of WAG Field Experience. Society of Petroleum Engineers, International Petroleum Conference and Exhibition of Mexico, Villahermosa, Mexico, 1998.
[8] J.J. Taber, F. Martin, R. Seright, EOR screening criteria revisited-Part 1: introduction to screening criteria and enhanced recovery field projects, SPE Reservoir Eng. 12 (1997) 189–198.
[9] P.H. Lowry, H.H. Ferrell, D.L. Dauben, A Review and Statistical Analysis of Micellar-Polymer Field Test Data. Report No. DOE/BC/10830-4, National Petroleum Technology Office, US Department of Energy, Tulsa, OK, 1986.
[10] Bachu, S., Shaw, J.C., 2004. CO_2 Storage in Oil and Gas Reservoirs in Western Canada: Effect of Aquifers, Potential for CO_2-Flood Enhanced Oil Recovery and Practical Capacity, Alberta Geological Survey, Alberta Energy and Utilities Board, Edmonton, Alberta, Canada, 1–77.
[11] T. Gamadi,, J. Sheng,, M. Soliman,, H. Menouar,, M. Watson,, H. Emadibaladehi,, An experimental study of cyclic CO_2 injection to improve shale oil recovery., SPE Improved Oil Recovery Symposium., Society of Petroleum Engineers, 2014.
[12] R. Hadlow,, Update of industry experience with CO_2 injection., SPE Annual Technical Conference and Exhibition., Society of Petroleum Engineers., 1992.
[13] Heinrich, J.J., Herzog, H.J., Reiner, D.M., 2003. Environmental assessment of geologic storage of CO_2. In: Second National Conference on Carbon Sequestration. pp. 5–8.
[14] J. Clancy, R. Gilchrist, L. Cheng, D. Bywater, Analysis of nitrogen-injection projects to develop screening guides and offshore design criteria, J. Pet. Technol. 37 (1985) 1097–1104.
[15] R.D. Tewari,, S. Riyadi,, C. Kittrell,, F.A. Kadir,, M. Abu Bakar,, T. Othman,, et al., Maximizing the oil recovery through immiscible water alternating gas (IWAG) in mature offshore field, SPE Asia Pacific Oil and Gas Conference and Exhibition, Society of Petroleum Engineers, 2010.
[16] D.W. Green, G.P. Willhite, Enhanced Oil Recovery, Henry L. Doherty Memorial Fund of AIME, Society of Petroleum Engineers, Richardson, TX, 1998.
[17] M. Leach, W. Yellig, Compositional model studies-CO_2 oil-displacement mechanisms, Soc. Pet. Eng. J. 21 (1981) 89–97.
[18] P. Zanganeh, S. Ayatollahi, A. Alamdari, A. Zolghadr, H. Dashti, S. Kord, Asphaltene deposition during CO_2 injection and pressure depletion: a visual study, Energy Fuels 26 (2012) 1412–1419.
[19] P. Wylie,, K.K. Mohanty,, Effect of wettability on oil recovery by near-miscible gas injection., SPE/DOE Improved Oil Recovery Symposium., Society of Petroleum Engineers, 1998.
[20] F.I. Stalkup,, Displacement behavior of the condensing/vaporizing gas drive process, SPE Annual Technical Conference and Exhibition, Society of Petroleum Engineers, 1987.
[21] R. Johns, B. Dindoruk, F. Orr Jr, Analytical theory of combined condensing/vaporizing gas drives, SPE Adv. Technol. Ser. 1 (1993) 7–16.
[22] P. Oren, J. Billiotte, W. Pinczewski, Mobilization of waterflood residual oil by gas injection for water-wet conditions, SPE Form. Eval. 7 (1992) 70–78.
[23] S.C. Ayirala,, D.N. Rao,, Comparative evaluation of a new MMP determination technique, SPE/DOE Symposium on Improved Oil Recovery, Society of Petroleum Engineers, 2006.

[24] J.-N. Jaubert, L. Avaullee, C. Pierre, Is it still necessary to measure the minimum miscibility pressure? Ind. Eng. Chem. Res. 41 (2002) 303–310.
[25] A.M. Elsharkawy, F.H. Poettmann, R.L. Christiansen, Measuring CO_2 minimum miscibility pressures: slim-tube or rising-bubble method?, Energy Fuels 10 (1996) 443–449.
[26] D.N. Rao, A new technique of vanishing interfacial tension for miscibility determination, Fluid Phase Equilib. 139 (1997) 311–324.
[27] D.N. Rao, J. Lee, Application of the new vanishing interfacial tension technique to evaluate miscibility conditions for the Terra Nova Offshore Project, J. Pet. Sci. Eng. 35 (2002) 247–262.
[28] D.N. Rao, J.I. Lee, Determination of gas–oil miscibility conditions by interfacial tension measurements, J. Colloid Interface Sci. 262 (2003) 474–482.
[29] A. Hemmati-Sarapardeh, S. Ayatollahi, M.-H. Ghazanfari, M. Masihi, Experimental determination of interfacial tension and miscibility of the CO_2–crude oil system; temperature, pressure, and composition effects, J. Chem. Eng. Data 59 (2013) 61–69.
[30] A. Firoozabadi, Thermodynamics of Hydrocarbon Reservoirs, McGraw-Hill, New York, 1999.
[31] X. Liao, Y.G. Li, C.B. Park, P. Chen, Interfacial tension of linear and branched PP in supercritical carbon dioxide, J. Supercrit. Fluids 55 (2010) 386–394.
[32] A. Zolghadr, M. Escrochi, S. Ayatollahi, Temperature and composition effect on CO_2 miscibility by interfacial tension measurement, J. Chem. Eng. Data 58 (2013) 1168–1175.
[33] Cronquist, C., 1978. Carbon dioxide dynamic miscibility with light reservoir oils. In: Proc. Fourth Annual US DOE Symposium, Tulsa. pp. 28–30.
[34] J. Lee, Effectiveness of Carbon Dioxide Displacement Under Miscible and Immiscible Conditions, Report RR-40, Petroleum Recovery Inst., Calgary, 1979.
[35] W. Yellig, R. Metcalfe, Determination and prediction of CO_2 minimum miscibility pressures (includes associated paper 8876), J. Pet. Technol. 32 (1980) 160–168.
[36] F. Orr Jr, C. Jensen, Interpretation of pressure–composition phase diagrams for CO_2/crude-oil systems, Soc. Pet. Eng. J. 24 (1984) 485–497.
[37] R. Alston, G. Kokolis, C. James, CO_2 minimum miscibility pressure: a correlation for impure CO_2 streams and live oil systems, Soc. Pet. Eng. J. 25 (1985) 268–274.
[38] H. Sebastian, R. Wenger, T. Renner, Correlation of minimum miscibility pressure for impure CO_2 streams, J. Pet. Technol. 37 (1985) 2076–2082.
[39] D. West, S. Shatynski, Ternary Equilibrium Diagrams, American Society of Mechanical Engineers, New York, 1982.
[40] D. West, Ternary Equilibrium Diagrams, Springer Science & Business Media, Berlin, Germany, 2012.
[41] T. Monger, J. Coma, A laboratory and field evaluation of the CO_2 Huff 'n' Puff process for light-oil recovery, SPE Reservoir Eng. 3 (1988) 1168–1176.
[42] W.D. McCain, The Properties of Petroleum Fluids, PennWell Books, Houston, Texas, United States, 1990.
[43] E.J. Manrique, V.E. Muci, M.E. Gurfinkel, EOR field experiences in carbonate reservoirs in the United States, SPE Reservoir Eval. Eng. 10 (2007) 667–686.
[44] Christensen, J.R., Stenby, E.H., Skauge, A., 1998. Review of WAG field experience. International Petroleum Conference and Exhibition of Mexico, Society of Petroleum Engineers.
[45] M. Stein, D. Frey, R. Walker, G. Pariani, Slaughter estate unit CO_2 flood: comparison between pilot and field-scale performance, J. Pet. Technol. 44 (1992) 1026–1032.
[46] Johns, R.T., Dindoruk, B., 2013. Gas flooding. In: Enhanced Oil Recovery Field Case Studies. pp. 1–22, Gulf Professional Publishing, Houston, TX.
[47] J. Lawrence,, N. Maer,, D. Stern,, L. Corwin,, W. Idol,, Jay nitrogen tertiary recovery study: managing a mature field, Abu Dhabi International Petroleum Exhibition and Conference, Society of Petroleum Engineers, 2002.
[48] J. Burger, K. Mohanty, Mass transfer from bypassed zones during gas injection, SPE Reservoir Eng. 12 (1997) 124–130.
[49] A. Caruana, R. Dawe, Experimental studies of the effects of heterogeneities on miscible and immiscible flow processes in porous media, Trends Chem. Eng. 3 (1996) 185–203.

[50] Y.M. Al-Wahaibi, First-contact-miscible and multicontact-miscible gas injection within a channeling heterogeneity system, Energy Fuels 24 (2010) 1813–1821.
[51] R. Giordano,, S. Salter,, K. Mohanty,, The effects of permeability variations on flow in porous media, SPE Annual Technical Conference and Exhibition, Society of Petroleum Engineers, 1985.
[52] Pande, K., Sheffield, J., Emanuel, A., Ulrich, R., De Zabala, E., 1995. Scale-up of near miscible gas injection processes: integration of laboratory measurements and compositional simulation. In: IOR1995-8th European Symposium on Improved Oil Recovery.
[53] D. Kjonsvik,, J. Doyle,, T. Jacobsen,, A. Jones,, The effects of sedimentary heterogeneities on production from a shallow marine reservoir—what really matters? SPE Annual Technical Conference and Exhibition, Society of Petroleum Engineers, 1994.
[54] P. Ballin, P. Clifford, M. Christie, Cupiagua: modeling of a complex fractured reservoir using compositional upscaling, SPE Reservoir Eval. Eng. 5 (2002) 488–498.
[55] D. Ajose,, K. Mohanty,, Compositional upscaling in heterogeneous reservoirs: effect of gravity, capillary pressure, and dispersion, SPE Annual Technical Conference and Exhibition, Society of Petroleum Engineers, 2003.
[56] Saner, W.B., Patton, J.T., CO_2 Recovery of Heavy Oil: Wilmington Field Test, *Journal of Petroleum Technology*, 38 (07), 1986, 769–776.
[57] Sayegh, S.G., Maini, B.B., Laboratory Evaluation of The CO Huff-N-Puff Process For Heavy Oil Reservoirs, Journal of Canadian Petroleum Technology, 23 (03), 1984, 29–36.
[58] Z. Dai, R. Middleton, H. Viswanathan, J. Fessenden-Rahn, J. Bauman, R. Pawar, et al., An integrated framework for optimizing CO_2 sequestration and enhanced oil recovery, Environ. Sci. Technol. Lett. 1 (2013) 49–54.
[59] L. James,, N. Rezaei,, I. Chatzis,, VAPEX, Warm VAPEX, and hybrid VAPEX-the state of enhanced oil recovery for in situ heavy oils in Canada, Canadian International Petroleum Conference, Petroleum Society of Canada, 2007.
[60] K.K. Mohanty,, C. Chen,, M.T. Balhoff,, Effect of reservoir heterogeneity on improved shale oil recovery by CO Huff-n-Puff, SPE Unconventional Resources Conference-USA, Society of Petroleum Engineers, 2013.
[61] Wan, T., 2013. Evaluation of the EOR Potential in Shale Oil Reservoirs by Cyclic Gas Injection, Doctoral dissertation, Texas Tech University, Broadway, Lubbock.
[62] F. Torabi, A.Q. Firouz, A. Kavousi, K. Asghari, Comparative evaluation of immiscible, near miscible and miscible CO_2 Huff-n-Puff to enhance oil recovery from a single matrix–fracture system (experimental and simulation studies), Fuel 93 (2012) 443–453.
[63] F. Torabi, K. Asghari, Effect of operating pressure, matrix permeability and connate water saturation on performance of CO_2 huff-and-puff process in matrix-fracture experimental model, Fuel 89 (2010) 2985–2990.
[64] R. Simon, D. Graue, Generalized correlations for predicting solubility, swelling and viscosity behavior of CO_2–crude oil systems, J. Pet. Technol. 17 (1965) 102–106.
[65] J.S. Miller,, R.A. Jones,, A laboratory study to determine physical characteristics of heavy oil after CO_2 saturation, SPE/DOE Enhanced Oil Recovery Symposium, Society of Petroleum Engineers, 1981.
[66] V. Sankur, J. Creek, S. Di Julio, A. Emanuel, A laboratory study of Wilmington Tar Zone CO_2 injection project, SPE Reservoir Eng. 1 (1986) 95–104.
[67] F.T. Chung, R.A. Jones, H.T. Nguyen, Measurements and correlations of the physical properties of CO_2–heavy crude oil mixtures, SPE Reservoir Eng. 3 (1988) 822–828.
[68] R. Azin,, R. Kharrat,, C. Ghotbi,, S. Vossoughi,, Applicability of the VAPEX process to Iranian heavy oil reservoirs., SPE Middle East Oil and Gas Show and Conference, Society of Petroleum Engineers, 2005.
[69] I.J. Mokrys, R.M. Butler, The rise of interfering solvent chambers: solvent analog model of steam-assisted gravity drainage, J. Can. Pet. Technol. 32 (1993) 26–36.
[70] R.M. Butler, I.J. Mokrys, A new process (VAPEX) for recovering heavy oils using hot water and hydrocarbon vapour, J. Can. Pet. Technol. 30 (1991) 97–106.

[71] V. Pathak, T. Babadagli, N. Edmunds, Mechanics of heavy-oil and bitumen recovery by hot solvent injection, SPE Reservoir Eval. Eng. 15 (2012) 182–194.
[72] I.J. Mokrys, Vapor extraction of hydrocarbon deposits, in, Google Patents, 2001.
[73] R. Butler, I. Mokrys, Closed-loop extraction method for the recovery of heavy oils and bitumens underlain by aquifers: the VAPEX process, J. Can. Pet. Technol. 37 (1998) 41–50.
[74] Butler, R., Mokrys, I., Das, S., 1995. The solvent requirements for VAPEX recovery. SPE International Heavy Oil Symposium. Society of Petroleum Engineers.
[75] CH, E., 2016. Introduction to Chemical Engineering Thermodynamics, Pennsylvania State University, State College, PA.
[76] S. Upreti, A. Lohi, R. Kapadia, R. El-Haj, Vapor extraction of heavy oil and bitumen: a review, Energy Fuels 21 (2007) 1562–1574.
[77] U.E. Guerrero Aconcha,, A. Kantzas,, Diffusion of hydrocarbon gases in heavy oil and bitumen, Latin American and Caribbean Petroleum Engineering Conference, Society of Petroleum Engineers, 2009.
[78] C.-L. Chang, M. Chang, Non-iteration estimation of thermal conductivity using finite volume method, Int. Commun. Heat Mass Transfer 33 (2006) 1013–1020.
[79] T. Wan,, X. Meng,, J.J. Sheng,, M. Watson,, Compositional modeling of EOR process in stimulated shale oil reservoirs by cyclic gas injection, SPE Improved Oil Recovery Symposium, Society of Petroleum Engineers, 2014.
[80] J. Sheng, Enhanced Oil Recovery Field Case Studies, Gulf Professional Publishing, Houston, TX, 2013.
[81] J. Shaw, S. Bachu, Screening, evaluation, and ranking of oil reservoirs suitable for CO_2-flood EOR and carbon dioxide sequestration, J. Can. Pet. Technol. 41 (2002) 51–61.
[82] O. Rivas, S. Embid, F. Bolivar, Ranking reservoirs for carbon dioxide flooding processes, SPE Adv. Technol. Ser. 2 (1994) 95–103.
[83] A.N. Carcoana,, Enhanced oil recovery in Rumania, SPE Enhanced Oil Recovery Symposium, Society of Petroleum Engineers, 1982.
[84] J.J. Taber,, Technical screening guides for the enhanced recovery of oil, SPE Annual Technical Conference and Exhibition, Society of Petroleum Engineers, 1983.
[85] Klins, M.A., 1984. Carbon Dioxide Flooding: Basic Mechanisms and Project Design, U.S. Department of Energy Office of Scientific and Technical Information, Oak Ridge, TN.
[86] D. Lopez, T. Perez, S. Simison, The influence of microstructure and chemical composition of carbon and low alloy steels in CO_2 corrosion. A state-of-the-art appraisal, Mater. Des. 24 (2003) 561–575.
[87] H.K. Sarma,, Can we ignore asphaltene in a gas injection project for light-oils? SPE international improved oil recovery conference in Asia Pacific, Society of Petroleum Engineers, 2003.
[88] J.P. Meyer, Summary of carbon dioxide enhanced oil recovery (CO_2 EOR) injection well technology, Am. Pet. Inst. 54 (2007) 1–54.
[89] L. Koottungal, 2008 Worldwide EOR survey, Oil Gas J. 106 (2008) 47.
[90] G. Benge,, E. Dew,, Meeting the challenges in design and execution of two high rate acid gas injection wells, SPE/IADC Drilling Conference, Society of Petroleum Engineers, 2005.
[91] M.E. Parker, J.P. Meyer, S.R. Meadows, Carbon dioxide enhanced oil recovery injection operations technologies (poster presentation), Energy Procedia 1 (2009) 3141–3148.
[92] R.J. Larkin,, P.G. Creel,, Methodologies and solutions to remediate inner-well communication problems on the SACROC CO_2 EOR project: a case study, SPE Symposium on Improved Oil Recovery, Society of Petroleum Engineers, 2008.
[93] J.J. Brnak,, B. Petrich,, M.R. Konopczynski,, Application of SmartWell technology to the SACROC CO_2 EOR project: a case study, SPE/DOE Symposium on Improved Oil Recovery, Society of Petroleum Engineers, 2006.
[94] P. Behnoud far, M.J. Ameri, M. Orooji, A novel approach to estimate the variations in stresses and fault state due to depletion of reservoirs, Arabian J. Geosci. 10 (2017) 397.
[95] P. Behnoud far, A.H. Hassani, A.M. Al-Ajmi, H. Heydari, A novel model for wellbore stability analysis during reservoir depletion, J. Nat. Gas Sci. Eng. 35 (Part A) (2016) 935–943.

[96] A. Gholilou, P. Behnoud far, S. Vialle, M. Madadi, Determination of safe mud window considering time-dependent variations of temperature and pore pressure: Analytical and numerical approaches, J. Rock Mech. Geotech. Eng. 9 (5) (2017) 900–911.
[97] Strazisar, B., Kutchko, B., 2006. Degradation of wellbore cement due to CO_2 injection-effects of pressure and temperature. In: 2006 International Symposium on Site Characterization for CO_2 Geological Storage. March 2006, Berkeley, CA.
[98] J. Zhang, Y. Wang, M. Xu, Q. Zhao, Effect of carbon dioxide corrosion on compressive strength of oilwell cement, J. Chin. Ceram. Soc. 37 (2009) 642–647.
[99] Z. Krilov,, B. Loncaric,, Z. Miksa,, Investigation of a long-term cement deterioration under a high-temperature, sour gas downhole environment, SPE International Symposium on Formation Damage Control, Society of Petroleum Engineers, 2000.
[100] J. Wang,, J. Abiazie,, D. McVay,, W.B. Ayers,, Evaluation of reservoir connectivity and development recovery strategies in Monument Butte Field, Utah, SPE Annual Technical Conference and Exhibition, Society of Petroleum Engineers, 2008.
[101] M. Zahoor, M. Derahman, M. Yunan, WAG process design—an updated review, Braz. J. Pet. Gas 5 (2011) 109–121.
[102] C.K. Ho, S.W. Webb, Gas Transport in Porous Media, Springer, New York, 2006.
[103] Skauge, A., Dale, E.I., 2007. Progress in immiscible WAG modelling. SPE/EAGE Reservoir Characterization and Simulation Conference. Society of Petroleum Engineers.
[104] J.R. Fanchi, Principles of Applied Reservoir Simulation, Gulf Professional Publishing, Houston, TX, 2005.
[105] Faisal, A., Bisdom, K., Zhumabek, B., Zadeh, A.M., Rossen, W.R., 2009. Injectivity and gravity segregation in WAG and SWAG enhanced oil recovery. SPE Annual Technical Conference and Exhibition. Society of Petroleum Engineers.
[106] Esmaiel, T., Fallah, S., van Kruijsdijk, C., 2004. Gradient based optimization of the WAG process with smart wells. In: ECMOR IX-9th European Conference on the Mathematics of Oil Recovery.
[107] J.C. Heeremans,, T.E. Esmaiel,, C.P. Van Kruijsdijk,, Feasibility study of WAG injection in naturally fractured reservoirs, SPE/DOE Symposium on Improved Oil Recovery, Society of Petroleum Engineers, 2006.
[108] Moghanloo, R.G., Lake, L.W., 2010. Simultaneous water–gas-injection performance under loss of miscibility. SPE Improved Oil Recovery Symposium. Society of Petroleum Engineers.
[109] L. Lo,, D. McGregor,, P. Wang,, S. Boucedra,, C. Bakhoukhe,, WAG pilot design and observation well data analysis for Hassi Berkine South field, SPE Annual Technical Conference and Exhibition, Society of Petroleum Engineers, 2003.
[110] M.D. Jackson, P.H. Valvatne, M.J. Blunt, Prediction of wettability variation within an oil/water transition zone and its impact on production, SPE J. 10 (2005) 185–195.
[111] J.M. Schembre, G.-Q. Tang, A.R. Kovscek, Interrelationship of temperature and wettability on the relative permeability of heavy oil in diatomaceous rocks (includes associated discussion and reply), SPE Reservoir Eval. Eng. 9 (2006) 239–250.
[112] S. Chen, H. Li, D. Yang, P. Tontiwachwuthikul, Optimal parametric design for water-alternating-gas (WAG) process in a CO_2-miscible flooding reservoir, J. Can. Pet. Technol. 49 (2010) 75–82.
[113] P.L. Bondor,, J.R. Hite,, S.M. Avasthi,, Planning EOR projects in offshore oil fields, SPE Latin American and Caribbean Petroleum Engineering Conference, Society of Petroleum Engineers, 2005.
[114] M. Garcia Quijada, Optimization of a CO_2 Flood Design Wesson Field-west Texas, Texas A&M University, College Station, TX, 2006.
[115] A. Ettehadtavakkol, L.W. Lake, S.L. Bryant, CO_2-EOR and storage design optimization, Int. J. Greenhouse Gas Control 25 (2014) 79–92.
[116] Z. Dai, H. Viswanathan, R. Middleton, F. Pan, W. Ampomah, C. Yang, et al., CO_2 Accounting and risk analysis for CO_2 sequestration at enhanced oil recovery sites, Environ. Sci. Technol. 50 (2016) 7546–7554.
[117] M.O. Eshkalak,, E.W. Al-shalabi,, A. Sanaei,, U. Aybar,, K. Sepehrnoori,, Enhanced gas recovery by CO_2 sequestration versus re-fracturing treatment in unconventional shale gas reservoirs, Abu Dhabi International Petroleum Exhibition and Conference, Society of Petroleum Engineers, 2014.

[118] M. Celia, S. Bachu, J. Nordbotten, K. Bandilla, Status of CO_2 storage in deep saline aquifers with emphasis on modeling approaches and practical simulations, Water Resour. Res. 51 (2015) 6846–6892.

[119] R.C. Reid, J.M. Prausnitz, T.K. Sherwood, The Properties of Gases and Liquids, McGraw-Hill, New York, 1977.

[120] J. Welker, Physical properties of carbonated oils, J. Pet. Technol. 15 (1963) 873–876.

[121] W. Shu, A viscosity correlation for mixtures of heavy oil, bitumen, and petroleum fractions, Soc. Pet. Eng. J. 24 (1984) 277–282.

[122] F. Gozalpour, S. Ren, B. Tohidi, CO_2 EOR and storage in oil reservoir, Oil Gas Sci. Technol. 60 (2005) 537–546.

[123] M. Goodfield, C. Woods, Potential UKCS CO_2 Retention Capacity from IOR Projects, DTI SHARP Programme, UK, 2001.

[124] S. Bachu,, J.C. Shaw,, R.M. Pearson,, Estimation of oil recovery and CO_2 storage capacity in CO_2 EOR incorporating the effect of underlying aquifers, SPE/DOE Symposium on Improved Oil Recovery, Society of Petroleum Engineers, 2004.

CHAPTER FIVE

Thermal Recovery Processes

Forough Ameli[1], Ali Alashkar[1] and Abdolhossein Hemmati-Sarapardeh[2]
[1]School of Chemical Engineering, Iran University of Science and Technology, Tehran, Iran
[2]Department of Petroleum Engineering, Shahid Bahonar University of Kerman, Kerman, Iran

5.1 INTRODUCTION

The recovery processes of oil are increased due to dependency of the industry to them and the requirements to petrochemical products. This leads to oil extraction from unconventional reservoirs to compensate the possible deficiency between production and demand, and oil with low API gravity. The total amount of heavy oils and bitumen is about 9 trillion barrels. The characteristics of these oils include low API gravity, high viscosity, and asphaltene content [1]. One technique for increasing the displacement which leads to enhancement of heavy oil recovery is viscosity reduction [2]. The fluid resistance to flow is called viscosity. Reduction of this quantity leads to increasing the mobility value as temperature is increased [3]. This fact states the significance of enhanced oil recovery (EOR) (thermal EOR) processes, in which the generated heat at the surface or in situ from steam or hot water is injected through the porous media [4].

5.2 VARIOUS THERMAL ENHANCED OIL RECOVERY PROCESSES

Various methods of thermal EOR are applied in different ranges of viscosity. For example, steam flooding is effective for heavy oil extraction, cyclic steam stimulation (CSS) is applicable for extra-heavy oil, and steam-assisted gravity drainage (SAGD) process is introduced for the recovery of bitumen [5]. This process leads to generation of greenhouse gases as a result of burning the fuel which may cause environmental problems. Solvent steam process was then introduced to resolve this environmental issue by generating less amount of steam. This leads to reduction in amount of emitted greenhouse gases [6,7]. If solvents for instance normal alkanes, CO_2, and CH_4 are also added to this stream, viscosity reduction would be increased

Fundamentals of Enhanced Oil and Gas Recovery from Conventional and Unconventional Reservoirs.
DOI: https://doi.org/10.1016/B978-0-12-813027-8.00005-9

due to miscibility of the solvent in oil. Howbeit, some heavy fractions of the crude oil, namely, asphaltenes are immiscible in such solvents [8]. Another alternative to steam injection processes is in situ combustion (ISC), which is applied for highly viscous oils. Oil displacement in this technique leads to 95% recovery. The generated heat front is so difficult to control. This leads to just few successful field experiences [9]. The nature of combustion reactions including cracking and oxidation, which happen in heterogeneous reservoirs, increases the complexity of this process [10]. In tight reservoirs, using electromagnetic methods is incumbent for increasing the recovery, although using various thermal EOR techniques would not lead to full recovery of the system as a result of fluid channeling in heterogonous media or heat losses occurring in underburden or overburden of thin reservoirs [11]. Another leading technique is application of this method in a specified part of the reservoir. In this method, contorting the wave's penetration and their absorption is of paramount importance.

To deeply understand the mechanism of thermal EOR processes, it is necessary to review various mechanisms of heat transfer including conduction, convection, and radiation. The momentum equation reveals the dependency of flow to viscosity. Variations of interfacial tension and the phase change lead to application of mass transfer equations and complicated reactions which occur in thermal EOR processes. This complexity and difficulty in prediction of the reservoir behavior may lead to reviewing of thermal EOR processes. In this section, various methods of thermal recovery are represented and compared to each other.

5.2.1 Steam Flood and Steam-Assisted Gravity Drainage

To increase the recovery of heavy oil and bitumen, the SAGD technique was developed. In this technique, two uneven horizontal wells are drilled. The upper wellbore is occupied with continuous stream of high pressure gas. This heat leads to oil viscosity reduction and moves the heated oil from upper well to the lower one and pumps it out. As heat transfer has occurred in this process, the steam which is injected creates a "steam chamber." Steam and other gases are accumulated within the upper well due to their lower density in comparison to oil and fill its empty space left by the oil. The associated gas forms an insulating space over the steam. No vapor is produced in the lower well [12]. A countercurrent flow of oil and water is produced by gravity drainage in the lower well. This fluid is pumped to the surface using cavity pump which is appropriate for viscous fluids containing suspended solids.

ISC process was first introduced as forward dry combustion which starts with ignition of crude oil downhole using an air stream to initiate the combustion. By propagating the front of generated flame, much energy is lost. To reduce this phenomenon, the process is reversed in which the injection of air stream occurs in another well and the stated well is ignited. In other words, the air stream and flame move in opposite directions. Of course, in the absence of oxygen supply, the flame might shut off. This is an economical process while the oil price is within 30–35 $/bbl.

The difference between boiling point of water at the lower well pressure and the temperature of the producer is called subcool. The producer temperature is lowered by increasing the liquid level in upper well which yields to higher subcool. As the actual systems are highly heterogeneous, there would not be a uniform subcool through the whole length of the well. In practice, a portion of steam is kept in the producer to keep the bitumen warm and lower its viscosity which leads it to flow toward colder regions of the reservoir. As process has a long shut-in time or start-up, the steam is circulated in the lower well. This process is called partial SAGD. From thermal stand point, low value of subcool would not be beneficial. This includes reduction in the rate of steam injection. Low temperature would actually result in viscosity increase and decrease in bitumen mobility [13]. Another point is that as the subcool is very high, the steam pressure would be low, and maintaining the chamber would not be possible leading the steam chamber to collapse. The condensed steam would prevent the chamber development. If the processes of injection and production are continuously operated at reservoir pressure, this would remove the instability issue in this process. The output of SAGD process would be 70%–80% recovery of the OIP for suitable reservoirs. This process is not affected by vertical barriers for fluid flow and steam. By heating the rock and considering the conduction mechanism of heat transfer, fluid and steam flow into the production well. This mechanism would lead to 60%–70% oil recovery even with many shale barriers [13].

This technique was first proposed by Roger Butler, the engineer at Imperial Oil at 1970s. Then, he worked as the director of technical programs of Alberta Oil Sands Technology and Research Authority (AOSTRA) which developed new technique for increasing the recovery of heavy oil and oil sand. They welcomed SAGD technique [14]. This process as the improvement to steam injection technique was applied for Kern River Oil Field in California [15]. As this field was produced using CSS technique, the oil was recovered from some specific portions of oil sands, for instance, Cold Lake oil sands. This approach was not efficient for bitumen production from deeper layers in oil sand, namely, Athabasca and Peace River oil sands. However, most of the reserves lie in that area. This led to developing the SAGD process in order to increase the oil recovery with cooperation of AOSTRA and industry partners with Bulter [16].

A number of geological formations that apply SAGD include Clearwater Formation, Lloydminster Sand of the Mannville Group, General Petroleum Sand, McMurray Formation, Grand Rapids Formation, a Stratigraphic range in the Western Canadian Sedimentary Basin Canada which is now the largest supplier of the United States oil with over 35% supply capacity. This value is more than the contributed share of Venezuela or Saudi Arabia and also the OPEC countries [17]. SAGD technology

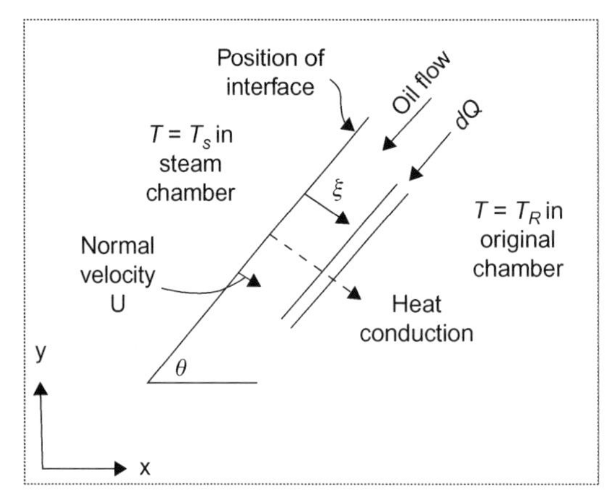

Figure 5.1 Representing Darcy law in SAGD process. *SAGD*, steam-assisted gravity drainage.

plays an important role in recovery of Alberta's oil sand deposits. For shallow bitumen reservoirs, strip-mining technique is mainly applied. However, for the large deep deposits that surround the shallow ones, application of SAGD method is more suitable. This technology is expected to be the primer for oil recovery from Canadian oil sands [18]. On the other hand, application of surface mining techniques consumes more than 20 times water quantity in comparison to other common techniques.

Darcy flow for oil (per unit thickness) in SAGD process is formulated as the following equation (Fig. 5.1):

$$dq = \frac{K(d\xi \times 1)\left(\rho_o - \rho_g\right)g\sin\theta}{\mu_o}d\xi \tag{5.1}$$

where dq is the incremental oil flow rate, K is the permeability, $d\xi$ is the incremental oil thickness, θ is the interface inclination.

SAGD-Darcy flow of mobilize oil is formulated as follows:

$$dq = \frac{Kg\sin\theta}{\nu_o}d\xi \text{ (heated reservoir)} \tag{5.2}$$

$$dq_r = \frac{Kg\sin\theta}{\nu_r}d\xi \text{ (cold reservoir)} \tag{5.3}$$

If the only mechanisms of heat transfer beyond the interface (ξ-direction) is conduction, then

$$\left(\frac{T(\xi) - T_R}{T_s - T_R}\right) = e^{\left(-U\xi/\alpha\right)} \tag{5.4}$$

$$dq = dq - dq_r = kg\sin\theta \int_0^{\infty} \left(\frac{1}{\nu} - \frac{1}{\nu_R}\right) d\xi \tag{5.5}$$

The other possibility is to have integral limit up to $\xi_{\max}$, beyond which no drainage is practically happening.

$$\frac{\nu_s}{\nu} = \left(\frac{T - T_R}{T_s - T_R}\right)^m \quad \mathrm{m} = 3 - 4(\text{typically}) \tag{5.6}$$

$$\int_0^{\infty} \left(\frac{1}{\nu} - \frac{1}{\nu_R}\right) d\xi = \frac{\alpha}{U} \frac{1}{m\nu_s} \tag{5.7}$$

$$q = \frac{Kg\alpha\sin\theta}{m\nu_s U} \tag{5.8}$$

5.2.1.1 SAGD-Material Balance

The flow into an incremental element is less than that leaving the element; the difference determines the rate of interface advancement.

$$dq_t = \phi \Delta S_o \left(\frac{dy}{dt}\right)_x dx \tag{5.9}$$

$$\left(\frac{dq}{dx}\right)_t = \phi \Delta S_o \left(\frac{dy}{dt}\right)_x \tag{5.10}$$

$(dy/dt)_x$ interface velocity U and inclination angle θ.

Determining interface velocity in SAGD:

$$U = -\cos\theta \left(\frac{dy}{dt}\right)_x \tag{5.11}$$

$$\left(\frac{dy}{dt}\right)_x < 0 \tag{5.12}$$

$$\tan\theta = \frac{dy}{dx} \tag{5.13}$$

SAGD-interface advancement velocity is calculated as

$$q = \frac{Kg\alpha\sin\theta}{m\nu_s U} \tag{5.14}$$

$$q = -\frac{Kg\alpha\sin\theta}{m\nu_s\cos\theta\left(\frac{\partial y}{\partial t}\right)} \tag{5.15}$$

$$q = -\frac{Kg\alpha\left(\frac{\partial y}{\partial x}\right)}{m\nu_s\left(\frac{\partial y}{\partial t}\right)} \tag{5.16}$$

$$U = -\cos\theta\left(\frac{\partial y}{\partial t}\right)_x \tag{5.17}$$

$$\tan\theta = \frac{\partial y}{\partial x} \tag{5.18}$$

$$\left(\frac{\partial q}{\partial x}\right)_t = \phi\Delta S_o\left(\frac{\partial y}{\partial t}\right)_x \tag{5.19}$$

$$q = -\frac{Kg\alpha\phi\Delta S_o\left(\partial y/\partial q\right)_t}{m\nu_s} \tag{5.20}$$

SAGD-oil production rate from 1/2 chamber is as follows:

$$\int_0^q q dq = \int_0^{h-y}\frac{Kg\alpha\phi\Delta S_o}{m\nu_s}dy \tag{5.21}$$

$$q = \sqrt{\frac{2Kg\alpha\phi\Delta S_o(h-y)}{m\nu_s}} \tag{5.22}$$

$$q^{\dagger} = \sqrt{\frac{2Kg\alpha\phi\Delta S_o h}{m\nu_s}} \tag{5.23}$$

$$f(x, y, t) = 0 \ \text{(At SAGD interface)} \tag{5.24}$$

$$\left(\frac{\partial x}{\partial t}\right)_y\left(\frac{\partial t}{\partial y}\right)_x\left(\frac{\partial y}{\partial x}\right)_t = -1 \ \text{(shape of interface vs time)} \tag{5.25}$$

$$\left(\frac{\partial x}{\partial t}\right)_y = \frac{-1}{\left(\partial y/\partial x\right)_t\left(\partial t/\partial y\right)_x} = \frac{-\left(\partial y/\partial t\right)_x}{\left(\partial y/\partial x\right)_t} \tag{5.26}$$

$$q = \frac{-Kg\alpha\left(\partial y/\partial x\right)_t}{m\nu_s\left(\partial y/\partial t\right)_x} \tag{5.27}$$

$$\frac{Kg\alpha}{qm\nu_s} = \frac{-\left(\partial y/\partial t\right)_x}{\left(\partial y/\partial x\right)_t} \tag{5.28}$$

$$\left(\frac{\partial x}{\partial t}\right)_y = \sqrt{\frac{Kg\alpha}{2\phi\Delta S_o m\nu_s(h-y)}} \tag{5.29}$$

As it is clear from the following equation, horizontal velocity is a function of y but not t.

$$x = \left(\frac{\partial x}{\partial t}\right)_y t + x_0 \tag{5.30}$$

If SAGD interface develops vertically above the wells ($x_0 = 0$),

$$x = t\sqrt{\frac{Kg\alpha}{2\phi\Delta S_o m\nu_s(h-y)}} \tag{5.31}$$

$$y = h - \frac{Kg\alpha}{2\phi\Delta S_o m\nu_s}\left(\frac{t}{x}\right)^2 \tag{5.32}$$

SAGD-dimensionless interface shape is defined as follows:

$$X = \frac{x}{h} Y = \frac{y}{h} t' = \frac{t}{h}\sqrt{\frac{Kg\alpha}{\phi\Delta S_o m\nu_s h}} \tag{5.33}$$

$$Y = 1 - \frac{1}{2}\left(\frac{t'}{X}\right)^2 \tag{5.34}$$

5.2.2 Cyclic Steam Stimulation Technique (Huff-and-Puff)

CSS is a technique for enhanced recovery of heavy oils at primary production phase. The steam would assist the heavy oil to flow more easily through the formation into injection or production wells. A specified amount of steam is injected to the drilled well. Then process is stopped to give a chance to steam for heating the formation around the well. Finally, the wells are allowed to produce and heat is exhausted with the produced fluid. This process is called "huff-and-puff" and is repeated until there would be a considerable amount of produced water. This process is then continued to heat the oil and compensate the pressure decline in the reservoir to continue the production. In this technique, some injection wells may convert to production ones and the total number of production wells would increase.

This process is recommended due to its high rate of success and high investment rate of return. However, from thermal point of view, SAGD process is two times more efficient in comparison to CSS technique. Fewer damages would occur due to lower pressures in comparison to CSS, and ultimately, SAGD is more economic for thick reservoirs, in comparison to cyclic steam processes [13]. More recent studies in this regard are centralized on optimization of fracture design, additives, and geomechanical solutions to poroelastic effects.

5.2.2.1 Underlying Technology

This process is composed of three stages, namely, injection, soaking, and production. These are continuously repeated until oil production becomes economical without gasification [2,19]. In this technique, the value of the residual oil is decreased by application of various methods. Namely, reducing the viscosity, wettability changes, and solution gas expansion [20] The schematic representation of this process is represented in Fig. 5.2.

Moreover, many other products are generated due to chemical reactions including hydrogen sulfide, carbon dioxide, hydrogen during steam injection [22]. These are the result of decarboxylation of oil, conversion of S to H_2S, and production of H_2, CO, CH_4, and CO_2 that yields from reaction between crude oil, water, and the produced CO_2 as a result of carbonate decomposition and further reactions [2]. These produced gases create an additional driving force that is called gas drive. Moreover, oil viscosity is reduced by increasing its mobility [2,23]. The results of Hongfu et al. [22] study reveal that viscosity was reduced 28%−42% by applying cyclic steam injection (CSI).

This process works best for formations thicker than 30 ft, and reservoirs do not deepen than 3000 ft with porosity and oil saturation more than 0.3% and 40%, respectively. The geological structure of the near wellbore is very important in this process.

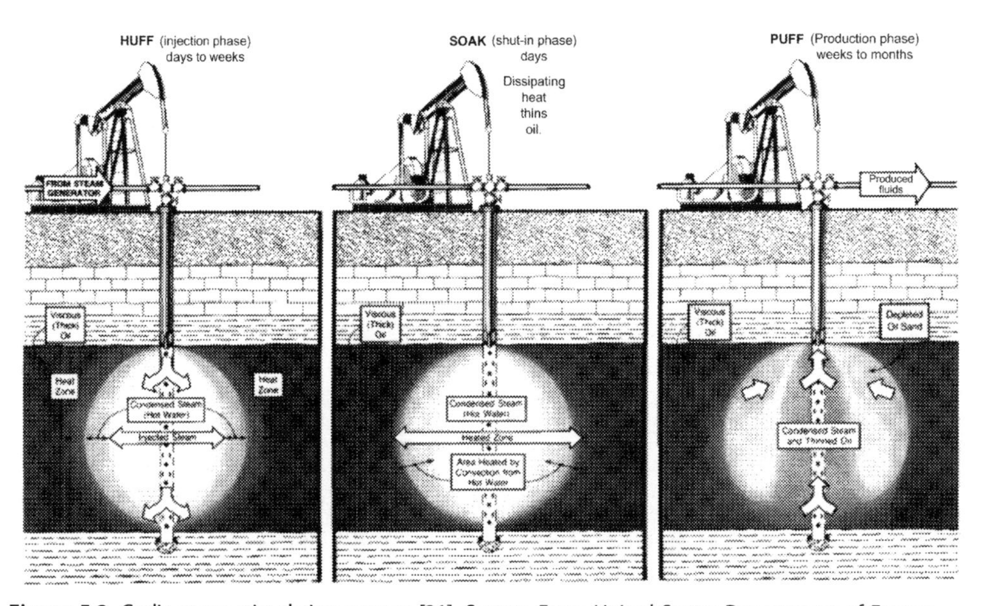

Figure 5.2 Cyclic steam stimulation process [21]. *Source: From United States Department of Energy.*

This would affect oil mobilization and steam distribution. The rock should be of moderate strength with low content of clay. The oil with API gravity more than 10 and viscosity within the range of 1000–4000 cP is favorable. The permeability of the reservoir should be more than 100 mD [13,19].

5.2.2.2 Reservoir Properties Changes With CSI

It is important to study the effects of EOR techniques on reservoir properties. For instance, the induced heat in CSI technique leads to creating tension and changing the structure of formations. This may lead to alteration in reservoir permeability and mobility of water [24]. The change in pore volume and permeability is a function of three parameters: (1) change in mean effective stress, (2) change in temperature, and (3) shear stress alterations. As temperature increases, the sand structure is expanded. The latter parameter was studied in Cold Lake field in Canada which revealed that as steam is injected into the reservoir and the pores are pressurized, the effective stress would be decreased [24].

In another study, for Clearwater formation in Canada, the expansion effect of CSI process was transferred to the surface and different areas in the reservoir [25]. This is reflected by changing the level of the well which is mostly observed in shallow reservoirs. Walters [26] also studied changing the pressure in the isolated aquifer which has sealed the Clearwater formation as an outcome of poroelastic effect. On the other hand, sand deformation and geomechanical changes would lead to initial injectivity of formation, supplying the driving energy for production [27]. The shear may be enlarged due to hot fluid injection in CSI [28]. Permeability changes as a function of shear dilation were reported by Wong et al. [28]. Yale et al. [29] confirmed that the most sensible changes occur in relative water permeability. On the other hand, as water is condensed by moving in the front of the hot steam, the pressure of the reservoir increases. This mechanism leads to saving the driving energy of the reservoir and supplying it by dilation. Gronseth [30] studied on streamline distribution during the fluid injection into Clearwater formation and confirmed that as injection rate of the fluid is more than its diffusion into matrix, the reservoir volume is justified by that of injected fluid. Increase in volume would lead to pressure increase. As the production initiates, the effective stress increases and pressure would reduce. This leads to contraction of the reservoir whereby a portion of increase in the initial steps of the process is compensated [30]. The reservoir deformation is studied using various techniques. The results of these studies are applied for optimization of production parameters including injection rate, well length, and well spacing. Migration of the steam and changes in the formation are recorded using tilt-meter and inclinometer [31]. The accuracy of tilt-meters is also greater one order of magnitude in comparison to inclinometer [32].

5.2.2.3 CSS Aziz and Gontijo Model

Aziz and Gontijo developed a model for vertical radial flows in the well. In this model, gravity and potential were introduced as the driving force. The steam zone was considered conical. To derive this model, the following assumptions were made:

1. There should be considered an initial saturation for water and oil phases.
2. By injection of the steam, a conical volume is formed.
3. The heat transfer is ignored in the phase of steam injection and the reservoir mean temperature is equal to steam temperature.
4. Oil is moved below a thick layer at oil–steam interface.
5. The heat transfer mechanism from steam to oil zone is conduction.
6. The flow regime is pseudosteady state.
7. By oil production, the steam occupation zone is increased.
8. A combination of pressure drop and gravity leads to flow through the reservoir.
9. The mean temperature of the heated zone is determinative for the steam pressure and consequently pressure draws down.

Using the above assumptions, heat transfer equation is as follows:

$$q_o = 1.87R_x\sqrt{\frac{k_o\varphi\Delta S_o\alpha\Delta\varnothing}{m_o v_{avg}\left[\left(\ln R_x/R_w\right) - 0.5\right]}} \tag{5.35}$$

where

$$R_x = \sqrt{h_t^{\,2} + R_h^{\,2}} \tag{5.36}$$

$$\Delta S_o = S_{oi} - S_{ors} \tag{5.37}$$

$$\Delta\varnothing = \Delta Hg\sin(\theta) + \frac{P_s - P_{wf}}{\rho_o} \tag{5.38}$$

$$\sin(\theta) = \frac{h_t}{R_x} \tag{5.39}$$

$$\Delta h = h_t - h_{st} \tag{5.40}$$

θ represents the interface and reservoir base angle. The pressure is determined using the following equation:

$$P_s = \left[\frac{T_s}{115.95}\right]^{4.4543} \tag{5.41}$$

The thickness of the steam zone is calculated using Van Lookeren [33] equation as follows:

$$h_{st} = 0.5h_t A_{RD} \tag{5.42}$$

In the above equation, A_{RD} represents a dimensionless number for scaling the steam zone.

$$A_{\mathrm{RD}} = \sqrt{\frac{(350)(144)Q_s\mu_{\mathrm{st}}}{6.328\pi(\rho_o - \rho_{\mathrm{st}})h_t^2 k_{\mathrm{st}}\rho_{\mathrm{st}}}} \tag{5.43}$$

The values of steam viscosity and density are calculated using the following equations (5.37):

$$\rho_{\mathrm{st}}(\mathrm{lb/ft^3}) = \frac{P_s^{0.9588}}{363.9};\ P_s \text{ in psia} \tag{5.44}$$

$$\mu_{\mathrm{st}}(\mathrm{cP}) = 10^{-4}(0.2T_s + 82);\ T_s \text{ in } ^\circ\mathrm{F} \tag{5.45}$$

The radius and volume of the steam zone are calculated according to the following equations:

$$R_h = \sqrt{\frac{v_s}{\pi h_{\mathrm{st}}}} \tag{5.46}$$

$$v_s = \frac{Q_s t_{\mathrm{inj}} \rho_w Q_i + H_{\mathrm{last}}}{v(T_s - T_R)} \tag{5.47}$$

The previous method for estimation of the volume represented by Parts (2) in which the remaining heat of previous cycles that accumulated in the reservoir was considered. The above equation is a modification to it. The value of the injected heat per unit mass of steam is calculated as follows:

$$Q_i = C_w(T_s - T_R) + L_{\mathrm{vdh}}f_{\mathrm{sdh}} \tag{5.48}$$

To calculate the value of water enthalpy, latent heat of the steam, and isobaric volumetric heat capacity of the steam, the following equations are represented [2,34]:

$$h_w = 68\left[\frac{T_s}{100}\right]^{1.24};\ T \text{ in } ^\circ\mathrm{F} \tag{5.49}$$

$$L_{\mathrm{vdh}} = 94(705 - T_s)^{0.38};\ T_s \text{ in } ^\circ\mathrm{F} \tag{5.50}$$

$$(\rho c)_t = (1 - \varphi)M_o + \varphi[(1 - S_{wi})M_o + S_{wi}M_w \tag{5.51}$$

As the initial condition, the occupied heat in the reservoir is set to zero. Volume of the steam and the mean temperature in each cycle are the basis for calculations.

$$H_{\mathrm{last}} = V_s(\rho c)_t(T_{\mathrm{avg}} - T_R) \tag{5.52}$$

To calculate the mean temperature, use the following correlation proposed by Boberg and Lantz [35]. It is emphasized that this equation is an approximation for our media and is actually represented for the cylindrical shape volumes.

$$T_{avg} = T_R + (T_s - T_R)[f_{HD}f_{VD}(1 - f_{PD}) - f_{PD}] \tag{5.53}$$

f_{HD}, f_{VD}, and f_{PD} are dimensionless parameters indicating radial loss, vertical loss, and exhausted energy from the fluids, respectively. These parameters are function of time. They were introduced by Boberg and Lantz [35] graphically in terms of dimensionless time or as error and gamma functions. To simplify calculations, the following equations are represented:

$$f_{HD} = \frac{1}{1 + 5t_{DH}} \tag{5.54}$$

$$t_{DH} = \frac{\alpha(t - t_{inj})}{R_h^2} \tag{5.55}$$

$$f_{VD} = \frac{1}{\sqrt{1 + 5t_{DV}}} \tag{5.56}$$

$$t_{DV} = \frac{4\alpha(t - t_{inj})}{h_t^2} \tag{5.57}$$

To calculate the amount of the energy removed by the fluid, this equation is introduced:

$$f_{PD} = \frac{1}{2Q_{max}} \int_0^t Q_P dt \tag{5.58}$$

Q_{max} is the maximum amount of heat transfer to the reservoir. This parameter is calculated as follows:

The value of the heat loss to overburden minus the summation of the heat remaining in the reservoir and the value of heat injected to the reservoir in the present time step.

$$Q_{max} = H_{inj} + H_{last} - \pi R_h^2 K_R (T_s - T_R) \sqrt{\frac{T_{soak}}{\pi a}} \tag{5.59}$$

The amount of H_{last} is substituted from the last time step. The value of heat injection in each cycle is calculated as follows:

$$H_{inj} = 350 Q_i Q_s t_{inj} \tag{5.60}$$

To calculate the rate of heat transfer, use the following equation (5.2):

$$Q_P = 5.615(q_o M_o + q_w M_w)(T_{avg} - T_R) \tag{5.61}$$

This problem is solved explicitly in which the mean temperature of the last time step is applied for the present time. The integral is converted to

$$f_{\mathrm{PD}}^{n}=f_{\mathrm{PD}}^{n-1}+\Delta f_{\mathrm{PD}} \tag{5.62}$$

where n represents the time step and Δf_{PD} is calculated as

$$\Delta f_{PD}=\frac{5.615(q_o M_o+q_w M_w)(T_{avg}^{n-1}-T_R)\Delta t}{2Q_{\max}} \tag{5.63}$$

The main steps for solving this problem are summarized as follows:

1. Initialize the model by inserting the fluid, reservoir, and operational data.
2. Calculate the initial values for the radius or thickness of the zone, fluid properties, and saturations.
3. Using small time steps, calculate the water and oil production cumulatively or in each time step. Determine the mean temperature for each cycle and finally determine the oil in place by checking the cumulative production in each cycle.
4. Check the requirements for the end of cycle by the number of time steps and then continue.
5. Determine the amount of the remained heat and water in the reservoir.
6. It may require continuing calculations for another cycle, then go to step 2. Otherwise this is end of calculations.

5.2.2.4 CSS – Boberg–Lantz Model [36]

The most important assumptions of this model include

- Boberg–Lantz model uses Marx–Langenheim model to calculate radius of the heated zone.
- The reservoir is assumed to heat to T_s instantaneously.
- Initially, the entire heated zone is at T_s while (remaining of) o/u and reservoir are at T_R.
- Heat loss and production of hot fluid are considered.
- Although cold fluids enter the hot region, it has not been considered in energy balance.

$$\frac{k_h}{r}\frac{\partial}{\partial r}\left(r\frac{\partial T}{\partial r}\right)+k_h\frac{\partial^2 T}{\partial z^2}=\rho C_p\frac{\partial T}{\partial t} \tag{5.64}$$

The initial conditions are reported as follows and represented in Fig. 5.3.

$$\text{at } t=t_i,\ T=T_s \text{ for } (r_w\le r\le r_s \text{ and } 0\le z\le h)$$

$$\text{at } t=t_i,\ T=T_R \text{ for } (r>r_s \text{ and } z<0 \text{ and } z>h)$$

The boundary conditions for r and z direction include

$$\text{at } r=0,\ \frac{\partial T}{\partial r}=0 \tag{5.65}$$

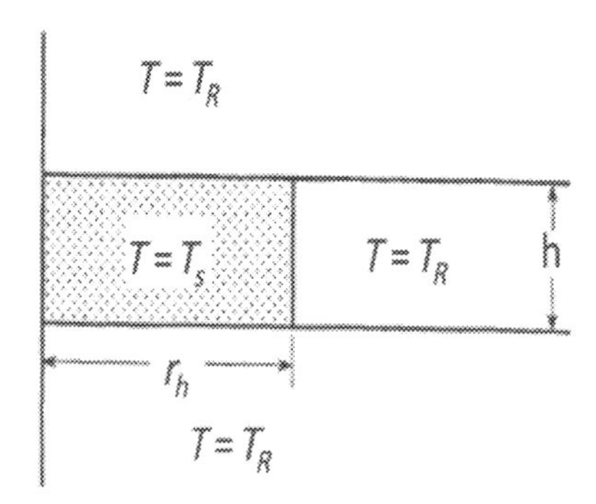

Figure 5.3 Boberg–Lantz model.

$$\text{at } r \to \infty, \frac{\partial T}{\partial r} = 0 \tag{5.66}$$

$$\text{at } z = \frac{h}{2}, \frac{\partial T}{\partial r} = 0 \tag{5.67}$$

$$\text{at } z \to \infty, \frac{\partial T}{\partial r} = 0 \tag{5.68}$$

Using the superposition principle in this model:

$$T(r, z) = T(r)T(z) \tag{5.69}$$

$$\overline{T}(r) \text{ for } 0 \leq r \leq r_h \tag{5.70}$$

$$\overline{T}(z) \text{ for } 0 \leq z \leq h \tag{5.71}$$

$$\overline{T}_D = \overline{T}_{Dr}\overline{T}_{Dz} \tag{5.72}$$

Introducing dimensionless parameters in Boberg–Lantz model:
Temperature:

$$\overline{T}_{Dr} = \frac{\overline{T}(r) - T_R}{T_s - T_R} \tag{5.73}$$

$$\overline{T}_{Dz} = \frac{\overline{T}(z) - T_R}{T_s - T_R} \tag{5.74}$$

Time:

$$t_{Dr} = \frac{\alpha(t - t_i)}{{r_h}^2} \tag{5.75}$$

$$t_{Dz} = \frac{\alpha(t - t_i)}{(h/2)^2} \tag{5.76}$$

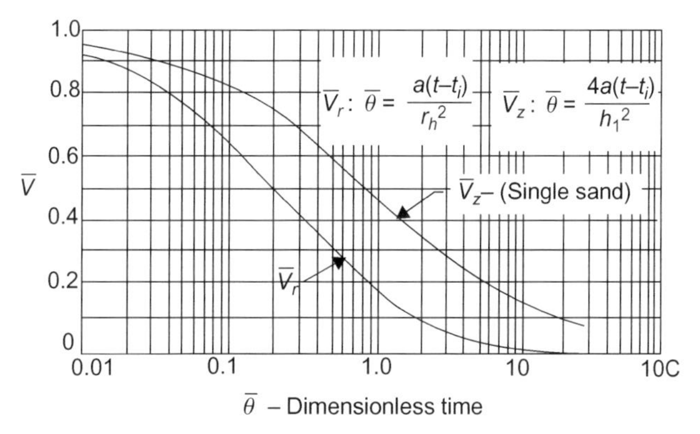

Figure 5.4 Temperature–time correlation.

The solution of the equation in r and z directions results in the following equations. This has been schematically represented in terms of dimensionless numbers in Fig. 5.4.

r-direction for $t_D \leq 10$

$$\overline{T}_{Dr} = \sqrt{\frac{t_{Dr}}{\pi}}\left(2 - \frac{t_{Dr}}{2} - \frac{3}{16}{t_{Dr}}^2 - \frac{15}{64}{t_{Dr}}^3 - \frac{525}{1024}{t_{Dr}}^4 - \ldots\right) \tag{5.77}$$

for $t_D > 10$

$$\overline{T}_{Dr} = \left(\frac{1}{4t_{Dr}} - \frac{1}{16{t_{Dr}}^2} + \frac{5}{384{t_{Dr}}^3} - \frac{1}{439{t_{Dr}}^4} + \frac{7}{20,480{t_{Dr}}^5} - \ldots\right) \tag{5.78}$$

For z direction:

$$\overline{T}_{Dz} = \text{erf}\left(\frac{1}{\sqrt{t_{Dz}}}\right) - \frac{\sqrt{t_{Dz}}}{\pi}\left(1 - e^{-1/t_{Dz}}\right) \tag{5.79}$$

Two adjustments to Boberg–Lantz made for this model include

1. T distribution in the o/u shale exists.

   A hypothetical length (z) is added to the pay zone ($h + z$ at T_s) to account for heat loss to o/u.

2. The average temperature does not consider removal of heated fluids.

A dimensionless parameter (δ) is defined to adjust the temperature to account for produced energy. Energy in a disk of radius r_h and thickness $z + h$ is calculated as follows:

$$m_s H_s = \pi {r_h}^2 M(T_s - T_R)(z + h) \tag{5.80}$$

$$z = \left(\frac{m_s H_s}{\pi r_h^2 M(T_s - T_R)}\right) - h \tag{5.81}$$

$$t_{Dz} = \frac{\alpha(t - t_i)}{\left[(z+h)/2\right]^2} \tag{5.82}$$

Removal of hot fluids in Boberg–Lantz model:

$$\dot{Q}_p = 5.615 q_{oh}\left(\rho_o C_o + F_{\mathrm{WOR}} \rho_w C_w\right)\left(\overline{T}_p - T_R\right) \tag{5.83}$$

Q_p is the rate of energy removal (Btu/d), T_p is the average T of heated region, adjusted for removal of hot fluids (°F), F_{WOR} is the water-to-oil ratio, energy removed by hot fluid in $(t - t_i)$:

$$\delta = \frac{1}{2}\int_{t_i}^{t} \frac{\dot{Q}_p d\lambda}{m_s H_s} \tag{5.84}$$

$$T_{Dp} = \frac{\overline{T}_p - T_R}{T_s - T_R} = \overline{T}_{Dr}\overline{T}_{Dz}(1 - \delta) - \delta \tag{5.85}$$

$$\overline{T}_p = T_R + (T_s - T_R)\left[\overline{T}_{Dr}\overline{T}_{Dz}(1 - \delta) - \delta\right] \tag{5.86}$$

Gas–water production in Boberg–Lantz model is as follows:

$$\dot{Q}_p = q_{oh}\left(H_{\mathrm{ogv}} + H_{\mathrm{wrv}}\right) \tag{5.87}$$

$$H_{\mathrm{ogv}} = \left(5.615 \rho_o C_o + F_{\mathrm{GOR}} C_g\right)\left(\overline{T}_p - T_R\right) \tag{5.88}$$

$$H_{\mathrm{wrv}} = 5.615\left[F_{\mathrm{WOR}}\left(H_{\mathrm{wT}} - H_{wr}\right) + F_{\mathrm{wv}}\lambda_s\right] \tag{5.89}$$

where H_{wrv} is the enthalpy of produced water (Btu/bbl), H_{ogv} is the enthalpy of produced oil and gas (Btu/stb), H_{wT} is the enthalpy of saturated water at T (Btu/lbm), F_{GOR} is the GOR (scf/stb), F_{WOR} is the WOR (bbl/stb), F_{WV} is the for water condensed from produced vapor (bbl/stb).

The amount of water condensed from vapor is determined by:

$$F_{\mathrm{wv}} = (0.0001356)\left(\frac{p_{\mathrm{wv}}}{p_w - p_{wr}}\right) F_{\mathrm{GOR}} \tag{5.90}$$

where F_{wv} is the WOR for water condensate from produced gas (bbl w@60 F/stb), p_{wv} is the vapor pressure of water at $\langle T \rangle$, p_{w} is the BHP, p_{wr} is the vapor pressure of water at T_R, F_{GOR} is the GOR (scf/stb).

$$F_{\mathrm{wv}} = (0.0001356)\left(\frac{p_{\mathrm{wv}}}{p_w - p_{wr}}\right) F_{\mathrm{GOR}} \tag{5.91}$$

- There will always be water vapor produced if gas is being produced.
- If $p_{wv} > p_w$, water will be flashed to vapor at production well. In this case, the above equation fails.
- Boberg–Lantz model need parameter tuning for any formation to account for WOR over time.

Example 5.1: Calculation of mean temperature at a zone shut-in after steam stimulation

Steam with temperature of 400°F is injected to the reservoir to give a heated zone of 25 ft. The reservoir temperature, thickness, and conductivity are 120°F, 40 ft, and 1.6 Btu/h ft^2 °F/ft. The mean heat capacity for overburden and formation is 33 Btu/ft^3 °F. Calculate the mean temperature of the heated zone for 100,200,300 days after reaching the reservoir temperature to 400°F. The reservoir has no production.

α should be determined to calculate $\overline{T}_{Dr}$ and $\overline{T}_{Dz}$

$a = k_h/M = (1.6\ \text{Btu/h ft}^2\ °\text{F/ft})/(33\ \text{Btu/ft}^3\ °\text{F}) = 0.0484\ \text{ft}^2/\text{h} = 1.163\ \text{ft}^2/\text{D}$

For radial dimensionless temperature component,

$$\overline{T}_{Dr} = \overline{T}_{Dr}(t_{Dr})$$

$$(t_{Dr}) = [\alpha(t - t_i)]/r_h{}^2 = [(1.163\ \text{ft}^2/\text{D})(t - t_i)]/(25\ \text{ft})_2 = 0.00186(t - t_i)$$

At $(t - t_i) = 100$ days and $t_{Dr} = 0.186$ $\overline{T}_{Dr} = 0.66$; at $(t - t_i) = 200$ days $t_{Dr} = 0.372$ and $\overline{T}_{Dr} = 0.53$; and at $(t - t_i) = 300$ days, $t_{Dr} = 0.558$ $\overline{T}_{Dr} = 0.45$. For the thickness-averaged temperature, $\overline{T}_{Dz}$:

$$\overline{T}_{Dz} = [\alpha(t - t_i)](h/2)^2 = [(1.163\ \text{ft}^2/\text{D})(t - t_i)]/(20\ \text{ft})^2 = 0.0029(t - t_i)$$

At $(t - t_i) = 100$ days and $t_{Dz} = 0.29$ $\overline{T}_{Dz} = 0.78$; at $(t - t_i) = 200$ days $t_{Dz} = 0.58$ and $\overline{T}_{Dz} = 0.64$; and at $(t - t_i) = 300$ days, $t_{Dz} = 0.87$ and $\overline{T}_{Dz} = 0.56$

Average heated zone temperature after shut-in (°F)

$(t - t_i)$Days	$\overline{T}_{Dr}$	$\overline{T}_{Dz}$	$\overline{T}_D$	$\overline{T}$(°F)
100	0.66	0.78	0.48	251.1
200	0.53	0.64	0.31	206.2
300	0.45	0.56	0.23	184.2

5.2.3 Fire Flood and In Situ Combustion

The process of ISC includes exothermic reactions which lead to increasing the final temperature of the reservoir. For instance, the temperature increase may reach to 300–400°C which leads to phase change. Though there are complex reactions embedded in this process, the engineers study on the mechanisms due to advantages of this process.

5.2.3.1 Description of the Method

In this method, air is injected to crude oil reservoir. After ignition, the generated heat keeps the combustion front moving toward the production well. Combustion front burns all the fuel in its way. Usually 5%–10% of the crude oil is used as a fuel and the rest is going to be produced in the production well. The heat of reaction vaporizes initial water and also the light components of oil at the advance of combustion front. The steam is condensed while distancing from the hot region. This method is also applicable for light oil reservoirs. The viscosity of the fluid would decrease which leads to more recovery of the reservoir. Combustion causes decomposition of asphaltene and other heavy fractions to lighter compounds, flue gases, and heat. A stable steam front is generated as a result of water condensation that under-effects the crude oil mobility. As the gases are miscible in oil, the process of miscible gas injection also occurs.

5.2.4 Toe-to-Heel Air Injection

A newly developed thermal process in the category of ISC is called toe-to-heel air injection (THAI). As in common ISC processes, the producer and injectors are vertical, the sweeping efficiency is limited which is due to overrunning or channeling. This problem is resolved by using horizontal producers or toe-to-heel well construction, to control the flow regime in the reservoir. This technology is applicable for two well configurations, namely, direct line drive and staggered line drive. In the first configuration, the position of the vertical injector is in front of horizontal producer toe. For the second approach, the horizontal producer toe is closed to the vertical injector. This process includes minimum one vertical air or an air/water injector. The position of this injector is at the upper part of the oil layer and its toe is faced to vertical injector. The schematic of this process is represented in Fig. 5.5. Short-distance oil

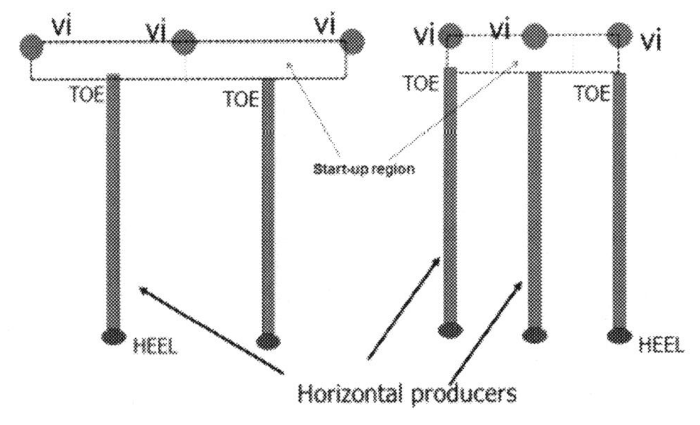

Figure 5.5 SLD and DLD THAITM schematics (VI) [37]. *SLD*, staggered line drive; *DLD*, direct line drive; *VI*, vertical injector.

displacement (SDOD) process is Toe-to-Heel Air Injection (THAITM) which is stable in gravity. The priority of SDOD process is that it occurs ahead of ISC front. The simulation results and laboratory scale tests revealed that this may occur as a result of blockage near the wellbore area of horizontal wells. This leads to avoiding oxygen short-circuiting in this process.

5.2.4.1 Benefits of THAI Process

- In this process, the front propagation is more under control in comparison to ISC in which the breakthrough occurs at the toe and the advancement happens at the heel.
- As in ISC process, some lab tests are required for this process.
- There are more choices available to optimize the process.
- If there is an initial mobility for the heavy oil, the process is run simpler; otherwise, heating of the reservoir may be inevitable.
- This process is run easily at horizontal wells which are near the oil bottom layer.
- The sensitivity of this process to formation heterogeneity is low.

5.2.4.2 Criteria for THAI Application

It is important for the selected reservoirs to have the following specifications:

- If there exists a bottom water zone, its thickness should not be more than 30% of the oil zone thickness.
- There should not exist a natural or hydraulic fracturing.
- There should be a sandstone or sand formation.
- Pay zone thickness must be more than 6 m.
- The oil viscosity and density should be more than 200 mPa s and 900 kg/m^3, respectively.
- The horizontal and vertical permeability of the reservoir should be more than 200 and 50 mD, respectively, and the value of $K_V/K_H > 0.25$.
- Water cut should be less than 70%.

If the reservoir permeability is increased moving to downhole, the last two conditions may not be important. The final decision on starting this process is achieved by analyzing the reservoir simulation outputs.

5.2.5 THAI With Catalyst (THAI–CAPRI)

The THAI process was developed using a catalyst to increase the recovery of the crude oil. This process calling Toe-to-Heel Air Injection (THAI)-Catalytic Peteroleum Recovery Institute (CAPRI) was first introduced in 1998. Actually, this process is combination of the previously described, ISC, production from a horizontal well and catalytic cracking processes that leads to a light product with no requirement to surface upgrading. However, other thermal recovery processes including SAGD, steam flooding, CSS, and ISC need upgrading at the surface.

One drawback for processes containing catalyst is its deactivation as a result of coke, heavy metals, and asphaltene deposition in catalyst pores. To study the mechanism of catalyst fouling in oil well, a fixed bed microreactor was applied. The main issues regarding this process include

- Produces oils with commercial grade which is suitable for refinery applications.
- Increases the global need of the energy.
- Decreases the volume of light oil in reservoirs along with increasing its temperature.
- Increases in the recovery of heavy oil and bitumen sources which are managed to be exploited. The number of reservoirs which is forecasted for heavy oil and bitumen is eight trillion which could be applied as a source of energy in the future.

This process starts with combustion of a portion of oil in the reservoir by reacting the injected oil with heavy oil. This process is accompanied by creating high temperature oxidation (HTO) [39]. As depicted in Fig. 5.6, this process starts at toe position and there is a continuous front which propagates to the production well's heel [41]. The mobilization of heavy oil at the advance of the front is due to the gravity into producer well. Application of the catalyst in this integrated process leads to conversion of the products to the light oil in horizontal well and elimination of the upgrading process at the surface. This idea was first accomplished by cooperation of Petroleum Recovery Institute, Calgary and University of Bath Improved Oil Recovery group. They equipped the existing system with active layer of catalyst which is placed in slotted liners of the production well [42,43].

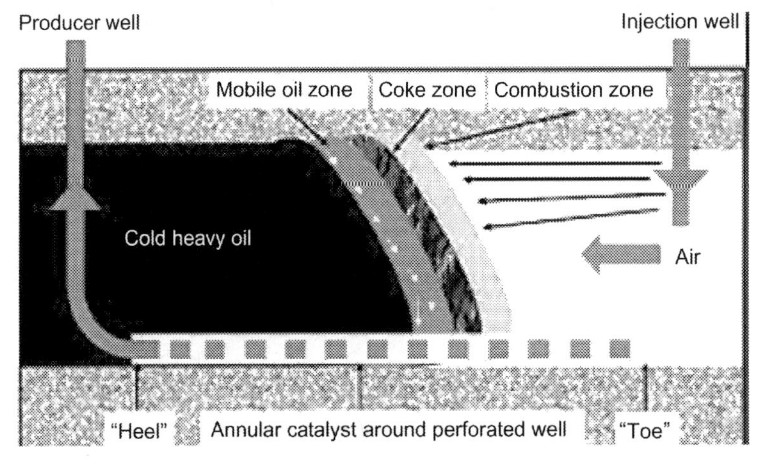

Figure 5.6 Schematic representation of the THAI–CAPRI process [40]. *THAI*, toe-to-heel air injection.

This process could be applied in high pressure shallow reservoirs. The temperature of 400–600°C is required to run the CAPRI process [42,43]. As stated earlier, the combustion front moves from toe of the well to its heel. This leads to occurring coke lay-down. The deposited coke produces the required heat for the process by combustion. The generated heat leads to reduction of heavy oil viscosity and its movement to mobile oil zone in horizontal well. The following equations represent various combustion reactions which occurs in ISC process [42,43]:

1. Thermal cracking (or pyrolysis):

$$\text{Heavy residue} \rightarrow \text{Light oil} + \text{Coke}$$

2. Oxidation of coke (high temperature oxidation, HTO):

$$\text{Coke} + O_2 \rightarrow CO + CO_2 + H_2O$$

3. Oxidation of heavy residue:

$$\text{Heavy residue} + O_2 \rightarrow CO + CO_2 + H_2O$$

The process of oil upgrading is composed of two chemical reactions, namely, addition of hydrogen and carbon rejection [44]. The latter causes thermal cracking equation in THAI process. This reaction is a function of reservoir pressure and temperature.

4. Carbon rejection:

$$CH_x \rightarrow CH_{x1} + C \quad (x_1 > x)$$

The next step includes heavy oil pyrolysis accompanied by catalytic hydrogenation using hydrotreating catalyst in CAPRI process as the following reaction:

5. Hydrogen addition:

$$CH_x + H_2 \rightarrow CH_{x1} \quad (x_1 > x)$$

Hydrogen containing products are produced during water–gas shift reaction and/or hydrocarbon gasification [45].

6. Gasification of hydrocarbon:

$$CH_x \rightarrow C + \frac{x}{2H_2}$$

$$C + H_2O \text{ (steam)} \rightarrow CO + H_2$$

$$C + CO_2 \rightarrow CO$$

7. Water–gas shift:

$$CO + H_2O \rightarrow CO_2 + H_2$$

As it is clear, heavy oil is a mixture of cycloparaffins, paraffins, and aromatic compounds. Catalytic reactions continue by *B*-scission and intermediates by chain reactions [46]. The latter include three stages, namely, initiation, propagation, and termination. The first step consists of carbenium ions formation which occurs during protonation of catalysts acid sites. There may be other routes that lead to producing carbenium ions including (1) paraffins hydride abstraction, (2) olefins protonation, (3) protolytic cracking [46]. Carbenium ions are produced by hydride ion transfer from carbocation and oil molecules in a chain reaction. It is possible to generate stable secondary or tertiary carbenium ions by alkyl or hydride shift [47]. The generated carbeniums are unstable and are converted to smaller hydrocarbons by cracking, isomerization, ring opening, alkylation, etc. Through the last stage, namely, termination, protons are separated from carbenium ions and change to lighter hydrocarbons, hydrogen, and tricoordinated carbenium ions [46].

5.2.6 Steam/Solvent-Based Hybrid Processes

Vapor extraction (VAPEX) consists of injecting a solvent into heavy oil reservoir to reduce the oil viscosity. The solvent is injected within the upper well and is produced from the lower one by gravity drainage mechanism. The first studies on this issue was performed by Mokrys and Butler [49] using a similar solvent to SAGD process. In this study, toluene was applied for extraction of two oils, namely, Suncor Coker and Athabasca. However, the initial idea of proposing this technique was introduced by Allen [50] as he changed "huff-and-puff" process by changing the solvents of butane and propane. Moreover, liquid solvents accompanied by a noncondensable gas were applied for injection through the reservoir [51]. Pure gas and a mixture of gases were also applied while the injection pressure is less than the vapor pressure to increase the recovery of heavy crude oils. As CH_4 and CO_2 were inexpensive and available, they were selected for heavy oil recovery by Dunn et al. [52]. The drawback of using these solvents was low production rates which caused this process never be introduced in field scale. The idea of horizontal wells leads to revival of solvent injection after 10 years of interruption. In this period, some lab-scale efforts were focused on porous media and nonporous media models [53]. Moreover, the VAPEX technology was upgraded by introducing hybrid VAPEX and warm VAPEX in which the solvent was heated. This causes heat transfer to the VAPEX interface and in situ condensation occurs in heavy oil. In Hybrid or wet VAPEX, steam is injected into solvent. In order to increase the rate of viscosity decrease, these techniques combine the effects of heat and mass transfer to optimize the rate of production. Another technique represented by Farouq Ali and Snyder [54] and Awang and Farouq Ali [55], involved application of solvents in hot miscible displacement. The effect of high temperature on this process is represented by Butler and Jiang [56] and Karmaker and Maini [57]. In Butler and Jiang [56] experiments, a packed bed model of glass bead was saturated with 870 mPa s viscose oil. Results revealed that choosing propane as solvent increased the production rate by 21.5% as temperature increased from 21 to 27°C. Enhancement in oil production rate was 18% as temperature raised in the range of 10−19°C [58]. The method of hybrid and warm VAPEX was compared to normal VAPEX by Frauenfeld et al. [48]. He declared that application of heat lead to faster communication from injection to production wells by more reduction in viscosity in the area near the well.

Hybrid VAPEX process was also studied, in which steam and solvent were alternatively or simultaneously injected. Farouq Ali [6] revealed this idea in 1976 and Butler [59] compared SAGD with solvent-added SAGD which lead to decrease in steam requirement up to 30% and the recovery of propane up to 99%. As dew point of water is higher in comparison to light hydrocarbons, Mokrys and Butler [59] reported steam trapping which results in reduction of steam-to-oil ratio and energy consumption. A group of researchers at Texas A&M University studied on the effects of propane injection into a limited region of the reservoir contained with 160−170°C temperature steam. Results indicated that the steam injectivity increases by adding the solvent into steam. The starting time and energy consumption decrease and overall recovery increases. This process is simulated using STARS module by Deng [60] and Mamora et al. [61] to consider heat effects in VAPEX process. The results agreed well with experimental values and confirm it as a hybrid process. Moreover, Zhao [62] developed a combined SAGD−VAPEX process with alternatively injection of solvent and steam. The properties of the system were as SAGD. This was similar to Allen [50] in which "huff-and-puff" process was studied with injection of solvents in definite cycles. Zhao [62] studied on steam alternating solvent (SAS) process and compared it to SAGD process. The results of his studies revealed that for the same production rate, the required energy decreased 18% for SAS. The results of his studies were also simulated using STARS and CMG for a typical Cold Lake reservoir conditions and compared it to SAGD process. He showed that SAS process lead to higher production rates in comparison to SAGD.

5.2.6.1 Comparison of Steam/Solvent-Based Hybrid Processes

There are a number of advantages of solvent-aided processes in comparison to SAGD, namely, high energy efficiency, low operating and capital costs, and more oil recovery. In comparison of VAPEX to SAGD, it should be noted that VAPEX needs less energy due to lower latent heat value of hydrocarbon in comparison to water. The temperature in this process is also lower. Singhal et al. [63] found that energy consumption for VAPEX is 3% of the same project done by SAGD process. One deficiency of such processes is transfer of the generated heat to the solid structure, overburden and underburden, and connate water of the well in addition to the heavy oil in place. The temperature of the reservoir in steam applications rises significantly. However in VAPEX, the temperature increase is 5−10°C [63]. As Das [64] reported for 1 kg of the produced oil, 0.5 kg of solvent is needed, whereas 3 kg of steam is required. Such estimations have also been reported by some other researchers [65]. For the processes in which solvent and steam are used simultaneously, the cost would be higher in comparison to VAPEX and less energy is required in comparison to SAGD. Some operating costs include solvent and water purchase and their handling, separation cost of water or solvent from oil. The required amount of steam for SAGD process with high steam-to-oil ratio is reduced but there would be difficulty in separating oil and water [60,63]. A flash vaporizer is applied to separate light hydrocarbons in low temperature. This process leads to solvent recovery of about 90%.

Application of VAPEX and warm VAPEX increases the chance of asphaltene elimination. Asphaltene is removed as the equilibrium condition for its solution wipes out. This leads to reduction of heavy oil viscosity. If asphaltene content of the crude oil is reduced from 16% to zero, the viscosity would reduce 20-order of magnitude [66]. If asphaltene is present in the solution, it would increase the operating costs regarding to oil upgrading. This would be from thermal and catalytic operation points of view. On the other hand, as permeability of the system is reduced, the sweeping efficiency would also be affected. In reservoirs, where aquifer in the bottom or top water layer exists, thermal operations could not be run [67]. Hydrocarbon solvents are slightly soluble in water. A bottom aquifer leads to better connection of injection and production wells. The production rate is increased due to changing the flow mechanism from gravity drainage to countercurrent flow regime [59,68,69]. It could be concluded that VAPEX is the only appropriate process for reservoirs with overlying layers and bottom aquifers [58,69]. The heat loss is also the case for tight and shallow reservoirs. SAGD process is also inappropriate for reservoirs with high clay content which is due to clay swelling as a result of water condensations. This actually happens for reservoirs with clay contents more than 10% [63]. The recovery of such reservoirs could be enhanced using VAPEX process.

5.2.7 Formation Heating by Hot Fluid Injection

By injection of a hot fluid into porous media, heat transfer occurs in the rock matrix and the contained fluid in it. The same thing happens at overburden and underburden. The mechanism of heat transfer includes convection and conduction. If phase change occurs in this media, equations would become more complicated [70]. For hot fluid injection process, the mechanism of heat transfer to fluid and matrix include convection and conduction. This fluid causes the movement of water, gas, and oil in place and heating them by conduction and convection mechanisms. The porous media is also warm up by conduction. Of course, the equilibrium rate is a function of injected fluid properties, namely, viscosity and density. For modeling heat transfer processes, it is assumed that the rock and fluid temperatures are the same. As heat transfer coefficient for the condensed steam is higher than hot water, its sweeping efficiency becomes lower. There is always a vertical temperature gradient in formation which is normally ignored; this is called infinite vertical thermal conductivity. Although there is temperature gradient in direction of the heat transfer injection, this also is ignored and the steam temperature (T_S) is assumed to suddenly change into reservoir temperature (T_R). This approximation using the step function may be useful, as the injected fluid in the porous media leads to conductive heat transfer from the sand to overburden and underburden. Steam would also lead to modification of temperature distribution along the formation. This is emphasized that the fluid front moves more rapidly in comparison to heat front [70].

5.2.8 Steam Generation

To generate the steam, usually flue gas is used as the source of energy in generation plants. Furnaces consist of boilers, soot blowers to transfer the mixture of steam—water to furnace, burners and combustion air systems, a pressure system for emission of flue gases, and pressurized air system for sealing the system to avoid flue gas emission. The boiler tubes are placed between steam distribution drums and water collectors at the bottom of boiler. A super heater is placed before steam distribution system.

5.2.8.1 Heater Fuel

A combination of natural and refinery gas, coal, and fuel oil are applied as heater fuel. A combination of LPG, natural gas, and off-gas from the process units make the refinery off-gas stream. Fuel oil consisting of a mixture of straight-run and residuals provides the fuel for the system at the required pressure and temperature. The duties of balance drum include providing fuel with stable heat content, constant pressure, and recovery of gas vapors from liquid. It also prevents carrying condensates through the system. The fuel is heated under control to flow within the unit. It is filtered before burning. Sometimes these fuels are applied in various units. For instance, heat recovery from catalytic cracking unit is provided in carbon monoxide boiler. It is then converted to carbon dioxide via combustion. The units of waste heat recovery provide steam from the flue gas.

5.2.8.2 Steam Distribution

In this system, a number of fittings, pipes, valves, and connections are provided. The steam-required pressure is determined by the process units or electrical generators that use it. This pressure is declined as steam enters turbines for driving the compressors and pumps. The steam is condensed to water by traveling through the heat exchangers. The condensates are then recycled to boiler or are transported to waste water treatment unit. The steam which is applied in generators must be produced at pressures higher than the steam pressure required for the process.

5.2.8.3 Feed Water

This is a significant parameter for steam production. A huge amount of input water is needed. If there is any impurity in the system, the operation would be affected. Dissolved minerals which lead to corrosion and make deposits on turbine blade are precipitated and filtered using soda ash or lime. Moreover, insoluble materials such as oil and silt that lead to scale formation are filtered. Oxygen and carbon dioxide are deaerated, as they cause corrosion in the boiler. The cooling water which is recirculated is also treated. The most challenging operation for generating steam is starting up the heater. There may generate a flammable air and gas mixture. All operating systems must be equipped with emergency procedure for purging and also a start-up process. As the water flow rate is low and there exists a dry boiler, the tubes would not run correctly. On the other hand, entering the excess water into the distribution system would damage the turbine. Boilers should equip with blowdown system for removing the remaining water to avoid scale formation on tubes and blades of the turbine. The unit is equipped with a knockout pot to eliminate the liquid from fuel gas. There should always be an alternative source of fuel to be applied for emergency conditions.

5.2.9 Heat Loss Rate From Distribution Lines

Insulation materials including aluminum cover on calcium silicate are applied for insulating distribution lines. Heat is transferred from insulation to aluminum using conduction heat transfer mechanism. Moreover, natural and forced convection along with radiation are responsible for heat transfer to the surroundings. Heat loss due to the deposited scale as well as conduction through the steel are not accounted in calculations. As the condensation heat transfer coefficient has a large value, distribution line temperature is equal to steam temperature. Schematic representation of the system is illustrated in Fig. 5.7.

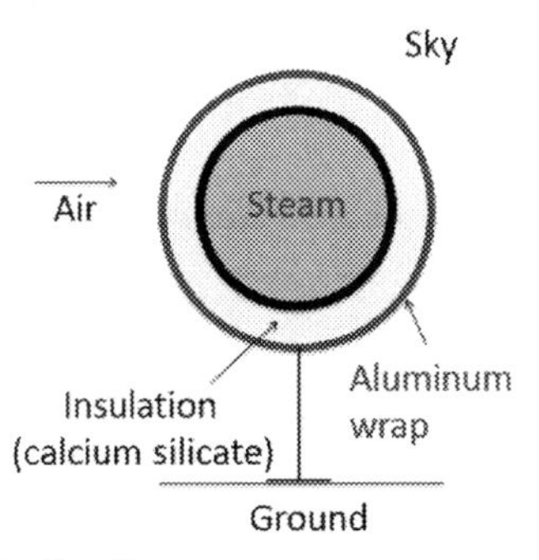

Figure 5.7 Q_{loss} from steam distribution lines.

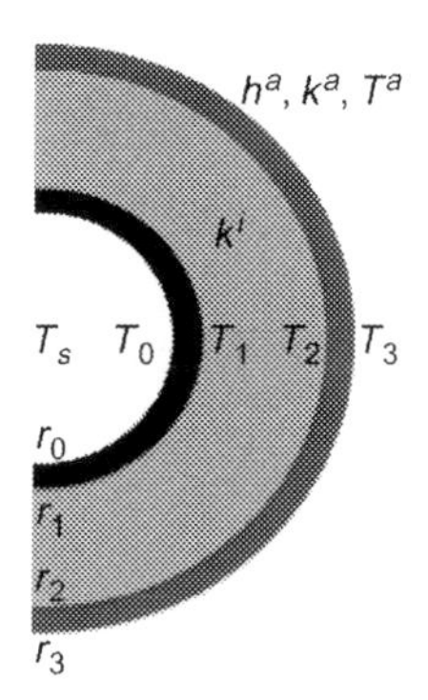

Figure 5.8 Heat transfer through insulation.

To calculate the heat loss through steam line some assumptions are considered:

- Negligible heat loss from tube and scale.
- Temperature of distribution line is equal to steam temperature

There are three mechanisms for heat transfer from insulation:

- Convection
- Radiation
- Natural convection

5.2.9.1 Heat Transfer Through Insulation/L

The conduction heat transfer in insulation is represented in Fig. 5.8.

$$\dot{Q}_i = \frac{2\pi K^i(T_1 - T_2)}{\ln(r_2/r_1)} \tag{5.92}$$

where Q_i is the heat transfer through insulation (heat loss) and k^i is the thermal conductivity of insulation.

5.2.9.2 Rate of Heat Loss – Distribution Lines

Heat transfer from insulation to the surroundings by a combination of radiation and convection mechanisms is represented as follows:

$$Q_i = Q_{Ir} + Q_{Ic} \tag{5.93}$$

The heat is lost by convection or radiation to environment: Q_{Ir} is the heat loss through radiation and Q_{Ic} is the heat loss through convection.

For high velocities of wind, the dominated heat transfer mechanism would be forced convection.

5.2.9.3 Forced Convection /L (Normal to Tube)

The value of heat transfer by convection is calculated as follows:

$$Q_{lc} = 2\pi r_3 h_c (T_3 - T_a) \tag{5.94}$$

where h_c is the convective heat transfer coefficient based on outer tube diameter.

$$1000 \le N_{Re} \le 50,000$$

To calculate the convective heat transfer coefficient, there are correlations that are applicable for covering a wide-spreading range of wind velocities. The following correlation represented by McAdams [71] may be used for forced convection as a dominated mechanism of heat transfer.

$$h_{fc} = \frac{0.12 k_{ha}}{r_3} N_{Re}{}^{0.6} \tag{5.95}$$

where k_{ha} is the thermal cond. air (Btu/h ft °F), h_{fc} is the forced conv. htc (Btu/h ft °F).

To calculate Reynolds number for air flowing into the pipe, use the following correlation:

$$N_{Re} = 4365 \frac{r_3 v_a \rho_a}{\mu_a} \tag{5.96}$$

where ρ_a is the air density at Ta (lbm/ft^3), μ_a is the air viscosity (cP), v_a is the air velocity normal to pipe (mi/h), k_{ha} and μ_a are calculated at film $T = (T_3 + T_a)/2$. As T_3 is unknown, the solution technique would be trial and error.

5.2.9.4 Radiation Heat Loss/L

To calculate the value of surface heat transfer by the mechanism of radiation (Fig. 5.9), use the following equation:

$$Q_{lr} = \pi r_3 \epsilon \sigma [(T_3{}^4 - T_{sky}{}^4) + (T_3{}^4 - T_g{}^4)] \tag{5.97}$$

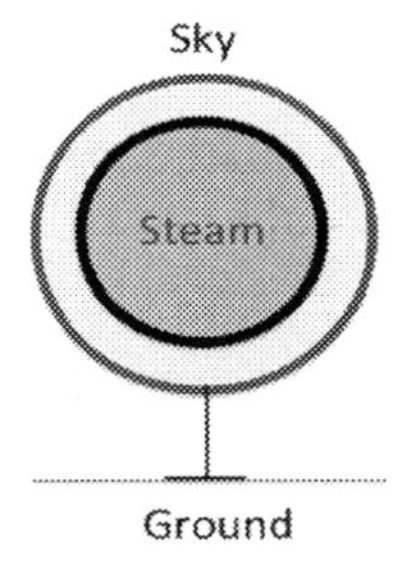

Figure 5.9 Radiation heat loss.

where ε is the emissivity of surface, σ is the Stefan–Boltzmann constant ($1.713e^{-9}$ Btu/h ft^2 °R^4), T_{sky} is the absolute sky T °R (°F + 460), T_g is the ground temperature beneath pipe °R (sky $T \cong 414-515$°R → in calc = 460°R), $T_3 \approx T_2$.

5.2.10 Heat Loss Rate From Wellbore

A well completion design is represented in Fig. 5.10 for the process of steam injection. The tubing is placed on a packer above the injection section. For initial stages of the process, the annulus is boiled dry and is filled with a mixture of vapor and air. Heat losses occur through the tubing, from annulus, and casing and cement by the mechanisms of radiation, convection, and conduction, respectively. The temperature distribution is illustrated in Fig. 5.11.

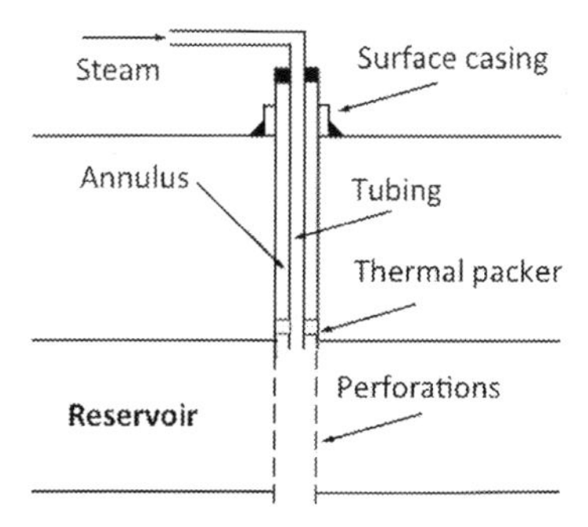

Figure 5.10 Well completion for steam injection.

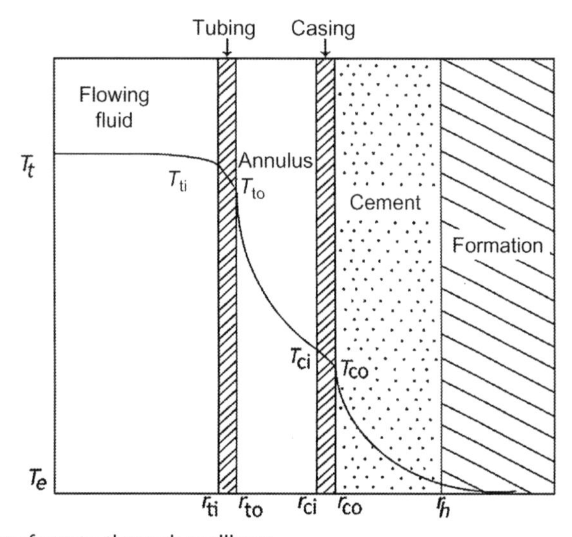

Figure 5.11 Heat-transfer rate through wellbore.

Ramey [72] represented that rate of heat transfer at the radius of drill hole is a transient process and is calculated using the following equation:

$$\dot{Q}_1 = \frac{[2\pi k_{\text{hf}}(T_h - T_e)]}{f(t)} \tag{5.98}$$

where k_{hf} is the k of formation (Btu/h ft °F), T_h is the T at cement-formation interface (°F), T_e is the stabilized T (°F), $f(t)$ is the transient dimensionless time function.

If injection time is long enough, the transient function would be applied. Dimensionless time function:

$$(t) = \ln\left(2\sqrt{\frac{a_f t}{r_{hd}}}\right) - 0.29 \text{ Ramey for } t > 1\text{week} \tag{5.99}$$

where α_f is the thermal diffusivity $(k/\rho c)$ of formation (ft^2/h) t = time (hour), r_{hd} is the radius of hole drill (ft).

Integration of the model considering the quasisteady states as a series of steady-state mechanisms would be

$$Q_1 = 2\pi r_{\text{to}} \dot{U}_{\text{to}}(T_s - T_h) \tag{5.100}$$

where U_{to} is the overall heat transfer coefficient fluid–cement/formation interface (Btu/h ft^2 °F) based on outside tube surface area, assuming $\Delta T = T_s - T_h$

$$T_h = \frac{\left[T_s f(t) + T_e\left(k_{hf} r_{\text{to}}/U_{\text{to}}\right)\right]}{\left[f(t) + \left(k_{hf} r_{\text{to}}/U_{\text{to}}\right)\right]} \tag{5.101}$$

In short time, $f(t)$ is a function of U_{to}.

5.2.10.1 Overall Heat Transfer Coefficient

To calculate the overall heat transfer coefficient in an annulus, use the following correlation represented by Willhite [73]:

$$U_{\text{to}} = \left[\frac{r_{\text{to}}}{r_{\text{ti}} h_f} + \frac{r_{\text{to}} \ln\left(r_{\text{to}}/r_{\text{ti}}\right)}{k_{\text{htub}}} + \frac{1}{h_{\text{nc}} + h_r} + \frac{r_{\text{to}} \ln\left(r_{\text{co}}/r_{ci}\right)}{k_{\text{hcas}}} + \frac{r_{\text{to}} \ln\left(r_{\text{hd}}/r_{\text{co}}\right)}{k_{\text{hcem}}}\right]^{-1} \tag{5.102}$$

where h_f is the film htc between flowing fluid and inside tube, h_{nc} is the natural convection htc in annulus based on r_{to} and $T_{\text{to}} - T_{\text{ci}}$, h_r is the radiation htc based on r_{to} and $T_{\text{to}} - T_{\text{ci}}$, k_{hcas} is the thermal conductivity of casing based on average casing T, k_{hcem} is the thermal conductivity of cement based on average cement T,P; k_{htub} is the thermal conductivity of tubing; h is calculated in Btu/h ft^2 °F, and k is calculated in Btu/h ft °F.

Approximate U_{to} by the following equation.

$$U_{to} = \left[\frac{r_{to}}{r_{ti}h_f} + \frac{r_{to}\ln(r_{to}/r_{ti})}{k_{htub}} + \frac{1}{h_{nc}+h_r} + \frac{r_{to}\ln(r_{co}/r_{ci})}{k_{hcas}} + \frac{r_{to}\ln(r_{hd}/r_{co})}{k_{hcem}}\right]^{-1} \tag{5.103}$$

In most cases, h_f, k_{htub}, and k_{hcas} are so large that previous equation can be approximated by

$$U_{to} = \left[\frac{1}{(h_{nc}+h_r)} + \frac{r_{to}\ln(r_{hd}/r_{co})}{k_{hcem}}\right]^{-1} \tag{5.104}$$

5.2.10.2 Radiation Heat-Transfer Rate/L

Evaluation of U_{to} with approximated equation requires estimation of h_r and h_c. Heat transfer coefficient for radiation, h_r, is given by

$$h_r = \sigma F_{tci}(T_{to}{}^2 + T_{ci}{}^2)(T_{to} + T_{ci}) \tag{5.105}$$

where

$$\frac{1}{F_{tci}} = \frac{1}{\varepsilon_{to}} + \frac{r_{ti}}{r_{ci}}\left(\frac{1}{\varepsilon_{ci}} - 1\right) \tag{5.106}$$

where ε_{to} is the emissivity of external tubing surface, (−), ε_{ci} is the emissivity of internal casing surface, (−), T is the absolute T (°R), σ is the Stefan–Boltzmann constant ($=1.713 \times 10^{-9}$ Btu/h ft^2 °R^4).

5.2.10.3 Heat Transfer-Rate Through Wellbore/L

$$T_{ci} = T_h + \frac{r_{to}U_{to}}{k_{hcem}}\ln\left(\frac{r_{hd}}{r_{co}}\right)(T_s - T_h) \tag{5.107}$$

To calculate h_r, the values of T_{to} and T_{ci} are required.

T_{ci} is correlated to T_h and T_s using the above equation. It should be mentioned that T_{ci} and T_h are dependent on U_{to} and also time.

5.2.10.4 Natural Convection Heat-Transfer Rate

The heat transfer coefficient for natural convection, h_c, is given by the following equation (Applicable $5 \times 10^4 < \mathbf{N_{Gr}N_{Pr}} < 7.2 \times 10^8$):

$$h_{nc} = \frac{k_{nc}}{\left(r_{to}\ln(r_{ci}/r_{to})\right)} \tag{5.108}$$

where

$$\frac{k_{hc}}{k_{ha}} = 0.049(N_{Gr}N_{Pr})^{0.333}N_{Pr}^{\ 0.074} \tag{5.109}$$

$$N_{Gr} = \frac{(r_{ci} - r_{to})g\beta\rho_{an}^{\ 2}(T_{to} - T_{ci})}{\mu_{an}^{\ 2}} \tag{5.110}$$

$$N_{Pr} = \frac{C_{an}\mu_{an}}{k_{ha}} \tag{5.111}$$

5.2.10.5 Unit Definitions in h_{nc} *Term*

N_{Gr} = Grashof number (−)
N_{Pr} = Prandtl number $= \nu/\alpha = (\mu/\rho)/(k/\rho c_P)$ (−)
C_{an} = heat capacity of fluid in annulus at average annulus temperature (Btu/lbm °F)
k_{ha} = thermal conductivity of air in the annulus at average T,P in annulus (Btu/h ft °F)
K_{hc} = equivalent thermal conductivity of fluids in the annulus with natural convection effects at average T,P in annulus (Btu/h ft °F)
β = thermal expansion coefficient of fluids in the annulus at average T,P in annulus (1/°R)
g = gravity acceleration ($=4.17 \times 10^8$ ft/h^2)
μ_{an} = viscosity of fluids in the annulus at average T,P in annulus (1/°R)
$\langle T_{annulus}\rangle = (T_{ci} + T_{to})/2$.

The step-by-step procedure to calculate heat loss to wellbore:

1. Give an initial guess for U_{to} based on T_s or T_{to} (depending on well completion)
2. Determine $f(t)$
3. Calculate T_h
4. Calculate T_{ci}
5. Guess h_r and h_c
6. Calculate the updated U_{to}
7. Compare the calculated and the initial value for U_{to}
8. Determine the value of heat loss while the error criterion for U_{to} is met.

Example 5.2: Heat loss from wellbore

Consider tubing with 3.5″ in which gas in injected at temperature of 600°F. The casing is N-80. The air pressure is 14.7 psia. The casing is cemented in a hole of 12″ diameter. The depth of the reservoir is 3000 ft. The packer size is 9.625″. The average temperature of subsurface is 100°F. Determine the overall heat transfer coefficient, the mean temperature of the casing, the value of heat loss from the wellbore after 21 days of injection.

(**Data:** r_{to}, r_{ci}, r_{co}, r_h, α_f, k_{hf}, ε_{to}, ε_{ci}, k_{hcem})

$$U_{to} = \left[\frac{r_{to}}{r_{ti}h_f} + \frac{r_{to}\ln\left(r_{to}/r_{ti}\right)}{k_{htub}} + \frac{1}{h_{nc}+h_r} + \frac{r_{to}\ln\left(r_{co}/r_{ci}\right)}{k_{hcas}} + \frac{r_{to}\ln\left(r_{hd}/r_{co}\right)}{k_{hcem}}\right]^{-1}$$

Approximation:

h_f, k_{htub}, and k_{hcas} are large:

$$U_{to} = \left[\frac{1}{(h_{nc}+h_r)} + \frac{r_{to}\ln\left(r_{hd}/r_{co}\right)}{k_{hcem}}\right]^{-1}$$

Data (known parameters):

Injecting steam

$T_s = 600$ (°F)

Well completion

$r_{to} = 0.146$ (ft), $r_{co} = 0.4$ (ft), $r_{ci} = 0.355$ (ft), and $r_h = 0.5$ (ft)

Stagnant air at 14.7 psia in annulus

Formation characteristics

$T_e = 100$ (°F), $L = 3000$ (ft),

Heat transfer parameters

$k_{hf} = 1$ (Btu/h ft °F), $k_{hcem} = 0.2$ (Btu/h ft °F),

$\varepsilon_{to} = \varepsilon_{to} = 0.9$

$\alpha_f = 0.0286$ (ft^2/h)

Mill scale: Iron oxides

1. Estimate U_{to} from T_{to}
2. Calc. $f(t)$ (for $t > 1$ week)

$$f(t) = \ln\left(\frac{\sqrt{\alpha_f t}}{r_{hd}}\right) - 0.29$$

$$= \ln\left(\frac{\sqrt{(0.0286)(504)}}{0.5}\right) - 0.29$$

$$= 2.43$$

3. T_h,

$$T_h = \frac{T_s f(t) + T_e\left(k_{hf}/r_{to}U_{to}\right)}{f(t) + \left(k_{hf}/r_{to}U_{to}\right)}$$

$$= \frac{(600)(2.43) + 100\left(1/((0.146)(4.05))\right)}{(2.43) + \left(1/((0.146)(4.05))\right)} = 395°\text{F}$$

4. Calc. T_{ci}

$$T_{ci} = T_h + \frac{r_{to}U_{to}}{k_{hcem}}\ln\left(\frac{r_{hd}}{r_{co}}\right)(T_s - T_h)$$
$$= 395 + \frac{(0.146)(4.05)}{0.2}\ln\frac{0.5}{0.4}(600 - 395)$$
$$= 530$$

5. a. Estimate h_r

$$F_{tci} = \left(\frac{1}{\varepsilon_{to}} + \frac{r_{ti}}{r_{ci}}\left(\frac{1}{\varepsilon_{ci}} - 1\right)\right)^{-1}$$
$$= \left(\frac{1}{0.9} + \frac{0.146}{0.355}\left(\frac{1}{0.9} - 1\right)\right)^{-1}$$
$$= 0.865$$

$$T_{to} = (600 + 460) = 1060°\text{R}$$
$$T_ci = (530 + 46) = 990°\text{R}$$

$$h_r = \sigma F_{tci}\left(T_{to}^{\ 2} + T_{ci}^{\ 2}\right)(T_{to} + T_{ci})$$
$$= \left(1.713 \times 10^{-9}\right)(0.865)\left(1060^2 + 990^2\right)(1060 + 990)$$
$$= 6.39\left(\frac{\text{Btu}}{\text{hr ft}^2}\right)$$

b. Estimate h_{nc}

$$\overline{T}_a = \frac{T_{to} + T_{ci}}{2} = \frac{600 + 530}{2}$$
$$= 565\ °\text{F}$$

The air properties in this temperature:
$\rho_{an} = 0.0388\ \text{lb}_m/\text{ft}^3$, $\mu_{an} = 0.069\ \text{lb}_m/\text{ft h}$, $C_{an} = 0.245\ \text{Btu/lb}_m\ °\text{F}$, $k_{ha} = 0.0255\ \text{Btu/h ft °F}$

$$\beta = \frac{-1}{\rho_{an}}\left(\frac{\partial\rho_{an}}{\partial T}\right)_P \quad \text{for ideal gas } \beta$$
$$= \frac{1}{\overline{T}_{an}} = \frac{1}{565 + 460}$$
$$= 9.75 \times 10^{-4}°\text{R}^{-1}$$

c. Estimate h_{nc} (cont'd)

$$N_{Pr} = \frac{C_{an}\mu_{an}}{k_{ha}} = \frac{(0.245)(0.069)}{0.0255} = 0.66$$

$$N_{Gr} = \frac{(r_{ci}-r_{to})^3 g\beta\rho_{an}{}^2(T_{to} - T_{ci})}{\mu_{an}{}^2}$$

$$= \frac{(0.355-0.146)^3(4.17 \times 10^8)(9.75 \times 10^{-4})(0.0388)^2(600 - 530)}{(0.069)^2}$$

$$= 8.26 \times 10^4$$

$$\frac{k_{hc}}{k_{ha}} = 0.049(N_{Gr}N_{Pr})^{0.333}N_{Pr}{}^{0.074}$$

$$= 0.049\left(8.26 \times 10^4 \times 0.66\right)^{0.333}(0.66)^{0.074} = 1.81$$

$$k_{hc} = k_{ha} \times 1.81 = 0.046 \text{ (Btu/h ft °F)}$$

d. Estimate h_{nc} (cont'd)

$$h_{nc} = \frac{k_{hc}}{\left(r_{to}\ln\left(r_{ci}/r_{to}\right)\right)} = \frac{(0.046)}{\left((0.146)\ln\left(0.355/0.146\right)\right)}$$

$$= 0.36\left(\text{Btu}/(\text{h ft}^2\ °\text{F})\right)$$

6. Calculate U_{to}

$$U_{to} = \left[\frac{1}{(h_{nc}+h_r)} + \frac{r_{to}\ln\left(r_{hd}/r_{co}\right)}{k_{hcem}}\right]^{-1}$$

$$= \left[\frac{1}{(0.36+6.39)} + \frac{(0.146)\ln\left(0.5/0.4\right)}{0.2}\right]^{-1}$$

$$= 3.22\left(\text{Btu}/(\text{h ft °F}^2)\right)$$

7. Checking for U_{to} tolerance, we started with $U_{to}^{old} = 4.05$ (Btu/h ft^2 °F), $U_{to}^{new} = 3.22$
If this tolerance is not acceptable, go back to Slide No. 10 using

$$U_{to} = U_{to}^{new}$$

	Estimate	Calculated				
Trial	U_{to}	T_h (°F)	T_{ci} (°F)	h_r	h_{nc}	U_{to}
1	4.05	395	530	6.39	0.36	3.22
2	3.22	367	487	6.00	0.42	3.15
3	3.15	364	485	5.97	0.42	3.14

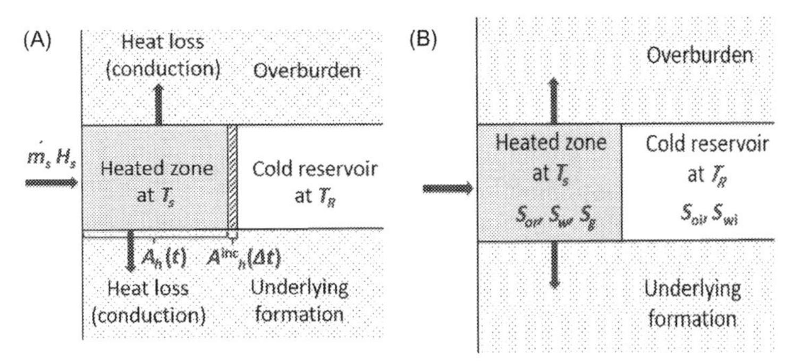

Figure 5.12 Advancement of the heated zone in the reservoir.

Overall heat loss from wellbore:

$$\begin{aligned}\dot{Q}_{\text{lwb}} &= 2\pi r_{\text{to}} U_{\text{to}}(T_s - T_h)L \\ &= 2\pi(0.146)(3.14)(600 - 364)(3000) \\ &= 2040,000 \ \text{Btu/h}\end{aligned}$$

5.2.11 Reservoir Heating by Steam Injection Using Marx–Langenheim Model

By heating the reservoir, a considerable amount of energy is lost to the formation. The modeling procedure for reservoir was introduced by Marx–Langenheim [70] in which the rock and fluid properties are considered constant. The distribution of steam is vertical in the selected region which leads to constant temperature in vertical direction. No segregation of condensate and steam occurs. The advancement of the heated zone in the reservoir is characterized by step function, as depicted in Fig. 5.12A and B.

- This process which carries a part of oil, gas, or water through the reservoir also carries heat.
- The mechanisms of heat transfer include conduction and convection. This is the case also for the displaced fluids
- For nonmoving fluids, the mechanism of heat transfer is conduction. This leads to equality of fluid and solid temperature.
- Determinative parameter to reach the equilibrium is heat transfer coefficients.
- As heat conduction occurs in surrounding sands, the heated sand area considering o/u is greater than that of invaded area by steam.
- The speed of heat front is less than the transferred heat by any injection mechanism.

5.2.11.1 The Assumptions of Marx–Langenheim Model

1. The temperature through the reservoir has a constant value of T_s.
2. The reservoir thickness is uniform.

3. The fluid and rock properties are uniform.
4. Consider no gravity segregation.
5. No heat is lost ahead of heated area.

The energy to heat unit volume of formation to temperature of T is formulated as follows:

$$Q = M_R(T - T_R) \tag{5.112}$$

$$M_R = (1-\phi)\rho_r C_r + \phi\left[S_o\rho_o C_o + S_w\rho_w C_w + S_g\rho_s C_s\right] \tag{5.113}$$

where M_R is the average heat capacity of reservoir (Btu/ft^3 °F) or (kJ/m^3 °C), C is the mean specific heat capacity (Btu/lb °F) or (kJ/kg °C), S is the saturation, ϕ is the effective porosity, ρ is the density (lb/ft^3) or (kg/m^3), r, o, w, s are the rock, oil, water, and steam, respectively.

Heat balance over heated area is as follows:

$$\dot{Q}_{\text{in}} - \dot{Q}_{\text{loss}} = \frac{dQ_R}{dt} = M_R(T_s - T_R)\frac{hdA_h}{dt} \tag{5.114}$$

where Q_{in} is the rate of energy input by steam inj., Q_{loss} is the rate of heat loss by conduction to o/u, dQ_R/dt is the rate of change of energy in reservoir, dA_h/dt is the rate of increase in heated area, M_R is the heat capacity of reservoir, T_R is the reservoir T, T_s is the steam T, t is the time, h is the reservoir thickness.

The input heat by steam injection is

CF = (350/24) lb/h when bbl/d (1 bbl = 159 L)

$$\dot{Q}_{\text{in}} = \dot{m}_s(C_w(T_s - T_R) + f_s\lambda_s) \tag{5.115}$$

where Q_{in} is the rate of energy in by steam inj. (Btu/h) or (kJ/s), m_s is the (CWE) stem inj. rate (lb/h) or (kg/s), f_s is the steam quality (=wt fraction of saturated steam as gas in the mix), λ_s is the latent heat of vaporization of water at T_s (Btu/lb) or (kJ/kg), C_w is the specific heat capacity of water (Btu/lb °F) or (kJ/kg °C).

5.2.11.2 Heat Loss to O/U

In this case, the lost heat to the boundaries is calculated using conduction heat mechanism in a semiinfinite medium and T designated by a complimentary error function. This case is represented in Fig. 5.13.

Temperature distribution for heat conduction in semiinfinite media is as follows:

$$\frac{T - T_R}{T_s - T_R} = \text{erfc}\left(\frac{z}{2\sqrt{\alpha t}}\right) \tag{5.116}$$

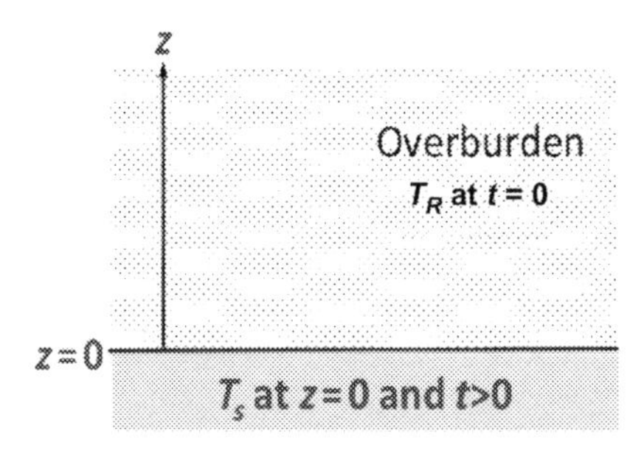

Figure 5.13 Heat loss to o/u.

$$\operatorname{erfc}(x) = 1 - \operatorname{erf}(x) = 1 - \frac{2}{\sqrt{\pi}}\int_0^x e^{-t^2}dt = \frac{2}{\sqrt{\pi}}\int_x^\infty e^{-t^2}dt \tag{5.117}$$

Heat conduction to semiinfinite media is calculated as

$$q|_{z=0} = -k_h\left(\frac{\partial T}{\partial z}\right)_{z=0} = \frac{k_h(T_s - T_R)}{\sqrt{\pi\alpha t}} \tag{5.118}$$

$$\dot{Q}_{\text{loss}} = 2\int_0^{A_h} q(t-t^*)\,dA_h \tag{5.119}$$

where t is the process time (injection), t^* is the time for the heated interface to advance to a specific location, $t-t^*$ is the time that o/u has been in contact with the heated zone.

Ordinary differential equation (ODE) equation in terms of A_h is as follows:

$$\dot{Q}_{\text{in}} = M_R(T_s - T_R)\frac{\text{hd}A_h}{dt} + 2\int_0^t \frac{k_h(T_s - T_R)}{\sqrt{\pi\alpha(t-t^*)}}\left(\frac{dA_h}{dt^*}\right)dt^* \tag{5.120}$$

$$t_D = 4\left(\frac{M}{M_R}\right)^2\left(\frac{\alpha t}{h^2}\right) \tag{5.121}$$

where t_D is the dimensionless time, M is the volumetric heat capacity of o/u, M_R is the heat capacity of reservoir, α is the thermal diffusivity of o/u.

Heated area as a function of time is

$$A_h = \frac{\dot{m}_s H_s M_R h}{4(T_s - T_R)\alpha M^2}G(t_D) \tag{5.122}$$

$$G(t_D) = \left(e^{t_D}\cdot\operatorname{erfc}\left(\sqrt{t_D}\right) + 2\sqrt{\frac{t_D}{\pi}} - 1\right) \tag{5.123}$$

Rate of growth for the heated zone is

$$\frac{dA_h}{dt} = \left(\frac{\dot{m}_s H_s}{M_R(T_s - T_R)h}\right) G_1(t_D) \tag{5.124}$$

$$G_1(t_D) = e^{t_D}\mathrm{erfc}\left(\sqrt{t_D}\right) \tag{5.125}$$

Example 5.3: Radius of steam zone for constant injection rate

The steam with the quality of 80% at pressure of 500 psig is injected to the reservoir at rate of 500 BWPD CWE. The reservoir has 25% porosity. The initial saturations of oil and water is $S_{oi} = 0.2$ and $S_{wi} = 0.8$. The volume occupied by the steam is 40% of the PV.

$k_h = 1.5$ Btu/h ft °F, $\alpha = 0.0482$ ft^2/h, $M_R = 32.74$ Btu/ft^3 °F, $M(=k_h/\alpha) =$ 31.12 Btu/ft^3 °F, $T_R = 80$°F.

Determine the heated area radius while 14 days is passed from injection [36].

At 500 psig, $T_s = 470.9$°F

$H_{wTR} = 77$ Btu/lb$_m$ (at 80°F)

$H_{wTs} = 452.9$ Btu/lb$_m$ (at 470.9°F)

$\lambda_s = 751.4$ Btu/lb$_m$

$m_s = (500$ bbl/d$)(350$ lb$_m$/bbl$)(24$ h/d$) = 7292$ lb$_m$/h

$$A_h = \frac{\dot{m}_s H_s M_R h}{4(T_s - T_R)\alpha M^2} G(t_D)$$

$$H_s = \left(H_w^{T_s} - H_w^{T_R}\right) + f_s\lambda_s$$

$$t_D = 4\left(\frac{M}{M_R}\right)^2\left(\frac{\alpha t}{h^2}\right)$$

$$M = \frac{k_h}{\alpha}$$

$$t_D = 4\left(\frac{M}{M_R}\right)^2\left(\frac{\alpha t}{h^2}\right)$$

$$= 4\left(\frac{31.12}{32.74}\right)^2\left(\frac{(0.0482)(14\text{d}\times 24\ \text{h/d})}{20^2}\right)$$

$$= 0.146$$

$G(t_D) = 0.113$

$H_s = (452.9 - 77) + 0.8 \times (751.4) = 977$ Btu/lb$_m$

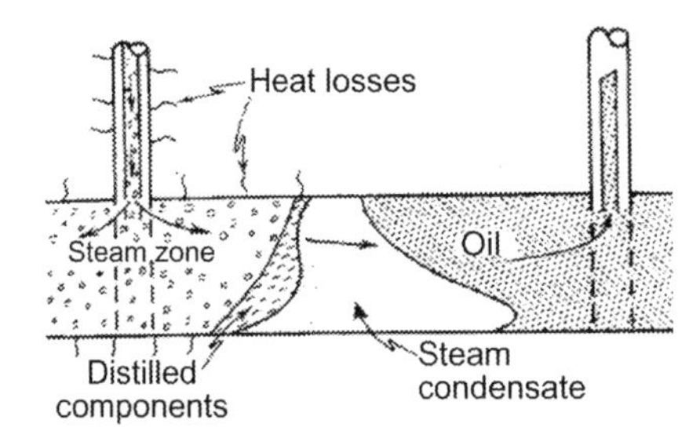

Figure 5.14 Steam-drive oil recovery mechanism.

$$A_h = \frac{\dot{m}_s H_s M_R h}{4(T_s - T_R)\alpha M^2} G(t_D)$$

$$= \frac{(7292)(977)(32.74)(20)}{4(470.9 - 80)(0.0482)\left(31.12^2\right)} 0.113$$

$$= 7222 \text{ ft}^2$$

$$r_h \approx \sqrt{\frac{A_h}{\pi}} = \sqrt{\frac{7222}{\pi}} = 47.94 \text{ ft}$$

5.2.12 Steam Drive Oil Recovery Mechanism

The assumption of this technique is occupation of the whole area by steam. After formation of the hot water zone, it invades to steam zone of the formation. The other assumption is preceding the oil to each of the zones. This mechanism is illustrated in Fig. 5.14. Willman et al. [74] analyzed various experimental data to determine the main mechanisms of displacement by the tests conducted on long and short cores. Tests were done at 330 and 250°F and the pressure of 800 psig. The other output was that the required number of pore volumes of injected fluid is more in the case of hot and cold water injection in comparison to steam injection. The main issues that resulted in increased recovery of hot water include thermal expansion and viscosity reduction. The residual saturation of oil in water-drive systems is a weak function of temperature. This leads to application of fractional-flow concept in hot water-drive systems for determining the performance of waterflood by displacement calculations.

5.2.12.1 Steam Distillation

- The presence of an immiscible phase (water/steam) will lower the temperature for the volatile organic phase to resolve. Vapor phase is in contact with two immiscible liquids.

- Ignoring nonidealities at T of system distillation starts when sum of vapor pressure equals total pressure:
  $P = P_V + P_W$ (V, volatile; W, water/steam)
  $P_V V_V = n_V RT$ and $P_W V_W = n_W RT$ ($V_v = V_W$)

5.2.12.2 Myhill and Stegemeier Model (MS Model) [75]

Myhill and Stegemeier [75] introduced an average thermal efficiency of combined steam-hot water zones, resulting in a uniform displacement model.

$$\overline{E}_h = \frac{1}{t_D}$$
$$\left\{ G(t_D) + \frac{(1-f_{h,v})U(t_D - t_{cD})}{\sqrt{\pi}} \left[2\sqrt{t_D} - 2(1-f_{h,v})\sqrt{t_D - t_{cD}} - \int_0^{t_{cD}} \frac{e^u \operatorname{erfc}(\sqrt{u})}{\sqrt{t_D - u}} du - \sqrt{\pi} G(t_D) \right\} \right. \tag{5.126}$$

$U = 0$ for $t_D > t_{cD}$ and $U = 1$ before critical time $t_D \le t_{cD}$, $f_{h,v}$ = fraction of inj. energy that is latent heat.

5.2.12.3 MS Uniform Model Limitations

The coefficient $(1 - f_{h,v})$ is empirical. It is the fraction of injected energy that is in the form of sensible heat. MS uniform model fails at low-quality steam drive ($f_{sd} < 0.2$) because it does not predict the oil displacement of HW zone adequately.

5.2.12.4 Capture Factor and Steam-to-Oil Ratio (SOR)

In reality, more oil is displaced than produced. To account for this, capture factor (E_c = 70%–100% in field) is introduced:

$$F_{SOR} = \frac{W_{s,eq}}{N_p} \tag{5.127}$$

where F_{SOR} is the steam-to-oil ratio (SOR), $W_{s,eq}$ is the vol. of steam inj. N_p is the vol. oil produced.

5.2.12.5 Steam-to-Oil Ratio

In the definition of $W_{s,eq}$, it is assumed that steam leaving boiler has 1000 Btu/lbm energy, so

$$\left(W_{s,eq}\right)\left(62.4\ \text{lbm/ft}^3\right)\left(350\ \text{lbm/bbl}\right)\left(1000\ \text{Btu/lb}\right) = \dot{m}_s(H_{ws} - H_{wA} + f_{sb}\lambda_s) \tag{5.128}$$

$$\left(W_{s,eq}\right) = 2.854 \times 10^{-6}(\dot{m}_s H_{sbA}) \tag{5.129}$$

where H_{sbA} is the enthalpy of steam at boiler outlet relative to feeder water temperature (T_A), f_{sb} is the steam quality, after boiler.

5.2.12.6 Oil-Production Rate

$$q_o = \left(\frac{7758\ \text{bbl}}{\text{ac ft}}\right)\phi\left(\frac{S_{oi}}{B_{o1}} - \frac{S_{ors}}{B_{os}}\right)\frac{dV_s}{dt} \quad (5.130)$$

where q_o is the oil-flow rate (bbl/day), t is the time (day), V_s is the volume of steam zone (ac ft).

$$q_o = \left(\frac{7758\ \text{bbl}}{\text{ac ft}}\right)\phi\left(\frac{S_{oi}}{B_{o1}} - \frac{S_{ors}}{B_{os}}\right)\frac{hdA_s}{dt} \quad (5.131)$$

$$\frac{dA_h}{dt} = \left(\frac{\dot{m}_s H_s}{M_R(T_s - T_R)h}\right) G_1(t_D)\ (\text{for } t_D \leq t_{cD}) \quad (5.132)$$

$$q_o = \left[\left(\frac{7758\ \text{bbl}}{\text{ac ft}}\right)\phi\left(\frac{S_{oi}}{B_{o1}} - \frac{S_{ors}}{B_{os}}\right)\right] \times \left(\frac{\dot{m}_s H_s}{M_R(T_s - T_R)}\right) G_1(t_D)\ (\text{for } t_D \leq t_{cD}) \quad (5.133)$$

$$q_{ohw} = \left[\left(\frac{7758\ \text{bbl}}{\text{ac ft}}\right)\phi\left(\frac{S_{oi}}{B_{o1}} - \frac{S_{orhw}}{B_{ohw}}\right)\right] \times \left(\frac{\dot{m}_s H_s}{M_R(T_s - T_R)}\right) G_1(t_D)\ (\text{for } t_D > t_{cD}) \quad (5.134)$$

5.2.12.7 Oil-Production Rate From Steam Zone

$$q_{os} = \left[\left(\frac{7758\ \text{bbl}}{\text{ac ft}}\right)\phi\left(\frac{S_{orw}}{B_{ow}} - \frac{S_{ors}}{B_{os}}\right)\right] \times \left(\frac{\dot{m}_s H_s}{M_R(T_s - T_R)}\right)\frac{dG_1(t_{Ds})}{dt}\ (\text{for } t_D > t_{cD}) \quad (5.135)$$

Example 5.4: Calculation of SOR using Myhill and Stegemeier Model

Use the following steam-drive data that were reported by Myhill and Stegemeier [75]. The injection rate and pressure are 850 B/D and of 200 psig, respectively. The temperatures of the reservoir and input water are 110 and 70°F, respectively. The quality of the output steam is 0.8. This value is decreased to 0.7 at injection sandface. Determine the value of SOR using the following information:

$Ps = 215$ psia

$T_s = 387.9$°F

$L_{vdh} = 837.4$ Btu/lbm

$f_{sd} = 0.7$

$H_{wr} = 77.94$ Btu/lbm at 110°F

$h = 32$ ft

$H_{wT} = 361.91$ Btu/lbm at 387.9°F
$\varnothing = 0.30$
$H_{wA} = 38$ Btu/lbm at 70°F
$\Delta S_o = 0.31$
$M_R = 35$ Btu/ft^3 °F
$M_s = 42$ Btu/ft^3 °F
$\alpha = 1.2$(Btu/h ft °F)/42 (Btu/ft^3 °F)
$= 0.0286$ ft^2/h
$= 0.6857$ ft^2/day
$K_h = 1.2$ Btu/h ft °F.

Solution: When time is in days,
$T = 4\ (M_S/M_R)^2(\alpha/h^2)t = 4(42/35)^2[0.6857/(32\text{ ft})^2]t = 3.857 \times 10^{-3}$ t
When time is in years,
$t_D = 1.408$ t, so that at $t = 4.5$ years, $t_D = 6.335$,
$H_s = 361.91 + (0.7)\ (837.4)\ - 77.94 = 870.15$ Btu/Ibm

$$f_{h,v} = \frac{((0.7)(837.4))}{870.15} = 0.674$$

As we know $\alpha = K_h/M$,

$$\overline{E}_{h,s} = 0.33$$

The volume of oil displaced from the steam zone:

$$V_s = \left[\frac{m_s H_s t}{M_R(T_s - T_r)}\right]\overline{E}_{h,s}(t_D)$$
$$= [(850\ \text{B/D})(350.4\ \text{lbm/bbl})(870.15\ \text{Btu/lbm})$$
$$\times (4.5\text{years})(365\ \text{D/year})]/[(35\ \text{Btu/ft}^3\ °\text{F})$$
$$\times (387.9 - 110°\text{F})(43,560\ \text{ft}^2/\text{ac ft}](0.33)$$
$$= 331.5\text{n ac ft.}$$

$N_{Ps} = (7758\text{ bbl/ac ft})\ \varnothing\frac{h_n}{h_t}\left(\frac{S_{oi}}{B_{oi}} - \frac{S_{ors}}{B_{ors}}\right)V_s = (7758\text{ bbl/ac ft})\ \ (0.30)\ \ (1.0)\ \ (0.31)$ (331.5 ac ft) = 239,197 stb.

The equivalent volume of water injected is determined.

The energy content of the steam relative to the feed water temperature and the steam leaving the boiler is computed below.

$H_{wA} = H_{ws} - H_{wA} + f_{sb}\ L_{vs} = 361.9 - 38 + 0.8(837.4) = 993.83$ Btu/lbm.

$W_t = (850\text{ B/D})\quad (5.615\text{ ft}^3/\text{bbl})\quad (62.4\text{ lbm/ft}^3) \times (4.5\text{ years})\quad (365\text{ D/year}) = 489.17 \times 10^6$ lbm.

$W_{s,\ eq} = (2.854 \times 10^{-6})\ (489.17 \times 10^{6})\ (993.83) = 1.388 \times 10^{6}$ bbl.
$F_{os} = 239198/1387500 = 0.172$ bbl oil/bbl steam,
$F_{so} = 5.81$ bbl steam/bbl oil.

PROBLEMS

P1. Determine the heat loss from the wellbore from a 2-in. tube after 80 days of injection. The depth of the formation is 900 ft.The steam quality is 83%. The feed-water rate is 1000 B/D and the reservoir temperature is 390°F.

1. Determine the heat-loss rate from the tubing.
2. Calculate the heat-injection rate at surface.
3. How much heat is heat loss from the wellbore?
4. Determine the quality of the saturated steam applied for injection (Table TP.1).

P2. The distribution line in Fig. P.1 is 3-in. [3.5-in. outer diameter (OD)]. The insulation thickness is 2 in. of magnesia silicate covered by a layer of aluminum ($\varepsilon_{A1} = 0.77$). The rate, pressure, and temperature of the stream injected are 350 B/D, 1650 psia, and 620°F. Mean ambient temperature is 100°F. The velocity of wind is 10 mi/h, and temperature of the subsurface is 70°F.

1. Determine the heat-loss rate of the distribution line.
2. Calculate the steam quality at wellhead as steam-generator quality is 0.85.

Table TP.1 Steam Injection Conditions

Mean subsurface temperature (°F)	70
Geothermal gradient (°F/ft)	0.02
Overall heat-transfer coefficient (Btu/D ft °F)	33
Tubing ID (in.)	2
Drill hole diameter (in.)	7
Thermal conductivity of Earth (Btu/D ft °F)	36
Thermal diffusivity of Earth (ft^2/D)	0.96

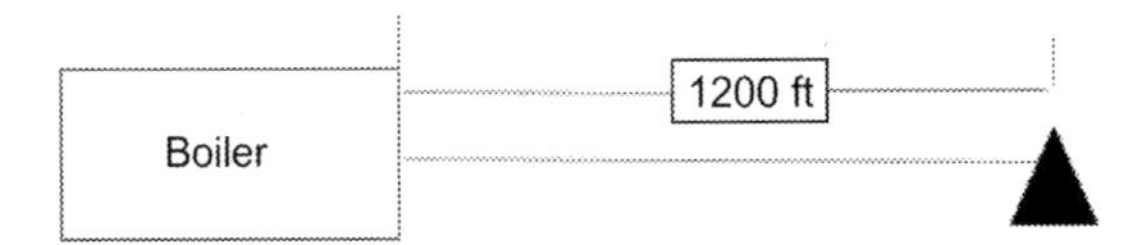

Figure P.1 Steam distribution system.

Table TP.2 Properties of Steam Injection Well

Injection tubing (ID, in.)	2.99
Injection tubing (OD, in.)	3.50
Casing (OD, in.)	7.0
Casing (ID, in.)	6.276
Hole (in.)	10.75
Depth (in.)	3500
Injection tubing (Btu/h ft °F)	24.84
Casing (Btu/h ft °F)	24.84
Cement (Btu/h ft °F)	0.3
Formation (Btu/h ft °F)	1.4
Thermal diffusivity of formation (0.04 ft^2/h)	0.04
Mean subsurface temperature (°F)	77
Geothermal gradient (°F/ft)	0.017

P3. Calculate the steam quality at the wellhead in Problem **P2** without insulation of the production line.

P4. The steam with pressure and temperature of 419°F and 350 psia is injected to the reservoir by rate of 300 B/D CWE. The characteristic of the casing is 7-in. (J-55, 26 lbm/ft cemented to surface in a $10^{3/4}$-in. drill hole). The earth conductivity and mean heat capacity are 1.0 Btu/(h ft °F) and 35 Btu/ft Oe, thermal gradient is 0.015°F/ft. The average temperature of subsurface is 100°F. The cement conductivity is 0.6 Btu/(h ft^2 °F/ft). The wellhead-stem quality is 0.75. Determine the heat loss of the wellbore after 5 months of injection.

P5. Steam with characteristics of Table TP.2 indicates injected steam properties with the rate of 1000 B/D CWE. Determine the rate of heat loss and casing temperature. The annulus of the casing is soil dry through the injection.

1. Determine heat-loss rate and temperature of the casing after 30 days of injection with pressure of 2500 psi. The quality of steam at wellhead is 89.7%.
2. Recalculate the steam quality at the sandface as pressure change in the tubing is zero.

REFERENCES

[1] R.F. Meyer, E.D. Attanasi, P.A. Freeman, Heavy oil and natural bitumen resources in geological basins of the world. Report No.: 2331 -1258.

[2] M. Prats, Thermal Recovery, Society of Petroleum Engineers, United States, 1982. Web.

[3] J. Raicar, R. Procter (Eds.), Economic considerations and potential of heavy oil supply from Llodminster -Alberta, Canada. In: The Future of Heavy Oil and Tar Sands, Second Internal Conference. McGraw-Hill, New York, NY, 1984.

[4] P.S. Sarathi, D.K. Olsen, Practical Aspects of Steam Injection Processes: A Handbook for Independent Operators, National Inst. for Petroleum and Energy Research, Bartlesville, OK, 1992.

[5] R. Butler, D. Stephens, The gravity drainage of steam-heated heavy oil to parallel horizontal wells, J. Can. Pet. Technol. 20 (02) (1981) 90–96.
[6] S. Farouq Ali, B. Abad, Bitumen recovery from oil sands, using solvents in conjunction with steam, J. Can. Pet. Technol. 15 (03) (1976) 80–90.
[7] O.E. Hernández, S. Ali (Eds.), Oil recovery from Athabasca tar sand by miscible-thermal methods. in: Annual Technical Meeting, Petroleum Society of Canada, 1972.
[8] B. Hascakir, How to select the right solvent for solvent-aided steam injection processes, J. Pet. Sci. Eng. 146 (2016) 746–751.
[9] A. Turta, J. Lu, R. Bhattacharya, A. Condrachi, W. Hanson (Eds.), Current status of the commercial in situ combustion (ISC) projects and new approaches to apply ISC, in: Canadian International Petroleum Conference, Petroleum Society of Canada, 2005.
[10] J.G. Burger, Chemical aspects of in-situ combustion-heat of combustion and kinetics, Soc. Pet. Eng. J. 12 (05) (1972) 410–422.
[11] A. Chhetri, M. Islam, A critical review of electromagnetic heating for enhanced oil recovery, Pet. Sci. Technol. 26 (14) (2008) 1619–1631.
[12] K. Holdaway, Harness Oil and Gas Big Data with Analytics: Optimize Exploration and Production with Data Driven Models, John Wiley & Sons, New York, 2014.
[13] J.G. Speight, The Chemistry and Technology of Petroleum, CRC Press, Boca Raton, FL, 2014.
[14] Deutsch, C., McLennan, J., 2005. Guide to SAGD (steam assisted gravity drainage) reservoir characterization using geostatistics. Centre for Computational Excellence (CCG), Guidebook Series, vol. 3. University of Alberta, April 2003.
[15] D.S. Law, A New Heavy Oil Recovery Technology to Maximize Performance and Minimize Environmental Impact, SPE International, Bethel, CT, 2011.
[16] M.R. Carlson, Practical Reservoir Simulation: Using, Assessing, and Developing Results, PennWell Books, Houston, TX, 2003.
[17] D. Glassman, M. Wucker, T. Isaacman, C. Champilou, The Water-Energy Nexus: Adding Water to the Energy Agenda, World Policy Institute, New York, 2011. 1.
[18] Q. Jiang, B. Thornton, J. Russel-Houston, S. Spence, Review of thermal recovery technologies for the clearwater and lower grand rapids formations in the cold lake area in Alberta, J. Can. Pet. Technol. 49 (09) (2010) 2–13.
[19] S. Thomas, Enhanced oil recovery-an overview, Oil Gas Sci. Technol. -Rev. l'IFP 63 (1) (2008) 9–19.
[20] M. Prats, A current appraisal of thermal recovery, J. Pet. Technol. 30 (08) (1978) 1–129.
[21] J. Alvarez, S. Han, Current overview of cyclic steam injection process, J. Pet. Sci. Res. 2 (3) (2013) 116–127.
[22] F. Hongfu, L. Yongjian, Z. Liying, Z. Xiaofei, The study on composition changes of heavy oils during steam stimulation processes, Fuel 81 (13) (2002) 1733–1738.
[23] H. Pahlavan, I. Rafiqul, Laboratory simulation of geochemical changes of heavy crude oils during thermal oil recovery, J. Pet. Sci. Eng. 12 (3) (1995) 219–231.
[24] J. Scott, S. Proskin, D. Adhikary, Volume and permeability changes associated with steam stimulation in an oil sands reservoir, J. Can. Pet. Technol. 33 (07) (1994) 44–52.
[25] Poroelastic effects of cyclic steam stimulation in the Cold Lake Reservoir, in: D. Walters, A. Settari, P. Kry, (Eds.), SPE/AAPG Western Regional Meeting, Society of Petroleum Engineers, 2000, pp. 1–10.
[26] D. Walters, A. Settari, P. Kry, Coupled geomechanical and reservoir modeling investigating poroelastic effects of cyclic steam stimulation in the Cold Lake reservoir, SPE Reservoir Eval. Eng. 5 (06) (2002) 507–516.
[27] Geomechanics for the thermal stimulation of heavy oil reservoirs-Canadian experience, in: Y. Yuan, B. Xu, B. Yang, (Eds.), SPE Heavy Oil Conference and Exhibition, Society of Petroleum Engineers, 2011.
[28] R. Wong, Y. Li, A deformation-dependent model for permeability changes in oil sand due to shear dilation, J. Can. Pet. Technol. 40 (08) (2001) 37–44.
[29] Geomechanics of oil sands under injection, in: D.P. Yale, T. Mayer, J. Wang (Eds.), 44th US Rock Mechanics Symposium and 5th US-Canada Rock Mechanics Symposium, American Rock Mechanics Association, 2010.

[30] Geomechanics monitoring of cyclic steam stimulation operations in the clearwater formation, in: J. Gronseth, ISRM International Symposium, International Society for Rock Mechanics, 1989.
[31] Mapping reservoir volume changes during cyclic steam stimulation using tiltmeter based surface deformation measurements, in: J. Du, S.J. Brissenden, P. McGillivray, S.J. Bourne, P. Hofstra, E.J. Davis, et al. (Eds.), SPE International Thermal Operations and Heavy Oil Symposium, Society of Petroleum Engineers, 2005.
[32] M. Dusseault, L. Rothenburg, Analysis of deformation measurements for reservoir management, Oil Gas Sci. Technol. 57 (5) (2002) 539–554.
[33] J. Van Lookeren, Calculation methods for linear and radial steam flow in oil reservoirs, Soc. Pet. Eng. J. 23 (03) (1983) 427–439.
[34] S.M. Farouq Ali, No. CONF-820316- Steam Injection Theories: A Unified Approach, Society of Petroleum Engineers, 1982.
[35] T.C. Boberg, Calculation of the production rate of a thermally stimulated well, J. Pet. Technol. 18 (12) (1966) 1–613.
[36] D.W. Green, G.P. Willhite, Enhanced Oil Recovery: Henry L. Doherty Memorial Fund of AIME, Society of Petroleum Engineers Richardson, TX, 1998.
[37] ⟨http://www.insitucombustion.ca/newprocesses.htm⟩.
[38] A. Hart, The novel THAI–CAPRI technology and its comparison to other thermal methods for heavy oil recovery and upgrading, J. Pet. Explor. Product. Technol. 4 (4) (2014) 427–437.
[39] Greaves M, Turta AT. Oilfield in-situ combustion process. Google Patents; 1997.
[40] Abarasi Hart, et al., Optimization of the CAPRI process for heavy oil upgrading: effect of hydrogen and guard bed, Ind. Eng. Chem. Res. 52 (44) (2013) 15394–15406.
[41] M. Greaves, T. Xia, Downhole upgrading of Wolf Lake oil using THAI–CAPRI processes-tracer tests, Prepr. Pap. Am. Chem. Soc. Div. Fuel Chem. 49 (1) (2004) 69–72.
[42] Downhole upgrading Athabasca tar sand bitumen using THAI-SARA analysis, in: T. Xia, M. Greaves, (Eds.), SPE International Thermal Operations and Heavy Oil Symposium, Society of Petroleum Engineers, 2001.
[43] Xia, T., Greaves, M. (Eds.), 2001. 3-D physical model studies of downhole catalytic upgrading of Wolf Lake heavy oil using THAI. In: Canadian International Petroleum Conference. Petroleum Society of Canada.
[44] J.G. Weissman, Review of processes for downhole catalytic upgrading of heavy crude oil, Fuel Process. Technol. 50 (2–3) (1997) 199–213.
[45] Hydrogen generation during in-situ combustion, in: L. Hajdo, R. Hallam, L. Vorndran, (Eds.), SPE California Regional Meeting, Society of Petroleum Engineers, 1985.
[46] J.-H. Gong, L. Jun, Y.-H. Xu, Protolytic cracking in Daqing VGO catalytic cracking process, J. Fuel Chem. Technol. 36 (6) (2008) 691–695.
[47] J.H. Lee, S. Kang, Y. Kim, S. Park, New approach for kinetic modeling of catalytic cracking of paraffinic naphtha, Ind. Eng. Chem. Res. 50 (8) (2011) 4264–4279.
[48] Frauenfeld, T., Deng, X., Jossy, C. (Eds.), 2006. Economic analysis of thermal solvent processes. In: Canadian International Petroleum Conference, Petroleum Society of Canada.
[49] I.J. Mokrys, R.M. Butler, The rise of interfering solvent chambers: solvent analog model of steam-assisted gravity drainage, J. Can. Pet. Technol. 32 (03) (1993) 26–36.
[50] Allen JC, Woodward CD, Brown A, Wu CH. Multiple solvent heavy oil recovery method. Google Patents; 1976.
[51] Allen JC, Redford DA. Combination solvent-noncondensible gas injection method for recovering petroleum from viscous petroleum-containing formations including tar sand deposits. Google Patents; 1978.
[52] S. Dunn, E. Nenniger, V. Rajan, A study of bitumen recovery by gravity drainage using low temperature soluble gas injection, Can. J. Chem. Eng. 67 (6) (1989) 978–991.
[53] R.M. Butler, I.J. Mokrys, A new process (VAPEX) for recovering heavy oils using hot water and hydrocarbon vapour, J. Can. Pet. Technol. 30 (01) (1991) 97–106.
[54] S. Farouq Ali, S. Snyder, Miscible thermal methods applied to a two-dimensional, vertical tar sand pack, with restricted fluid entry, J. Can. Pet. Technol. 12 (04) (1973) 20–26.

[55] Awang, M., Farouq Ali, S. (Eds.), 1980. Hot-solvent miscible displacement. In: Annual Technical Meeting. Petroleum Society of Canada.
[56] R. Butler, Q. Jiang, Improved recovery of heavy oil by VAPEX with widely spaced horizontal injectors and producers, J. Can. Pet. Technol. 39 (01) (2000) 48–56.
[57] Experimental investigation of oil drainage rates in the VAPEX process for heavy oil and bitumen reservoirs, in: K. Karmaker, B.B. Maini, (Eds.), SPE Annual Technical Conference and Exhibition, Society of Petroleum Engineers, 2003.
[58] Applicability of vapor extraction process to problematic viscous oil reservoirs, in: K. Karmaker, B.B. Maini, (Eds.), SPE Annual Technical Conference and Exhibition, Society of Petroleum Engineers, 2003.
[59] In-situ upgrading of heavy oils and bitumen by propane deasphalting: the VAPEX process, in: I. Mokrys, R. Butler (Eds.), SPE Production Operations Symposium, Society of Petroleum Engineers, 1993.
[60] Recovery performance and economics of steam/propane hybrid process, in: X. Deng, SPE International Thermal Operations and Heavy Oil Symposium, Society of Petroleum Engineers, 2005.
[61] Experimental and simulation studies of steam-propane injection for the Hamaca and Duri fields, in: D. Mamora, J. Rivero, A. Hendroyono, (Eds.), SPE Annual Technical Conference and Exhibition, Society of Petroleum Engineers, 2003.
[62] Steam alternating solvent process, in: L. Zhao, SPE International Thermal Operations and Heavy Oil Symposium and Western Regional Meeting, Society of Petroleum Engineers, 2004.
[63] Screening of reservoirs for exploitation by application of steam assisted gravity drainage/VAPEX processes, in: A. Singhal, S. Das, S. Leggitt, M. Kasraie, Y. Ito (Eds.), International Conference on Horizontal Well Technology, Society of Petroleum Engineers, 1996.
[64] S.K. Das, VAPEX: An efficient process for the recovery of heavy oil and bitumen, SPE J. 3 (03) (1998) 232–237.
[65] Effect of drainage height and grain size on the convective dispersion in the VAPEX process: Experimental study, in: B.B. Maini, SPE/DOE Symposium on Improved Oil Recovery, Society of Petroleum Engineers, 2004.
[66] Effects of asphaltene content and solvent concentration on heavy oil viscosity, in: P. Luo, Y. Gu (Eds.), SPE International Thermal Operations and Heavy Oil Symposium, Society of Petroleum Engineers, 2005.
[67] R. Butler, Some recent developments in SAGD, J. Can. Pet. Technol. 40 (01) (2001) 18–22.
[68] R. Butler, I. Mokrys, Closed-loop extraction method for the recovery of heavy oils and bitumens underlain by aquifers: The VAPEX process, J. Can. Pet. Technol. 37 (04) (1998) 41–50.
[69] Countercurrent extraction of heavy oil and bitumen, in: S. Das, R. Butler, (Eds.), International Conference on Horizontal Well Technology, Society of Petroleum Engineers, 1996.
[70] J.W. Marx, R.H. Langenheim, Reservoir Heating by Hot Fluid Injection, Society of Petroleum Engineers, 1959.
[71] W.H. McAdams, third ed., Heat Transmission, 742, McGraw-Hill, *New York*, 1954.
[72] H. Ramey Jr, Wellbore heat transmission, J. Pet. Technol. 14 (04) (1962) 427–435.
[73] G.P. Willhite, Over-all heat transfer coefficients in steam and hot water injection wells, J. Pet. Technol. 19 (05) (1967) 607–615.
[74] B. Willman, V. Valleroy, G. Runberg, A. Cornelius, L. Powers, Laboratory studies of oil recovery by steam injection, J. Pet. Technol. 13 (07) (1961) 681–690.
[75] N. Myhill, G. Stegemeier, Steam-drive correlation and prediction, J. Pet. Technol. 30 (02) (1978) 173–182.

CHAPTER SIX

Chemical Flooding

Mohammad Ali Ahmadi
Department of Chemical and Petroleum Engineering, University of Calgary, Calgary, AB, Canada

6.1 INTRODUCTION

Historically, oil production from oil reservoirs has been divided into three phases. Initially, oil is produced by the native energy of the reservoir (such as the dissolved gas drive, or the natural water-drive aquifer), and this period is called primary production. Primary production results in 5%–30% original oil in place (OOIP) recovery [1]. Since water is the cheapest fluid available, water flooding is carried out to increase the oil production beyond that of primary production and this stage is called secondary production. The water pushes the oil in front of it toward the production wells and helps to increase the total recovery to 40%–60% OOIP. The process continues until the water–oil ratio at the production wells becomes very high and reaches the economic level at which the oil production is not cost-effective anymore. At this stage, there is a significant amount of oil (40%–60% OOIP) still left in the reservoir due to many factors including unfavorable wettability conditions, heterogeneity of reservoir rock, and capillary-trapped oil. In order to recover this residual oil and increase the ultimate oil recovery of the reservoir, enhanced oil recovery (EOR) methods are utilized. Since these methods often follow the secondary production, they are sometimes called tertiary oil recovery methods. The EOR processes can be divided into three main categories [2–6]:

- Chemical methods
- Miscible methods
- Thermal methods

6.2 CHEMICAL-BASED ENHANCED OIL RECOVERY METHOD

In chemical EOR methods, an agent that is not normally present in the reservoir is injected to enhance the oil displacement. Examples of the chemical processes are gel polymer and polymer flooding aimed to shutoff the high-permeability areas of

Fundamentals of Enhanced Oil and Gas Recovery from Conventional and Unconventional Reservoirs.
DOI: https://doi.org/10.1016/B978-0-12-813027-8.00006-0

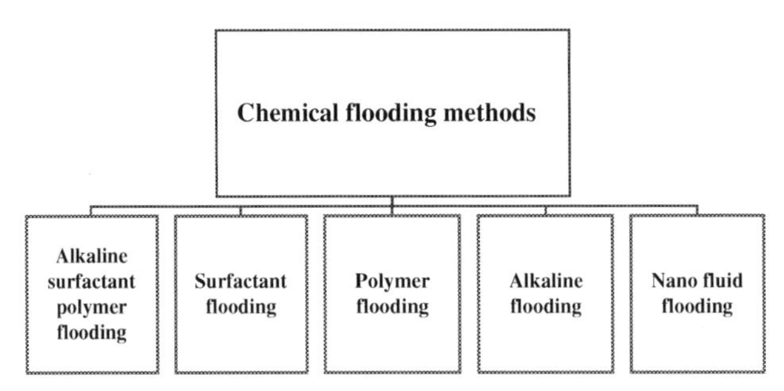

Figure 6.1 Classification of chemical EOR methods. *EOR*, enhanced oil recovery.

the reservoir [7,8], and to increase the viscosity of the injected water to increase the sweep areas in the reservoir [9], and alkaline and surfactant flooding to create low oil–water interfacial tension (IFT) and hence remobilizing the trapped oil [10,11]. It is also possible to enhance the oil production through wettability alteration of the reservoir rock during a surfactant flooding [10–14]. Fig. 6.1 depicts the classification for chemical EOR methods.

As an example of chemical EOR, surfactant flooding is to lower the IFT, which causes lower capillary pressure and enhances the imbibition mechanism by gravity drive in oil-wet reservoir [13]. The concept behind the chemical-based EOR methods is increasing the capillary number; the number is dimensionless representing the ratio of viscous forces over capillary forces. The capillary number can be expressed through the following equation [15–17]:

$$N_C = \frac{v\mu}{\sigma} \tag{6.1}$$

where v represents the fluid velocity, μ denotes the fluid viscosity, and σ is the IFT. There are different relations between the residual oil saturation and capillary number in the depleted oil reservoir. Fig. 6.2 illustrates the typical variation of residual oil saturation versus capillary number reported in literature [17].

6.2.1 Surfactant Flooding

Ahmadi and Shadizadeh [19] carried out comprehensive series of core-flooding experiments on real carbonate rocks to examine the efficiency of a new natural surfactant for EOR goals. They conducted core-displacement test along with IFT measurement to determine the efficiency of the surfactant in terms of oil recovery factor. They discussed that the surfactant extracted from *Ziziphus spina-christi* could considerably increase the oil recovery factor. Besides this advantage, there is no environmental issue using such a surfactant [19].

Ahmadi et al. [20] proposed an environment-friendly surfactant extracted from mulberry leaves. They conducted different experiments under reservoir condition to

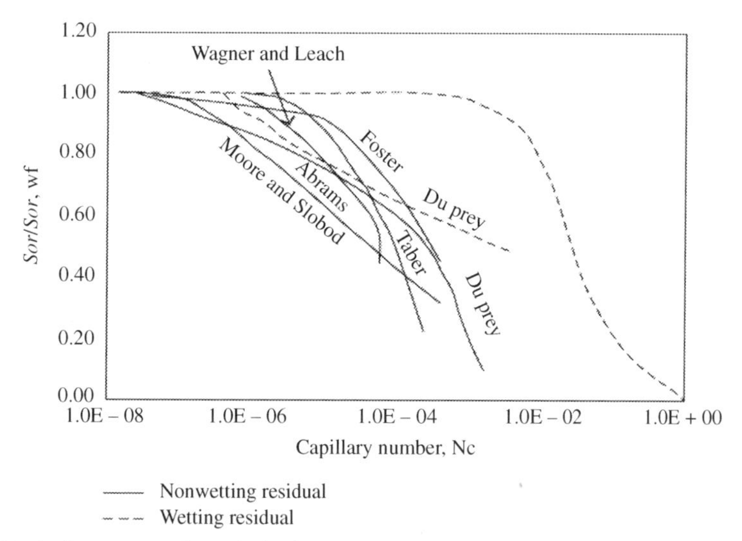

Figure 6.2 Typical variation of residual oil saturation versus capillary number [17,18].

Figure 6.3 Structure of a surfactant molecule [21].

evaluate the performance of the surfactant in such conditions. Their experiments were core displacement as well as interfacial measurements. They concluded that this surfactant could be a good option for further investigations as an EOR agent in field scale [20].

6.2.1.1 Type of Surfactant

Surface active agents (surfactants) are amphiphilic materials with a characteristic chemical structure consists of one molecular component that will have little attraction for the surrounding phase, normally called lyophobic group, and a chemical component that have a strong attraction for the surrounding phase, the lyophilic group [21]. In the standard surfactant terminology, the soluble component lyophilic is called "head" group and the lyophobic group called "tail." A schematic of a surfactant molecule structure is illustrated in Fig. 6.3 [22,23].

The simplest classification of surfactants is determined by the nature of the hydrophilic group, and the subgroups are based on the nature of the hydrophobic groups. Myers in (1999) [21] classified surfactants into four general groups nonionic, ionic, cationic, and zwitterionic.

6.2.1.1.1 Nonionic Surfactant

In the case of nonionic surfactants, the hydrophile has no charge but it is water soluble due to the presence of highly polar groups, such as polyoxyethylene $(-CH_2CH_2O-)_n-H$, where n is the number of ethylene oxide units, or polyols. Nonionic surfactants include alcohol ethoxylates, alkylphenol ethoxylates, and polysorbates [22–24]. Many of these surfactants are based on polyoxyethylene and a typical example is, dodecyl hexaoxyethylene glycol monoether $[C_{12}H_{25}(OCH_2CH_2)_6OH]$, often abbreviated to $C_{12}E_6$ to denote hydrocarbon chain length and ethylene oxide chain length. It should also be noted that with this type of compound the head-group is larger than the hydrocarbon chain [24]. Nonionic surface active agents with small head-groups also exist, some examples being, dodecyl sulphinyl ethanol $(C_{12}H_{25}SOCH_2CH_2OH)$ and decyl dimethyl amine oxide [22–24].

6.2.1.1.2 Ionic Surfactant

The hydrophilic group acquires a negative charge (anion) when it is dissolved in water. Anionic hydrophilic groups include sulfates $(ROSO^{3-})$, sulfonates (RSO^{3-}), carboxylates $(RCOO^-)$, and phosphates (RPO_4^-). Soaps are sodium or potassium fatty acid carboxylates, which have been implemented by man for more than 2000 years [25]. One of the common anionic surfactants is sodium dodecyl sulfate $(C_{12}H_{25}SO_4^-Na^+)$, a synthetic material. The name is often abbreviated to sodium dodecyl sulfate (SDS) [24]. Sodium dodecanoate $(C_{11}H_{23}COO^-Na^+)$, a material that can be synthesized is another one. There is also a large group of materials of this type formed from natural sources, e.g., the hydrolysis of triglycerides [22–24].

6.2.1.1.3 Cationic Surfactant

In this category, the charge of the hydrophilic group is positive (cation). Although cationic surfactants are more expensive than anionic ones, their economic importance has increased greatly in recent years. Typical cationic surfactants studied in petroleum industry are quaternary ammonium salts with one long alkyl chain (carbon number from 8 to 22) and methyl or hydroxyethyl groups in the head group positions [25]. Some usual examples of these quaternary ammonium salts (R_4N^+) are dodecyl trimethyl ammonium bromide $(C_{12}H_{25}N^+Me_3Br^-)$, the name is often abbreviated to $C_{12}TAB$ or DTAB and hexadecyl trimethyl ammonium bromide $(C_{16}H_{25}N^+Me_3Br^-)$ abbreviated to $C_{16}TAB$ or HTAB [22–24].

6.2.1.1.4 Zwitterionic Surfactant

The hydrophile's charge can be either positive or negative depending on the pH of the solution, or it can have both charges simultaneously (these are also known as amphoteric surfactants). Compounds of this type include imidazoline derivatives, amino acid derivatives, and lecithins. Zwitterionic surfactants represent 1% or less of the worldwide surfactant production [25]. A simple example of this type of material is 3-dimethyldodecylamine propanesulphonate; within this group, there are a number of important naturally occurring materials, known as triglycerides, a good example being lecithin, which occurs in the membranes of many animal cells [22–24].

6.2.1.2 Concerns Associated With Surfactant Flooding

One of the main concerns associated with surfactant flooding is surfactant adsorption from injected fluid onto the reservoir rock. This means that the concentration of surfactant decreases during the injection process, and this issue results in reduction of performance of surfactant in oil recovery. Because the ability of surfactant in IFT reduction or wettability alteration of the rock surface highly depends on the surfactant content of the injected fluid.

Ahmadi and Shadizadeh [26] examined adsorption of a plant-based surfactant on to real carbonate reservoir sample. In their experiments, increasing surfactant content of the solution until a certain point resulted in increasing the amount of adsorption.

Ahmadi and Shadizadeh [27] conducted different core-displacement tests as well as static adsorption experiments at different temperatures to determine the adsorption phenomenon in both static and dynamic conditions for a specific natural surfactant. Also, they employed different adsorption kinetic models besides adsorption isothermal methods to determine the adsorption behavior from kinetic viewpoint. They figured out increasing the temperature could reduce the amount of adsorbed surfactant on to the rock surface. Also, they pointed out that in the case of dynamic adsorption, it is much lower than static ones. They proposed that the main adsorption mechanism was electrostatic attraction between positive charge of a shale sandstone rock and negative charge of hydroxyl head of surfactant as shown in Fig. 6.4 [27].

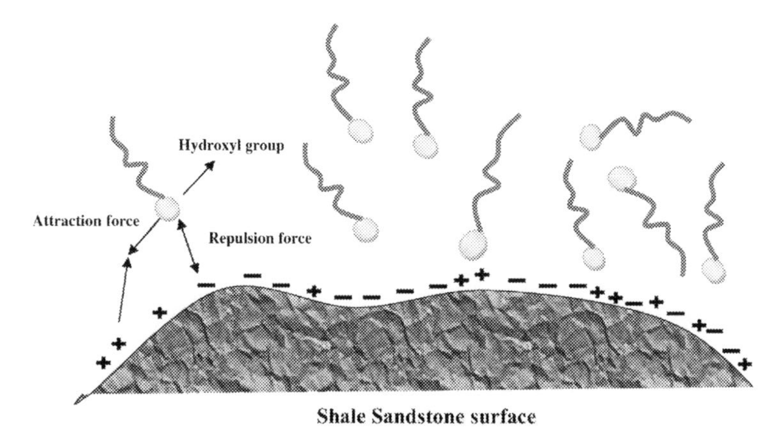

Figure 6.4 Adsorption process on to a shale sandstone rock [27].

Another issue that might occur during surfactant flooding is surfactant partitioning from injected fluid into the reservoir oil.

The last but not the least concern regarding surfactant flooding is surfactant stability in terms of thermal stability and salinity stability. In the case of surfactant flooding in high temperature or oil reservoirs with high value of salinity, surfactant instability could damage the reservoir permanently and reduce the oil production drastically. As a result, before any field application, different experiments should be done to figure out the stability of the surfactant versus temperature and salinity.

6.2.2 Alkaline Flooding

When an alkaline solution is injected in a reservoir, it reacts with the acid component of the crude oil and a surfactant (called soap to differentiate it from injected synthetic surfactants) is generated in situ. Therefore, most of surfactant-related mechanisms such as reduction of IFT apply to alkaline flooding. Injected alkalis can react with divalents so that insoluble precipitates are generated. The precipitates reduce permeability; thus sweep efficiency is improved [28]. Other mechanisms include emulsification, oil entrainment, bubble entrapment, and wettability reversal [28–30].

Mayer et al. summarized the research and laboratory results on the alkaline reactions with oil, water, and rocks in terms of fundamental theories and mechanisms. Here new or important findings are summarized [31]. One important alkali–oil reaction is to generate soap. It can be understood that if a high enough concentration or enough amount of alkali is injected, all the acid components in the crude oil will be converted to soap, as assumed by Delshad et al. and Karpan et al. in their numerical simulation models [32,33]. However, Sheng did a simulation study and found that only 25% of acids are converted into soap at a practical injection concentration of 2 wt.%. In other words, under practical conditions, not all acids can be converted into soap [4]. This result was confirmed by Wang and Gu's experimental data [34]. Sheng's results also showed that the amount of generated soap was small. From a particular model, the soap concentration was only 0.1%. To be able to generate soap, pH must be at least higher than 9.5 [4]. Multivalent cations (such as Ca^{2+} and Mg^{2+}) cause soap to form greasy, water-insoluble soap curds that are ineffective in recovering oil. However, the precipitates generated from alkali–water reaction may divert water flow into less-permeability zones, thus improving sweep efficiency [28]. If this mechanism is designed to be the main objective of alkaline injection, a high divalent content is beneficial. For this purpose, sodium silicate is better than sodium carbonate or sodium hydroxide, because the reaction between sodium silicate and calcium or magnesium will generate highest amount of insoluble precipitants and lead to the highest permeability reduction. Alkalirock reactions are complex depending on minerals and alkalis [4,29,35].

The acid number or total acid number is the mass of potassium hydroxide (KOH) in milligrams that is required to neutralize 1 g of crude oil. It is intuitive that if a crude oil has a higher acid number, more soap will be generated. Then more oil can be recovered. Therefore, for alkaline flooding to be effective, some minimum acid numbers may exist. Cooke et al. reported that a number of crude oils and synthetic oils had been flooded by alkaline water in the laboratory [36]. No oil with an acid number less than about 1.5 had been successfully flooded to a residual oil saturation much below that for a normal water flood. Therefore, Cooke et al. proposed a minimum acid number of 1.5 [36]. Sheng proposed a minimum acid number of 0.3 mg KOH/g oil [37]. However, Ehrlich and Wygal reported that crude oils with acid numbers higher than 0.1–0.2 mg KOH/g oil or IFT at 0.1% NaOH less than 0.5 mN/m gave significant caustic-flood oil [38]. There was no further correlation of increased oil at higher acid numbers or lower IFTs. Sheng calculated the reduction in residual oil saturation by alkaline flood at different acid numbers from the data presented by Ehrlich and Wygal [4,38]. The alkali used was 0.1% NaOH. The calculated data show that these two variables were not correlated. The current conclusions are that there is no minimum acid number for a successful alkaline flooding, and there is no relationship between the oil recovery and acid number [29].

No simple relationship has been observed between the amount of oil recovered and the measured value of IFT [36,38]. Cooke et al. pointed out that low IFT is a necessary but not sufficient condition for a successful alkaline flood [36]. However, the alkaline flooding results from Castor et al. showed that the recovery efficiencies could be better correlated with the stability of emulsions and wettability alteration rather than with the IFT [39]. Interestingly, Li's data showed that even when the acid number was zero, the IFT decreased with alkaline concentration [40]. The currently accepted statement is that oil recovery is not correlated to IFT for alkaline flooding, because other factors like wettability alteration may play the role [29].

The average values of the parameters that are important to an alkaline process from the real projects are also listed. Each parameter is analyzed using the method of rank and percentile, and the average value (median) is taken at 50% percentile. The most important parameters are formation of water divalent contents, clay contents, oil acid number, and oil viscosity [29].

6.2.3 Polymer Flooding

The polymer flooding process is defined as an addition of polymer with injected water to increase the viscosity of the injected water and reduce the mobility ratio between oil and injected water. It is well known that when the viscosity of polymer solution is increased, the sweep efficiency is improved by reducing viscous fingering. According to the fractional flow equation [45],

$$f_w = \frac{1}{1 + (k_{ro}/k_{rw})(\mu_w/\mu_o)} = \frac{1}{1 + (\lambda_o/\lambda_w)} \tag{6.2}$$

where, μ_w is water viscosity, μ_o denotes the oil viscosity, k_{rw} represents the water relative permeability, k_{ro} represents the oil relative permeability, λ_o is oil mobility and λ_w is water mobility [46].

A secondary mechanism is related to polymer viscoelastic behavior. Because of polymer viscoelastic properties, there is normal stress between oil and polymer solution. Thus, polymer exerts a larger pull force on oil droplets or oil films. Oil is "pushed and pulled" out of dead-end pores. Thus, residual oil saturation is decreased [45,47,48].

Presence of salt in a polymer aqueous solution results in decreasing of the solution viscosity. Flory-Huggins model [49] can be used to evaluate the behavior of polymers in the presence of salt in an aqueous solution [50].

Polymer solution viscosity may decrease as the pH is increased owing to the increased salt effect of an alkali [51,52]. However, Mungan [53] reported that the HAPM viscosity at 50 seconds^{-1} shear rate significantly decreased when lowering pH. And Szabo [54] reported the increase in the viscosity of AM/2-acrylamido-2-methyl propane sulfonate copolymer solution when NaOH was added. Those observations are probably related to the early-time hydrolysis effect. The effect of pH is complex considering different hydrolysis effect and salt effect [45].

One economic impact of polymer flooding that has been less discussed is the reduced amount of water injected and produced, compared with water flooding [4]. Because polymer improves the sweep efficiency, less water is produced and injected. In some situations, like an offshore environment and a desert area, water and the treatment of water could be costly [45].

6.2.4 Alkaline–Surfactant–Polymer Flooding

An important mechanism of alkaline–surfactant–polymer (ASP) flooding is the synergy between in situ generated soap and an injected surfactant. Generally, the optimum salinity for the soap is unrealistically low. To satisfy the low optimum salinity, the injected alkaline concentration must be so low that the injected alkali is lower than the amount of consumption; thus, the alkali cannot propagate forward. To solve this problem, a synthetic surfactant is added because the optimum salinity for a surfactant is high. When the soap and the surfactant are mixed, the optimum salinity range in which IFT reaches its low values is increased and is widened [55].

Table 6.1 Screening Criteria for ASP Flooding [45]

Reference	K (mD)	T (°C)	Lithology	Oil Viscosity (cP)	Water Saturation (Fraction)	Aquifer	Gas Cap	API	Depth (ft)
Lake et al. [56]				< 200					
Taber et al. [41,42]	> 10	< 93.3	Sandstone	< 35	> 0.35			> 20	< 9000
Al-Bahar et al. [43]	> 50	< 70	Sandstone	< 150	0.35	No	No		
Dickson et al. [44]	> 100	< 93.3		< 35	> 0.45				500–9000

API, American Petroleum Institute. *ASP*, alkaline–surfactant–polymer.

Screening criteria for EOR processes were discussed by several researchers, e.g., Lake et al. [56], Taber et al. [41,42], Al-Bahar et al. [43], Dickson et al. [44], and Al Adasani and Bai [57]. Some of the screening criteria for ASP are summarized in Table 6.1 [45].

6.2.4.1 Concerns Associated With Surfactant–Polymer Flooding

This section discusses issues resulting from ASP applications, including produced emulsion, chromatographic separation, precipitation and scaling, and others [45,58].

Emulsification is an important mechanism in alkaline flooding [30,45]. In other words, emulsion in ASP flooding could improve oil recovery. Cheng et al. [59] reported that emulsification increased the oil recovery factor by about 5% in their core floods; however, the main drawback of such a method is difficulty in oil separation from emulsion as well as increasing in injection pressure. oil/water separation.

To overcome the emulsion obstacles in ASP projects, using demulsifier is highly recommended. The ability of demulsifiers is related to how well it is absorbed at the oil/water interface, how it spreads toward the interface to form a film, and how much it can affect the interfacial intensity [45,60]. Wylde et al. [61] tested other demulsifiers. They found a mixtured demulsifier of Diep oxide, amine polyester, amine block-polymer, and a noni acid catalyzed resin worked best for an ASP project in a heavy oil reservoir [45].

Fig. 6.5 depicts a typical graph for the relative concentration of surfactant, alkaline, and polymer in an ASP injection scenario. This graph just shows the ratio of concentration at output to the concentration at injection of each component for one slug. As shown in Fig. 6.5, surfactant broke through later than alkaline and polymer. Also, the relative concentration of polymer and alkaline is higher than one for surfactant. In general, actual effluent concentrations and breakthrough times depend on their individual balance between the injection concentration and the retention or consumption [45].

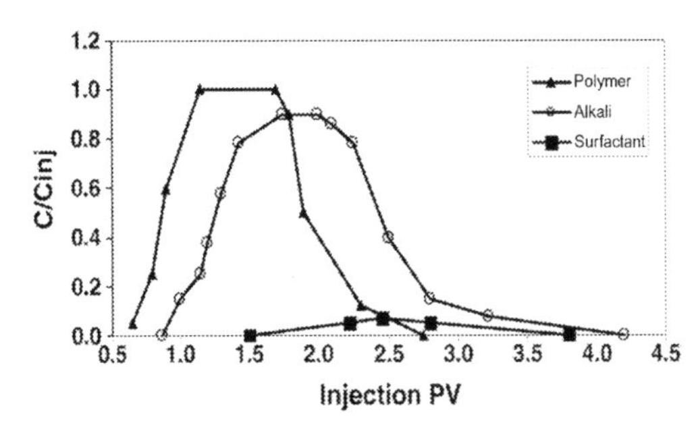

Figure 6.5 Effluent concentration histories of polymer, alkali, and surfactant [45].

Although ASP outperformed any other combinations of alkaline, surfactant, and polymer flooding, the problems with produced emulsions, scaling, maturation at the bottom of injection tank, pump vibration due to oscillation of injection velocity, and corrosion have led the industry to seek alkaline-free options like surfactant polymer (SP) process [45,62–65].

Other important problems include produced water cleanup and polymer degradation. Polymer degradation could be due to mechanical shear through pumps, perforation, pore throats, and so on. Pang et al. [66] found that the viscosity losses due to mechanic shearing at the high-pressure metering pump, transportation pump, and filter were about 5%, 2%, and 1%, respectively [45,67].

To prevent the polymer degradation by oxygen, the polymer makeup water, and dissolution equipment need to be kept under a nitrogen blanket [64,67–69]. Luo et al. [70] reported that the combination of thiourea and cobalt salt could prevent oxidation reduction more effectively than using the individual alone. A typical preservation package includes isopropyl alcohol and thiourea, which mitigates degradation of polymer due to oxygen, ions such as iron, and H_2S [71].

Wang et al. [72] combined biodegradation and filtration for removal of oil and suspended solids in polymer-containing produced water. Zhang et al. [73] treated produced water by a combined method of hydrolysis acidification-dynamic membrane bioreactor-coagulation process. Jiang et al. [74] presented a design of three cubed curve hydrocyclone tube, and Liu et al. [75] proposed to use a double-cone air-sparged hydrocyclone to treat produced water from polymer flooding [45].

Wu et al. [76] synthetized a new demulsifier that was a mixture of nonionic and reverse demulsifier; they called the output as SP1002. They carried out several experiments to evaluate the performance of that material in flocculation and coalescence of the droplets of oil in crude oil emulsion.

Some companies developed a new type of technology to remove the dispersed oil from emulsions; they used magnetic filtration and high-rate magnetic ballasted clarification combined with chemical flocculant and coagulant [77–78].

6.2.5 Application of Nanoparticles in Enhanced Oil Recovery Schemes

Ahmadi and Shadizadeh [79] studied first the effect of nanoparticles on the adsorption of surfactant on to reservoir rock, particularly carbonates. They pointed out that adding nanoparticles could reduce the surfactant loss due to adsorption on to reservoir rock; they concluded that the performance of hydrophobic nanosilica was higher than hydrophobic ones. The main mechanism was hydrophobic between hydrophobic groups in nanosilica and surfactant [79]. Ahmadi and Shadizadeh [80] examined the effect of nanoparticles on the adsorption behavior of a surfactant on to reservoir rock. They employed hydrophilic and hydrophobic nanosilica in their adsorption experiments. They concluded that hydrophilic nanosilica could significantly reduce the surfactant adsorption on to sandstones; the main mechanism behind this reduction was hydrogen bond between hydroxyl groups in nanosilica and head of surfactant as depicted in Fig. 6.6.

Ehtesabi et al. [81] conducted core-displacement experiments, contact angle measurements, scanning electron microscopy, and energy-dispersive spectrometry on titanium oxide nanoparticles to figure out its performance on oil recovery. According

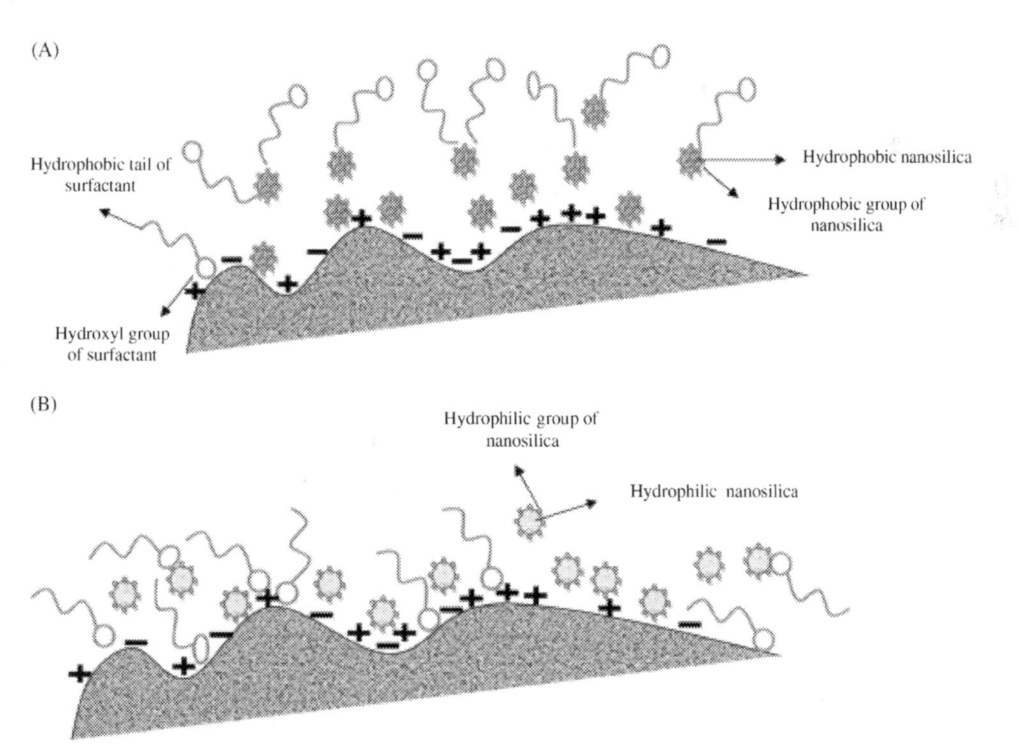

Figure 6.6 Adsorption mechanism of a natural surfactant onto a sandstone rock in presence of (A) hydrophobic nanosilica and (B) hydrophilic nanosilica [80].

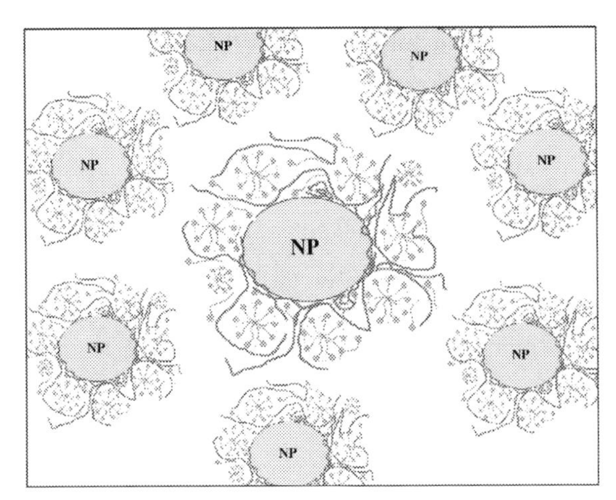

Figure 6.7 Effect of nanoparticles and polymer in foam stabilization [87].

to their results, TiO_2 nanoparticles might be an EOR agent; however, a concern regarding nanoparticle deposition is associated with [81]. Hendraningrat et al. [82–85] examined experimentally the application of nanofluid flooding as a branch of chemical flooding for EOR purposes. They figured out lipophilic–hydrophilic nanoparticles could reduce the IFT between water and oil phases ; however, in the case of illite core sample, using such nanoparticles could damage core sample in terms of reduction in both porosity and permeability [82–85]. Ahmadi and Shadizadeh [86] evaluated the ultimate oil recovery factor of nanofluid flooding in carbonate oil reservoirs using core-displacement experiments. They concluded that adding hydrophilic nanosilica could considerably increase viscosity of water; increasing viscosity of water means that sweep efficiency of water is improved. They pointed out the efficiency of nanofluid increases with nanoparticle content until a certain point and higher concentrations could not significantly increase the ultimate recovery factor [86].

Kumar and Mandal [87] experimentally studied the performance of carbon dioxide foam in terms of stability in presence of different nanoparticles as well as different type of surfactants. They employed nonionic, cationic, and ionic surfactants to figure out the impact of ionic strength on CO_2 foam stability. Experimental results revealed that adding nanoparticles could significantly improve the CO_2 foam stability. Also, they concluded that adding alcohol and polymer resulted in higher stability in comparison with conventional CO_2 foam system [87]. They presented a possible phenomenon, as shown in Fig. 6.7, to justify their conclusion.

Kumar et al. [88] investigated experimentally the impact of nanoparticles in performance of emulsions in terms of thermal stability and viscosity stability. Moreover, they carried out core-displacement experiments to find the performance of emulsion flooding. They concluded that nanoparticles improve the performance of emulsions in terms of both thermal and viscoelastic stability. Also, Pickering emulsion injection resulted in

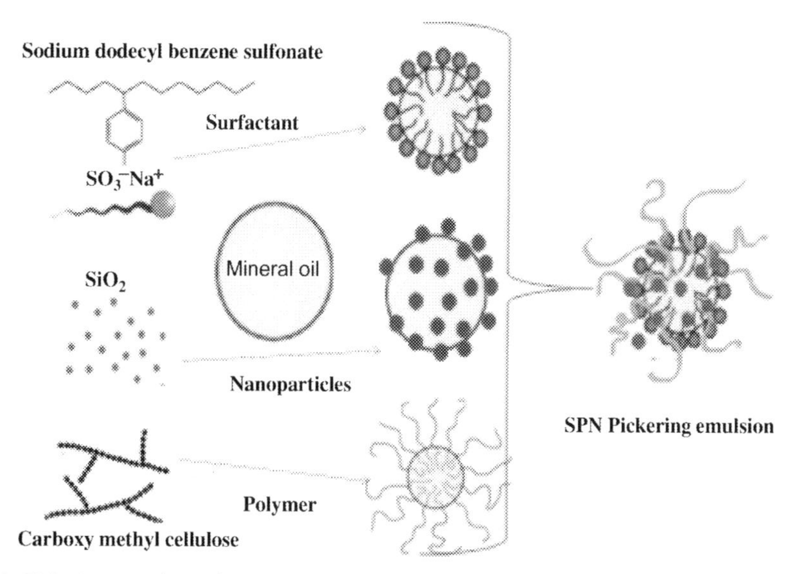

Figure 6.8 Pickering emulsion formation in presence of nanoparticles and polymer [88].

considerably higher oil recovery compared to conventional water injection [88]. Fig. 6.8 depicts the graphical demonstration of Pickering emulsion generated by surfactant, nanoparticles, and polymer. Afzali Tabar et al. [89] proposed a new method for preparing Pickering emulsions using graphene/silica nanoparticles. Their experimental results showed that graphene nanoparticles could improve the performance of Pickering emulsion in terms of rheological behavior in comparison with other conventional nanoparticles. However, they did not use real oil sample in conduction of experiments.

Another application of nanoparticles for EOR purposes is improving the performance of surfactant in wettability alteration of oil reservoirs, particularly carbonates, from oil-wet toward water-wet or neutral-wet conditions. Nwidee et al. [90] studied the effect of nanoparticle and surfactant on the rock wettability using contact angle measurement, imbibition test as well as morphological experiments. They concluded that combination of nanoparticle and surfactant could be able to change reservoir rock wettability; take this advantage for improving the oil recovery factor or employing in decontamination of soils [90].

Ahmadi [91] conducted batch adsorption tests along with core-flooding experiments using an ionic surfactant (SDS) and nanosilica. He employed real sandstone core samples in his experiments. Based on the experimental results he concluded that adding nanosilica could improve considerably the oil recovery factor of SDS; he maintain this incremental oil is due to reduction of surfactant adsorption [91]. Ahmadi and Sheng [92] studied experimentally the flooding efficiency of nanosurfactant in carbonate core samples. They used nanosilica and SDS in their tests. They concluded that adding nanosilica increases the efficiency of SDS in oil recovery due to reduction of

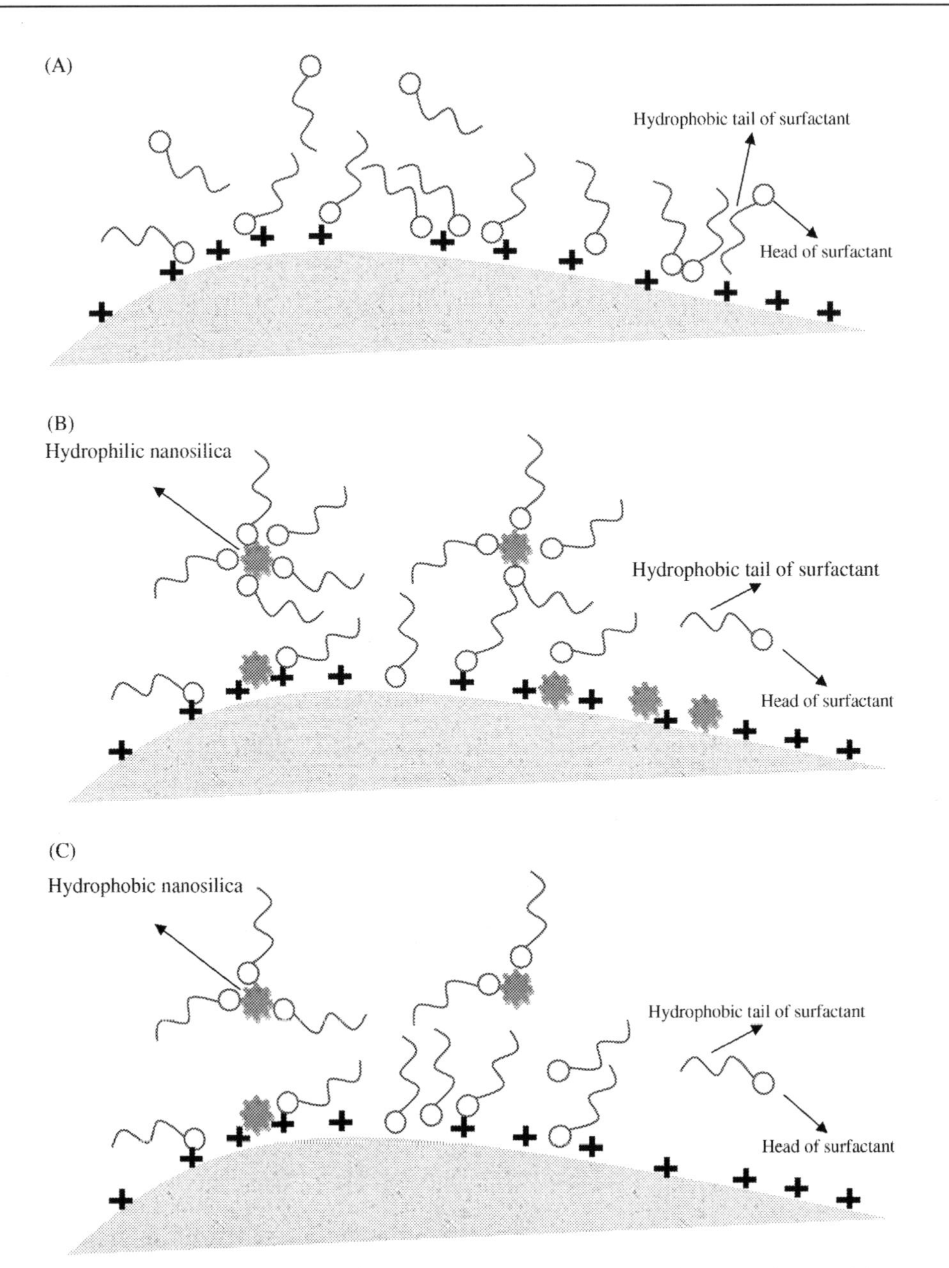

Figure 6.9 Schematic of adsorption of SDS on to carbonates: (A) standalone surfactant, (B) in presence of hydrophilic nanosilica, and (C) in presence of hydrophobic nanosilica [92]. *SDS*, sodium dodecyl sulfate.

SDS adsorption on to carbonate surface. They proposed phenomena behind this adsorption process as illustrated in Fig. 6.9. Moreover, Fig. 6.10 shows a graphical illustration about aggregation of SDS around nanosilica particles, especially hydrophilic ones. Aggregation of SDS around nanosilica resulted in reduction critical

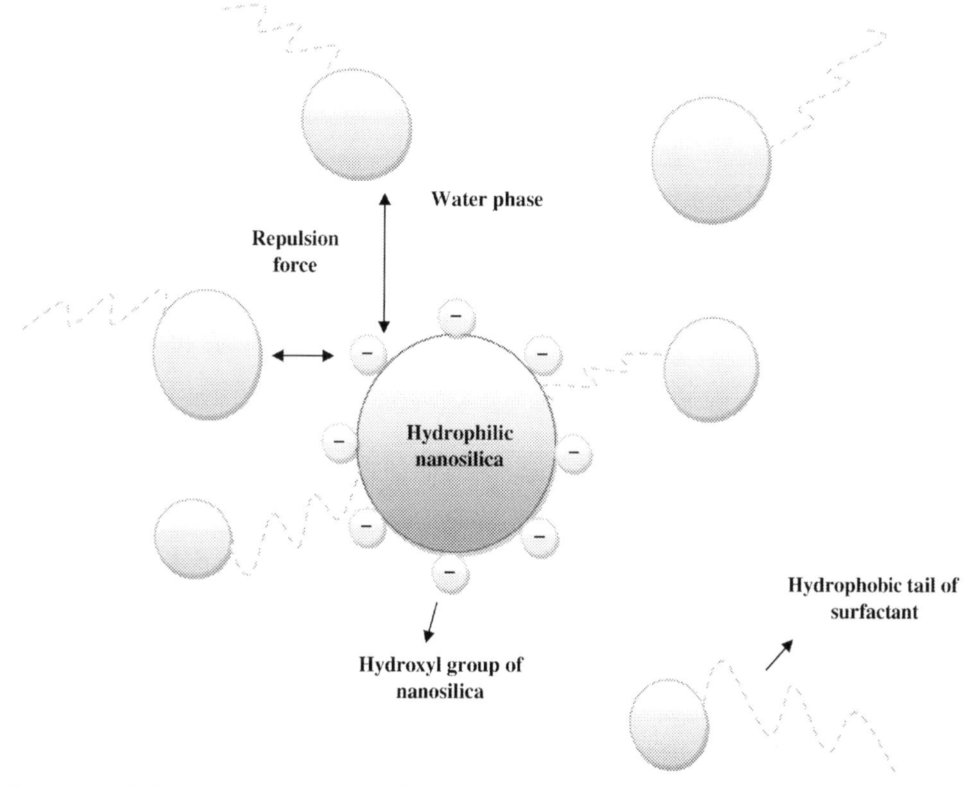

Figure 6.10 Schematic of aggregation of SDS around nanosilica particles [92]. *SDS*, sodium dodecyl sulfate.

micelle concentration of SDS in aqueous solution. Ahmadi and Shadizadeh [93] carried out mechanistic study on displacement efficiency of nanosurfactant system in oil recovery using nanosilica and a natural surfactant derived from leaves of *Z. spina-christi*. According to the experimental results, they concluded that adding nanosilica could not significantly increase the oil recovery compared to surfactant flooding alone.

REFERENCES

[1] S. Farouq-Ali, C. Stahl, Increased oil recovery by improved waterflooding, Earth Miner. Sci. (United States) 39 (1970).
[2] Lake, L.W., 1989. Enhanced Oil Recovery.
[3] Lake, L.W., Johns, R.T., Rossen, W.R., Pope, G.A., 2014. Fundamentals of Enhanced Oil Recovery.
[4] J. Sheng, Modern Chemical Enhanced Oil Recovery: Theory and Practice, Gulf Professional Publishing, New York, United States, 2010.
[5] D.W. Green, G.P. Willhite, Enhanced Oil Recovery, Henry L. Doherty Memorial Fund of AIME, Society of Petroleum Engineers, Richardson, TX, 1998.

[6] S. Thomas, Enhanced oil recovery—an overview, Oil Gas Sci Technol—Revue de l'IFP 63 (2008) 9–19.
[7] R. Seright, J. Liang, A survey of field applications of gel treatments for water shutoff, SPE Latin America/Caribbean Petroleum Engineering Conference, Society of Petroleum Engineers, 1994.
[8] R. Sydansk, G. Southwell, More than 12 years of experience with a successful conformance-control polymer gel technology, SPE/AAPG Western Regional Meeting, Society of Petroleum Engineers, 2000.
[9] M. Bavière, Basic Concepts in Enhanced Oil Recovery Processes, Springer, Netherlands, 1991.
[10] T. Austad, B. Matre, J. Milter, A. Saevareid, L. Øyno, Chemical flooding of oil reservoirs 8. Spontaneous oil expulsion from oil- and water-wet low permeable chalk material by imbibition of aqueous surfactant solutions, Colloids Surf., A 137 (1998) 117–129.
[11] H. Chen, L. Lucas, L. Nogaret, H. Yang, D. Kenyon, Laboratory monitoring of surfactant imbibition using computerized tomography, SPE International Petroleum Conference and Exhibition in Mexico, Society of Petroleum Engineers, 2000.
[12] O. Wagner, R. Leach, Improving oil displacement efficiency by wettability adjustment, Trans. AIME 216 (1959) 65–72.
[13] T. Austad, J. Milter, Spontaneous imbibition of water into low permeable chalk at different wettabilities using surfactants, International Symposium on Oilfield Chemistry, Society of Petroleum Engineers, 1997.
[14] E. Spinler, D. Zornes, D. Tobola, A. Moradi-Araghi, Enhancement of oil recovery using a low concentration of surfactant to improve spontaneous and forced imbibition in chalk, SPE/DOE Improved Oil Recovery Symposium, Society of Petroleum Engineers, 2000.
[15] L. Treiber, W. Owens, A laboratory evaluation of the wettability of fifty oil-producing reservoirs, Soc. Pet. Eng. J. 12 (1972) 531–540.
[16] G.V. Chilingar, T. Yen, Some notes on wettability and relative permeabilities of carbonate reservoir rocks, II, Energy Sources 7 (1983) 67–75.
[17] H. Guo, M. Dou, W. Hanqing, F. Wang, G. Yuanyuan, Z. Yu, et al., Proper use of capillary number in chemical flooding, J. Chem. 2017 (2017) 1–11.
[18] R.A. Fulcher Jr, T. Ertekin, C. Stahl, Effect of capillary number and its constituents on two-phase relative permeability curves, J. Pet. Technol. 37 (1985) 249–260.
[19] M.A. Ahmadi, S.R. Shadizadeh, Implementation of a high-performance surfactant for enhanced oil recovery from carbonate reservoirs, J. Pet. Sci. Eng. 110 (2013) 66–73.
[20] M.A. Ahmadi, Y. Arabsahebi, S.R. Shadizadeh, S.S. Behbahani, Preliminary evaluation of mulberry leaf-derived surfactant on interfacial tension in an oil-aqueous system: EOR application, Fuel 117 (2014) 749–755.
[21] D. Myers, Surfaces, Interfaces and Colloids, Wiley-VCH, New York, NY etc, 1990.
[22] O. Guertechin, Surfactant: classification, handbook of detergents. Part A—Properties, Surfactant Sci. Ser. 82 (1999) 40.
[23] D. Myers, Surfactant Science and Technology, John Wiley & Sons, 2005.
[24] O.A. de Swaan, Analytic solutions for determining naturally fractured reservoir properties by well testing, Soc. Pet. Eng. J. 16 (1976) 117–122.
[25] M.J. Rosen, J.T. Kunjappu, Surfactants and Interfacial Phenomena, John Wiley & Sons, Hoboken, United States, 2012.
[26] M.A. Ahmadi, S.R. Shadizadeh, Experimental investigation of adsorption of a new nonionic surfactant on carbonate minerals, Fuel 104 (2013) 462–467.
[27] M.A. Ahmadi, S.R. Shadizadeh, Experimental investigation of a natural surfactant adsorption on shale–sandstone reservoir rocks: Static and dynamic conditions, Fuel 159 (2015) 15–26.
[28] Sarem, A., 1974. Secondary and tertiary recovery of oil by MCCF (mobility-controlled caustic flooding) process. In: Paper SPE, 4901.
[29] J.J. Sheng, Status of alkaline flooding technology, J. Pet. Eng. Technol. 5 (2015) 44–50.
[30] C. Johnson Jr, Status of caustic and emulsion methods, J. Pet. Technol. 28 (1976) 85–92.
[31] E. Mayer, R. Berg, J. Carmichael, R. Weinbrandt, Alkaline injection for enhanced oil recovery—a status report, J. Pet. Technol. 35 (1983) 209–221.

[32] M. Delshad, C. Han, F.K. Veedu, G.A. Pope, A simplified model for simulations of alkaline–surfactant–polymer floods, J. Pet. Sci. Eng. 108 (2013) 1–9.
[33] Karpan, V., Farajzadeh, R., Zarubinska, M., Stoll, M., Dijk, H., Matsuura, T., 2011. Selecting the "right" ASP model by history matching core flood experiments. In: IOR 2011-16th European Symposium on Improved Oil Recovery.
[34] W. Wang, Y. Gu, Experimental studies of detection and reuse of the produced chemicals in alkaline–surfactant–polymer floods, SPE Reservoir Eval. Eng. 8 (2005) 362–371.
[35] K. Cheng, Chemical consumption during alkaline flooding: a comparative evaluation, SPE Enhanced Oil Recovery Symposium, Society of Petroleum Engineers, 1986.
[36] C. Cooke Jr, R. Williams, P. Kolodzie, Oil recovery by alkaline waterflooding, J. Pet. Technol. 26 (1974) 1,365–361,374.
[37] J. Sheng, Enhanced Oil Recovery Field Case Studies, Gulf Professional Publishing, New York, United States, 2013.
[38] R. Ehrlich, R.J. Wygal Jr, Interrelation of crude oil and rock properties with the recovery of oil by caustic waterflooding, Soc. Pet. Eng. J. 17 (1977) 263–270.
[39] T. Castor, W. Somerton, J. Kelly, Recovery mechanisms of alkaline flooding, Surface Phenomena in Enhanced Oil Recovery, Springer, Netherlands, 1981, pp. 249–291.
[40] H. Li, Advances in Alkaline–Surfactant–Polymer Flooding and Pilot Tests, Science Press, China, 2007.
[41] J.J. Taber, F. Martin, R. Seright, EOR screening criteria revisited—Part 1: Introduction to screening criteria and enhanced recovery field projects, SPE Reservoir Eng. 12 (1997) 189–198.
[42] J. Taber, F. Martin, R. Seright, EOR screening criteria revisited—Part 2: Applications and impact of oil prices, SPE Reservoir Eng. 12 (1997) 199–206.
[43] M.A. Al-Bahar, R. Merrill, W. Peake, M. Jumaa, R. Oskui, Evaluation of IOR potential within Kuwait, Abu Dhabi International Conference and Exhibition, Society of Petroleum Engineers, 2004.
[44] J.L. Dickson, A. Leahy-Dios, P.L. Wylie, Development of improved hydrocarbon recovery screening methodologies, SPE Improved Oil Recovery Symposium, Society of Petroleum Engineers, 2010.
[45] J.J. Sheng, A comprehensive review of alkaline–surfactant–polymer (ASP) flooding, Asia-Pac. J. Chem. Eng. 9 (2014) 471–489.
[46] K.S. Sorbie, Polymer-Improved Oil Recovery, Blackie and Son Ltd, Glasgow and London, 1991.
[47] D. Wang, J. Cheng, Q. Yang, G. Wenchao, L. Qun, F. Chen, Viscous-elastic polymer can increase microscale displacement efficiency in cores, SPE Annual Technical Conference and Exhibition, Society of Petroleum Engineers, 2000.
[48] D. Wang, J. Cheng, H. Xia, Q. Li, J. Shi, Viscous-elastic fluids can mobilize oil remaining after water–flood by force parallel to the oil–water interface, SPE Asia Pacific Improved Oil Recovery Conference, Society of Petroleum Engineers, 2001.
[49] P.J. Flory, Principles of Polymer Chemistry, Cornell University Press, United States, 1953.
[50] R.B. Bird, W.E. Stewart, E.N. Lightfoot, Transport Phenomena, John Wiley & Sons, New York, NY, 1960, p. 413.
[51] D. Sheng, P. Yang, Y. Liu, Effect of alkali–polymer interaction on the solution properties, Pet. Explor. Dev. 21 (1994) 81–85.
[52] W. Kang, Study of Chemical Interactions and Drive Mechanisms in Daqing ASP Flooding, Petroleum Industry Press, Beijing, 2001.
[53] N. Mungan, Rheology and adsorption of aqueous polymer solutions, J. Can. Pet. Technol. 8 (1969) 45–50.
[54] M.T. Szabo, An evaluation of water-soluble polymers for secondary oil recovery—Parts 1 and 2, J. Pet. Technol. 31 (1979) 553–570.
[55] R. Nelson, J. Lawson, D. Thigpen, G. Stegemeier, Cosurfactant-enhanced alkaline flooding, SPE Enhanced Oil Recovery Symposium, Society of Petroleum Engineers, 1984.
[56] L.W. Lake, P.B. Venuto, A niche for enhanced oil recovery in the 1990s, Oil Gas J. 88 (1990) 62–67.

[57] A. Al Adasani, B. Bai, Analysis of EOR projects and updated screening criteria, J. Pet. Sci. Eng. 79 (2011) 10–24.
[58] Weatherill, A., 2009. Surface development aspects of alkali–surfactant–polymer (ASP) flooding. In: International Petroleum Technology Conference, International Petroleum Technology Conference.
[59] J. Cheng, G. Liao, Z. Yang, Q. Li, Y. Yao, D. Xu, Overview of Daqing ASP pilots, Pet. Geol. Oilfield Dev. Daqing 20 (2001) 46–49.
[60] W. Kang, D. Wang, Emulsification characteristic and de-emulsifiers action for alkaline/surfactant/polymer flooding, SPE Asia Pacific Improved Oil Recovery Conference, Society of Petroleum Engineers, 2001.
[61] J.J. Wylde, J.L. Slayer, V. Barbu, Polymeric and alkali–surfactant polymer enhanced oil recovery chemical treatment: chemistry and strategies required after breakthrough into the process, SPE International Symposium on Oilfield Chemistry, Society of Petroleum Engineers, 2013.
[62] H. Wang, G. Liao, J. Song, Combined chemical flooding technologies, In: Shen, P.P. (Editor), Technological Developments in Enhanced Oil Recovery, Petroleum Industry Press (2006) 126–188.
[63] D. Wang, H. Dong, C. Lv, X. Fu, J. Nie, Review of practical experience by polymer flooding at Daqing, SPE Reservoir Eval. Eng. 12 (2009) 470–476.
[64] J.J. Sheng, B. Leonhardt, N. Azri, Status of polymer-flooding technology, J. Can. Pet. Technol. 54 (2015) 116–126.
[65] W. Demin, J. Youlin, W. Yan, G. Xiaohong, W. Gang, Viscous-elastic polymer fluids rheology and its effect upon production equipment, SPE Prod. Facil. 19 (2004) 209–216.
[66] Pang, Z., Li, J., Liu, H., Li, Y., 1998. Polymer viscosity loss in injection and production processes. In: Q.-L. Gang, et al. (Eds.), Chemical Flooding Symposium–Research Results during the Eighth Five-Year Period (1991–1995). pp. 385–394.
[67] H. Liu, Y. Wang, Y. Liu, Mixing and injection techniques of polymer solution, in: Y.-Z. Liu, et al. (Eds.), Enhanced Oil Recovery–Polymer Flooding, Petroleum Industry Press, Beijing, 2006, pp. 157–181.
[68] D.C. Standnes, I. Skjevrak, Literature review of implemented polymer field projects, J. Pet. Sci. Eng. 122 (2014) 761–775.
[69] H.L. Chang, H. Hou, F. Wu, Y. Gao, Chemical EOR injection facilities—from pilot test to field-wide expansion, SPE Enhanced Oil Recovery Conference, Society of Petroleum Engineers, 2013.
[70] J. Luo, Y. Liu, P. Zhu, Polymer solution properties and displacement mechanisms, Enhanced Oil Recovery–Polymer Flooding, Petroleum Industry Press, Beijing, 2006, pp. 1–72.
[71] S.C. Ayirala, E. Uehara-Nagamine, A.N. Matzakos, R.W. Chin, P.H. Doe, P.J. van den Hoek, A designer water process for offshore low salinity and polymer flooding applications, SPE Improved Oil Recovery Symposium, Society of Petroleum Engineers, 2010.
[72] Z. Wang, B. Lin, G. Sha, Y. Zhang, J. Yu, L. Li, A combination of biodegradation and microfiltration for removal of oil and suspended solids from polymer-containing produced water, SPE Americas E&P Health, Safety, Security, and Environmental Conference, Society of Petroleum Engineers, 2011.
[73] Y. Zhang, B. Gao, L. Lu, Q. Yue, Q. Wang, Y. Jia, Treatment of produced water from polymer flooding in oil production by the combined method of hydrolysis acidification-dynamic membrane bioreactor–coagulation process, J. Pet. Sci. Eng. 74 (2010) 14–19.
[74] M. Jiang, F. Li, L. Zhao, Y. Zhang, The design of three cubed curve hydrocyclone tube, The Sixteenth International Offshore and Polar Engineering Conference, International Society of Offshore and Polar Engineers, 2006.
[75] S. Liu, X. Zhao, X. Dong, B. Miao, W. Du, Experimental research on treatment of produced water from a polymer-flooding process using a double-cone air-sparged hydrocyclone, SPE Projects Facil. Constr. 2 (2007) 1–5.
[76] Wu, D., Meng, X., Zhao, F., Lin, S., Jiang, N., Zhang, S., Qiao, L., Song, H., 2013. Dual function reverse demulsifier and demulsifier for the improvement of polymer flooding produced water treatment. In: IPTC 2013: International Petroleum Technology Conference.

[77] Raney, K.H., Ayirala, S.C., Chin, R.W., Verbeek, P., 2011. Surface and subsurface requirements for successful implementation of offshore chemical enhanced oil recovery. In: Offshore Technology Conference, Offshore Technology Conference.

[78] Barnes, J.R., Dirkzwager, H., Dubey, S.T., Reznik, C., 2012. A new approach to deliver highly concentrated surfactants for chemical enhanced oil recovery. SPE Annual Technical Conference and Exhibition. Society of Petroleum Engineers.

[79] M.A. Ahmadi, S.R. Shadizadeh, Adsorption of novel nonionic surfactant and particles mixture in carbonates: enhanced oil recovery implication, Energy Fuels 26 (2012) 4655–4663.

[80] M.A. Ahmadi, S.R. Shadizadeh, Induced effect of adding nano silica on adsorption of a natural surfactant onto sandstone rock: experimental and theoretical study, J. Pet. Sci. Eng. 112 (2013) 239–247.

[81] H. Ehtesabi, M.M. Ahadian, V. Taghikhani, M.H. Ghazanfari, Enhanced heavy oil recovery in sandstone cores using TiO_2 nanofluids, Energy Fuels 28 (2013) 423–430.

[82] L. Hendraningrat, S. Li, O. Torsæter, A coreflood investigation of nanofluid enhanced oil recovery, J. Pet. Sci. Eng. 111 (2013) 128–138.

[83] L. Hendraningrat, S. Li, O. Torsater, A coreflood investigation of nanofluid enhanced oil recovery in low-medium permeability Berea sandstone, SPE International Symposium on Oilfield Chemistry, Society of Petroleum Engineers, 2013.

[84] L. Hendraningrat, S. Li, O. Torsaeter, Enhancing oil recovery of low-permeability Berea sandstone through optimised nanofluids concentration, SPE Enhanced Oil Recovery Conference, Society of Petroleum Engineers, 2013.

[85] L. Hendraningrat, S. Li, O. Torsater, Effect of some parameters influencing enhanced oil recovery process using silica nanoparticles: an experimental investigation, SPE Reservoir Characterization and Simulation Conference and Exhibition, Society of Petroleum Engineers, 2013.

[86] M.-A. Ahmadi, S.R. Shadizadeh, Nanofluid in hydrophilic state for EOR implication through carbonate reservoir, J. Dispersion Sci. Technol. 35 (2014) 1537–1542.

[87] S. Kumar, A. Mandal, Investigation on stabilization of CO_2 foam by ionic and nonionic surfactants in presence of different additives for application in enhanced oil recovery, Appl. Surf. Sci. 420 (2017) 9–20.

[88] N. Kumar, T. Gaur, A. Mandal, Characterization of SPN Pickering emulsions for application in enhanced oil recovery, J. Ind. Eng. Chem. 504 (October 25) (2017) 304–315.

[89] M. Afzali Tabar, M. Alaei, M. Bazmi, R.R. Khojasteh, M. Koolivand-Salooki, F. Motiee, et al., Facile and economical preparation method of nanoporous graphene/silica nanohybrid and evaluation of its Pickering emulsion properties for chemical enhanced oil recovery (C-EOR), Fuel 206 (2017) 453–466.

[90] L.N. Nwidee, M. Lebedev, A. Barifcani, M. Sarmadivaleh, S. Iglauer, Wettability alteration of oil-wet limestone using surfactant–nanoparticle formulation, J. Colloid Interface Sci. 54 (October 15) (2017) 334–345.

[91] M.A. Ahmadi, Use of nanoparticles to improve the performance of sodium dodecyl sulfate flooding in a sandstone reservoir, Eur. Phys. J. Plus 131 (2016) 435.

[92] M.A. Ahmadi, J. Sheng, Performance improvement of ionic surfactant flooding in carbonate rock samples by use of nanoparticles, Pet. Sci. 13 (2016) 725–736.

[93] M.A. Ahmadi, S.R. Shadizadeh, Nano-surfactant flooding in carbonate reservoirs: a mechanistic study, Eur. Phys. J. Plus 132 (2017) 246.

CHAPTER SEVEN

Waterflooding

Mohammad Ali Ahmadi
Department of Chemical and Petroleum Engineering, University of Calgary, Calgary, AB, Canada

7.1 INTRODUCTION

Darcy's law governs reservoir flow, which is the fundamental flow equation for fluids through a porous media in the steady state. The fractional flow equation together with the derivation of solutions to predict the performance of displacing water system started from Darcy equation. Many researchers have ventured into the development of waterflooding designs to understand the performance of water injection. This has led to diverse quantitative approaches in analyzing waterflooding based on different geometries [1].

According to literature, the most important advances in understanding multiphase flow in porous media started back in the 1920s [2] and majorly in the 1940s [3]. Reservoirs have transition zones between water and oil phase; the true oil zone sees the presence of connate water with respect to the true water zone, which is essentially 100% water. During production from a completed well, water production is from the true water zone whereas only oil is produced from the true oil zone [7]. However, at the transition zone, both oil and water are produced which is dependent on the saturation of oil and water at that point. As time passes, the saturation becomes a multiple valued function of the distance coordinate (x), solved by material balance consideration. When the initial saturation in the flow system is uniform, a simple graphic approach developed by Welge [4] can be used to determine the abrupt saturation front without difficulty. Sheldon and Cardwell [5] solved the Buckley–Leverett problem with the method of characteristics.

7.2 DERIVATION OF CONTINUITY EQUATION FOR DISPLACEMENT FRONT OF LINEAR DISPLACEMENT SYSTEM

Assume that the total flow rate is the same at all the medium cross-section. Neglect capillary and gravitational forces may be acting. Let the oil be displaced by

Fundamentals of Enhanced Oil and Gas Recovery from Conventional and Unconventional Reservoirs.
DOI: https://doi.org/10.1016/B978-0-12-813027-8.00007-2

water from left to right. The rate the water enters to the medium element from left-hand side is

$$q_t \times f_w = \text{water flow rate entering the element}$$

The rate of water leaving element from the right-hand side is

$$q_t \times \left\langle f_w + \Delta f_w \right\rangle = \text{water flow rate leaving the element}$$

The change in water flow rate across the element is found by performing a mass balance. The movement of mass for an immiscible, incompressible system gives

$$\begin{aligned} \text{Change in water flow rate} &= \text{water entering} - \text{water leaving} \\ &= q_t \times f_w - q_t \left\langle f_w + \Delta f_w \right\rangle = -\, q_t \times \Delta f_w \end{aligned} \tag{7.1}$$

This is equal to the change in element water content per unit time (see Fig. 7.1). Let S_w be the water saturation of the element at time t. Then, if oil is being displaced from the element, at the time $(t + \Delta t)$, the water saturation will be $(S_w + \Delta S_w)$. Therefore, water accumulation in the element per unit time is

$$\text{water accummulation per unit time} = \frac{\Delta S_w \times A \times \phi \times \Delta x}{\Delta t} \tag{7.2}$$

where ϕ is porosity; equating Eqs. (7.1) and (7.2) results

$$\frac{\Delta S_w \times A \times \phi \times \Delta x}{\Delta t} = -\, q_t \times \Delta f_w \rightarrow \frac{\Delta S_w}{\Delta t} = \frac{-\, q_t \times \Delta f_w}{A \times \phi \times \Delta x} \tag{7.3}$$

In the limit as $\Delta t \rightarrow 0$ and $\Delta x \rightarrow 0$ (for the water phase):

$$\left(\frac{\Delta S_w}{\Delta t}\right)_x = \frac{-\, q_t}{A \times \phi} \left(\frac{\mathrm{d} f_w}{\mathrm{d}x}\right)_t \tag{7.4}$$

The subscript x on the derivative indicates that this derivative is different for each element. It is not possible to solve for the general distribution of water saturation $S_w(x, t)$ in most realistic cases because of the nonlinearity of the problem. For example, water fractional flow is usually a nonlinear function of water saturation. It is, therefore, necessary to consider a simplified approach to solving Eq. (7.4).

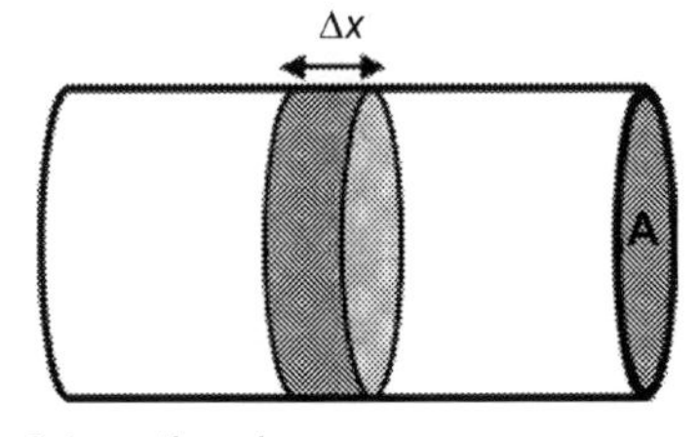

Figure 7.1 Horizontal bed containing oil and water.

For a given rock, the fraction of flow for water f_w is a function only of the water saturation S_w, as indicated by Eq. (7.4), assuming constant oil and water viscosities. The water saturation, however, is a function of both time and position, which may be express as $f_w = F(S_w)$ and $S_w = G(t, x)$. Then,

$$dS_w = \left(\frac{\partial S_w}{\partial t}\right)_x dt + \left(\frac{\partial S_w}{\partial x}\right)_t dx \tag{7.5}$$

$$\frac{dS_w}{dt} = \left(\frac{\partial S_w}{\partial t}\right)_x + \left(\frac{\partial S_w}{\partial x}\right)_t \frac{dx}{dt} \tag{7.6}$$

Now, there is interest in determining the rate of advance of a constant saturation plane or front $(\partial x/\partial t)_{S_w}$, where S_w is constant and $dS_w = 0$. So, from Eq. (7.5),

$$\frac{dx}{dt} = \frac{(\partial S_w/\partial t)_x}{(\partial S_w/\partial x)_t} \tag{7.7}$$

Substituting Eqs. (7.5) and (7.6) into Eq. (7.7) gives the Buckley–Leverett frontal advance equation:

$$\left(\frac{dx}{dt}\right)_{S_w} = \frac{-q_t}{A\phi}\left(\frac{df_w}{dS_w}\right)_{Sw} \tag{7.8}$$

The derivative $(df_w/dS_w)_{S_w}$ is the slope of the fractional flow curve and derivative $(dx/dt)_{S_w}$ is the velocity of the moving plane with water saturation S_w. Because the porosity, area, and flow rate are constant and because for any value of S_w, the derivative $(df_w/dS_w)_{S_w}$ is a constant, then the rate dx/dt is constant.

This means that the distance a plane of constant saturation, S_w, advances is proportional to time and the value of the derivative $(df_w/dS_w)_{S_w}$ at that saturation, or

$$X_{S_w} = \frac{-q_t}{A\phi}\left(\frac{df_w}{dS_w}\right)_{S_w} \tag{7.9}$$

where X_{S_w} is the distance traveled by a particular S_w contour and q_t is the cumulative water injection at reservoir conditions.

In field units,

$$X_{S_w} = -\frac{5.615 q_t}{A\phi}\left(\frac{df_w}{dS_w}\right)_{Sw} \tag{7.10}$$

Fig. 7.2 shows the linear flow through a body of constant cross-section as well as series and parallel flow in linear bed. Consider displacement of oil by water in a system with dip angle α.

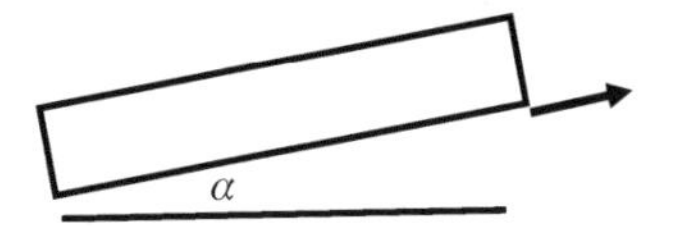

Figure 7.2 An inclined linear reservoir bed.

Darcy's equation for oil and water flow rate with respect to dip angle is given as

$$q_o = -\frac{kk_{ro}A}{\mu_o}\left(\frac{\partial P_o}{\partial x} + P_o g \sin \alpha\right) \tag{7.11}$$

$$q_w = -\frac{kk_{rw}A}{\mu_w}\left(\frac{\partial P_w}{\partial x} + P_w g \sin \alpha\right) \tag{7.12}$$

Replacing water pressure by $P_w = P_o - P_{cow}$ so that

$$q_w = -\frac{kk_{rw}A}{\mu_w}\left(\frac{\partial P_w}{\partial x} + P_w g \sin \alpha\right) \tag{7.13}$$

After rearranging, the equations may be written as

$$-q_o\frac{\mu_o}{kk_{ro}A} = \frac{\partial P_o}{\partial x} + P_o g \sin \alpha \tag{7.14}$$

$$-q_w\frac{\mu_w}{kk_{rw}A} = \frac{\partial P_o}{\partial x} - \frac{\partial P_{cow}}{\partial x} + P_w g \sin \alpha \tag{7.15}$$

Subtracting the first equation from the second one, we get

$$-\frac{1}{kA}\left(q_w\frac{\mu_w}{k_{rw}} - q_o\frac{\mu_o}{k_{ro}}\right) = -\frac{\partial P_{cow}}{\partial x} + P_w g \sin \alpha \tag{7.16}$$

Substituting for

$$q_t = q_w + q_o \tag{7.17}$$

and

$$f_w = \frac{q_w}{q_t} \tag{7.18}$$

Also, solving for a fraction of water flowing, we obtain the expression for a fraction of water flowing:

$$f_w = \frac{1 + \left(kk_{ro}A/q_t\mu_o\right)\left(\left(\partial P_{cow}/\partial x\right) - \Delta\rho g \sin \alpha\right)}{1 + \left(k_{ro}\mu_w/\mu_o k_{rw}\right)} \tag{7.19}$$

For the simplest case of horizontal flow, with negligible capillary pressure, the expression reduces to

$$f_w = \frac{1}{1 + \left(k_{ro}\mu_w / \mu_o k_{rw}\right)} \tag{7.20}$$

7.3 DERIVATION OF CONTINUITY EQUATION FOR DISPLACEMENT FRONT OF RADIAL DISPLACEMENT SYSTEM

Fig. 7.3 depicts a circular reservoir with the planar and lateral view of the radial flow system. Darcy's equation for estimating oil and water flow rates can be calculated using

$$q_o = \frac{kk_{ro}}{\mu_o}\frac{\partial\,(AP_o)}{\partial r} \tag{7.21}$$

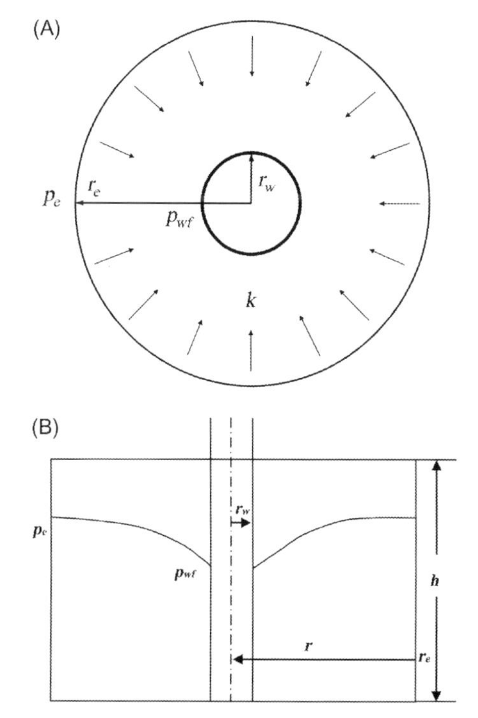

Figure 7.3 Reservoir geometry in (A) plan views and (B) lateral view [1].

$$q_w = \frac{kk_{rw}}{\mu_w}\frac{\partial(AP_w)}{\partial r} \tag{7.22}$$

where A is the flow area, k is the reservoir permeability, k_{ro} and k_{rw} are relative permeability to oil and water, respectively, P_o and P_w are pressure of oil and water, respectively, q_o and q_w are flow rate for oil and water, respectively, r is the radius of the wellbore, μ_w and μ_o are oil and water viscosities.

Recall that capillary pressure is defined as

$$P_c = P_o - P_w \tag{7.23}$$

where P_c is the capillary pressure.

Hence, we have $P_w = P_o - P_c$

Substituting P_w into Eq. (7.22), we have

$$q_w = \frac{kk_{rw}}{\mu_w}\frac{\partial[A(P_o - P_c)]}{\partial r} \tag{7.24}$$

In terms of pressure gradient, Eqs. (7.22) and (7.24) become

$$\frac{\partial(AP_o)}{\partial r} = \frac{\mu_o}{kk_{ro}}q_o \tag{7.25}$$

$$\frac{\partial[A(P_o - P_c)]}{\partial r} = \frac{\mu_w}{kk_{rw}}q_w \tag{7.26}$$

Subtracting Eq.(7.26) from Eq. (7.25), we have

$$-\frac{\partial[A(P_c)]}{\partial r} = \frac{\mu_w}{kk_{rw}}q_w - \frac{\mu_o}{kk_{ro}}q_o \tag{7.27}$$

The total liquid flow rate (q_t) is defined as

$$q_t = q_o + q_w \tag{7.28}$$

whereas the fractional flow (f_w) can be defined in terms of both oil and water. For water, the expression applies

$$f_w = \frac{q_w}{q_t} \tag{7.29}$$

Substituting Eq. (7.27) into Eq. (7.29) yields

$$f_w = \frac{1 - \left(\partial[A(P_c)]/\partial r\right)\left(kk_{ro}/q_t\mu_o\right)}{1 + \left(k_{ro}\mu_w/k_{rw}\mu_o\right)} \tag{7.30}$$

with flow area $A = 2\pi rh$, where h is the reservoir thickness; hence, Eq. (7.30) becomes

$$f_w = \frac{1 - \left(2\pi hkk_{ro}/q_t\mu_o\right)\left(\left(r\partial P_c/\partial r\right) + P_c\right)}{1 + \left(k_{ro}\mu_w/k_{rw}\mu_o\right)} \tag{7.31}$$

Water saturation is a function of time, t, and position, r; we can express

$$\mathrm{d}S_w = \frac{\partial S_w}{\partial t}\mathrm{d}t + \frac{\partial S_w}{\partial r}\mathrm{d}r \tag{7.32}$$

At the displacement front, the water saturation is constant, thus provides a boundary condition for us.

$$\mathrm{d}S_w = \frac{\partial S_w}{\partial t}\mathrm{d}t + \frac{\partial S_w}{\partial r}\mathrm{d}r = 0 \rightarrow \frac{\partial S_w}{\partial t} = -\frac{\partial S_w}{\partial r}\frac{\mathrm{d}t}{\mathrm{d}r} \tag{7.33}$$

Recall that water fraction is function of water saturation, $f_w(S_w)$, and partial differential equation result for change of fluid density governing equation:

$$-\frac{\mathrm{d}f_w}{\mathrm{d}S_w}\frac{\partial S_w}{\partial r} = \frac{(2r_e - 2r)\pi h\phi}{q_t}\frac{\partial S_w}{\partial t} \tag{7.34}$$

Substituting Eq. (7.33) into Eq. (7.34)

$$-\frac{\mathrm{d}f_w}{\mathrm{d}S_w}\left(-\frac{\partial S_w}{\partial r}\frac{\mathrm{d}t}{\mathrm{d}r}\right) = \frac{(2r_e - 2r)\pi h\phi}{q_t}\frac{\partial S_w}{\partial t} \rightarrow \frac{\mathrm{d}f_w}{\mathrm{d}S_w}\mathrm{d}t = \frac{(2r_e - 2r)\pi h\phi}{q_t}\mathrm{d}r \tag{7.35}$$

Integrating Eq. (7.35) yields an equation for displacement front position, r_f.

$$r_f^2 - 2r_e r_f + \frac{tq_t}{\pi h\phi}\left(\frac{\mathrm{d}f_w}{\mathrm{d}S_w}\right)_f = 0 \tag{7.36}$$

where r_f is the displacement front position in radial system; there are two solutions to Eq. (7.36), which are

$$r_f = r_e \pm \sqrt{r_e^2 - \frac{tq_t}{\pi h\phi}\left(\frac{\mathrm{d}f_w}{\mathrm{d}S_w}\right)_f} \tag{7.37}$$

For Eq. (7.37), only one solution is correct to match the physical phenomenon. Considering at the beginning of the displacement as $t \rightarrow 0$, we have $r_f \rightarrow 0$; therefore, we can ignore the solution

$$r_f = r_e + \sqrt{r_e^2 - \frac{tq_t}{\pi h\phi}\left(\frac{\mathrm{d}f_w}{\mathrm{d}S_w}\right)_f} \tag{7.38}$$

Therefore, the correct solution is

$$r_f = r_e - \sqrt{r_e^2 - \frac{tq_t}{\pi h\phi}\left(\frac{\mathrm{d}f_w}{\mathrm{d}S_w}\right)_f} \tag{7.39}$$

In field unit, the expression is

$$r_f = r_e - \sqrt{r_e^2 - \frac{5.615tq_t}{\pi h\phi}\left(\frac{\mathrm{d}f_w}{\mathrm{d}S_w}\right)_f} \tag{7.40}$$

There are other extensions, generalizations, and improvements to Buckley–Leverett theory developed to obtain and enhance understandings of the complicated flow behavior of multiple phases in porous media. In particular, the Buckley–Leverett fractional flow theory has been generalized and applied by various researchers to study enhanced oil recovery (EOR) [7], surfactant flooding [8], polymer flooding [9], mechanisms of chemical methods [10], and alkaline flooding [11].

More recently, studies have extended the Buckley–Leverett solution to flow in a composite, one-dimensional heterogeneous, composite-reservoir system [12], to non-Newtonian fluid flow [13–17] and the non-Darcy displacement of immiscible fluids in porous media [18–22]. Fundamentals of the physics of flow of multiphase fluids in porous media have been understood through laboratory experiments, theoretical analysis, mathematical modeling, and field studies [23–25]. Analysis of porous medium flow processes relies traditionally on Darcy's law-based approaches, and application of such analysis has provided quantitative methodologies and modeling tools for many related scientific and engineering disciplines.

Fayers and Sheldon [26] described the Frontal advance theory as an application of the law of conservation of mass. Flow through a small volume element with length Δx and cross-sectional area "A" can be expressed in terms of total flow rate q_t as

$$q_t = q_w + q_o \tag{7.41}$$

$$q_w = q_t \times f_w \tag{7.42}$$

$$q_o = q_t \times f_w = q_t \times (1 - f_w) \tag{7.43}$$

where q denotes volumetric flow rate at reservoir conditions and subscripts $\{o,w,t\}$ refer to oil, water, and total rate, respectively, and f_w and f_o are fractional flow to water and oil (or water cut and oil cut), respectively.

$$q_o = \frac{kk_{ro}}{\mu_o}\frac{\partial(AP_o)}{\partial r} \tag{7.44}$$

$$q_w = \frac{kk_{rw}}{\mu_w}\frac{\partial(AP_w)}{\partial r} \tag{7.45}$$

$$f_w = \frac{q_w}{q_o + q_w} = \frac{\left(k_{rw}A/\mu_w\right)\left(\mathrm{d}p/\mathrm{d}x\right)}{\left(k_{ro}A/\mu_o\right)\left(\mathrm{d}p/\mathrm{d}x\right) + \left(k_{rw}A/\mu_w\right)\left(\mathrm{d}p/\mathrm{d}x\right)} = \frac{k_{rw}/\mu_w}{\left(k_{ro}/\mu_o\right) + \left(k_{rw}/\mu_w\right)}$$
$$= \frac{1}{1 + \left(k_{ro}\mu_w/k_{rw}\mu_o\right)} \tag{7.46}$$

$\left(k_o/k_w\right)$ is a function of saturation; hence, for constant viscosity, f_w is just a function of saturation [20]. Wu [20] also considered studying the Buckley–Leverett flow in a one-dimensional radial system, where a fluid flowing radially toward (or away from) a vertical well in a radially symmetric manner as illustrated in Fig. 7.4.

The mass balance equation for the fluid phase is given as

$$pq|_r - \left(pq|_r + \frac{\partial(\rho q)}{\partial t}\mathrm{d}r\right) = 2\pi rh\mathrm{d}r\frac{\partial}{\partial t}(\varnothing\rho) \tag{7.47}$$

Wu [20] suggested that if Darcy's law for multipurpose flow is applied for radial flow in the above equation, then we have

$$q_o = -2\pi h\frac{kk_{ro}}{\mu_o}\frac{\partial P_o}{\partial r} \tag{7.48}$$

$$q_w = -2\pi h\frac{kk_{rw}}{\mu_w}\frac{\partial P_w}{\partial r} \tag{7.49}$$

Then, one can obtain the corresponding fractional flow equation to water in a radial flow system as

$$f_w = \frac{1 - \left(2\pi hkk_{ro}/q_t\mu_o\right)\left(r\partial P_c/\partial r\right)}{1 + \left(k_{ro}\mu_w/k_{rw}\mu_o\right)} \tag{7.50}$$

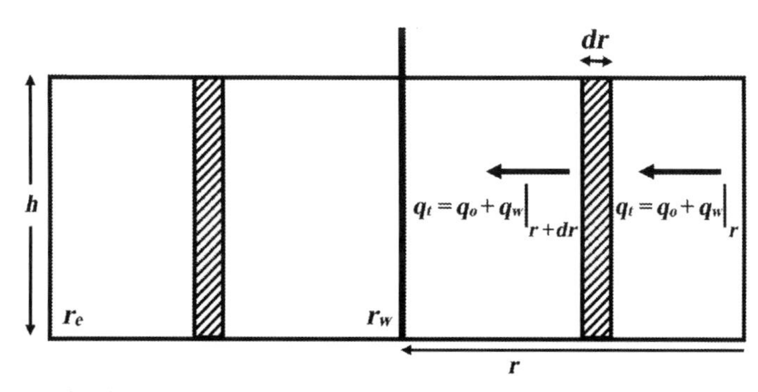

Figure 7.4 A control volume in a circular reservoir with a well located in the center [1].

The equation above is the water fractional flow equation for the displacement of oil by water in a one-dimensional radial system [20]. If we neglect the capillary-pressure gradient along the radii, the fractional flow equation is reduced to

$$f_w = \frac{1}{1 + \left(k_{ro}\mu_w / k_{rw}\mu_o\right)} \tag{7.51}$$

This expression is practically the same equation as the expression for fractional flow in a linear system. This tells us categorically that the fractional flow in a one-dimensional radial flow system is also a function of water saturation through the saturation dependence on relative permeability. When fluid viscosities are constant, the mass conservation of a one-dimensional flow and displacement in a radial system is rewritten as follows:

$$-\frac{1}{r}\frac{\partial f_w}{\partial r} q_t = 2\pi h\phi \frac{\partial s_w}{\partial t} \tag{7.52}$$

For Buckley–Leverett solution in one-dimensional linear flow, an expression $\left(\partial S_w / \partial t\right)$ is obtained and solved as

$$r_{s_w}^2 = r_w^2 + \frac{W_i}{\pi h\phi} \cdot \frac{\mathrm{d}f_w}{\mathrm{d}S_w}\bigg|_{S_w} \tag{7.53}$$

It is important to note that Buckley–Leverett solution in a radial flow system depends on but not solely on the assumption that the effect of the capillary-pressure gradient $\left(\partial P_c / \partial r\right)$ is small and negligible.

7.4 IMPORTANCE AND CAPABILITY OF FRACTIONAL FLOW IN RADIAL FLOW SYSTEM

The fractional flow is of great importance, because with the fractional flow, at any given point in time in the reservoir, we can calculate both the oil and water flow rates, and it depicts reservoir flow conditions. With the help of Buckley–Leverette who developed the frontal advance equation, we can determine water cut and recovery after breakthrough. Also, with the fractional flow cure, we can determine the mobility ratio (which is the ratio of the relative permeability of the rock to the viscosity of the fluid), bearing in mind that the lower the mobility ratio, the higher the recovery efficiency. Low mobility ratio also gives a good sweep efficiency. These two parameters enhance the oil recovery process.

Paul and Franklin [27] understood the importance of the above subject and therefore modified the work of Stiles [28] by developing equations based on the radial flow. Stiles was able to develop equations for fractional recovery, as well as that for water cut in terms of the fraction of a fraction of capacity, thickness, and permeability with the assumption that the flow was linear, which still holds. However, Paul makes us understand that as water is injected into the reservoir, the direction of flow occurs in three phases: the initial direction, which is that of a radial flow, transition from radial to linear, and then the linear flow. Paul and Franklin [27] then developed the equations of water cut and fractional recovery by assuming that fraction of recovery and production rates are proportional to the volumes created.

Radial fractional flow is an important factor to be considered in the waterflood design because it gives a good unit displacement efficiency with a minimal fraction of water flowing, thereby giving a good prediction to the recovery factor. This can be seen as described by Singh and Kiel [29], and the unit displacement efficiency is gotten by plotting a fractional flow curve against water saturation.

Ekwere [30] showed that fractional flow is important because it helps predict a stable frontal displacement at all mobility ratios. Millian and Parker [31] also helped validate that with fractional flow analysis, the waterflooding process is successful and it helps maintain the reservoir pressure, thereby increasing oil recovery.

The radial displacement method helps improve the prediction of reservoir performance unlike the Buckley–Leverette theory, which results in the much lower recovery process, and so, this process is, therefore, an important supplement to the Buckley–Leverette method as the process shows a short process of water breakthrough [32].

Example 7.1: A pilot-scale injection is performed on field alpha to ascertain the distinction between linear and radial displacement systems. Field alpha is a sandstone formation with no existing waterflooding scheme. The following parameters in Table 7.1 are utilized for model analysis.

Field alpha assumed a radial displacement scenario. The comparison analysis used the same inputs for the correlations for both displacement systems and the impact on results was critically observed.

First, the plot of relative permeability of oil and water (K_{ro} and K_{rw}) with increasing water saturation was generated to ascertain if the reservoir follows the conventional trends. Fig. 7.5 depicts the plot, which shows a normal trend.

The linear system pioneered by Buckley–Leverett assumed that capillary pressure is negligible. For the radial system, capillary-pressure effect is included in fractional flow calculations. It is important to note that for the total negligence of capillary pressure for both systems, the Buckley–Leverett equations are valid for both systems. Bearing this in mind, and considering incorporating the effect of capillary pressure into our pilot-scale analysis, the major challenge was how best to calculate the effect of capillary pressure accurately.

Table 7.1 Reservoir Characterization of Field Alpha

Reservoir Data	
Drainage radius (*re*)	1320 ft
Wellbore radius (*rw*)	0.25 ft
Porosity (φ)	20%
Absolute permeability	3 mD
Formation thickness	50 ft
Dip angle (Θ)	0
Connate water saturation (S_{wc})	20%
Initial water saturation (S_{wi})	35%
Residual oil saturation (S_{or})	20%
Oil FVF (B_o)	1.25 bbl/STB
Water FVF (B_w)	1.02 bbl/STB
Oil viscosity	2.0 cp
Water viscosity	1.0 cp
Total injection rate	250BWPD

FVF formation volume factor.

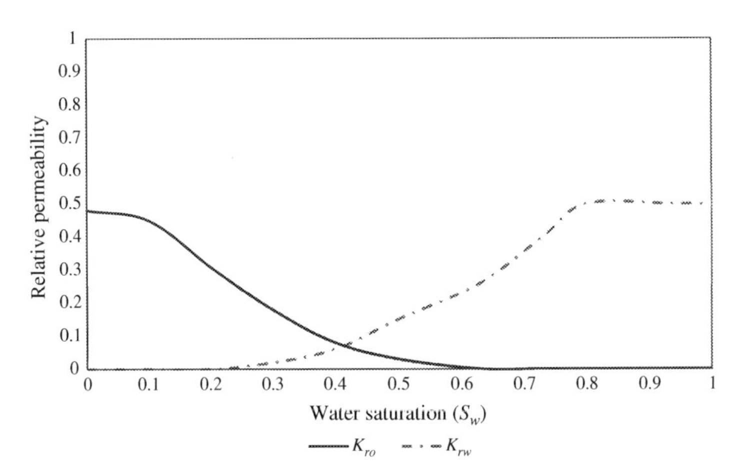

Figure 7.5 A plot of relative permeability and water saturation.

According to Brooks and Corey [33], where they develop a model to estimate the capillary pressure, based on rock property, threshold pressure, water saturation, and connate water saturation as following as:

$$P_c = P_d\left(\frac{S_w - S_{wi}}{1 - S_{wi}}\right)^{-1/\lambda} \tag{7.54}$$

where S_{wi} is the irreducible water saturation, P_d is the threshold pressure, λ is the rock property parameter.

This correlation was applied in our analysis and assuming $\left(\partial P_c/\partial r\right)$ is negligible, various capillary pressure was estimated for increasing water saturation. Note that the free water level (FWL) was assumed at a height of 20 ft; hence we could predict our threshold pressure to be

$$P_d = \Delta \rho g h \tag{7.55}$$

Assuming water density of 1005 kg/m^3 and oil density of 900 kg/m^3 with g taken as 9.81 m/s^2. P_d was calculated to be 0.91 psi. Brooks and Corey related the rock parameter λ to the distribution of pore sizes. For narrow distribution λ is >2 and for wide distribution λ is <2. For this analysis, a normal distribution was assumed and $\lambda = 2$. A plot for both systems were generated as shown in Fig. 7.6.

The plot above shows that the effect of capillary pressure drastically reduces the water cuts with respect to water saturation. Buckley–Leverett equation for linear systems showed a deviation of about 23% for a water saturation of 50%. However, it was observed for the absence of capillary pressure; the water cuts were the same for both systems.

Similarly, the waterfront correlations for both systems were also analyzed. For a duration of 100 days, the location of the waterfront populated with respect to waterflooding time enabled the development of the plot in Fig. 7.7 which showed that linear system had higher distances invaded as compared to radial system.

From the chart, one can see the discrepancies of waterfront for both systems. If the linear displacing mechanism is applied for waterflooding scenarios that follow a radial displacement, then error would be introduced into the analysis. For example, at 70 days of flooding, linear system estimated the waterfront to be at 17 ft, whereas the radial system estimated 11.5 ft; thus, it has been overestimated with a deviation of nearly 48%.

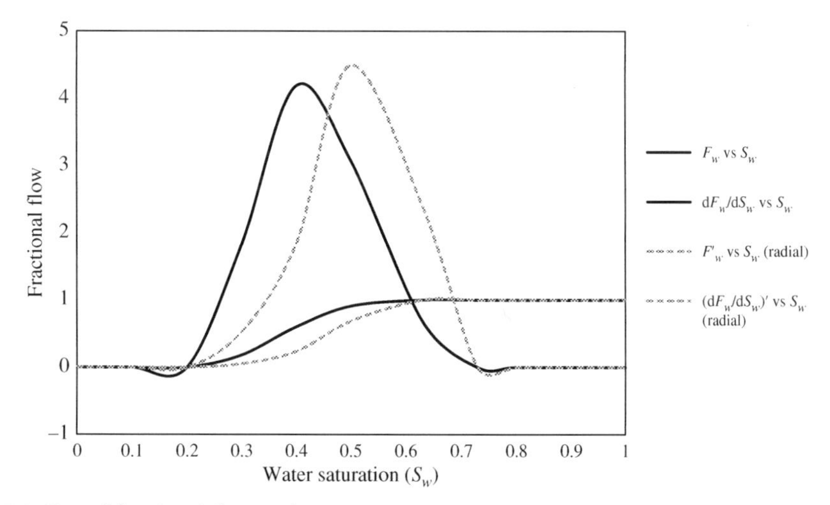

Figure 7.6 Plot of fractional flow and water saturation.

The pilot-scale analysis was further evaluated to estimate the likely recovery factor for both the linear and radial system based on water saturation. Fig. 7.8 shows the average saturation behind the fluid front determined by the intersection between the tangent line and $f_w = 1$.

Hence, at the water breakthrough, the oil recovery factor can be estimated by the expression

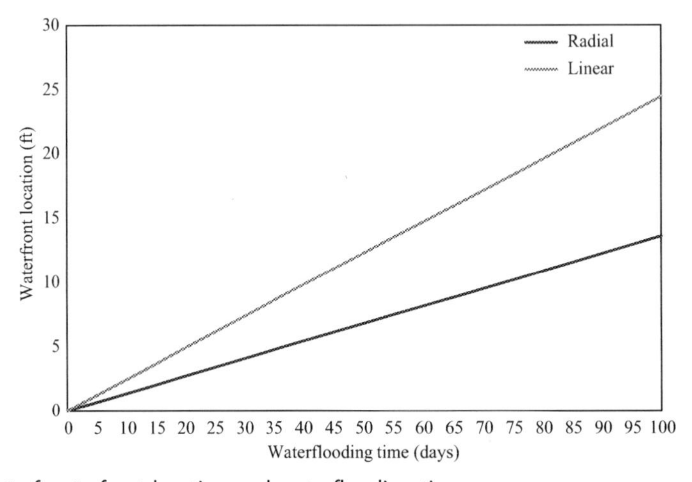

Figure 7.7 Plot of waterfront location and waterflooding time.

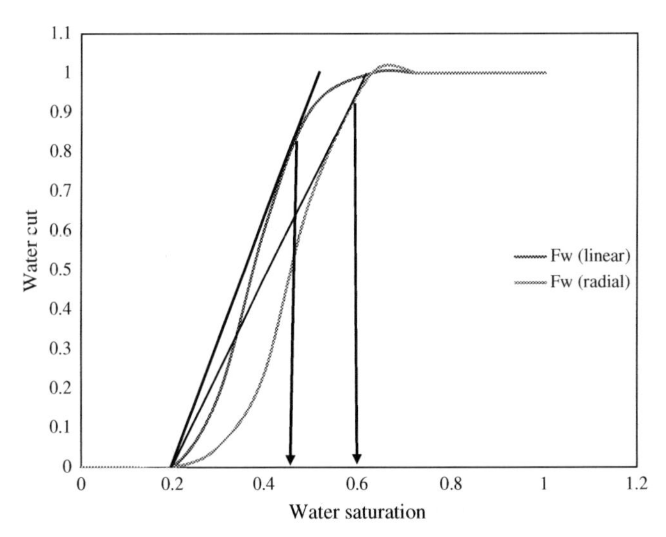

Figure 7.8 Average water saturation for both linear and radial systems.

$$RF = \frac{\overline{S}_w - S_{wir}}{1 - S_{wir}}$$

From the chart below, the linear system showed an average saturation $\overline{S}_w = 0.52$ and that for the radial system was $\overline{S}_w = 0.63$. Inputs of the extrapolated values into the expression above, it was observed that the recovery factor was much higher for the radial displacement system with a recovery of about 53% and that of the linear system was calculated to be about 40%. This has shown that radial displacing mechanism is much more practicable in terms of performance for radial waterflooding scenarios.

The pilot-scale analysis showcased the importance of water cut for reservoir performance analysis. This can give engineers an understanding of their unique flooding design with forecasted reports on the location of invading water with respect to production while preventing negative impacts on economic benefits. A major drawback for this work is how best capillary pressure can be estimated. Linear displacement assumes negligible capillary pressure. However, the inclusion of capillary pressure for the radial system contributed to lower water cut as observed in Fig. 7.6. If higher water cuts are observed, there is a tendency for early water breakthrough at the production well which isn't the best scenario in terms of economics. However, this can be resolved by increasing the viscosity of water during flooding with additives.

7.5 APPLICATION OF BUCKLEY–LEVERETT THEORY AND FRACTIONAL FLOW CONCEPT

The performance analysis of waterflooding is often based on reservoir flow systems. The two geometries of a system widely used are the linear displacement system and the radial displacement system [6]. The oil recovery factor for both systems would be estimated using the average saturation of both systems. At water breakthrough, the average water saturation would be determined and used to estimate the recovery factor for comparison of both systems. The Buckley–Leverett theory estimates the rate at which an injected water moves through a porous medium [6]. The approach applies fractional flow theory and assumes that

- Flow is linear and horizontal
- Water is injected into an oil reservoir
- Oil and water are both incompressible
- Oil and water are immiscible
- Gravity and capillary-pressure effects are negligible

Also, the following assumptions have been made for developing Buckley–Leverett approach:

- A circular reservoir with constant height
- Reservoir is homogeneous in all rock properties
- The dip angle of formation is zero
- Oil and water two-phase flow in reservoir, no gas present in the reservoir
- Compressibilities of oil and water are negligible
- Constant reservoir temperature is applied
- All rock properties do not change as pressure changes
- Constant oil and water viscosities during displacement

7.6 LOW-SALINITY WATERFLOODING

Injection of low-salinity (LS) water has widely been practiced because the water sources are available and relatively cheaper among other practical advantages. However, the EOR potential was not recognized until Morrow et al. [34–40] found that changing the composition of the injected water results in changing the oil recovery factor.

Yildiz and Morrow [40] supported that the injection water composition could affect the oil recovery factor; however, the maximum recoverable oil using water injection process occurs in a particular conditions of brine/rock/oil system.

There are different studies that have been done to evaluate the machanisms contributed in low salinity water injection [41–44].

7.6.1 Effect of Rock and Fluid Properties on Low-Salinity Waterflooding Performance

7.6.1.1 Effect of Connate Water Saturation

Tang and Morrow [37] carried out experimental tests to evaluate the effect of initial water saturation on both high salinity and low salinity water flooding processes. They concluded that in a case of zero connate water saturation the oil recovery achieved from both high salinity and low salinity water flooding was almost the same. Zhang and Morrow [46] realized that the more oil can be produce using low salinity water injection in a case of connate water existence; as a comparison, presence of initial saturation is crucial for success of such an EOR method.

7.6.1.2 Effect of the Salinity of Connate Water

Sharma and Filoco [47] found that the salinity of connate water was the primary factor controlling the oil recovery. They concluded that decreasing in salinity of connate water results in increasing the amount of recovered oil [35]. Different researchers supported this result that lower salinity of connate water results in more oil production, for instance, McGuire et al. [44] data and Zhang and Morrow's [46] data.

7.6.1.3 *Effect of Injection Water Salinity*

According to the data in open literature a higher oil recovery was obtained when the salinity of injection water is lower than that of connate water. Moreover, the injection water salinity was low enough [48,49]. Zhang et al. [48] observed that the LS of 1500 ppm which was about 5% formation water salinity resulted in sharp increase in the tertiary recovery and in the differential pressure.

7.6.1.4 *Effect of Wettability*

Jadhunandan and Morrow [35] found that the wettability related to initial water saturation in the cores. With higher initial water saturation, the cores showed more water-wet. Moreover, the oil recovery increased from strongly water-wet to a maximum close to neutral wet, which was agreed by Sharma and Filoco [47]. For the cores with high initial salinity, LS injection will make the cores more water-wet and result in a higher oil recovery [45].

7.6.2 Mechanisms Behind Low-Salinity Waterflooding

Seventeen mechanisms of LS waterflooding have been proposed in the literature, as follows [45]:

1. fine migration [37]
2. mineral dissolution [50]
3. limited release of mixed-wet particles [50]
4. increased pH effect and reduced interfacial tension (IFT) [44]
5. emulsification/snap-off [44]
6. saponification [44]
7. surfactant-like behavior [44]
8. multicomponent ion exchange (MIE) [51]
9. double layer effect [52]
10. particle-stabilized interfaces/lamella [50,53]
11. salt-in effects [54]
12. osmotic pressure [50]
13. salinity shock [50]
14. wettability alteration (more water-wet) [50]
15. wettability alteration (less water-wet) [50]
16. viscosity ratio [50]
17. end effects [50]

Most of these mechanisms are related to each other. In this section, we will discuss major mechanisms and their working conditions [45,55].

7.6.2.1 Fine Mobilization

Martin [56] and Bernard [57] observed that clay swelling and dispersion accompanied by increased differential pressure and incremental oil recovery resulted. Kia et al. [58] reported that when sandstones were previously exposed to sodium-salt solutions, flooding these sandstones by fresh water resulted in the release of clay particles and a drastic reduction in permeability. The permeability reduction was lessened, however, when calcium ions were also present in the salt solution.

However, Lager et al. [51] reported that no fine migration or significant permeability reductions were observed during numerous LS core displacement experiments under reduced conditions and full reservoir conditions, although these core floods had all shown increased oil recovery. Valdya and Fogler [59] reported that a gradual reduction in salinity kept the concentration of fines in the flowing suspension low, with formation damage minimized or avoided. Soraya et al. [60] conducted several low salinity water injection displacement experiments; they concluded that low salinity did not have effect; however, they reported sand production in their experiments.

7.6.2.2 Limited Release of Mixed-Wet Particles

Tang and Morrow [37] found and described the mechanism of limited removal of mixed-wet particles from locally heterogeneous pores in terms of wettability. They concluded that when the salinity is reduced, the electrical double layer in the aqueous phase between particles is expanded, and the tendency to strip fines is increased. The stripped fines migrate and aggregate so that the oil coalesces and consequently, the oil recovery factor increased. This mechanism combines the Derjaguin, Landau, Verwey, and Overbeek (DLVO) theory and fine migration [45].

7.6.2.3 Increased pH and Reduced IFT Similar to Alkaline Flooding

McGuire et al. [44] proposed that LS mechanisms could be due to increased pH and reduced IFT similar to alkaline flooding. This increase in pH is due to exchange of hydrogen ions in water with adsorbed sodium ions [61]. Another related mechanism is that a small change in bulk pH can impose a great change in the zeta potential of the rock. When pH is increased, organic materials will be desorbed from the clay surfaces [62–64].

Austad et al. [63] proposed a hypothesis that cation exchange resulted in local pH increase close to clay surfaces. Zhang et al. [48] reported that after LS brine injection, a slight rise and drop in pH were observed. However, no clear relationship between effluent pH and recovery was observed [45].

It has been observed from many field projects that the level of improved oil recovery from alkaline flooding is low. Based on analysis of the data reported by Mayer et al. [65], the incremental oil recovery factor over waterflooding was 1%−2% in most of the projects, and 5%−6% in a few projects [45].

7.6.2.4 Multicomponent Ion Exchange

Different affinities of ions on rock surfaces results in the Multicomponent Ion Exchange (MIE) in which multivalent or divalent such as Ca^{2+} and Mg^{2+} strongly adsorbed on rock surfaces until the saturation point of the rock. Polar compounds present in the oil phase and multivalent cations at clay surfaces bonded together and form organometallic complexes; this mechanism yields the rock surface toward oil-wet condition. During the injection of LS brine, MIE will take place, removing organic polar compounds and organometallic complexes from the surface and replacing them with uncomplexed cations [45,66]. The suggested mechanism of MIE is supported by the pore-scale model proposed by Sorbie and Collins [66].

Meyers and Salter [67] carried out several adsorption experiments; in their tests, they observed that the steady-state effluent concentrations of calcium and magnesium were observed to be slightly greater than the injected concentrations. These excess concentrations increased as the injection concentrations decreased. When NaCl brine was injected into the cores, "residual" calcium and magnesium concentrations were still observed in the effluent. However, Valocchi et al. [68] injected fresh water in a brackish water aquifer and noticed that the concentration of Ca^{2+} and Mg^{2+} in different control wells were lower than the invading water and the connate brine [45].

7.6.2.5 Double Layer Effect

The double layer theory combines the effects of the electrostatic repulsion and van der Waals attraction owing to the so-called double layer of counter ions. LS brine using the mechanism of the expansion of the electric double layer to reduce clay−clay attraction. Indirect interactions between oil, brine, and rock highly affect the discharge of clay particles; this mechanism normally occurs in kaolinite plates and involve in their charge distribution [46]. LS water makes water film more stable owing to this expanded double layer effect, resulting in more water-wet on clay surfaces and more oil is detached; conversely, adsorption of divalent at water/sand and water/oil interfaces changes the wettability from water-wet state to oil-wet condition [45,70,71].

7.6.2.6 Salt-in Effect

The solubility of the organic material in water can be drastically decreased by adding salt to the solution, that is, the salting-out effect, and the solubility can be increased by removing salt from the water, that is, the salting-in effect [54]. Therefore, a decrease in salinity below a critical ionic strength can increase the solubility of the organic material in the aqueous phase, so that oil recovery is improved [45].

7.6.2.7 Osmotic Pressure

Sandengen and Arntzen [71] demonstrated using experiments that oil droplets acted as semipermeable membranes; oil droplets could move under an osmotic pressure gradient. They proposed that such osmotic gradient relocate oil by expanding an otherwise inaccessible aqueous phase in a porous rock medium. This mechanism cannot explain the need for the existence of crude (polar) oil and clays [45].

7.6.2.8 Wettability Alteration

As mentioned earlier, brine films are more stable at a lower salinity. This suggests that LS water will cause cores to become mixed-wet (less water-wet). Mixed-wet cores show lower residual oil saturation or higher oil recoveries than strongly water-wet or oil-wet cores [72,73]. Buckley et al. [70] explained wettability alteration as a result of the interaction between crude oil and reservoir rock. Berg et al. [74] experimentally showed that LS water could achieve wettability alteration. Nasralla et al. [75] showed that LS water could decrease contact angles. Yousef et al. [76] and Zekri et al. [77] reported that LS water injection could change wettability to more water-wet in carbonates. Vledder et al. [78] even provided a proof of wettability alteration in a field scale [45].

Drummond and Israelachvili [79] showed that the wettability was altered from oil-wet to water-wet at $pH > 9$ and from water-wet to intermediate-wet at $pH < 9$, as shown in Fig. 7.9. Fig. 7.9 depicts the wettability map as a function of pH and Na^+ concentration. In LS waterflooding, pH is most likely below 9. This can also explain why connate water is needed for the LS effect because the existence of connate water makes water-wettability possible. The wettability alteration is the most frequently suggested mechanism [45,53].

7.6.3 Field Tests of Low-Salinity Waterflooding

Four different SWCTT were carried out in Alaska North slope; using low salinity water injection increased the oil production rate in all the tests [44,45].

Another field test of low salinity water injection was performed in Endicott offshore oil field located on the North Slope of Alaska. Field observations show the significantly reduction in water cut after using low salinity water injection process [80].

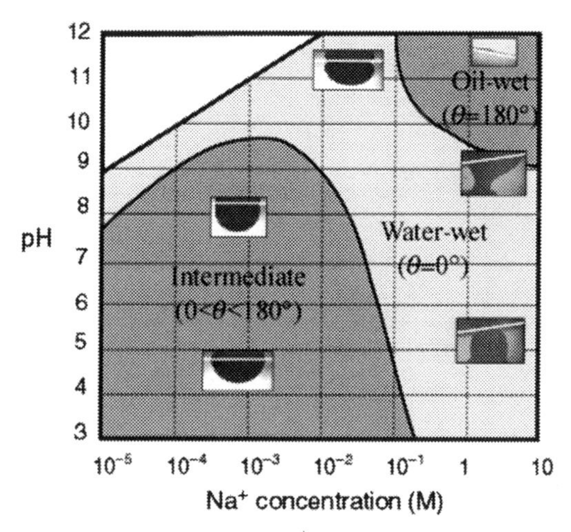

Figure 7.9 Wettability map in different pH and Na^+ concentration [45,79].

Robertson [81] compared three waterflooding field performances in Wyoming. The reservoir was pre-flushed using surfactant-polymer; With a lower salinity ratio, a higher oil recovery was obtained. preflush water should bring incremental oil if LS waterflooding worked. However, apparently, no oil rate increase was observed during the fresh water preflush in the North Burbank Unit surfactant–polymer pilot in Osage County, Oklahoma [82] and Loudon surfactant pilot [83]. Thyne and Gamage [84] evaluated the LS flooding effects in the fields in the Powder River Basin of Wyoming. They found no increase in recovery for the 26 fields where LS water was injected when they compared with the 25 fields where mixed water or formation water was injected [45].

Skrettingland et al. [85] evaluated LS flooding for the Snorre field. A SWCTT did not show a significant in oil recovery factor; they explained why low salinity water injection did not have a considerable effect because of the reservoir was in an optimum wettability condition. As a result, such an EOR method could not work for this reservoir [45].

REFERENCES

[1] K. Ling, Fractional flow in radial flow systems: a study for peripheral waterflood, J. Pet. Explor. Prod. Technol. 6 (2016) 441–450.
[2] Willhite, G.P., 1986. Waterflooding.
[3] S.E. Buckley, M. Leverett, Mechanism of fluid displacement in sands, Trans. AIME 146 (1942) 107–116.
[4] H.J. Welge, A simplified method for computing oil recovery by gas or water drive, J. Pet. Technol. 4 (1952) 91–98.
[5] Sheldon, J., Cardwell Jr, W., 1959. One-dimensional, incompressible, noncapillary, two-phase fluid flow in a porous medium.

[6] R.E. Terry, J.B. Rogers, B.C. Craft, Applied Petroleum Reservoir Engineering, Prentice Hall, NJ, United States, 2013.
[7] G.A. Pope, The application of fractional flow theory to enhanced oil recovery, So. Pet. Eng. J. 20 (1980) 191–205.
[8] R. Larson, Analysis of the physical mechanisms in surfactant flooding, Soc. Pet. Eng. J. 18 (1978) 42–58.
[9] J. Patton, K. Coats, G. Colegrove, Prediction of polymer flood performance, Soc. Pet. Eng. J. 11 (1971) 72–84.
[10] R.G. Larson, H. Davis, L. Scriven, Elementary mechanisms of oil recovery by chemical methods, J. Pet. Technol. 34 (1982) 243–258.
[11] E. DeZabala, J. Vislocky, E. Rubin, C. Radke, A chemical theory for linear alkaline flooding, Soc. Pet. Eng. J. 22 (1982) 245–258.
[12] Y.-S. Wu, K. Pruess, Z. Chen, Buckley–Leverett flow in composite porous media, SPE Adv. Technol. Ser. 1 (1993) 36–42.
[13] Y.-S. Wu, Theoretical studies of non-Newtonian and Newtonian fluid flow through porous media, Lawrence Berkeley National Laboratory, United States, 1990.
[14] Y.-S. Wu, K. Pruess, P. Witherspoon, Displacement of a Newtonian fluid by a non-Newtonian fluid in a porous medium, Transp. Porous Media 6 (1991) 115–142.
[15] Y. Wu, K. Pruess, P. Witherspoon, Flow and displacement of Bingham non-Newtonian fluids in porous media, SPE Reservoir Eng. 7 (1992) 369–376.
[16] Y.-S. Wu, K. Pruess, Flow of non-Newtonian fluids in porous media, Adv. Porous Media 3 (1996) 87–184.
[17] Y.-S. Wu, K. Pruess, A numerical method for simulating non-Newtonian fluid flow and displacement in porous media, Adv. Water Resources 21 (1998) 351–362.
[18] Y.-S. Wu, Numerical simulation of single-phase and multiphase non-Darcy flow in porous and fractured reservoirs, Transp. Porous Media 49 (2002) 209–240.
[19] Y.S. Wu, Non-Darcy displacement of immiscible fluids in porous media, Water Resources Res. 37 (2001) 2943–2950.
[20] Y.S. Wu, An approximate analytical solution for non-Darcy flow toward a well in fractured media, Water Resources Res. 38 (2002).
[21] Y.-S. Wu, B. Lai, J.L. Miskimins, P. Fakcharoenphol, Y. Di, Analysis of multiphase non-Darcy flow in porous media, Transp. Porous Media 88 (2011) 205–223.
[22] Y.-S. Wu, B. Lai, J.L. Miskimins, Simulation of non-Darcy porous media flow according to the Barree and Conway model, J. Comput. Multiphase Flows 3 (2011) 107–122.
[23] Scheidegger, A.E., 1974. The Physics of Flow Through Porous Media.
[24] R.E. Collins, Flow of Fluids Through Porous Media, Reinhold, New York, NY, 1961, p. 59.
[25] J. Bear, Dynamics of Flow in Porous Media, Dover, NY, 1972.
[26] Fayers, F., Sheldon, J., 1959. The effect of capillary pressure and gravity on two-phase fluid flow in a porous medium.
[27] P.S. Ache, L.A. Franklin, Inclusion of radial flow in use of permeability distribution in waterflood calculation, AIME Tech. Pap (1957) 935.
[28] W.E. Stiles, Use of permeability distribution in water-flood calculations, Trans. AIME 186 (1949) 9–13.
[29] S.P. Singh, O.G. Kiel, Waterflood design (pattern, rate, and timing), International Petroleum Exhibition and Technical Symposium, Society of Petroleum Engineers, January 1982.
[30] J.P. Ekwere, Scaling unstable immiscible displacements, SPE 12331 (1983) 1–6.
[31] H. Millian, A. Parker, Recapturing the Value of Fractional Flow Analysis in a Modern Day Waterflood, Society of Petroleum Engineers, SPE (2006). 101070.
[32] Zhang, H., Ling, K., Acura, H., 2013. New analytical equations of recovery factor for radial flow systems. In: North Africa Technical Conference and Exhibition. Society of Petroleum Engineers.
[33] R.H. Brooks, A.T. Corey, Hydraulic Properties of Porous Media. Hydrology Paper No. 3, Colorado State University, Fort Collins, Colorado, 1964, pp. 22–27.

[34] P.P. Jadhunandan, Effects of Brine Composition, Crude Oil, and Aging Conditions on Wettability and Oil Recovery, Department of Petroleum Engineering, New Mexico Institute of Mining & Technology, United States, 1990.
[35] P. Jadhunandan, N.R. Morrow, Effect of wettability on waterflood recovery for crude-oil/brine/rock systems, SPE Reservoir Eng. 10 (1995) 40–46.
[36] P. Jadhunandan, N. Morrow, Spontaneous imbibition of water by crude oil/brine/rock systems, In Situ (United States) 15 (1991).
[37] G.-Q. Tang, N.R. Morrow, Influence of brine composition and fines migration on crude oil/brine/rock interactions and oil recovery, J. Pet. Sci. Eng. 24 (1999) 99–111.
[38] G. Tang, N.R. Morrow, Salinity, temperature, oil composition, and oil recovery by waterflooding, SPE Reservoir Eng. 12 (1997) 269–276.
[39] Tang, G., 1999. Brine composition and waterflood recovery for selected crude oil/brine/rock systems.
[40] H.O. Yildiz, N.R. Morrow, Effect of brine composition on recovery of Moutray crude oil by waterflooding, J. Pet. Sci. Eng. 14 (1996) 159–168.
[41] Webb, K.J., Black, C.J.J., Tjetland, G., 2005. A laboratory study investigating methods for improving oil recovery in carbonates. In: International Petroleum Technology Conference.
[42] Webb, K., Black, C., Edmonds, I., 2005. Low salinity oil recovery—the role of reservoir condition corefloods. In: IOR2005-13th European Symposium on Improved Oil Recovery.
[43] Webb, K., Black, C., Al-Ajeel, H., 2003. Low salinity oil recovery—log–inject–log. In: Middle East Oil Show. Society of Petroleum Engineers.
[44] McGuire, P., Chatham, J., Paskvan, F., Sommer, D., Carini, F., 2005. Low salinity oil recovery: an exciting new EOR opportunity for Alaska's North Slope. In: SPE Western Regional Meeting. Society of Petroleum Engineers.
[45] J. Sheng, Critical review of low-salinity waterflooding, J. Pet. Sci. Eng. 120 (2014) 216–224.
[46] Zhang, Y., Morrow, N.R., 2006. Comparison of secondary and tertiary recovery with change in injection brine composition for crude-oil/sandstone combinations. In: SPE/DOE Symposium on Improved Oil Recovery. Society of Petroleum Engineers.
[47] M. Sharma, P. Filoco, Effect of brine salinity and crude-oil properties on oil recovery and residual saturations, SPE J. 5 (2000) 293–300.
[48] Zhang, Y., Xie, X., Morrow, N.R., 2007. Waterflood performance by injection of brine with different salinity for reservoir cores. In: SPE Annual Technical Conference and Exhibition. Society of Petroleum Engineers.
[49] Jerauld, G.R., Webb, K.J., Lin, C.-Y., Seccombe, J., 2006. Modeling low-salinity waterflooding. In: SPE Annual Technical Conference and Exhibition. Society of Petroleum Engineers.
[50] Buckley, J., Morrow, N., 2010. Improved oil recovery by low salinity waterflooding: a mechanistic review. In: 11th International Symposium on Evaluation of Wettability and Its Effect on Oil Recovery. Calgary, pp. 6–9.
[51] A. Lager, K.J. Webb, C. Black, M. Singleton, K.S. Sorbie, Low salinity oil recovery-an experimentalinvestigation1, Petrophysics 49 (2008).
[52] Ligthelm, D.J., Gronsveld, J., Hofman, J., Brussee, N., Marcelis, F., van der Linde, H., 2009. Novel waterflooding strategy by manipulation of injection brine composition. In: EUROPEC/EAGE Conference and Exhibition. Society of Petroleum Engineers.
[53] N. Morrow, J. Buckley, Improved oil recovery by low-salinity waterflooding, J. Pet. Technol. 63 (2011) 106–112.
[54] A. RezaeiDoust, T. Puntervold, S. Strand, T. Austad, Smart water as wettability modifier in carbonate and sandstone: a discussion of similarities/differences in the chemical mechanisms, Energy Fuels 23 (2009) 4479–4485.
[55] Boston, W., Brandner, C., Foster, W., 1969. Recovery of oil by waterflooding from an argillaceous oil-containing subterranean formation.
[56] Martin, J.C., 1959. The effects of clay on the displacement of heavy oil by water. In: Venezuelan Annual Meeting. Society of Petroleum Engineers.

[57] Bernard, G.G., 1967. Effect of floodwater salinity on recovery of oil from cores containing clays. In: SPE California Regional Meeting. Society of Petroleum Engineers.
[58] Kia, S., Fogler, H.S., Reed, M., 1987. Effect of salt composition on clay release in Berea sandstones. In: SPE International Symposium on Oilfield Chemistry. Society of Petroleum Engineers.
[59] R. Valdya, H. Fogler, Fines migration and formation damage: influence of pH and ion exchange, SPE Prod. Eng. 7 (1992) 325–330.
[60] Soraya, B., Malick, C., Philippe, C., Bertin, H.J., Hamon, G., 2009. Oil recovery by low-salinity brine injection: laboratory results on outcrop and reservoir cores. In: SPE Annual Technical Conference and Exhibition. Society of Petroleum Engineers.
[61] K.K. Mohan, H.S. Fogler, R.N. Vaidya, M.G. Reed, Water sensitivity of sandstones containing swelling and non-swelling clays, Colloids in the Aquatic Environment, Elsevier, Netherlands, 1993, pp. 237–254.
[62] T. Austad, Water-based, EOR in carbonates and sandstones: new chemical understanding of the EOR potential using smart water, Enhanced Oil Recovery Field Case Studies. (2013) 301–335.
[63] Austad, T., RezaeiDoust, A., Puntervold, T., 2010. Chemical mechanism of low salinity water flooding in sandstone reservoirs. In: SPE Improved Oil Recovery Symposium. Society of Petroleum Engineers.
[64] J. Sheng, Modern Chemical Enhanced Oil Recovery: Theory and Practice, Gulf Professional Publishing, Amsterdam, Netherlands, 2010.
[65] E. Mayer, R. Berg, J. Carmichael, R. Weinbrandt, Alkaline injection for enhanced oil recovery—a status report, J. Pet. Technol. 35 (1983) 209–221.
[66] Sorbie, K.S., Collins, I., 2010. A proposed pore-scale mechanism for how low salinity waterflooding works. In: SPE Improved Oil Recovery Symposium. Society of Petroleum Engineers.
[67] Meyers, K., Salter, S., 1984. Concepts pertaining to reservoir pretreatment for chemical flooding. In: SPE Enhanced Oil Recovery Symposium. Society of Petroleum Engineers.
[68] A.J. Valocchi, R.L. Street, P.V. Roberts, Transport of ion-exchanging solutes in groundwater: chromatographic theory and field simulation, Water Resources Res. 17 (1981) 1517–1527.
[69] Q. Liu, M. Dong, K. Asghari, Y. Tu, Wettability alteration by magnesium ion binding in heavy oil/brine/chemical/sand systems—analysis of electrostatic forces, J. Pet. Sci. Eng. 59 (2007) 147–156.
[70] J. Buckley, Y. Liu, S. Monsterleet, Mechanisms of wetting alteration by crude oils, SPE J. 3 (1998) 54–61.
[71] Sandengen, K., Arntzen, O., 2013. Osmosis during low salinity water flooding. In: IOR2013-17th European Symposium on Improved Oil Recovery.
[72] N.R. Morrow, Wettability and its effect on oil recovery, J. Pet. Technol. 42 (1990) 1,476–471,484.
[73] N.R. Morrow, G.-Q. Tang, M. Valat, X. Xie, Prospects of improved oil recovery related to wettability and brine composition, J. Pet. Sci. Eng. 20 (1998) 267–276.
[74] S. Berg, A. Cense, E. Jansen, K. Bakker, Direct experimental evidence of wettability modification by low salinity, Petrophysics 51 (2010).
[75] Nasralla, R.A., Bataweel, M.A., Nasr-El-Din, H.A., 2011. Investigation of wettability alteration by low salinity water. In: Offshore Europe. Society of Petroleum Engineers.
[76] A.A. Yousef, S.H. Al-Saleh, A. Al-Kaabi, M.S. Al-Jawfi, Laboratory investigation of the impact of injection-water salinity and ionic content on oil recovery from carbonate reservoirs, SPE Reservoir Eval. Eng. 14 (2011) 578–593.
[77] Zekri, A.Y., Nasr, M.S., Al-Arabai, Z.I., 2011. Effect of LoSal on wettability and oil recovery of carbonate and sandstone formation. In: International Petroleum Technology Conference, International Petroleum Technology Conference.
[78] Vledder, P., Gonzalez, I.E., Carrera Fonseca, J.C., Wells, T., Ligthelm, D.J., 2010. Low salinity water flooding: proof of wettability alteration on a field wide scale. In: SPE Improved Oil Recovery Symposium. Society of Petroleum Engineers.
[79] C. Drummond, J. Israelachvili, Surface forces and wettability, J. Pet. Sci. Eng. 33 (2002) 123–133.
[80] Seccombe, J., Lager, A., Jerauld, G., Jhaveri, B., Buikema, T., Bassler, S., et al., 2010. Demonstration of low-salinity EOR at interwell scale, Endicott field, Alaska. In: SPE Improved Oil Recovery Symposium. Society of Petroleum Engineers.

[81] E.P. Robertson, Low-Salinity Waterflooding to Improve Oil Recovery-Historical Field Evidence, Idaho National Laboratory (INL), United States, 2007.
[82] J. Trantham, H. Patterson Jr, D. Boneau, The North Burbank Unit, Tract 97 surfactant/polymer pilot operation and control, J. Pet. Technol. 30 (1978) 1068–1074.
[83] S. Pursley, R. Healy, E. Sandvik, A field test of surfactant flooding, Loudon, Illinois, J. Pet. Technol. 25 (1973) 793–802.
[84] Thyne, G.D., Gamage, S., Hasanka, P., 2011. Evaluation of the effect of low salinity waterflooding for 26 fields in Wyoming. In: SPE Annual Technical Conference and Exhibition, 30 October–2 November, Denver, Colorado, USA.
[85] K. Skrettingland, T. Holt, M.T. Tweheyo, I. Skjevrak, Snorre low-salinity-water injection—coreflooding experiments and single-well field pilot, SPE Reservoir Eval. Eng. 14 (2011) 182–192.

CHAPTER EIGHT

Enhanced Gas Recovery Techniques From Coalbed Methane Reservoirs

Alireza Keshavarz[1], Hamed Akhondzadeh[1], Mohammad Sayyafzadeh[2] and Masoumeh Zargar[1]

[1]School of Engineering, Edith Cowan University, Perth, WA, Australia
[2]Australian School of Petroleum, The University of Adelaide, Adelaide, SA, Australia

8.1 INTRODUCTION

Over the past recent years, coalbed methane (CBM) has been among the fastest growing unconventional reservoirs in the world. Although coal mining is well known and has been regarded as one of the most reliable fuel supplies for a long time, gas production from coal seams had not previously drawn oil and gas industries' attention. This is mainly attributed to the fact that the gas content in these reservoirs was not considered to be substantial before the occurrence of some major explosions in mining sites. Subsequent to perceiving the deep coal layers as potential unconventional gas resources, the production was not yet encouraging due to the unique mechanism of this reservoir type, rendering the gas companies disinterested in exploring CBM fields. However, owing to economic obstacles in heavy oil production as well as depletion of conventional reservoirs, in the past two decades, gas production feasibility from CBM reservoirs has been investigated extensively, and CBM has proven to be a promising unconventional gas resource, and such reservoirs are being extracted worldwide for the time being. The world cumulative CBM in place has been estimated to be over 8000 Tcf, with North America being the richest region, globally [1].

CBM reservoirs are, in essence, naturally fractured reservoirs, and the fluid bulk flow occurs inside the fractures (cleats) toward the wellbore. The fracture system in CBM reservoirs consists of two distinct sets of fracture, namely face cleats and butt cleats, which are normally perpendicular to the reservoir bedding layers. Face cleats are the well-developed cracks inside the reservoir that are fairly parallel and play the most significant role in transferring the fluids toward the production well. Butt cleats, on the other hand, comprise a set of less-developed parallel fractures that expectedly end at their interconnection with face cleats, almost vertically. Normally, the natural cleats in CBM reservoirs are initially filled with mobile water and contain negligible sorbed gas.

Fundamentals of Enhanced Oil and Gas Recovery from Conventional and Unconventional Reservoirs.
DOI: https://doi.org/10.1016/B978-0-12-813027-8.00008-4

The original gas content of the cleats might be zero in some coals, while in other coals there might be some initial gas available in the cleat system [2]. The uniqueness of these reservoirs roots in the fact that opposite to conventional reservoirs in which the gas is trapped in the reservoir's void space, the gas is sorbed on the coal rock's surface, and the production requires desorption of the gas. These reservoirs are referred to as CBM, because most of the reservoir gas content is accounted for by methane. The adsorption sites consist of micropore and mesopore spaces inside the matrix, while cleats behave as the fluid flow conduits. Desorption process is viable through depressurization of the reservoir, and upon desorption from the rock surface, the gas diffuses within the matrix porous media toward the cleats by gas concentration and partial pressure gradient.

The primary production is accomplished by declining the pressure of the reservoir through depletion, thereby facilitating the gas desorption process. Since the gas release from coal matrix into the cleat system in such reservoirs is controlled by the gas partial pressure gradient rather than reservoir pressure gradient, and given that methane partial pressure could not reach zero in cleats due to economic limitations, it is expected to obtain a recovery factor less than 50% in the natural depletion condition [3]. Therefore, in order to get a desirable gas production rate from the reservoir, we need to keep two pressure gradients at maximum: methane partial pressure gradient between matrix and cleats and reservoir pressure gradient between cleats and the production wellbore. As such conditions are not achievable in the natural depletion, the injection of a foreign gas to the CBM reservoir, namely enhanced coalbed methane (ECBM) recoveryprocess, was suggested in the early 1990s [4]. Another approach to improve gas production from CBM reservoirs is to improve reservoir conductivity. Hydraulic stimulation techniques, such as natural fracture stimulation and hydraulic fracturing, are the most common productivity enhancement techniques in CBM reservoirs.

In this chapter, CBM and its associated recovery and enhanced recovery processes are discussed in detail. First, the properties of CBM reservoirs including coal rank, macerals, porosity and permeability, coal density, and coal rock mechanical properties are explored. Subsequently, a typical production profile of CBM reservoirs is illustrated, and the flow mechanism in such reservoirs is scrutinized. In this section, the most unique feature of CBM reservoirs is introduced. Finally, the production enhancement approaches in CBM reservoirs, split into stimulation techniques and ECBM, are investigated.

8.2 COALBED METHANE RESERVOIR PROPERTIES

Due to the unique behavior of CBM reservoirs in terms of gas storage as well as production mechanism compared to conventional reservoirs, the key properties of these reservoirs differ to some extent. The most significant parameters in coals, playing key roles in reservoir studies, consist of coal rank, macerals, permeability, porosity, density, rock mechanical properties, and sorption properties. Therefore, in this section these features are briefly introduced.

8.2.1 Coal Rank

One of the main aspects of coals' classification is their rank, defined as the degree to which the coal is thermally mature or in other words, the degree of metamorphosis of the existing organic materials in the coal. Coal rank is one of the main criteria for gas content determination, because this factor indicates the adsorption capacity of the rocks. Coals are organic rocks that corresponding to their geological age, the level of impurity, and moisture content, reveal distinct complexities in different reservoirs. In fact, deeper coals are generally more mature, and due to subjection to more heat and pressure, these coals have expelled most of the water from the rock texture, resulting in a more mechanically integrated rock with its surface covered with longer chained gases as well as aromatic molecules. On the contrary, more recent coals with less burial depth are of lower rank and illustrate higher levels of moisture and impurity content. There are several standards classifying coal ranks based on different parameters, in all of which the coal rank major classifications in ascending order of maturity include peat, lignite, subbituminous, bituminous, and anthracite [5–8]. The process of coal rocks maturing from peat toward anthracite is called coalification, in which the volatile and moisture contents diminish, and the carbon content of the rock increases. Therefore, coals containing low amount of volatile matter and moisture content, and high carbon content are referred to as high-rank coals.

In coals, the pore structure (specific surface area, size distribution, volume, etc.) is of great importance, because the matrix pore surface provides the main sorption sites for the gas. Ji et al. examined the pore structure of some raw coals from anthracite and bituminous ranks in comparison with their residues [9]. They found out that in comparing the raw coals with their residues, the coal rank determines the changes in the size of specific surface areas and micropore volumes. Furthermore, Zhang et al. conducted an experimental study on coal samples to predict the specific surface fractal dimension. They concluded that coal rank is among the impressive factors on this parameter in a coal rock [10]. Hu et al. also proposed that coal rank is one of the influential parameters on the methane content of the coal [11]. It is also a fundamental parameter in CO_2 sequestration studies in coals, which will be discussed later in ECBM section [12,13].

Coals might sometimes be marked with their proximate analysis, which is based on their organic contents (fixed carbon and volatile matter) and inorganic contents (moisture and ash). The fixed carbon content of a coal rock increases with coal rank, while volatile and moisture constituents decrease. However, there is no clear-cut verdict regarding the ash content of a coal rock based on its rank. Indeed, ash is an index of purity level of coals and is dependent on the coal mineralization as well as burying environment.

8.2.2 Macerals

The organic equivalent of inherent minerals in sandstones is referred to as "macerals" in coals and is mainly composed of remaining fossilized plants in the coal. The macerals are manifest under microscopes. The macerals content of coals could be categorized into three groups: vitrinite, liptinite, and inertinite, which change independently as the coal rock progresses toward coalification [14]. These three macerals group are originated from different parts of a plant. Vitrinite group materials are originated from woody plants (mainly lignin) and make up the major part of macerals in a coal. Increasing vitrinite fraction in a coal macerals composition often correlates with the sorption capacity of the rock [15]. On the other hand, the derivation of liptinite is lipids as well as waxy plant substances, and inertinite group components are originated during peat formation stage from oxidized plant materials. The macerals composition of a coal might include all of the three types or might be accounted for by two or just one of these groups. The macerals composition of the three groups vary toward the same composition as the coal rank increases, and when the carbon content reaches 94% they become almost indistinguishable [16].

One of the approaches based on which the coal rank is determined is through interpreting the maceral composition of the coal rock. For this purpose, the reflected light (which is the reflection of a vertical ray of a prespecified wavelength) by the vitrinite maceral from a polished sample is quantitatively measured. This measurement is accomplished under oil and might vary up to 8% for a given sample based on the orientation of the sample. Subsequent to examining this value for many times in different sample orientations, the maximum vitrinite reflectance is reported. In fact, this measure increases with maturity of coal rocks; therefore, this value is normally considered among the criteria in determination of the coal rank.

8.2.3 Coal Porosity

Coals are among naturally fractured reservoirs for which a dual porosity system is considered, as they encompass a wide spectrum of pore sizes categorized into two distinct groups: primary (micropore and mesopore) and secondary (macropore and cleats) porosity systems. The primary porosity is the major sorption site for the gas and holds the greatest fraction of the containing gas, while secondary porosity system serves as the main conduits to transfer the fluid toward the production wellbore. In coals, micropore and mesopore sites are considered to be impermeable, and the gas flow occurs by diffusion through the porous media toward the cleats. The flow gradient in the primary porosity system, on the contrary to conventional reservoirs, is controlled by gas concentration gradient. The cleat porosity has been suggested to be dependent on the coal composition and rank [17]. Through X-ray CT scanning, Karacan and Mitchell recognized that coal microlithotypes determine the coal porosity [18]. Similarly, Mukhopadhyay and Hatcher suggested that coal porosity is related to both coal type and coal rank [19].

Coal porosity refers to the total void space in a coal rock. However, in some reservoir engineering studies, the mobile water porosity is considered instead of the total porosity. The mobile water porosity is defined as the space filled with water, which will be produced in dewatering stage (the first CBM production stage). It is obvious that in the latter porosity concept, the void space containing gas or immobile moisture is excluded from the porosity, and the coal porosity consists of only the mobile water space inside the cleats, macropore, and some mesopore porosities. In fact, the mobile water porosity is the major conduit of fluid flow toward the wellbore, for which the Darcy flow is applicable. The fracture (cleat) porosity inside a coal is the same as that in typical naturally fractured reservoirs, at about 1% or lower [20]. The estimation of matrix and cleat porosities differs in methodology; matrix porosities are estimated through laboratory experiments, while in order to determine the coal cleat porosity, conceptual models or simulation history matches are reliable tools. However, an experimental method termed "miscible tracer technique" in which the displacing fluid contains a traceable component is also an experimental approach for cleat porosity estimation using cores [21].

It is noteworthy that in coals, the porosity distribution is related to the fixed carbon content of the rock, such that the increase in carbon content would result in a greater fraction of coal porosity to be composed of micropores, which in turn roots in the consolidation of the coal rock as it progresses through coalification process. Therefore, as it is expected, in high-rank coals with the fixed carbon content of over 85%, the coal porosity is mostly accounted for by micropore structures [22]. There is a widely agreed classification of the coal pores with regard to coal rank [23]. This classification is resulted from high-resolution electron microscopy and is illustrated in Table 8.1 [24,25]. As is observed on the table, an increase in the coal rank (from lignite and subbituminous to the highest rank of bituminous and anthracite coals) results in the reduction in the pore size, and also the dominant pore structure changes from macropore in lignite to micropores in the highest ranked coals. Furthermore, it alludes that while in high-rank coal, adsorption is the main storage mechanism due to a high specific surface fractal dimension, low-rank coals might encompass considerable amount of free gas compressed in their void space, in the macropores.

Table 8.1 Relationship Between Coal Rank and Pore Size

Pore sizes	Coal rank (ASTM Designation D388-98a)
Micropores $d < 2$ nm	High-volatile bituminous coal A and higher
Mesopores 2 nm $< d <$ 50 nm	High-volatile bituminous coal (C + B)
Macropores $d > 50$ nm	Lignites + subbituminous

Source: After Rodrigues, C., De Sousa, M.L., 2002. The measurement of coal porosity with different gases. Int. J. Coal Geol. 48 (3), 245–251.

8.2.4 Coal Permeability

Coal permeability is among the most significant parameters in economic and technical viability of a CBM reservoir, and also it is a key factor in predicting CBM behavior in the state of natural depletion as well as ECBM [26–29]. In coals, the matrix system is considered to be impermeable, and the fluid flow is expected to occur through cleat system; thus, cleats serve as the permeability path for Darcy flow of gas and water toward the production well. Theoretically, the development of cleat system permeability is supposed to be dependent on the rank, grade, and type of the concerned coal. It is also affected by in situ stresses, matrix shrinkage during gas desorption, the frequency of natural fractures as well as their interconnection level, degree of fissure aperture opening, face and butt cleats' direction, reservoir depth, and initial water saturation [30]. The permeability value of a coal seam changes during the production profile of the reservoir and is one of the most difficult parameters to be estimated accurately. The gas relative permeability is further challenging to obtain during the reservoir depletion that roots in the changing of cleat aperture measure and the corresponding water saturation inside the cleats within the production of the reservoir. The typical coal seam permeability range is from impermeable to over 100 mD [30].

The permeability of coals could be defined as a function of cleat porosity and spacing when permeability is in millidarcies and cleat spacing is in millimeter square:

$$k = 1.0555(10)^5 \phi^3 a^2 \tag{8.1}$$

Considering the assumption of a stiff coal matrix, in which the cleat spacing during depletion changes negligibly, the permeability ratio is cubically related to the porosity ratio:

$$\frac{k}{k_i} = \left(\frac{\phi}{\phi_i}\right)^3 \tag{8.2}$$

However, this relationship between permeability and porosity is more comprehensively applicable to many naturally fractured reservoirs [2,31].

One of the most striking features of coal permeability, being unique to CBM reservoirs, is its variation mechanism within the production profile of the reservoir.

Coal permeability, similar to other reservoirs, is dependent on the effective stress onto the reservoir that is a function of reservoir depth and pressure differential. However, unique to coals, the gas desorption process is also influential in determination of the permeability value of the reservoir at any given time. In fact, these two effects function competitively. The effective stress increase due to the reservoir pressure drawdown obviously imposes an adverse effect on the cleat permeability through narrowing the fissures. On the other hand, the gas desorption as a response to reservoir pressure drawdown causes the reservoir matrix to shrink, thereby increasing the cleat permeability. Consequently, the interaction of these two phenomena governs the shape of the cleat system and determines permeability value at any given condition [32].

The outcome of this interaction might increase the permeability value up to 100 times, being the case in San Juan basin, USA. This occurrence is in a sharp contrast with conventional reservoirs in which the reservoir depletion leads to a decrease in absolute permeability [33]. In bituminous coals, the typical porosity is around 1%, and 99% of the reservoir volume is accounted for by matrix [2]. Under such condition, given the cubic relationship between permeability and porosity, an increase in cleat porosity from 1% to 2% in response to reservoir depletion and corresponding matrix shrinkage would result in an eightfold rise in permeability value of the reservoir. It goes without saying that for lower rank coals, in which the initial porosity is far more than 1%, the same increase in porosity (1%) results in smaller increase in permeability measure. Therefore, matrix shrinkage has a more significant effect on permeability measure in high-rank coals. Moreover, apart from the positive matrix-shrinkage effect on cleat absolute permeability, such phenomenon favors the relative permeability to gas, because for the same amount of water inside the cleats, an increase in porosity lowers the water saturation percentage.

There has been some models presented to describe the behavior of coal permeability with respect to the effect of mentioned influential factors on this parameter within the reservoir depletion. One of the most widely used permeability models, which is based on the cubic relationship of permeability and porosity, was suggested by Palmer and Mansoori and is shown below after solving for the axial modulus and the relationship between axial and bulk modulus [34]:

$$\frac{\phi}{\phi_i} = 1 + \frac{(1+\nu)(1-2\nu)}{(1-\nu)E\phi_i}(p - p_i) + \frac{c_o}{\phi_i}\frac{2(1-2\nu)}{3(1-\nu)}\left(\frac{p_i}{p_i + p_l} - \frac{p}{p + p_l}\right) \tag{8.3}$$

where ν is the Poisson ratio, E is the Young modulus, p is the pressure at any given time, p_i is the initial pressure, c_o is the volumetric strain coefficient, and p_l is the Langmuir pressure.

The second term on the right-hand side of the equation represents the effects of stress on the porosity, and the third term illustrates how porosity is affected by matrix shrinkage. While during natural depletion, the third term of this equation is always positive because gas desorption causes the matrix to shrink, the second term is negative since the depletion increases the effective stress, and this phenomenon in turn attempts to lower the cleats' width, thereby reducing porosity and permeability. Depending upon the mechanical characteristics and also sorption properties of the rock, either of the two terms might dominate. It should be noted that for ECBM recovery, in which the effective stress decreases with the injection pressure, and matrix swelling is the expected phenomenon rather than matrix shrinkage, the two terms of the above equation function much differently. It will be discussed in detail in the ECBM section.

Example 8.1: San Juan Basin (Fruitland coal)—permeability estimation using Palmer and Mansoori model.

Mavor and Vaughn in 1998 presented some required properties of Fruitland coal located in San Juan Basin for calculating the coal permeability, taking into account the matrix shrinkage as well as stress-related effects [34]. The well VC 32-1 had an average pressure of 956.7 psia in November 1990, which reduced to 527 psia through depletion by October 1994. The volumetric strain was 0.01266 and the initial porosity stood at 0.000457. Young's modulus, Langmuir pressure, and Poisson's ratio of the rock amounted at 521,000 psi, 368.5 psia, and 0.21, respectively.

Substituting the given measures in Eq. (8.3):

$$\frac{\phi}{\phi_i} = 1.22$$

It means that the porosity measure of the rock has been enhanced by 0.22 during the pressure drawdown. Considering the initial absolute permeability of 17.2 mD in Eq. (8.2):

$$\frac{k}{17.2} = \left(\frac{\phi}{\phi_i}\right)^3 = (1.22)^3$$

Therefore, the permeability value of the mentioned coal would be 31.2 mD at 527 psia according to Palmer and Mansoori model.

A couple of years later than the Palmer and Mansoori model, Shi and Durucan introduced a dynamic model for predicting the permeability changes taking into account the effects of stress and matrix shrinkage during the natural depletion of coals [27]. The permeability ratio in this model is exponentially related to the changes in effective horizontal stress normal to the cleats, as observed in Eq. (8.5).

$$\sigma - \sigma_0 = -\frac{\nu}{1-\nu}(p - p_0) + \frac{E\varepsilon_l}{3(1-\nu)}\left(\frac{p}{p + P_\varepsilon} - \frac{p_0}{p_0 + P_\varepsilon}\right) \tag{8.4}$$

$$\frac{k}{k_i} = \exp\left(-3c_f(\sigma - \sigma_i)\right) \tag{8.5}$$

where σ is the effective horizontal stress normal to cleats, σ_i is the initial effective horizontal stress normal to cleats, c_f is the cleat volume compressibility, P_ε is the matrix deformation Langmuir pressure, and ε_l is the matrix-shrinkage coefficient.

Based on their model, Shi and Durucan claimed that the matrix-shrinkage term is 1.5−3 times stronger than Palmer and Mansoori model. Indeed, they postulated that Palmer and Mansoori model has an extra multiplier in the matrix-shrinkage term causing the model to underestimate matrix-shrinkage term [27].

8.2.5 Coal Density

Coal density is among the important parameters for reservoir engineering purposes and is inserted as an input property in simulation studies. Coal density is typically less than that of conventional reservoirs and differs from seam to seam based on the given coal rank and purity [2]. The bulk density of a coal consists of the matrix and the void space, with the latter being expectedly filled with water. The dry coal density, however, is only made up of matrix system density. The coal density is supposed to increase with coalification, implying that high-rank coals are denser than low-rank coal rocks [16]. The bulk coal density was presented by Seidle as a function of the densities of entailed ash, moisture, and the organic rock with their corresponding weight percentage, with the assumption of no free or sorbed gas available in the cleat system [2].

$$\rho = \frac{1}{\left((1-a-w)/\rho_o\right) + \left(a/\rho_a\right) + \left(w/\rho_w\right)} \tag{8.6}$$

where ρ is the coal bulk density, ρ_o is the organic fraction density, ρ_a is the ash density, $\rho_w$is the water density.

It should also be noted that o, a, and w represent the weight percentage of organic, ash, and water content of the rock, respectively.

Eq. (8.6) shows that depending upon the organic and inorganic content of the coal rock, the rock density could vary substantially. In lack of the laboratory measurements for a given coal rock, the organic and ash densities are assumed 1.25 and 2.55 g/cm^3, respectively. These numbers allude that for a high-rank coal, being denser in essence, with a low ash content, the density might be lower than a low-rank coal with a vast presence of ash inside the coal rock. Additionally, the density of organic fraction of coal rock is further dependent on maceral composition of that particular coal.

8.2.6 Coal Rock Mechanical Properties

The mechanical properties of rocks are concerned with the rock physical changes as a result of subjection to force from its physical environment. The mechanics of reservoir rocks is of vital importance in naturally fractured reservoirs, particularly CBM, in which the fissures serve as the circuit to fluid flow and vary in response to changes in effective stress due to reservoir depletion. These properties are also among the main design criteria to assure a promising hydraulic fracturing in the reservoir [35]. Among key mechanical properties, the elastic properties mainly include Young's modulus and Poisson's ratio. These properties are obtainable either by triaxial laboratory examinations of representative samples or through analyzing the field data.

Young's modulus represents the stiffness level of the rock, or in other words, it quantifies the rock resistance against the compressive stress. Young's modulus, with pressure unit, could be calculated through dividing the tensile stress by the extensional strain which are given by

$$\sigma_x = \frac{F_n}{A_o} \tag{8.7}$$

$$\varepsilon_x = \frac{\Delta L}{L_o} \tag{8.8}$$

where σ_x is the tensile stress in the x direction, F_n is the normal force, A_o is the area, ε_x is the strain in the x direction, dimensionless, ΔL is the change of length, L_o is the initial length.

Equating (8.7) and (8.8) for Young's modulus, we will obtain

$$E = \frac{FL_o}{\Delta L A_o} \tag{8.9}$$

Eq. (8.9) vividly describes that for high Young's modulus values, being the property of stiff formations, a large force on the rock area would result in a minor change in the rock length. Therefore, for stiff rocks, the permeability reduction due to increase in effective stress during reservoir depletion is lower compared to softer rocks. The effect of this elastic property of coal rocks has been investigated in some works. Palmer and Mansoori suggested that for soft rocks of low Young's modulus, a considerable permeability reduction within the reservoir depletion is expected, since stress effects play the most important role compared to matrix shrinkage [36]. In fact, coals with smaller Young's modulus compress more in subjection to an increase in effective stress compared to coals with larger values of this property [37]. Geertsma and De Klerk proposed that the major effect of Young's modulus is on the width of the fractures in the coal rock. They claimed that the maximum width of a fracture near the wellbore is inversely proportional to the fourth power of Young's modulus; that is, the lower the value of Young's modulus, the higher the width of the fracture network [38].

Young's modulus for coals is typically in the range of 7000–35,000 bars [39]. High existence of fractures inside the coal rock decreases Young's modulus [39]. Therefore, since fissures inside samples are not authentic representatives of the fracture system in the reservoirs, exact estimation of Young's modulus in laboratory studies is much more challenging. Typically, the modulus of high-rank coals such as anthracite is higher compared to bituminous or lower rank coals [30]. The values of the Young modulus of coals are much lower than that of conventional rocks, and also compared to its surrounding rocks [30]. This difference in the modulus between a coal and adjacent rock, which could be up to an order of magnitude in some cases, contributes to fracture confinement inside the coal [30].

Another important elastic property of coals to be taken into account for reservoir engineering purposes is Poisson's ratio, defined as the ratio of transverse (lateral) strain to axial strain while subjected to uniaxial loads, or in other words, the measure of the lateral expansion versus longitudinal shrinkage for a longitudinally imposed load, Eq. (8.10) [40]. The range for Poisson's ratio for all rocks is between 0 and 0.5, while the typical range of this factor for coals is between 0.2 and 0.4, with its average being somewhere around 0.3 [2].

$$\nu = -\frac{d\varepsilon_{\text{lateral}}}{d\varepsilon_{\text{axial}}} \tag{8.10}$$

where ν is the Poisson ratio, dimensionless, $d\varepsilon_{\text{lateral}}$ is the lateral strain, $d\varepsilon_{\text{axial}}$ is the longitudinal (axial) strain.

Rogers suggested that Poisson's ratio for the reservoir rock and the surrounding rock affects the reservoir stress profile and the parameters defining fracture boundary as well as orientation [30]. Additionally, Poisson's ratio is among the influential parameters in cleat width determination [30]. In reservoir engineering studies, Young's modulus and Poisson's ratio are often interpreted concurrently to evaluate the elastic behavior of reservoir and cleat system in coals. Lu et al. suggested that during hydraulic fracturing, a larger effective disturbance zone is expected in the condition of a low Young's modulus and high Poisson's ratio. This was attributed to the better compressibility of coal in such condition, which in turn would result in a greater shear stress to facilitate disturbance zone expansion [41].

8.3 PRODUCTION PROFILE IN COALS

The production mechanism in coals is exclusively associated with this type of reservoir and differs substantially from that of conventional reservoirs. The production process in coals mostly initiates with extracting a considerable amount of water, being the mobile water originally filling the cleats. This stage of coal production is referred

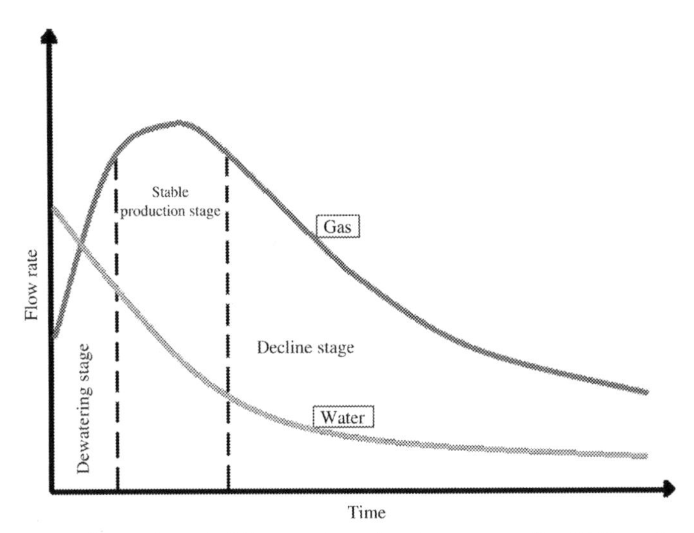

Figure 8.1 Schematic illustration of CBM typical production rate profile. *CBM*, coalbed methane.

to as the dewatering stage. In terms of economic viability of a CBM reservoir, the water removal stage is among the key criteria, because it often adds a remarkable cost to the process.

Generally, the decline curve of oil and gas reservoirs is a reliable source for estimating the reserves. This curve typically follows specific patterns (exponential, harmonic, or hyperbolic), providing the ability for reservoir engineers to predict the production rate profile as well as annual production until the abandonment. The decline curve has been given this name based on the decline that occurs in the production rates of conventional reservoirs, throughout the reservoir life time. However, one of the most striking features of CBM reservoirs, which is conspicuously different from conventional reservoirs, is the shape of this curve in the very early stages of the process. In coals, the gas production rate increases in the dewatering stage, until reaching a plateau after which negligible amount of water will be produced. Therefore, the beginning stage in a CBM reservoir production is termed "negative decline." Subsequent to this gas production rate increase, the curve starts declining and follows a decline pattern. Fig. 8.1 illustrates the production rate profile of a typical CBM reservoir, schematically.

8.4 GAS-FLOW MECHANISM IN COALS

Gas flow in coals, from the initial storage sites (micropores) toward the wellbore, consists of three stages: desorption, diffusion, and permeation. In the dewatering stage of CBM production, the reservoir pressure decreases, which in turn results in

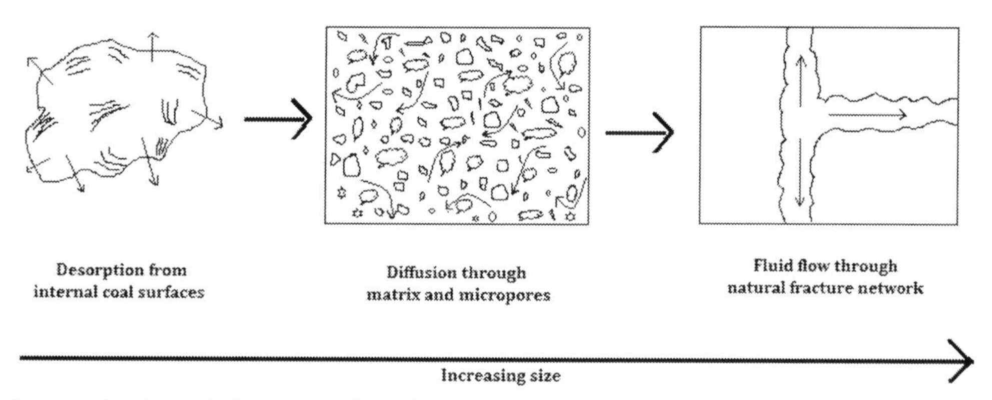

Figure 8.2 Schematic illustration of gas-flow stages in coals.

desorption of the gas molecules from the sorption sites. The desorbed gas from the micropores diffuses through the porous media (governed by Fick's law) and enters the cleat system in which the gas flows freely (Darcy flow). Therefore, these mechanisms are discussed in this section, focusing on sorption and diffusion. Fig. 8.2 schematically illustrates theses three steps associated with gas flow in coals.

8.4.1 Sorption

The main gas-storage mechanism in coals differs from that of conventional reservoirs in which the compressed gas fills the void space (pores) due to imposed pressure from overlying planes. In coals, a little amount of dissolved gas in water or free gas might exist in cleats, and the bulk volume of gas is stored onto the surface of micropores, a mechanism termed sorption or adsorption. Therefore, in order to extract gas from CBM reservoirs, gas desorption is the first step, accomplished through pressure draw-down in the reservoir, which is in turn viable by dewatering the reservoir. The adsorbed gas is attached to the coal rock by van der Waal's bonds between gaseous and solid molecules. Therefore, the affinity of a given rock for gas adsorption differs for different gases. CH_4 and CO_2 are two gases with a strong capability to bond with the coal rock surface. It should also be mentioned that most of the existing gas in coals in adsorbed state is at liquid-like density [42].

There is a critical point for the pressure in coal reservoirs, termed saturation pressure, over which no gas molecule is desorbed from or adsorbed on the rock surface during pressure alteration. However, for reservoir pressures lower than this point, any reduction in pressure (during reservoir depletion) would lead to gas desorption. The desorption rate is controlled by reservoir temperature, porosity distribution, gas composition, coal rank, and composition [43,44]. One of the most striking features of coals is their abnormal large pore surface area, providing the site on the rock to store enormous amount of gas in adsorbed state. A given reservoir volume of a coal might encompass two to three times more gas compared to the same volume of a sandstone having 25% porosity and 30% water saturation, with the same burial depth [30].

Corresponding to the unique gas-storage mechanism in coals, the estimation of original gas in place, at preliminary stages of field development and during the production profile, is not viable through conventional volumetric approaches for sandstone reservoirs. Therefore, in reserve estimation in coalbeds, a new set of equations considering adsorption must be taken into account. Yang suggested that for a given adsorbate and adsorbent, the amount of adsorption at equilibrium is a function of pressure and temperature [45]

$$V = F\,(p,\ T) \tag{8.11}$$

Therefore, at constant temperature, the rate of adsorption is only a function of pressure. Assuming negligible temperature change in the reservoir during production, the adsorption of gaseous phase on coal rocks is described by sorption isotherms. In fact, a sorption isotherm illustrates the relationship between the volume of adsorption of a given adsorbate on the surface of a specific adsorptive rock as a function of pressure, at a constant temperature. The adsorption of a gas on a solid has been categorized into five different types, each of which has its distinct curve on the pressure–adsorption graph [46]. The isotherm type I, as depicted schematically on Fig. 8.3, was found to be applicable in adsorption of gaseous phase onto the surface of microporous solids [46].

As observed on Fig. 8.3, under low-pressure conditions, an alteration in pressure results in a considerable amount of gas desorbed from or adsorbed on the rock surface. In higher ranges of pressure, however, the same pressure alteration leads to much lower amount of gas content change on the rock. This trend continues until the saturation point, beyond which no adsorption is expected with an increase in pressure. The condition of saturation implies that all of the adsorption sites (rock surface) is covered with a monolayer of the gas, or in other words, the rock's maximum capacity

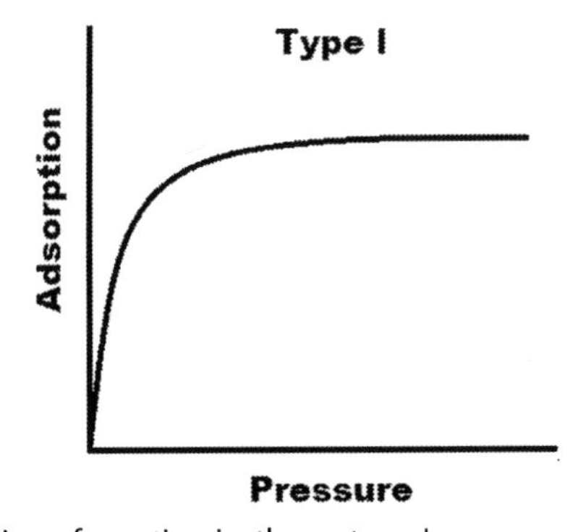

Figure 8.3 Schematic depiction of sorption isotherm type I.

for gas adsorption is reached. It should be noted that pressure increase occurs in laboratory investigations for the purpose of sorption characteristic determination, and in reservoir conditions during depletion, pressure decreases and desorption is the expected phenomenon.

The adsorption/desorption behavior of gases on coals could be described with type I isotherms. Among the most popular models presented to describe gas sorption on rocks is Langmuir and Freudlich equation that fits the type I isotherm curve. This equation closely matches the gas adsorption data on rocks in CBM reservoirs and is widely used to predict the sorption behavior of gaseous phase on coals [2].

$$V = V_L \frac{P}{P + P_L} \tag{8.12}$$

where V is the gas content (in adsorbed state), V_L is the Langmuir volume constant, P is the pressure, P_L is the Langmuir pressure constant.

Langmuir volume constant is actually the maximum capacity of a coal rock to adsorb gas onto its surface area. Eq. (8.12) clearly implies that the higher the reservoir pressure, the more gas adsorbed onto the coal, and at the infinite pressure, the maximum adsorption capacity of a coal is expected to be met. It also reveals that Langmuir pressure constant represents a pressure at which half of the Langmuir volume constant of the coal surface is covered with gas. Therefore, the Langmuir pressure constant indicates the half-saturation pressure.

There are several factors affecting the maximum capacity of a coal rock to adsorb gas. In coals, some of the potential gas adsorption sites might be preoccupied with ash or moisture, both of which obstruct the access of gas to these sites, hence lowering volume constant at a specific pressure. Therefore, for reservoir engineering purposes, the Langmuir volume constant needs to be replaced with the dry, ash-free (daf) Langmuir volume constant in order to minimize the estimation errors. Particularly, in low-rank coals where a considerable amount of water is available, and also the rock has higher affinity to water than methane in some sorption spots, the moisture content would play a significant role in determination of Langmuir volume constant. Therefore, isotherm curves for coal samples must be drawn taking into account the moisture and ash contents. Eq. (8.13) considers effects of moisture and ash content in Langmuir equation [2]:

$$V_{\mathrm{Ldaf}} = \frac{V_{\mathrm{Lis}}}{(1 - a - w)} \tag{8.13}$$

where V_{Ldaf} is the dry, ash-free Langmuir volume constant, V_{Lis} is the in situ Langmuir volume constant, a is the ash mass fracture, w is the equilibrium moisture mass fraction.

The sorption capacity of coal is also influenced by coal rank. Coals of higher ranks are typically more tortuous rocks presenting more complicated pore structures.

Consequently, high-rank coals provide larger surface area and more sorption sites compared to low-rank coals, thereby increasing the maximum adsorption capacity. Another parameter that could alter the sorption capacity of a given coal rock at a specific pressure is temperature. There is a negative correlation between the maximum sorption capacity and temperature of a given coal at a specific pressure. Although considerable variation in reservoir temperature occurs quite seldom in practice, this evaluation might be accomplished for some especial purposes such as extrapolation of the data obtained from a shallower taken sample in the exploration stage of a basin, to a deeper basin section that is higher in temperature [2].

In order to make sure regarding the accuracy of the sorption isotherm considered in reservoir engineering evaluations for a given coal seam or basin, a number of isotherms should be calculated and compared to find the most reliable and precise isotherm curve to characterize the reservoir. Seidle suggested that the number of required sorption isotherms depends on the scale of the reservoir section considered; therefore, this number is different for a single seam, multiple seams in a single well, a project, and a basin [2]. While the justifiable number of required sorption isotherms varies from three isotherms in good consistency for a single coal seam, more than a dozen isotherms are required to characterize the sorption behavior of a basin [2]. This is quite obvious that this number increases with complexity degree of coals. It should also be mentioned that the simplistic consideration of a single isotherm for a given seam or well might provide substantially erroneous results, and this assumption would be much worse for a given basin.

The initial gas content in coals is mainly accounted for by methane, while considerable amount of other gases such as CO_2, ethane, and nitrogen might also exist in coal seams. The sorption isotherm of a given coal differs for distinct gases, and the coals' affinity degree toward various gases is related to their boiling point [47]. Typically, coals have more affinity to adsorb CO_2 than methane and have the least affinity to nitrogen among these three gases [47]. Consequently, for a given coal, the Langmuir volume constant for CO_2 is higher and for nitrogen is lower than methane. Correspondingly, among the three gases, the Langmuir pressure constant is the lowest for CO_2 and the highest for nitrogen. Therefore, coal seams are potential sites for CO_2 sequestration, a process which is also aimed at enhancing methane recovery from CBM reservoirs.

For sorption characterization of a coal in the state of entailing a mixture of gases, the Langmuir equation should be modified to be capable of describing multicomponent sorption behavior. The modified Langmuir equation is referred to as extended Langmuir isotherm equation [2]:

$$V_j = V_{Lj} \frac{p y_j / p_{Lj}}{1 + \sum_{k=1}^{nc} p y_k / p_{Lk}} \tag{8.14}$$

where V_j is the gas content of component j, V_{Lj} is the Langmuir volume constant of component j, p_{Lj} is the Langmuir pressure constant of component j, y_j is the free gas mole fraction of component j, and p is the reservoir pressure.

It should be noted that in gas reserve estimation of a coal, the total amount of free gas, dissolved gas in water, and adsorbed gas should be considered to prevent any reserve underestimation.

Example 8.2: San Juan Basin (Fruitland coal)—methane content estimation using extended Langmuir equation.

Arri et al. provided sorption characteristics of Fruitland formation for some gases including CO_2 and methane [48]. The Langmuir pressures for these two gases are 204.5 psia and 362.3 psia, respectively. Additionally, the Langmuir volume constant of CO_2 and methane were estimated at 1350.1 scf/ton and 908.4 scf/ton. Assuming the coal entails 90% mole fraction methane and 10% CO_2, at 800 psia the methane content of coal is calculated using Eq. (8.14)

$$V_{\text{methane}} = 908.4 \times \frac{\left((800 \times 0.9)/362.3\right)}{1 + \left((800 \times 0.9)/362.3\right) + \left((800 \times 0.1)/204.5\right)}$$

Therefore, in such reservoir condition, there is 534.3 scf of methane per each ton of the coal rock.

8.4.2 Diffusion

Subsequent to desorption from internal coal surfaces, the gas diffuses through matrix and micropores to reach cleats. Therefore, gas diffusion in coal matrix system is the second step in gas movement, for describing its mechanism that some models have been suggested.

8.4.2.1 Unipore Model

One of the earliest models presented to describe the gas diffusion behavior in coal matrix is unipore diffusion model. This model has been developed based on the Fick's second law with the assumption of a spherical symmetric flow [44,49]. The unipore model, Eq. (8.15), assumes that microspores in coal are monosized; whereas, in practice, coal matrix has a variety of pore sizes. Therefore, the unipore model is accurate only for coals with homogenous pore structure.

$$\frac{\partial C}{\partial t} = D\left(\frac{\partial^2 C}{\partial r^2} + \frac{2}{r}\frac{\partial C}{\partial r}\right) \tag{8.15}$$

Eq. (8.15) reveals that rather than pressure gradient, diffusion is dependent on concentration gradient. Fick's second law could be solved in terms of fractional adsorption/desorption:

$$\frac{M_t}{M_\infty} = 1 - \frac{6}{\pi^2}\sum_{n=1}^{\infty}\frac{1}{n^2}\exp\left(-n^2\pi^2\frac{Dt}{r_p^2}\right) \tag{8.16}$$

where M_t is the amount of adsorbed/desorbed gas at time t, M_∞ is the total amount of adsorbed/desorbed gas at equilibrium condition, D is the diffusion coefficient, r_p is the mean radius of coal particle radius.

In order to estimate diffusion coefficient measure, D, at a given time, the experimental data is inserted into Eq. (8.16), thereby D is calculated. Since in this equation micropores are assumed to be monosized, this model is referred to as "unipore diffusion model." Although unipore models have been applied to coals, they were proved to fit data only over restricted time intervals [44,50,51]. Thus, predicting gas flow in heterogeneous pore structures requires the consideration of different pore sizes in diffusion models.

8.4.2.2 Bidisperse Model

Considering the fact that the coal structure is highly heterogeneous, the unipore model usually does not predict the diffusion coefficient precisely [52–54]. Ruckenstein et al. proposed a bidisperse diffusion model to more realistically describe the pore size distribution and consequently more accurately predict the diffusion behavior [55]. Bidisperse model limits pore size distribution to two sizes: macropore and micropore. In this model, the adsorbent contains microporous spherical particles separated by inter-particle macropores. The bidisperse model was applied to coal for the first time by Smith and Williams [56,57]. In this approach, coal matrix is assumed as a double porosity medium with two distinct pore sizes, macropores indicating fast diffusion [Eq. (8.17)] and micropores characterized by slow diffusion, as observed in Eq. (8.18) [58,59].

$$\frac{M_{at}}{M_{a\infty}} = 1 - \frac{6}{\pi^2}\sum_{n=1}^{\infty}\frac{1}{n^2}\exp\left(-n^2\pi^2\frac{D_a t}{r_{pa}^2}\right) \tag{8.17}$$

$$\frac{M_{it}}{M_{i\infty}} = 1 - \frac{6}{\pi^2}\sum_{n=1}^{\infty}\frac{1}{n^2}\exp\left(-n^2\pi^2\frac{D_i t}{r_{pi}^2}\right) \tag{8.18}$$

where M_{at} and M_{it} are the gas adsorption/desorption amount from macropores and micropores at time t, respectively; $M_{a\infty}$ and $M_{i\infty}$ are the total amount of adsorbed/desorbed gas in macropores and micropores at equilibrium condition, respectively; D_a and D_i are the macropores' and micropores' diffusion coefficients, respectively.

The total amount of gas uptake/desorption of both micro/macropores is calculated as follows:

$$\frac{M_t}{M_\infty} = \frac{M_{at} + M_{it}}{M_{a\infty} + M_{i\infty}} = (1-\alpha)\frac{M_{at}}{M_{a\infty}} + \alpha\frac{M_{it}}{M_{i\infty}} \tag{8.19}$$

where α is $\left(M_{i\infty}/(M_{a\infty} + M_{i\infty})\right)$.

Although bidisperse model fits the data more accurately compared to unipore model, due to a greater number of fitting variables, it yet does not include the whole range of pore sizes.

8.4.2.3 Pseudo Steady State Model

Another model to describe gas diffusion behavior in coals is pseudo steady state model. This model describes gas diffusion in coal matrix as follows [60,61]:

$$\frac{dM}{dt} = -\frac{[M_\infty - M_t]}{t_0} \tag{8.20}$$

where t_0 is a time constant indicating the required time for adsorption/desorption of 63.2% of the total amount of gas at equilibrium condition. This equation, when separating the variables and integrating both sides, would be rewritten as

$$\int_0^{M_t} \frac{dM}{[M_\infty - M_t]} = \int_0^t -\frac{dt}{t_0} \tag{8.21}$$

The solution of Eq. (8.21) results in

$$\frac{M_t}{M_\infty} = 1 - \exp\left(-\frac{t}{t_0}\right) \tag{8.22}$$

This explicit equation suggests that adsorption/desorption amount of a gas in coals is exponentially correlated to time.

Furthermore, an experimental fitting parameter, β, was added to Eq. (8.22) to get more accurate history match results [53,54,62]:

$$\frac{M_t}{M_\infty} = 1 - \exp\left[-\left(\frac{t}{t_0}\right)^\beta\right] \tag{8.23}$$

While this parameter is aimed at describing the spread in sorption times, the variable β varies between 0 and 1 based on coal characteristics. Drawing a comparison among unipore, bidisperse, and exponential models reveals that the kinetics phenomena of gas diffusion in coal, during adsorption/desorption process, is more accurately described by exponential model [Eq. (8.23)] than unipore and bidisperse models [53,54].

In spherical crushed core samples, the relationship between the sorption time constant, t_0, and diffusion coefficient, D, is presented in Eq. (8.24) [63]:

$$D = \frac{r_p^2}{t_0} \tag{8.24}$$

In this equation, r_p represents the mean radius of coal particles, and the term D/r_p^2 is referred to as diffusivity with the dimension of 1/time.

Example 8.3: Sorption time.

Keshavarz et al. conducted sensitivity analyses on diffusion coefficient measure on 18 Australian coal samples [64]. The sample No. 8 had the diffusivity measure of 0.098 and 0.53 hour^{-1} for CH_4 and CO_2, respectively. The sample also had the β value of 0.53 and 0.5 for methane and CO_2. The required time for this sample to release 90% of the adsorbed gas for each of the gases is calculated using Eq. (8.23).

For CH_4,

$$0.9 = 1 - \exp\left[-(0.098 t_{90\%})^{0.53}\right] \rightarrow t_{90\%} = 49 \text{ hour}$$

For CO_2,

$$0.9 = 1 - \exp\left[-(0.53 t_{90\%})^{0.5}\right] \rightarrow t_{90\%} = 10 \text{ hour}$$

The results of this example show that the gas diffusion happens much quicker for CO_2 compared to methane.

8.4.2.4 Upscaling From Laboratory to Reservoir Scale

As discussed above, pseudo steady state diffusion model, Eq. (8.18), could closely describe the gas flow from matrix system to cleats during desorption and from cleats to matrix system within adsorption process in CBM reservoirs, as observed on Fig. 8.4 [60,61]. However, there is a difference between the time constant, t_0, related to laboratory experiments to that of reservoir scale. Kazemi model suggests that the matrix–fracture interface area per unit bulk volume, σ, in naturally fractured reservoirs is defined as [65–67]

$$\sigma = 4\left(\frac{1}{a_x^2} + \frac{1}{a_y^2} + \frac{1}{a_z^2}\right) \tag{8.25}$$

where a_i represents the fracture spacing in the i direction.

Therefore, in reservoir scale, the adsorption/desorption time, τ, would be

$$\tau = \frac{1}{D\sigma} \tag{8.26}$$

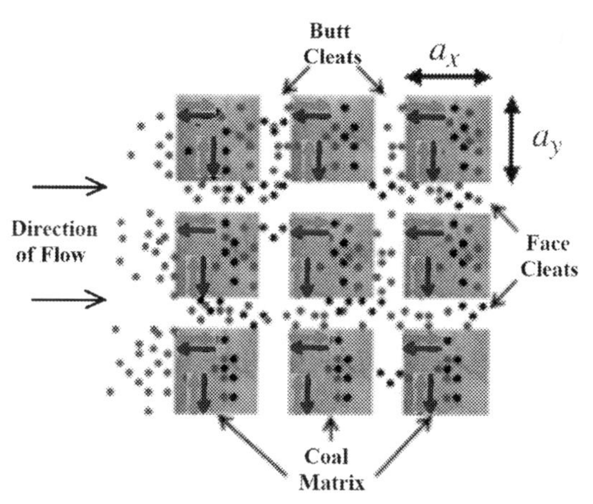

Figure 8.4 Gas transport in coal matrix and cleats. *Modified after Busch, A., Gensterblum, Y., 2011. CBM and CO_2-ECBM related sorption processes in coal: a review. Int. J. Coal Geol. 87 (2), 49–71.*

The desorbed gas that has diffused through the micropore and macropore media permeates through the cleat system toward the production well, a flow which is described by Darcy equation and is the third mechanism in gas flow in coal rocks.

8.5 COALBED METHANE PRODUCTIVITY AND RECOVERY ENHANCEMENT

Increasing gas production from CBM reservoirs is viable through two discrete approaches. The first method is productivity enhancement via hydraulic stimulation in which increasing the rock permeability is the main goal. The second method, termed ECBM recovery, is accomplished through injecting a foreign gas into the reservoir in order to maintain the pressure of the reservoir and also, in the case of injecting a more adsorbable gas than methane such as CO_2, to facilitate methane desorption from the rock surface.

8.5.1 Hydraulic Stimulation

One of the main challenges in gas production from CBM reservoirs is their low productivity index. Therefore, applying stimulation techniques can improve gas production from these type of reservoirs. Hydraulic fracturing and natural fracture stimulation techniques are the main productivity enhancement techniques in CBM reservoirs. In hydraulic-based stimulation techniques, high-pressure fluid is injected into the wellbore. The injected fluid improves conductivity of the existing cleat network in the coal seam if the injection pressure is below the fracturing pressure (natural fracture stimulation). The injected fluid can initiate and propagate new fractures if the injection pressure is above the fracturing pressure (hydraulic fracturing). In order to keep the conductivity of the fractures, after reproduction of the injected fluid, small propped agents (proppants) are added to the injection fluid. The proppants are placed inside the stimulated or induced fractures to mitigate fracture closure due to injected fluid withdrawal and hydrocarbon production.

8.5.1.1 Hydraulic Fracturing

Hydraulic fracturing is a common technique for productivity enhancement in conventional oil and gas reservoirs by inducing high-conductivity fractures from reservoir to wellbore. Hydraulic fracturing was applied successfully for the first time in the Hugoton gas field in 1947 [68]. Since that time, it has been recognized as one of the main stimulation techniques for productivity enhancement and has been conducted in many conventional oil and gas fields. Oil and gas industry has become more interested in hydraulic fracturing treatment recently because of its unique role in productivity enhancement from unconventional reservoirs (e.g., shale, CBM, tight sands, etc.). However, the design and performance of hydraulic fracturing treatment in unconventional reservoirs differs from that in conventional ones.

Hydraulic fracturing is the most common stimulation technique in CBM reservoirs and improves the gas productivity by connecting cleat network to wellbore [69]. However, fracture propagation is the main challenging issue as the injected fluid can divert into the natural cleats in the intersection of hydraulic fracture and the existing cleat system. In this case, short and noncontinuous cracks are created, instead of a single massive fracture, within a complex system of natural cleats and joints [70–75]. The induced complex fracture geometry can significantly decrease the efficiency of hydraulic fracturing treatment by accelerating fluid leak-off and causing ineffective proppant transport [76,77]. In addition, uncontrolled fracture propagation may cause some environmental issues such as contamination of nearby aquifers used for agriculture.

8.5.1.2 Natural Fracture Stimulation

The aim of natural fracture stimulation is to improve conductivity of the preexisting fracture network by injecting high-pressure fluids [78–80]. In this technique, injection pressure must not exceed the fracturing pressure. In natural fracture stimulation, the following mechanisms lead to conductivity enhancement of the fracture network.

1. *Improving conductivity of existing fracture network due to slip-dilation (shear dilation stimulation).*

   In this stimulation technique, injection of high-pressure fluid causes shear stress perturbation in the fracture network. Fracture surfaces slip to each other due to the shear stress perturbation and improve flow passes by offsetting the two fracture's rough surfaces [78]. Therefore, application of this stimulation method in CBM reservoirs can improve CBM productivity index because of mismatches and asperities in the cleat system created due to relocation of cleat's walls from their original place [76]. This is a long-term fracture stimulation as created asperities on the rough fracture walls resist against sliding back to their original locations after withdrawal of the injected fluid.

2. *Improving conductivity of preexisting fracture network by increasing the average of fractures' aperture due to high-pressure fluid injection.*

   In stress-dependent fractured reservoirs such as CBM reservoirs, the average of fractures' aperture in preexisting fracture network is proportional to the reservoir pressure. Therefore, increasing reservoir pressure can improve the conductivity of the fracture system by increasing fracture's aperture. In this stimulation technique, although injecting high-pressure fluid opens up preexisting fractures and improves fracture conductivity and connectivity, the opened fractures may get closed after withdrawal of the injected fluid. Therefore, the main challenge is to keep the fractures open after pressure decline.

8.5.1.3 Proppant Placement

The aim of hydraulic stimulation techniques, either hydraulic fracturing or natural fracture stimulation, is to improve productivity index by creating high conductive flow conduit from the reservoir toward the wellbore. This can be achieved by injecting high-pressure fluids into the reservoir. However, in order to keep the fractures open subsequent to pressure decline, injected fluids should be accompanied by rigid small propping agents (proppant) to be placed inside the open fractures in order to lessen the influence of pressure dissipation on fracture conductivity, due to withdrawal of the injected fluid.

Induced fractures are filled by multilayers of proppants in hydraulic fracturing treatments. The pack of proppants creates an artificial conductive porous media inside the fracture resisting against fracture closure in the production stage. Darin and Huitt reported that larger conductivity is achieved where a hydraulic fracture is filled by a partial monolayer of large-sized proppant particles rather than fully packing by multilayers of small-sized proppant particles [81]. The partial monolayer proppant placement technique can also be used to prop up natural fracture system using microsized proppant particles [82–84]. Two main parameters influencing fracture conductivity are confining stress and proppant concentration [82,85–88]. The higher the concentration of placed proppants in the fracture, the more barrier against flow conductivity in the fracture. The lower the concentration of placed proppants in the fracture, the higher the risk of fracture deformation and conductivity decline, as observed on

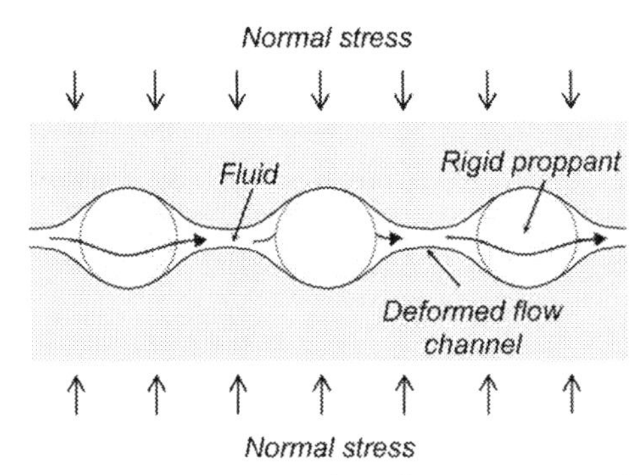

Figure 8.5 Effect of proppant concentration and confining stress on fracture conductivity. *After Khanna, A., Keshavarz, A., Mobbs, K., Davis, M., Bedrikovetsky, P., 2013. Stimulation of the natural fracture system by graded proppant injection. J. Pet. Sci. Eng. 111, 71–77.*

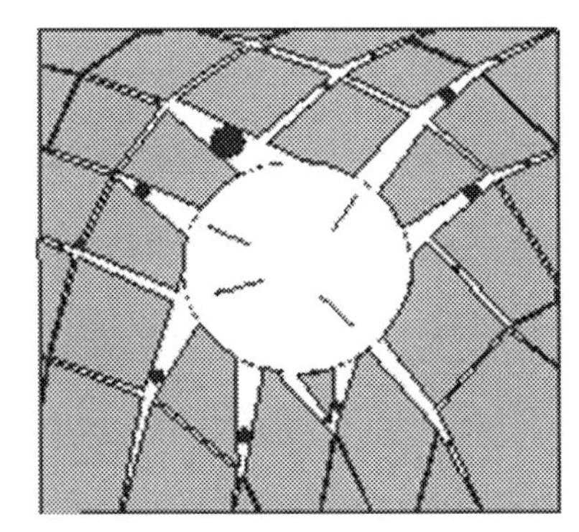

Figure 8.6 Schematic view of graded proppant injection into natural fracture system. *After Khanna, A., Keshavarz, A., Mobbs, K., Davis, M., Bedrikovetsky, P., 2013. Stimulation of the natural fracture system by graded proppant injection. J. Pet. Sci. Eng. 111, 71–77.*

Fig. 8.5. Therefore, for a given confining stress, there is an optimum concentration of placed particles in which fracture conductivity is maximized [76,83,84,89].

A novel technique, graded proppant injection, has been developed recently to stimulate natural cleats in CBM reservoirs using different sized micro proppants [76,84,89–95]. The graded proppant injection technique aims to improve cleat network conductivity by extending the stimulated zone around the wellbore. In fact, during the injection, pressure decreases along each fracture from well toward the reservoir. Hence, fracture aperture decreases too (Fig. 8.6). Therefore, the injection process is started by injecting small proppants to cover small fractures farther from the well and expand the stimulated area. Then, the intermediate-sized proppant particles are injected to be placed inside the cleats in the bulk of the drainage area. Gradually increasing the size of proppant particles during the injection stimulates the bigger fractures located near to the wellbore (Fig. 8.6).

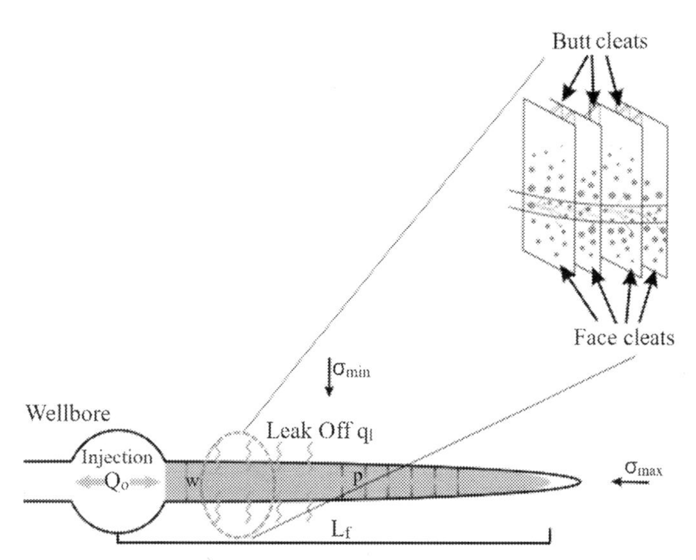

Figure 8.7 Schematic view of microsized graded proppant injection into the natural cleat network connected to the main hydraulic fracture. *After Keshavarz, A., Johnson, R., Carageorgos, T., Bedrikovetsky, P., Badalyan, A., 2016. Improving the conductivity of natural fracture systems in conjunction with hydraulic fracturing in stress sensitive reservoirs. In: SPE Asia Pacific Oil & Gas Conference and Exhibition. Society of Petroleum Engineers.*

Keshavarz et al. coupled the graded proppant injection technique with hydraulic fracturing treatment in CBM reservoirs to enhance postfrac conductivity of cleat network around the created hydraulic fracture [76,77]. In this method, microsized monolayer proppant particles are leaked-off into the natural cleat system at the leak-off pressure condition, as seen on Fig. 8.7 [96]. The proppants placed inside the cleat network maintain the cleat network conductivity and uphold the connectivity between natural cleat system and the main hydraulic fracture during the post fracturing production. Application of this technique can significantly improve the efficiency of hydraulic fracturing in CBM reservoirs by expanding the stimulated reservoir volume. In addition, the graded microsized proppants placed inside the cleat network decrease the fluid loss due to leak-off during the hydraulic fracturing operation, which cause the extension of the fracture half-length.

8.5.2 Enhanced Coalbed Methane Recovery

ECBM recovery is a secondary production mechanism to enhance the methane ultimate recovery and/or improve methane production rate, by means of injecting a foreign gas. The commonly used gases are nitrogen (N_2), carbon dioxide (CO_2), or a mixture of N_2 and CO_2. If the injected gas contains CO_2, the ECBM can be also beneficial from a carbon sequestration standpoint.

The natural depletion (primary production) of CBM resources may result in less than 50% methane recovery [3]. Besides, the methane depletion rate is usually marginal, due to one or a combination of different factors, such as (1) low initial pressure, (2) low cleats permeability and connectivity, (3) low gas diffusivity, and (4) high water production. In the natural depletion, at the beginning, mainly water is produced (in undersaturated conditions). Once the cleat pressure falls below the equilibrium pressure corresponding to the adsorbed gas content, the liberation of gas from the matrix starts. The equilibrium pressure depends on the coal matrix adsorption capacity, gas type, and total adsorbed gas. The liberated gas diffuses from the matrix to the cleats and then flows along with the remaining water, from the cleats to the wellbore. The gas rate usually increases, by the reduction of water-in-place, and the two-phase flow continues until the water saturation reaches the residual saturation or until the system energy (pressure) becomes insufficient for pushing water to the wellbore. Subsequent to the two-phase flow, there will be dominantly a single-phase flow of gas. The gas production continues until a point at which the desorption rate becomes insufficient for having an economically viable production, as a result of matrix gas content reduction and/or cleats closure.

To enhance the recovery and improve the production rate at the same time, (1) the reservoir pressure should be maintained at a reasonably high value (creating the pressure gradient for effective convection flow in cleats and also avoid cleats closure), and (2) methane partial pressure gradient, between cleats and matrix, should be maximized (accelerating desorption and diffusion process). A continuous injection of a foreign gas into the coalbed allows us to minimize methane partial pressure in cleats and simultaneously increase total cleat pressure. The injected gas sweeps methane from the cleats, resulting in methane partial pressure drop, while the injected mass into the system maintains the reservoir pressure, keeping the cleats open and boosting flow in cleats. The performance of ECBM is controlled by several factors, including the competitive sorption characteristics and the geomechanical and petrophysical properties of the coal.

One of the early studies on ECBM was conducted by Fulton et al.; they carried out laboratory analyses on five core samples from Pricetown mine in West Virginia [97]. By introducing the CO_2 to the coal samples, through a cyclic or single-stage injection, the recovery factor increased, compared to a natural depletion mechanism. Sinayuç and Gümrah performed a simulation study based on the data of Zonguldak coal basin [98]. In the study, an increment of 23% CH_4 production was observed by injecting CO_2. The first field application of CO_2-ECBM injection was piloted in Allison unit of San Juan basin, operated by Burlington Resources. CO_2 was injected for 6 years (1995–2001). The results revealed an increment of methane recovery, one additional volume of methane per three volume CO_2 injected [99].

An additional benefit of the CO_2-ECBM is the carbon sequestration, since the injected CO_2 is adsorbed on coal matrix [100]. Normally, coal affinity toward CO_2 is greater than its affinity toward CH_4 or N_2, making coalbeds a suitable host for carbon sequestration. The higher CO_2 affinity than methane can be explained by the geometrical, electrical, and physical–chemical properties of gas molecules and also the functional groups on the coal surface [101,102]. CO_2 molecules have a linear shape and are smaller than CH_4 molecules. These features facilitate CO_2 entrance to more restricted pore spaces and dislocate competitively preadsorbed CH_4 molecules out of micropore surfaces [101]. Moreover, CO_2 has a more favorable interaction enthalpy than methane, which increases CO_2 solubility in coals and may increase the CO_2 diffusivity and sorption capacity [102].

Due to this higher affinity, a larger amount of CO_2 is adsorbed, relative to the amount of desorbed methane. This leads to matrix swelling and may create operational problems in a CO_2-ECBM project. As the matrix is swollen, the cleat openings decrease, hence permeability decreases [103,104]. The injectivity decline was observed in the pilot study in Allison unit. A numerical study showed that the permeability might decrease by two orders of magnitude [103]. Mazzotti et al. conducted an experimental study on the coal matrix swelling effect for different types of gases [105]. In the study, the changes of coal volume subsequent to exposure to different gases (CO_2, N_2, CH_4, and He) were measured. The coal was swollen more severely by CO_2 in comparison with the other gases.

Nitrogen injection, known as N_2-ECBM, is another well-studied technique for methane recovery enhancement. In N_2-ECBM, the deterioration of well injectivity is not a concern, because N_2 is less adsorptive on coal than methane and CO_2. One of the first studies on N_2-ECBM is the paper written by Puri and Yee in which they investigated the impact of N_2 injection on methane recovery, through laboratory experiments and modeling [3]. They concluded that introducing N_2 to coal enhances methane recovery. N_2-ECBM was piloted for the first time in Tiffany unit of San Juan Basin, operated by BP America. N_2 was injected for 4 years (1998–2002) and resulted in methane recovery increment of one volume per 0.4 volume of injected N_2. In other studies, it was observed that the N_2 injection also boosts the production rate, rapidly and notably [106,107]. The main issue of N_2-ECBM is the early breakthrough of nitrogen. The low adsorbability of N_2 on the coal induces matrix shrinkage that increases the well injectivity. But most of the injected N_2 travels toward the producer(s), instead of being adsorbed on the coal matrix, which in turn causes an early breakthrough. The experimental study conducted by Zhou et al. confirmed the early breakthrough problem [108].

A more promising result can be obtained by injecting a mixture of different gases. Shi and Durucan carried out a micropilot study in the Fenn Big Valley. They argued that by injecting flue gas, more desirable results can be obtained [26]. Durucan and Shi investigated the overall performance of different N_2/CO_2 mixtures on production gas rate and the quality of produced gas [103]. It was seen that production rate increases, by an enriched N_2 mixture, although, the quality of the produced gas is still of concern due to the quick N_2 breakthrough. The best performance was obtained by a mixture of 13% CO_2 and 87% N_2. Sayyafzadeh et al. claimed that the performance of mixture gas injection can be improved by applying a varying composition strategy, throughout a continuous injection of N_2/CO_2 mixture [4]. A series of sensitivity analyses were performed to find an optimal scenario. The best scenario was the one that starts by injecting a mixture with less CO_2 and continues by a sequential rise in the CO_2 fraction.

The optimal gas composition depends on the petrophysical, geomechanical, and sorption characteristics of the coal. Hence, it is suggested to conduct a sensitivity analyses or an optimization using numerical simulation to find an operationally and economically viable scenario. Sayyafzadeh and Keshavarz used a genetic algorithm to optimize well controls and injectant composition to maximize the revenue from a semisynthetic coalbed model [109].

8.5.2.1 The Governing Equations for Modeling ECBM

A set of partial differential equations (PDEs) should be solved to model and simulate fluid flow in an enhanced coalbed recovery process. This allows us to predict adsorbed gas content (V_i), mole fraction in cleats (y_i), pressure (p_l), saturation (s_l), well flow rate (q_i), and well bottomhole pressure (p_{wf}) in space and time (x, y, z, t). i and l denote component and phase, respectively. The PDEs are derived based on the following laws and equations, including mass conservation law, Darcy's law, Fick's law, a sorption model (typically extended Langmuir isotherm), a permeability model and equations of state.

8.5.2.1.1 Mass Continuity Equations

In coalbed, there are typically two phases (water and oil). By writing molar mass balance for each component in each phase on a representative elementary volume (REV) through Eulerian formulation, the below equations for cleats will be achieved. The flow from matrix to cleats or from cleats to matrix ($\int_{\partial\Omega} j_i.\vec{n}ds$) can be seen as a sink/source term for the following equations in an isotherm state:

$$\int_{\partial\Omega} y_{i_g} b_g \left(\vec{u}_g \cdot \vec{n} dA\right) + \int_{\Omega} \frac{\partial}{\partial t}\left(y_{i_g} b_g \phi S_g\right) dV - \int_{\Omega} \tilde{q}_{i_g} dV - \int_{\partial\Omega} \vec{j}_i \cdot \vec{n} dA = 0 \tag{8.27}$$

$$\int_{\partial\Omega} y_{i_w} b_w (\vec{u}_w \cdot \vec{n} dA) + \int_{\Omega} \frac{\partial}{\partial t}\left(y_{i_w} b_w \phi S_w\right) dV - \int_{\Omega} \tilde{q}_{i_w} dV = 0 \tag{8.28}$$

where b_l is the molar density of each phase, y_{il} is the mole fraction of component i in phase l, $\tilde{q}_{il}$ is the molar rate production/injection of component i from phase l per unit volume, j_i is the molar flux rate of component i from matrix to cleats or from cleats to matrix, $\vec{u}_l$ is phase l fluid velocity, ϕ is cleats porosity, S_l is phase saturation,

Ω is the volume of REV, and $\partial\Omega$ is the surface enclosing Ω. $\vec{n}$ is a unit normal vector perpendicular to $\partial\Omega$.

These two equations should be written for each component. The unknowns are y_{i_g}, y_{i_w}, b_w, b_g, S_g, S_w, ϕ, j_i, $\vec{u}_g$, and $\vec{u}_w$. In modeling fluid flow in CBM, it is usually assumed that (1) the diffusion in cleats is negligible, compared to bulk flow, (2) gas is not dissolvable in water phase, (3) no water vapor exists in gas phase, e.g., for a four-component $y_{H_2O_g} = y_{CH_4w} = y_{N_2w} = y_{CO_2w} = 0$.

Therefore, the equations for cleats will be simplified as follows:

$$\int_{\partial\Omega} y_{i_g} b_g \left(\vec{u}_g \cdot \vec{n} dA\right) + \int_{\Omega} \frac{\partial}{\partial t}\left(y_{i_g} b_g \phi S_g\right) dV - \int_{\Omega} \tilde{q}_{i_g} dV - \int_{\partial\Omega} \vec{j}_i \cdot \vec{n} dA = 0 \tag{8.29}$$

$$\int_{\partial\Omega} b_w(\vec{u}_w \cdot \vec{n} dA) + \int_{\Omega} \frac{\partial}{\partial t}(b_w \phi S_w) dV - \int_{\Omega} \tilde{q}_{H_2O_w} dV = 0 \tag{8.30}$$

$$\sum_{l=1}^{N_p} S_l = 1 \tag{8.31}$$

$$\sum_{i=1}^{N_c - 1} y_{i_g} = 1 \tag{8.32}$$

where N_c is the number of components and N_p is the number of phases.

Eq. (8.29) should be written for all the components except water. For matrix, it is typically assumed that (1) there is no bulk flow, (2) fluid is in adsorbed state, and (3) there is no water in matrix. The following equation represents molar balance in coal matrix. It also should be written for all the components except water.

$$\int_{\Omega} \frac{\partial C_i}{\partial t} dV + \int_{\partial\Omega} \vec{j}_i \cdot \vec{n} dA = 0 \tag{8.33}$$

where C_i is the molar density of component i in the matrix.

Totally, there are $2N_c + 1$ equations and $3N_c + 4$ unknowns, e.g., for the four-component case, unknowns are y_{CH_4g}, y_{N_2g}, y_{CO_2g}, b_g, b_w, $\vec{u}_g$, $\vec{u}_w$, S_g, S_w, ϕ, C_{CH_4}, C_{CO_2}, C_{N_2}, j_{CH_4}, j_{CO_2}, and j_{N_4}.

A few constitutive equations are required to have the same number of equations and variables.

8.5.2.1.2 Darcy's Law

Using Darcy's law, flow rate $\vec{u}_l$ will be related to the phase pressure p_l

$$\vec{u}_l = -\frac{Kk_{rl}}{\mu_l}\left(\nabla p_l - \rho_l g\right) \tag{8.34}$$

where K is the permeability tensor, k_{rl} is the relative permeability of phase l which is a function of saturation, ρ_l is the density of phase l, μ_l is the viscosity of phase l, and g is the directional gravity acceleration.

The two pressures can be related to each other through capillary pressure (p_c) that is in turn a function of saturation.

$$p_g - p_w = p_c \tag{8.35}$$

8.5.2.1.3 Fick's Law

Using Fick's law, molar flux rate $\vec{j}_l$ will be related to the partial pressure (concentration) gradient between cleats and matrix:

$$\int_{\partial\Omega} \vec{j}_i \cdot \vec{n} dA \approx D_i \sigma \left(C_i - V_i \rho_{\text{coal}} \right) \Omega \tag{8.36}$$

where $V_i \rho_{\text{coal}}$ is the maximum molar density in matrix which is obtained in equilibrium condition corresponding to the partial pressure of component i at the cleats. V_i is calculated using a sorption model, ρ_{coal} is coal density, and D_i is the diffusion coefficient corresponding to component i. σ is the matrix–cleats interface area [Eq. (8.25)].

8.5.2.1.4 Sorption Model

As discussed in the sorption section [Eq. (8.14)], extended Langmuir model is a widely used equation for relating V_i to partial pressure $p_g y_{i_g}$ and coal characteristics (V_{Li} and p_{Li}).

8.5.2.1.5 Equation of State

Using equation of states, molar densities and viscosities can be related to partial pressure. Water viscosity is normally assumed constant. Water molar density is related to pressure using the compressibility factor (c_w) that can be assumed constant for water

$$b_w = b_{w0} e^{c_w \left(p_w - p^0 \right)} \tag{8.37}$$

where b_{w0} is the molar density at a reference pressure (p^0).

b_g can be related to pressure and gas composition using gas law.

$$b_g = \frac{p_g}{zRT} \tag{8.38}$$

where R is the gas constant, T is the temperature, and z is the gas compressibility that is a function of critical temperature and pressure.

Peng–Robinson model is one of the commonly used models for gas.

Gas viscosity is also pressure and composition dependent. Lorentz–Bray–Clark model is a typical model to relate gas viscosity to the composition and pressure [110].

$$\left[(\mu-\mu^{0})\xi+10^{-4}\right]=\left(\sum_{i=1}^{5}a_i b_r^{i-1}\right)^{4} \tag{8.39}$$

where a_i is a constant, ξ and μ^0 are the functions of composition of gas, molecular weights, critical temperature and pressure, and b_r represents the reduced molar density.

8.5.2.1.6 Porosity Model

The porosity of cleats and correspondingly permeability change with the alteration of pressure and/or the amount of adsorbed gas (which results in shrinkage or swelling). Palmer–Mansoori model is among the most popular models used to take this phenomenon into consideration [31].

$$\frac{\phi}{\phi_0}=\left\{1+\frac{c_m}{\phi_{c0}}(p-p_0)+\frac{1}{\phi_{c0}}\left(\frac{\mathcal{K}}{M}-1\right)\Delta\varepsilon\right\} \tag{8.40}$$

where ϕ_0 is the initial porosity at pressure p_0, $\mathcal{K}$ and M represent bulk modulus and axial modulus, respectively, and $\Delta\varepsilon$ represents the total volumetric strain calculated using the following equation:

$$\Delta\varepsilon=\sum_{k=1}^{N_c-1}\frac{\varepsilon_k\beta_k a_k p}{1+\sum_{j=1}^{N_c}\beta_j a_j p}-\sum_{k=1}^{N_c-1}\frac{\varepsilon_k\beta_k a_k p_0}{1+\sum_{j=1}^{N_c}\beta_j a_j p_0} \tag{8.41}$$

where β_k and ε_k are matching parameters, and a_k and c_m are calculated as follows:

$$a_k=\frac{V_k}{\sum_{j=1}^{N_c}V_j} \tag{8.42}$$

$$c_m=\frac{1}{M}-\left[\frac{K}{M}+f-1\right]\gamma \tag{8.43}$$

where γ is grain compressibility, and f is a fraction ranging between 0 and 1.

The cleat permeability is typically proportional to the cube of cleat porosity [2,20,75,82,87]. Incorporating all the mentioned equations, the number of unknowns and equations will be the same. Using numerical methods, e.g., finite volume with a backward Euler, the resulted PDEs can be solved.

REFERENCES

[1] Z. Dong, S. Holditch, D. McVay, W.B. Ayers, Global unconventional gas resource assessment, SPE Econ. Manage. 4 (04) (2012) 222–234.
[2] J. Seidle, Fundamentals of Coalbed Methane Reservoir Engineering, PennWell Corp., Tulsa, Okla, 2011, p. xiii. 401 p.
[3] R. Puri, D. Yee, Enhanced coalbed methane recovery, SPE Annual Technical Conference and Exhibition, Society of Petroleum Engineers, New Orleans, LA, 1990.
[4] M. Sayyafzadeh, A. Keshavarz, A.R.M. Alias, K.A. Dong, M. Manser, Investigation of varying-composition gas injection for coalbed methane recovery enhancement: a simulation-based study, J. Nat. Gas Sci. Eng. 27 (2015) 1205–1212.
[5] ASTM D388-05, 2005. Standard Classification of Coal by Rank. ASTM, Philadelphia.
[6] Speight, J.G., **1994**. The Chemistry and Technology of Coal. Chemical Industries. Marcel Dekker, New York, NY. Revised and Expanded.
[7] ISO-11760: 2005, 2005. Classification of Coals. ISO 11760:2005, Geneva.
[8] AS 1038.17: Part 17. 2000. Higher rank coal—moisture-holding capacity (equilibrium moisture). Coal and Coke–Analysis and Testing. S.A. Ltd., Sydney.
[9] H. Ji, Z. Li, Y. Peng, Y. Yang, Y. Tang, Z. Liu, Pore structures and methane sorption characteristics of coal after extraction with tetrahydrofuran, J. Nat. Gas Sci. Eng. 19 (2014) 287–294.
[10] S. Zhang, S. Tang, D. Tang, W. Huang, Z. Pan, Determining fractal dimensions of coal pores by FHH model: problems and effects, J. Nat. Gas Sci. Eng. 21 (2014) 929–939.
[11] X. Hu, S. Yang, X. Zhou, G. Zhang, B. Xie, A quantification prediction model of coalbed methane content and its application in Pannan coalfield, Southwest China, J. Nat. Gas Sci. Eng. 21 (2014) 900–906.
[12] A. Busch, Y. Gensterblum, B.M. Krooss, Methane and CO_2 sorption and desorption measurements on dry Argonne premium coals: pure components and mixtures, Int. J. Coal Geol. 55 (2) (2003) 205–224.
[13] E. Ozdemir, B.I. Morsi, K. Schroeder, CO_2 adsorption capacity of Argonne premium coals, Fuel 83 (7) (2004) 1085–1094.
[14] L.J. Thomas, L.P. Thomas, Coal Geology, John Wiley & Sons, London, 2002.
[15] M.N. Lamberson, R.M. Bustin, Coalbed methane characteristics of Gates Formation coals, northeastern British Columbia: effect of maceral composition, AAPG Bull. 77 (12) (1993) 2062–2076.
[16] N. Berkowitz, An Introduction to Coal Technology, Elsevier, London, 2012.
[17] Levine, J.R., **1993**. Coalification: The Evolution of Coal as Source Rock and Reservoir Rock for Oil and Gas. Hydrocarbons from coal: AAPG Studies in Geology, 38, 39–77.
[18] C.Ö. Karacan, G.D. Mitchell, Behavior and effect of different coal microlithotypes during gas transport for carbon dioxide sequestration into coal seams, Int. J. Coal Geol. 53 (4) (2003) 201–217.
[19] P.K. Mukhopadhyay, P.G. Hatcher, Composition of coal, Hydrocarbons from Coal. vol. 38. AAPG Studies in Geology, Tulsa, OK (1993) 79–118.
[20] L.H. Reiss, The Reservoir Engineering Aspects of Fractured Formations, vol. 3, Editions Technip, Paris, 1980.
[21] Gash, B.W., 1991. Measurement of "Rock Properties" in coal for coalbed methane production. SPE Annual Technical Conference and Exhibition. Society of Petroleum Engineers.
[22] H. Gan, S. Nandi, P. Walker, Nature of the porosity in American coals, Fuel 51 (4) (1972) 272–277.
[23] C. Rodrigues, M.L. De Sousa, The measurement of coal porosity with different gases, Int. J. Coal Geol. 48 (3) (2002) 245–251.
[24] D. Everett, Manual of symbols and terminology for physicochemical quantities and units, appendix II: definitions, terminology and symbols in colloid and surface chemistry, Pure Appl. Chem. 31 (4) (1972) 577–638.
[25] B. McEnaney, T. Mays, Porosity in carbons and graphites, Introduction to Carbon Science, *Butterworths, London*, 1989, pp. 153–196.
[26] H. Akhondzadeh, A. Keshavarz, M. Sayyafzadeh, A. Kalantariasl, Investigating the relative impact of key reservoir parameters on performance of coalbed methane reservoirs by an efficient statistical approach, J. Nat. Gas Sci. Eng. 53 (2018) 416–428.

[27] J.-Q. Shi, S. Durucan, A model for changes in coalbed permeability during primary and enhanced methane recovery, SPE Reservoir Eval. Eng. 8 (04) (2005) 291–299.
[28] Young, G., Paul, G., McElhiney, J., McBane, R., 1992. A parametric analysis of Fruitland coalbed methane reservoir producibility. SPE Annual Technical Conference and Exhibition. Society of Petroleum Engineers.
[29] Roadifer, R., Moore, T., Raterman, K., Farnan, R., Crabtree, B., 2003. Coalbed methane parametric study: what's really important to production and when? In: Paper SPE 84425 Presented at the 2003 SPE Annual Technical Conference and Exhibition. Denver.
[30] R.E. Rogers, K. Ramurthy, G. Rodvelt, M. Mullen, Coalbed Methane: Principles and Practices, third ed., Oktibbeha Publishing Co. (Halliburton Company), Mississippi, 2007.
[31] A. H. Alizadeh, A. Keshavarz, M. Haghighi, Flow rate effect on two-phase relative permeability in Iranian carbonate rocks. In: SPE Middle East Oil and Gas Show and Conference, 11–14 March 2007, Manama, Bahrain.
[32] S. Harpalani, G. Chen, Estimation of changes in fracture porosity of coal with gas emission, Fuel 74 (10) (1995) 1491–1498.
[33] Palmer, I.D., 2010. The permeability factor in coalbed methane well completions and production. In: SPE Western Regional Meeting. Society of Petroleum Engineers.
[34] M. Mavor, J. Vaughn, Increasing coal absolute permeability in the San Juan Basin fruitland formation, SPE Reservoir Eval. Eng. 1 (03) (1998) 201–206.
[35] Al-anazi, B.D., Algarni, M.T., Tale, M., Almushiqeh, I., 2011. Prediction of Poisson's ratio and Young's modulus for hydrocarbon reservoirs using alternating conditional expectation algorithm. In: SPE Middle East Oil and Gas Show and Conference. Society of Petroleum Engineers.
[36] S. Ian Palmer, J. Mansoori, How permeability depends on stress and pore pressure in coalbeds: a new model, SPE Reservoir Eval. Eng. 1 (1998) 539–544.
[37] F. Gu, R. Chalaturnyk, Sensitivity study of coalbed methane production with reservoir and geomechanic coupling simulation, J. Can. Pet. Technol. 44 (10) (2005) 23–32.
[38] J. Geertsma, F. De Klerk, A rapid method of predicting width and extent of hydraulically induced fractures, J. Pet. Technol. 21 (12) (1969) 1571–1581.
[39] Holditch, S., Ely, J., Carter, R., Semmelbeck, M., 1990. Coal Seam Stimulation Manual. Gas Research Institute, Chicago. Contract No. 5087-214-1469.
[40] D.J. Johnston, Geochemical logs thoroughly evaluate coalbeds, Oil Gas J. 88 (52) (1990) 45–51.
[41] P. Lu, G. Li, Z. Huang, S. Tian, Z. Shen, Simulation and analysis of coal seam conditions on the stress disturbance effects of pulsating hydro-fracturing, J. Nat. Gas Sci. Eng. 21 (2014) 649–658.
[42] Kovscek, A., Tang, G., Jessen, K., **2005**. Laboratory and simulation investigation of enhanced coalbed methane recovery by gas injection. In: Paper SPE95947 presented at SPE Annual Technical Conference and Exhibition.
[43] C. Bertard, B. Bruyet, J. Gunther, International Journal of Rock Mechanics and Mining Sciences & Geomechanics Abstracts Determination of Desorbable Gas Concentration of Coal (Direct Method), Elsevier, 1970. pp 43IN351-50IN465.
[44] C. Clarkson, R. Bustin, The effect of pore structure and gas pressure upon the transport properties of coal: a laboratory and modeling study. 2. Adsorption rate modeling, Fuel 78 (11) (1999) 1345–1362.
[45] R. Yang, Gas Separation by Adsorption Processes., Pergamon, Oxford, 1988.
[46] S. Brunauer, P.H. Emmett, E. Teller, Adsorption of gases in multimolecular layers, J. Am. Chem. Soc. 60 (2) (1938) 309–319.
[47] Testa, S., Pratt, T., 2003. Sample preparation for coal and shale gas resource assessment. In: Proceedings, International Coalbed Methane Symposium, Tuscaloosa, Alabama.
[48] Arri, L., Yee, D., Morgan, W., Jeansonne, M., 1992. Modeling coalbed methane production with binary gas sorption. In: SPE Rocky Mountain Regional Meeting. Society of Petroleum Engineers.
[49] J. Crank, 0198534116. The Mathematics of Diffusion, Clarendon Press, Oxford, GB, 1975.
[50] P.J. Crosdale, B.B. Beamish, M. Valix, Coalbed methane sorption related to coal composition, Int. J. Coal Geol. 35 (1) (1998) 147–158.

[51] G. Staib, R. Sakurovs, E.M. Gray, Kinetics of coal swelling in gases: influence of gas pressure, gas type and coal type, Int. J. Coal Geol. 132 (2014) 117–122.
[52] G. Staib, R. Sakurovs, E.M.A. Gray, A pressure and concentration dependence of CO_2 diffusion in two Australian bituminous coals, Int. J. Coal Geol. 116 (2013) 106–116.
[53] G. Staib, R. Sakurovs, E.M.A. Gray, Dispersive diffusion of gases in coals. Part I: model development, Fuel 143 (2015) 612–619.
[54] G. Staib, R. Sakurovs, E.M.A. Gray, Dispersive diffusion of gases in coals. Part II: An assessment of previously proposed physical mechanisms of diffusion in coal, Fuel 143 (2015) 620–629.
[55] E. Ruckenstein, A. Vaidyanathan, G. Youngquist, Sorption by solids with bidisperse pore structures, Chem. Eng. Sci. 26 (9) (1971) 1305–1318.
[56] D.M. Smith, F.L. Williams, Diffusion models for gas production from coals: application to methane content determination, Fuel 63 (2) (1984) 251–255.
[57] D.M. Smith, F.L. Williams, Diffusion models for gas production from coal: determination of diffusion parameters, Fuel 63 (2) (1984) 256–261.
[58] S. Bhatia, Modeling the pore structure of coal, AIChE J. 33 (10) (1987) 1707–1718.
[59] D. Van Krevelen, Coal: Typology-Physics-Chemistry-Composition., Elsevier, Amsterdam, 1993.
[60] Zuber, M., Sawyer, W., Schraufnagel, R., Kuuskraa, V., 1987. The use of simulation and history matching to determine critical coalbed methane reservoir properties. In: Low Permeability Reservoirs Symposium. Society of Petroleum Engineers.
[61] Paul, G., Sawyer, W., Dean, R., 1990. Validation of 3D coalbed simulators. In: SPE Annual Technical Conference and Exhibition. Society of Petroleum Engineers.
[62] E. Airey, International Journal of Rock Mechanics and Mining Sciences & Geomechanics Abstracts Gas Emission from Broken Coal. An Experimental and Theoretical Investigation, Elsevier, 1968pp. 475–494.
[63] D.J. Remner, T. Ertekin, W. Sung, G.R. King, A parametric study of the effects of coal seam properties on gas drainage efficiency, SPE Reservoir Eng. 1 (06) (1986) 633–646.
[64] A. Keshavarz, R. Sakurovs, M. Grigore, M. Sayyafzadeh, Effect of maceral composition and coal rank on gas diffusion in Australian coals, Int. J. Coal Geol. 173 (2017) 65–75.
[65] J. Warren, P.J. Root, The behavior of naturally fractured reservoirs, Soc. Pet. Eng. J. 3 (03) (1963) 245–255.
[66] H. Kazemi, Pressure transient analysis of naturally fractured reservoirs with uniform fracture distribution, Soc. Pet. Eng. J. 9 (04) (1969) 451–462.
[67] A. Busch, Y. Gensterblum, CBM and CO_2-ECBM related sorption processes in coal: a review, Int. J. Coal Geol. 87 (2) (2011) 49–71.
[68] M.J. Economides, T. Martin, Modern Fracturing: Enhancing Natural Gas Production, ET Publishing Houston, Houston, TX, 2007.
[69] Holditch, S., Ely, J., Semmelbeck, M., Carter, R., Hinkel, J., Jeffrey Jr, R., 1988. Enhanced recovery of coalbed methane through hydraulic fracturing. SPE Annual Technical Conference and Exhibition. Society of Petroleum Engineers.
[70] Jeffrey Jr, R., Brynes, R., Lynch, P., Ling, D., 1992. An analysis of hydraulic fracture and mineback data for a treatment in the German creek coal seam. In: SPE Rocky Mountain Regional Meeting. Society of Petroleum Engineers.
[71] Jeffrey, R., Settari, A., 1995. A comparison of hydraulic fracture field experiments, including mineback geometry data, with numerical fracture model simulations. In: SPE Annual Technical Conference and Exhibition. Society of Petroleum Engineers.
[72] Jeffrey, R., Vlahovic, W., Doyle, R., Wood, J., 1998. Propped fracture geometry of three hydraulic fractures in Sydney Basin coal seams. In: SPE Asia Pacific Oil and Gas Conference and Exhibition. Society of Petroleum Engineers.
[73] Jeffrey, R., Settari, A., 1998. An instrumented hydraulic fracture experiment in coal. In: SPE Rocky Mountain Regional/Low-Permeability Reservoirs Symposium. Society of Petroleum Engineers.
[74] Johnson Jr, R.L., Flottman, T., Campagna, D.J., 2002. Improving results of coalbed methane development strategies by integrating geomechanics and hydraulic fracturing technologies. In: SPE Asia Pacific Oil and Gas Conference and Exhibition. Society of Petroleum Engineers.

[75] Palmer, I.D., Lambert, S.W., Spitler, J.L., Coalbed Methane Well Completions and Stimulations. Hydrocarbons from coal: AAPG Studies in Geology, 38 (1993) 303–339.
[76] A. Keshavarz, A. Badalyan, T. Carageorgos, P. Bedrikovetsky, R. Johnson, Stimulation of coal seam permeability by micro-sized graded proppant placement using selective fluid properties, Fuel 144 (2015) 228–236.
[77] A. Keshavarz, A. Badalyan, R. Johnson, P. Bedrikovetsky, Productivity enhancement by stimulation of natural fractures around a hydraulic fracture using micro-sized proppant placement, J. Nat. Gas Sci. Eng. 33 (2016) 1010–1024.
[78] M. Hossain, M. Rahman, S. Rahman, A shear dilation stimulation model for production enhancement from naturally fractured reservoirs, SPE J. 7 (02) (2002) 183–195.
[79] M. Rahman, M. Hossain, S. Rahman, A shear-dilation-based model for evaluation of hydraulically stimulated naturally fractured reservoirs, Int. J. Numer. Anal. Methods Geomech. 26 (5) (2002) 469–497.
[80] A. Riahi, B. Damjanac, Numerical study of interaction between hydraulic fracture and discrete fracture network, Effective and Sustainable Hydraulic Fracturing, InTech, London, 2013.
[81] Darin, S., Huitt, J., Effect of a Partial Monolayer of Propping Agent on Fracture Flow Capacity. Petroleum Transactions, AIME, 1960, 219, 31–37.
[82] Fredd, C., McConnell, S., Boney, C., England, K., 2000. Experimental study of hydraulic fracture conductivity demonstrates the benefits of using proppants. SPE Rocky Mountain Regional/Low-Permeability Reservoirs Symposium and Exhibition. Society of Petroleum Engineers.
[83] A. Khanna, A. Kotousov, J. Sobey, P. Weller, Conductivity of narrow fractures filled with a proppant monolayer, J. Pet. Sci. Eng. 100 (2012) 9–13.
[84] A. Khanna, A. Keshavarz, K. Mobbs, M. Davis, P. Bedrikovetsky, Stimulation of the natural fracture system by graded proppant injection, J. Pet. Sci. Eng. 111 (2013) 71–77.
[85] Brannon, H.D., Starks, T.R., 2008. The effects of effective fracture area and conductivity on fracture deliverability and stimulation value. SPE Annual Technical Conference and Exhibition. Society of Petroleum Engineers.
[86] A. Gaurav, E. Dao, K. Mohanty, Evaluation of ultra-light-weight proppants for shale fracturing, J. Pet. Sci. Eng. 92 (2012) 82–88.
[87] Kassis, S.M., Sondergeld, C.H., 2010. Gas shale permeability: effects of roughness, proppant, fracture offset, and confining pressure. In: International Oil and Gas Conference and Exhibition in China. Society of Petroleum Engineers.
[88] Parker, M.A., Glasbergen, G., van Batenburg, D.W., Weaver, J.D., Slabaugh, B.F., 2005. High-porosity fractures yield high conductivity. In: SPE Annual Technical Conference and Exhibition. Society of Petroleum Engineers.
[89] A. Keshavarz, Y. Yang, A. Badalyan, R. Johnson, P. Bedrikovetsky, Laboratory-based mathematical modelling of graded proppant injection in CBM reservoirs, Int. J. Coal Geol. 136 (2014) 1–16.
[90] Bedrikovetsky, P.G., Keshavarz, A., Khanna, A., Mckenzie, K.M., Kotousov, A., 2012. Stimulation of natural cleats for gas production from coal beds by graded proppant injection. In: SPE Asia Pacific Oil and Gas Conference and Exhibition. Society of Petroleum Engineers.
[91] A. Keshavarz, A Novel Technology for Enhanced Coal Seam Gas Recovery by Graded Proppant Injection, PhD Thesis, The University of Adelaide, Adelaide, 2015.
[92] Keshavarz, A., Badalyan, A., Carageorgos, T., Bedrikovetsky, P., Johnson, R., 2015. Graded proppant injection into coal seam gas and shale gas reservoirs for well stimulation. In: SPE European Formation Damage Conference and Exhibition. Society of Petroleum Engineers.
[93] A. Keshavarz, A. Badalyan, R. Johnson, P. Bedrikovetsky, EUROPEC 2015 A New Technique for Enhancing Hydraulic Fracturing Treatment in Unconventional Reservoirs, Society of Petroleum Engineers, Madrid, 2015.
[94] Keshavarz, A., Badalyan, A., Carageorgos, T., Johnson, R., Bedrikovetsky, P., 2014. Stimulation of unconventional naturally fractured reservoirs by graded proppant injection: experimental study and mathematical model. In: SPE/EAGE European Unconventional Resources Conference and Exhibition.
[95] A. Keshavarz, A. Badalyan, T. Carageorgos, P. Bedrikovetsky, R. Johnson, Enhancement of CBM well fracturing through stimulation of cleat permeability by ultra-fine particle injection, APPEA J. 54 (1) (2014) 155–166.

[96] Keshavarz, A., Johnson, R., Carageorgos, T., Bedrikovetsky, P., Badalyan, A., 2016. Improving the conductivity of natural fracture systems in conjunction with hydraulic fracturing in stress sensitive reservoirs. In: SPE Asia Pacific Oil & Gas Conference and Exhibition. Society of Petroleum Engineers.

[97] Fulton, P.F., Parente, C.A., Rogers, B.A., Shah, N., Reznik, A., 1980. A laboratory investigation of enhanced recovery of methane from coal by carbon dioxide injection, In: SPE Unconventional Gas Recovery Symposium. Society of Petroleum Engineers.

[98] Ç. Sınayuç, F. Gümrah, Modeling of ECBM recovery from Amasra coalbed in Zonguldak Basin, Turkey, Int. J. Coal Geol. 77 (1) (2009) 162–174.

[99] Stevens, S.H., Spector, D., Riemer, P., 1998. Enhanced coalbed methane recovery using CO_2 injection: worldwide resource and CO_2 sequestration potential. In: SPE International Oil and Gas Conference and Exhibition in China. Society of Petroleum Engineers.

[100] Wong, S., Gunter, W., Mavor, M., 2000. Economics of CO_2 sequestration in coalbed methane reservoirs. In: SPE/CERI Gas Technology Symposium. Society of Petroleum Engineers.

[101] X. Cui, R.M. Bustin, Volumetric strain associated with methane desorption and its impact on coalbed gas production from deep coal seams, AAPG Bull. 89 (9) (2005) 1181–1202.

[102] J.W. Larsen, The effects of dissolved CO_2 on coal structure and properties, Int. J. Coal Geol. 57 (1) (2004) 63–70.

[103] S. Durucan, J.-Q. Shi, Improving the CO_2 well injectivity and enhanced coalbed methane production performance in coal seams, Int. J. Coal Geol. 77 (1) (2009) 214–221.

[104] S. Durucan, M. Ahsanb, J.-Q. Shia, Matrix shrinkage and swelling characteristics of European coals, Energy Procedia 1 (1) (2009) 3055–3062.

[105] M. Mazzotti, R. Pini, G. Storti, Enhanced coalbed methane recovery, J. Supercrit. Fluids 47 (3) (2009) 619–627.

[106] M. Perera, P. Ranjith, A. Ranathunga, A. Koay, J. Zhao, S. Choi, Optimization of enhanced coalbed methane recovery using numerical simulation, J. Geophys. Eng. 12 (1) (2015) 90.

[107] S. Reeves, A. Oudinot, The Tiffany Unit N2-ECBM Pilot: A Reservoir Modeling Study, Advanced Resources International, Incorporated, Houston, TX, 2004.

[108] F. Zhou, W. Hou, G. Allinson, J. Wu, J. Wang, Y. Cinar, A feasibility study of ECBM recovery and CO_2 storage for a producing CBM field in Southeast Qinshui Basin, China, Int. J. Greenhouse Gas Control 19 (2013) 26–40.

[109] M. Sayyafzadeh, A. Keshavarz, Optimisation of gas mixture injection for enhanced coalbed methane recovery using a parallel genetic algorithm, J. Nat. Gas Sci. Eng. 33 (2016) 942–953.

[110] J. Lohrenz, B.G. Bray, C.R. Clark, Calculating viscosities of reservoir fluids from their compositions, J. Pet. Technol. 16 (10) (1964) 1,171–1,176.

CHAPTER NINE

Enhanced Oil Recovery (EOR) in Shale Oil Reservoirs

Mohammad Ali Ahmadi
Department of Chemical and Petroleum Engineering, University of Calgary, Calgary, AB, Canada

9.1 INTRODUCTION

Oil production from shale and tight formations accounted for more than half of total US oil production in 2015 [1]. Such amount is expected to grow significantly as the active development of low-permeability reservoirs continues as shown in Fig. 9.1. Fig. 9.2 shows the unconventional oil fields across the world including oil shale, extra heavy oil and bitumen, and tight oil and gas. This figure implies the importance of shale oil reservoirs as promising energy resources for future life. Fig. 9.3 illustrates the most important light tight oil fields throughout the world and Fig. 9.4 describes global shale resource exploration and extraction momentum in selected countries [2]. It is worth to mention that several countries, especially in Europe, have banned the exploration of shale reservoirs; however, United States, Russia, and China as developed countries have explored their shale reservoirs [2,5].

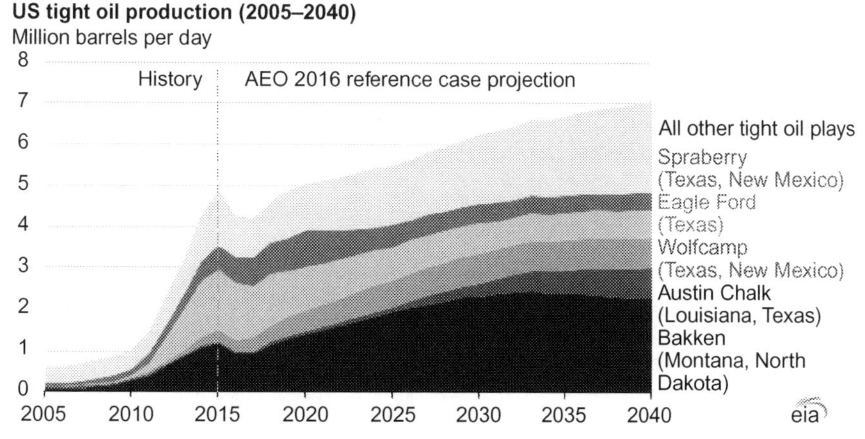

Figure 9.1 Projection of US petroleum and other liquid fuels production (Annual Energy Outlook 2016 www.eia.gov/aeo).

Fundamentals of Enhanced Oil and Gas Recovery from Conventional and Unconventional Reservoirs.
DOI: https://doi.org/10.1016/B978-0-12-813027-8.00009-6

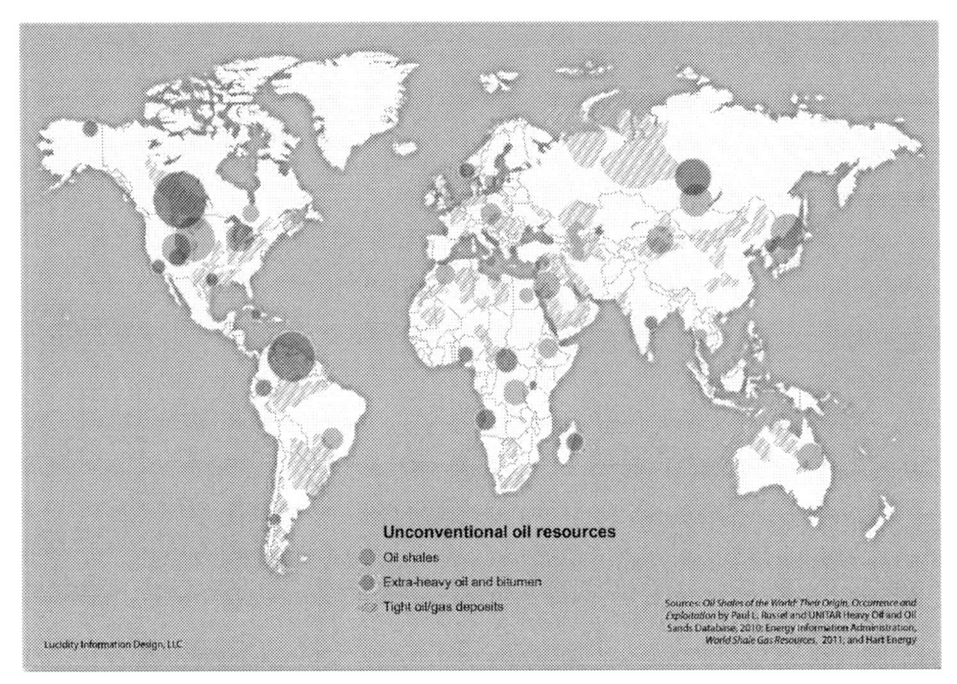

Figure 9.2 Distribution of shale oil reservoirs across worldwide [2,3]. Oil Shales of the World: Their Origin, Occurrence and Exploitation by Paul L. Russel; UNITAR Heavy Oil and Oil Sands Database, 2010; Energy Information Administration, World Shale Gas Resources, 2011; and Hart Energy.

Figure 9.3 Selected light tight oil plays worldwide [2,4].

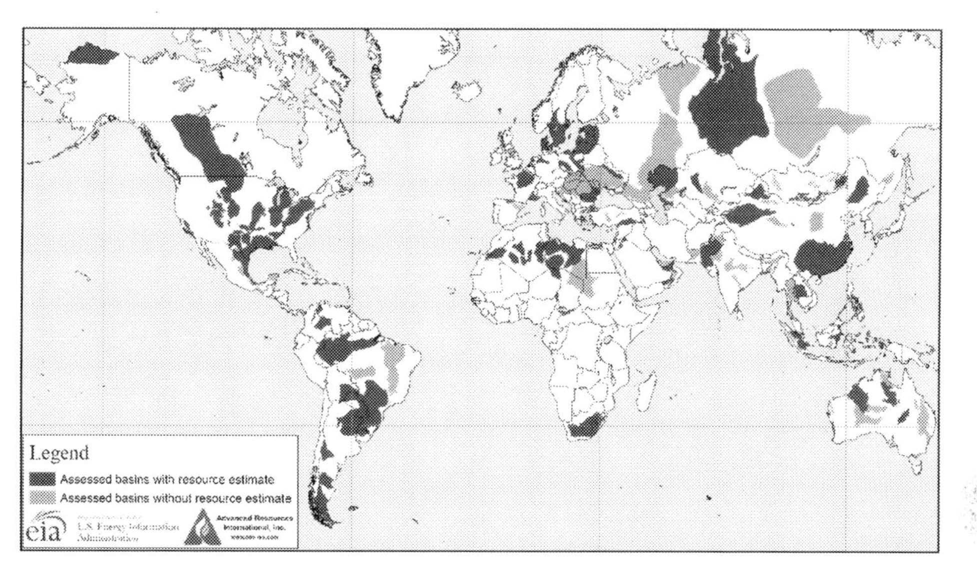

Figure 9.4 The map of assessments in terms of shale reserves across the world [2]. http://www.eia.gov/analysis/studies/worldshalegas/.

Fig. 9.5 depicts the portion of the shale oil in place and technically recoverable shale oil, individually. The graphs have been developed utilizing data reported by energy information administration (EIA) [6]. As clear be seen from these charts, Europe has the highest shale oil in place and producable shale oil reserves; conversely, Australia is the last in this ranking [2,6].

The main production method for producing shale oil from their tight reservoirs is using hydraulic fracturing in the horizontal wells. This method facilitate the oil production due to the reservoir depletion; however, the achievable recovery factor of such method in most of the cases is lower than 10% [7]. For instance, Clark [8] used different approaches to figure out the value of the shale oil recovery factor; according to the results the most probable recovery factor is almost 7%. This means that huge volume of oil remained in the shale oil reservoir and this will be a motivating force to innovate new technology to enhance production from such type of reservoirs [1].

The main and general rock property of tight and shale reservoir is drastically low permeability in comparison with other types of the oil/gas formations. For instance, the value of permeability in shale reservoirs varies from 0.001 to 0.0001 mD [9]; however, this value in most of the conventional oil/gas formations is 10000 times higher than that of the value in the shale and tight reservoirs. Practically speaking, in several rare cases the shale reservoir does have some micro-fractures that resulting in improvement of the effective permeability; in these cases the value of the effective permeability is much higher than that ot the shale matrix permeability without fractures. In addition to the permeability characterization, different common properties of

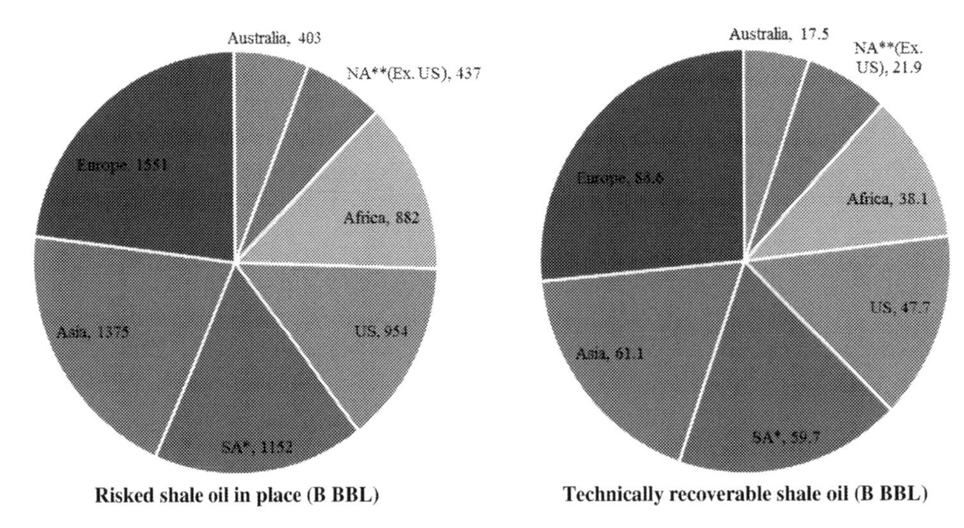

Figure 9.5 Continent-wise breakdown of risked in-place and technically recoverable shale oil [2]; *SA represents South America, **NA stands for North America (here excludes the USA). EIA [6].

the tight reservoirs reported in the literature including thermal maturity 0.6%–1.3%, porosity less than 10%, total organic carbon higher than 1%, and the API gravity higher than 40 [1,9]. Fig. 9.6 depicts the range of permeability for both conventional and unconventional oil/gas reservoirs.

It should be mentioned the definition of tight oil and shale oil reservoirs is different; to avoid any misunderstanding we should describe both terminologies. In technical speaking, the term of shale oil is used for reservoirs like mud shale rocks and source rocks; however, the term of tight oil reservoir is used in cases including low permeable carbonates, silty sandstones, and sandstones [1,10].

9.2 SHALE OIL AND OIL SHALE

In technical speaking, we do have two different terminologies that may cause a misunderstanding; these two terms are shale oil and oil shale. Oil shale defined as a rock comprises a solid organic materials called Kerogen; Kerogen is a compound of fossilized organic material. In other words, the rock containing the kerogen is not necessarily shale; apparently Kerogen is not a real crude oil. The terminology of shale oil is employed in case of trapping of oil in a very tight formation, for example, permeability is around 0.001 mD, the fluid cannot straightforwardly flow into the production well; hydraulic fracturing is employed to accelerate the fluency of the flow,

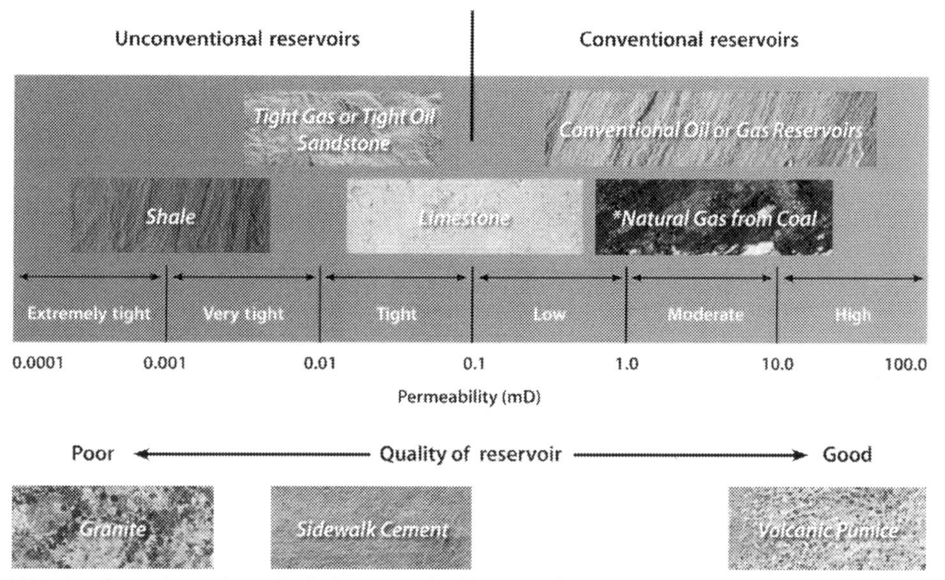

Figure 9.6 Quality of the oil and gas reservoirs in terms of permeability.

especially near the wellbore. On the other hand, to produce the oil from oil shale, very high temperature (almost 500°C) should be used. Such a process that the oil shale has been heated in a low oxygen condition is called retorting. Two techniques can be applied for heating the rock to extract the oil. The first technique comprises two step operation including mining the rock and heating the rock at surface. The second method is to heating the rock at the reservoir. Several oil companies including ExxonMobil and Shell have developed their technologies for heating the oil shale in the underground. Their technologies focused on the conversion of Kerogen into liquid oil using electric currents or electric heaters. Such methods are used to produce the oil from oil shale, on the other hand, horizontal wells with hydraulic fractures are employed as the main production technology in shale oil reservoirs [1].

9.3 EOR METHODS IN SHALE OIL AND GAS RESERVOIRS

Several enhanced oil recovery (EOR) techniques can be applied in the shale and tight oil reservoirs; these methods are water injection, gas injection, and surfactant injection. Following sections described the advantages, and drawbacks of such methods [1].

9.3.1 Gas Injection

Gas injection in shale reservoirs can be applied either in continuous mode or huff-n-puff mode [1]; these schemes described in following sections.

9.3.1.1 Continuous Gas Flooding

General speaking, continuous gas injection is most popular EOR scenario in conventional oil reservoirs compared to huff-n-puff gas injection method. According to the results gained by reservoir simulation studies, because of very low permeability of shale reservoirs the gas cannot simply propagate in a area between injection and production wells; this case happens when there is no fracture in the injector-producer well space [11]. Moreover, numerical simulation and experimental results reveal that the most portion of oil can be recover before the gas breakthrough time; this means that in a case of gas channeling through natural/induced fractures or fingering due to the heterogeneous nature of the shale formation, there is no significant oil production [12]. In practical speaking, the occurrence of gas channeling or gas fingering will make a big doubt on the applicability of such method in shale oil reservoirs. In Yu et al. carried out both experimental and numerical studies on the capability of N2 injection as an EOR method in shale oil reservoirs. Based on their results, they concluded that at certain conditions N2 injection could improve the oil production factor [12]. Kovscek et al. [13] and Vega et al. [14] performed several experiments on shale core samples with fracture to figure out the applicability of CO_2 injection in such reservoirs; core samples used in their studies had permeabilities in range of 0.2 to 1.3 mD [13–14].

Most of the research on tight and shale oil EOR scenarios have been done in the Bakken field. Wang et al. [15] performed numerical simulation method on the Bakken reservoir in the Saskatchewan province to assess the potential of CO_2 injection. They considered various injection plans including water injection alternating CO_2 injection, CO_2 injection, water injection, and CO_2 huff-n-puff injection. Based on their simulation results for this reservoir, continuous CO_2 injection provides better oil recovery factor compared to the other proposed scenarios. It is worth to mention that the configuration of injection and production wells in their simulations are not an actual huff-n-puff mode. Because they employed four injection wells and nine wells as continuous production wells. However, in a case of huff-n-puff scenario the injection and production well are the same. Another possible reason to make such a decision that continuous CO_2 injection has a higher oil recovery factor in comparison with other methods could be: (1) Using uncommon well configuration, especially, in CO_2 huff-n-puff scenario, (2) Considering long soaking time in that case (almost 5 years), (3) Rock properties, especially, permeability values used in their model are not ultra low as tight and shale reservoirs; permeability varies in their model from 0.04 to

2.5 mD. Considering these points can help to have a better evaluation and comparison between huff-n-puff CO_2 injection and continuous carbon dioxide flooding. According to their simulation results, enriching carbon dioxide improve the efficiency of the injection process. Also, the results revealed that the sweep efficiency of water flooding is much lower than CO_2 injection scenario. Their results confirmed those ones reported by Sheng and Chen [11], Joslin et al. [16], and Dong and Hoffman [17] for the Bakken Formation in the Sanish Field, North Dakota.

9.3.1.2 Huff-n-Puff Gas Injection

Sheng and Chen [11] proposed the use of CO_2 huff-n-puff scenario instead of continuous mode because of in a case of continuous injection the pressure in a near injection well region increases dramatically and near production well area the pressure decreases significantly due to the tightness of the rock. These points lead them to suggest the huff-n-puff scenario in such a reservoir [18] and they verified this recommendation with the experiments [19]. Furthermore, various experimental works have been done to evaluate this proposal [20–24]. In such a scheme there are different parameters should be optimized, such as number of cycles, injection time, soaking period, production period, and well configurations. In most of the researches, which have been performed in both lab scale and pilot/field scale, the shorter soaking time resulted in higher oil recovery factor; the best case scenario was zero soaking period in such a method [18,25–27]. It is worth to highlight that this observation works for a case of huff-n-puff with several cycles; it is obvious for a single cycle more soaking time yield more oil recovery factor [28,29]. In laboratory experiments the effect of soaking period when we dealing with gas condensate samples is ignorable [30,31].

The performance of carbon dioxide flooding coud be improved by enriching the composition of the injected gas; carbon dioxide is a promising EOR agent, especially in shale oil reservoirs [32]. Carbon dioxide injection scenario is employed broadly throughout world to improve oil recovery factor from different type of oil resources including naturally fractured reservoirs, deep and shallow conventional reservoirs, and tight reservoirs; however, in all these projects two drawbacks associated with carbon dioxide: (1) Availablity in huge volume, (2) Corrosion of the surface and downhole facilities in case of using conventional materials.

9.3.1.3 Advantages and Drawbacks of Gas Injection

Gas flooding is much more commonly used than huff-n-puff gas injection in conventional reservoirs. However, in shale or tight reservoirs, because of ultralow permeability and thus a significant pressure drop in the matrix, it is very difficult for the gas to drive oil from an injector to a producer. If a shale or tight reservoir has natural fracture networks or the hydraulic fractures connect an injector and a producer, gas will easily break through, resulting in a very low sweep efficiency [1,11]. There is no such a problem in a case of huff-n-puff injection mode. Wan et al. [33] compared the oil recovery from CO_2 flooding with that from huff-n-puff CO_2 injection and found that CO_2 huff-n-puff outperformed CO_2 flooding. Sheng and Chen [11] also compared gas flooding with huff-n-puff and their simulation results show that huff-n-puff oil recovery factors are higher than flooding ones. Yu et al. [34] performed experimental investigation to determine the performance of continuous gas injection compared to huff-n-puff CO_2 injection. They concluded that if the soaking period is short then the oil recovery factor gained by CO_2 huff-n-puff scenario will be higher than continuous CO_2 injection. Meng et al. [35] carried out different flooding experiments to evaluate the liquid condensate recovery in both huff-n-puff and continues schemes. The experimental results reveal that huff-n-puff is much higher than that the gas injection with continuous setting.

Shoaib and Hoffman [36] conducted different carbon dioxide injection scenarios including huff-n-puff, continuous carbon dioxide flooding in Elm Coulee Field in Richland County, United States. They employed various well configurations in their numerical simulation design. The reservoir formation of this field is Bakken formation; this formation is a main production zone in this oil field. Design of carbon dioxide injection in a huff-n-puff mode was 3 months for injection, 3 months as a soaking time, and 3 months for production time. Using such a scheme they gained 2.5% improvement in oil recovery factor from primary production scenario; in huff-n-puff scheme they achieved this value by injecting 0.19 pore volume (PV) of the reservoir. Their simulation results revealed that the continuous carbon dioxide flooding is much better than the huff-n-puff CO_2 injection scenario. Based on their results the continuous CO_2 injection improved oil recovery factor by 14-15%. Lower performance of CO_2 huff-n-puff scenario attributed to the parameters of this method which are not optimum value. Consequently, the performance of such a method lower than the continuous mode [7]. Also, very high permeability has a good contribution in the performance of continuous carbon dioxide flooding [11].

Wang et al. [15] carried out numerical reservoir simulation to evaluate the EOR potential of the Bakken reservoir in Saskatchewan. Their results reveal that continuous carbon dioxide injection works much better than that CO_2 huff-n-puff scenario in such a formation. The main reason for this conclusion is in their reservoir simulation they considered huff-n-puff as a single cycle with 10 years as an injection time, 5 years as a soaking period, and 5 years as a production period. As discussed in previous sections, in a cased of huff-n-puff injection mode there are lots of parameters that should be optimized. In their study, these parameters obviously not optimized and too large.

Kurtoglu [37] performed numerical simulation for carbon dioxide huff-n-puff scenario with three horizontal well configuration. In his model,all the wells are parallel and the central well was employed as an injection well; the two side wells were used as production wells. He also compared the results with the case of continuous CO_2

injection. In the huff-n-puff scenario he considered 60 days for injection, 10 days as a soaking time, and 120 days for production time in each cycle; he repeated this scenario for 6 cycles. According to his simulation outputs, the oil recovery factor improved significantly in a case of continuous carbon dioxide flooding; in both huff-n-puff and continuous CO_2 injection the oil production enhanced. One of the reasons come to mind for such a result is that in her simulation model the advantage of huff-n-puff was ignored because of using the central well as an injector; however, in a case of huff-n-puff model the injector and producer is exactly the same. Moreover, she did not incorporate the effect of molecular diffusion in the reservoir simulation.

Considering the points discussed above, in a case of CO_2 huff-n-puff injection scenario various factors should be taken into account. Also, all the main parameters of such a method should be optimized; these parameters are the number of cycles, injection time, soaking period, production time, and well configurations.

9.3.1.4 Field Test of Gas Injection

One of the field test of carbon dioxide injection in shale formation is immiscible CO_2 flooding in Bakken formation in Saskatchewan [38]. The top view of the Bakken field located in Saskatchewan is depicted in Fig. 9.7. The pilot project covered 1280 ac and was developed on a combination of 80 and 160-ac spacing. The length of the horizontal well was 1 mi; all the horizontal wells were stimulated using hydraulic

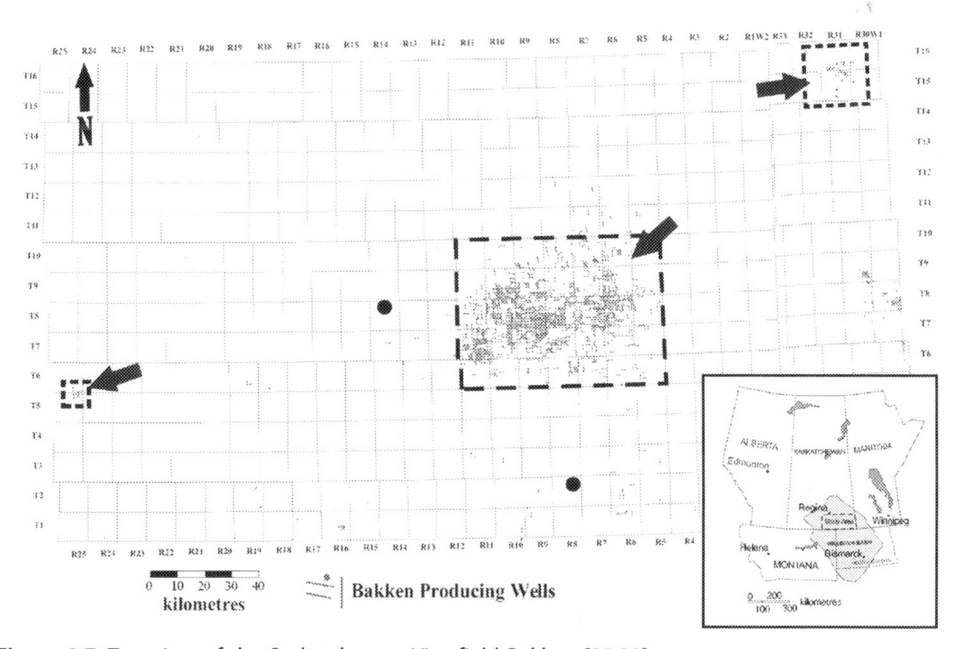

Figure 9.7 Top view of the Saskatchewan Viewfield Bakken [38,39].

fracturing operation. The well layout created a toe-to-heel injection pattern. The distance from the injector to the nearest hydraulic fracture of each offset producer was almost the same. When gas broke through at the toe end, the toe portions of the producer were plugged to alleviate gas cycling. The injected gas continued moving to the next port. This pattern enabled one injector to serve the gas requirement by nine producers. For the producers with heels close to the injector, a straddle packer system called "scab-liner" technique was applied at the immediate heel port [1].

The pilot test was started in December 2011; at the starting point of the project, the injection rate was 300 MSCF/day at the injection pressure of 500 psi. When compression was added in March 2012, the injection rate was increased to 1 MMSCF/day at 1000 psi. Immediately, gas broke through two pattern wells. The oil production rate decreased to 53 bbl/day by July 2012. After workovers, oil rates consistently increased in all of nine producers and the total rate climbed to 295 bbl/day. The average decline rate of the pattern wells decreased from 20% before gas injection to 15% after gas injection [1].

From this pilot test different valuable point could be learned. The first one is the producer to injector ratio highly affects the performance of the project from an economic viewpoint; this parameter is normally one to one in a case of conventional water flooding. The second point is very low investment costs of carbon dioxide injection compared to water injection scenario.

A CO_2 flooding pilot was performed in the Fuyang Layer in the Fang-48 fault block, Song-Fang-Dun Field, Daqing, China, starting in March 2003. The ratio of production well to injection well in this case is 5 to 1. At the end of two years from injection started the total cumulative injected carbon dioxide was 0.33 reservoir pore volume (PV). It should be noted that the injection well was not fractured in this pilot test. Injectivity of carbon dioxide in this case is much higher than the water injectivity; it is almost 6.3 times higher. The performance of CO_2 injection in this pilot test was promising.

One CO_2 huff-n-puff injection pilot was conducted in the Elm Coulee Field in the Bakken Formation in the North Dakota area in 2008 [40]. There was no injectivity problem associated with injection of CO_2 at 1 MMSCF/day; the injection period took 30 days. According to the production history before and after performing such an EOR method, there was no significant improvement on oil recovery factor from this test.

Another CO_2 huff-n-puff field test was performed in Richland County in the Montana part of Bakken reservoir in 2009. Well configuration for this test was horizontal well, which stimulated using hydraulic fracturing operation. The injection period took 45 days, in this time the cumulative injected carbon dioxide was 45 million cubic feet. After injection process, put the well in shut down model for 64 days; this time was a soaking time. After soaking time, the well started to production and the oil recovery rate increased by 44 bbl/day. However, this incremental oil recovery was not completely attributed to CO_2 injection; this is because the workover has been done in the well as well as some CO_2 effects in near wellbore region [40,41].

The third CO_2 huff-n-puff injection pilot test in Bakken formation was done in 2014; in this scenario, a vertical well drilled and completed in the Bakken formation. It was planned to inject CO_2 for 20-30 days by 300-500MSCF/day and the soaking time was 20 days. Also, an offset well was used to evaluate the CO_2 breakthrough time; after 1 day from injection started CO_2 broke through the offset well and the injection process was shut down [40]. This was mainly because huff-n-puff mode could not be successful when CO_2 find the way to break through offset well.

Another field test of CO_2 injection scenario was performed on the Parshall field, Mountrail County. In this pilot the huff-n-puff injection model was used in a horizontal well which completed in the Bakken formation. Injection time in this pilot test was 11 days and after this time, the oil production rate increased [1]. It is worth to mention that this field is a naturally fractured one and controlling the conformance is quite challenging in designing any type of EOR scenarios; mobility of carbon dioxide in such a reservoir was 304. Also, this reservoir has a local fracture network; it means that the connectivity of the fractures will affect the performance of the EOR scenario [41].

9.3.2 Water Injection

In US and Canadian shale formations, there have not been many water injection field projects. However, in China, large-scale water injection is carried in tight formations. This section provides descriptions for ongoing projects and pilots throughout the United States, China, as well as Canada.

9.3.2.1 Continuous Waterflooding

In shale and tight reservoirs, one main concern is water injectivity. It can be understood that water injection may have more injectivity issue than gas injection. Thus, the first objective of a water injection pilot is to check water injectivity. Surprisingly, the field tests conducted so far did not have the injectivity problem in shale reservoirs [40] and in many tight reservoirs in China.

It is commonly accepted that water—rock interaction causes permeability impairment. However, it was observed that the water may help to generate microfractures or open existing microfractures in shale formations if no confining pressure is applied [42—46].

However, Behnsen and Faulkner [47], Duan and Yang [48], and Faulkner and Rutter [49] reported that with isotropic confining pressure, a significant reduction was observed on clay-bearing rocks or montmorillonite sample permeability measured with water. In conventional propped hydraulic fracture treatments, water fractures rely on reactivation of natural fractures to induce permanent shear-induced dilation, which enhances reservoir permeability [50,51]. Hydraulic fracturing is performed where shear failure is anticipated to dominate [52] in shale under anisotropic stress. Hydration swelling due to water imbibition can weaken the mechanical strength of shale [53–55], and it can reduce the shear-induced fracture conductivity [1,56,57].

9.3.2.2 Huff-n-Puff Water Injection

One general mechanism for water huff-n-puff is that water preferentially invades in large pores and then imbibes into small pores to displace oil. Another important mechanism is the invaded water and imbibed water increase reservoir pressure and local pressure so that the drive energy is boosted. From the imbibition point of view, water-wet formation is preferred.

Yu and Sheng [57] carried out water flooding tests in a huff-n-puff mode. Their experimental results reveal that the oil recovery factor highly depends on the injection pressure. In their experiment increasing the soaking period resulted in increasing the oil recovery factor. However, the oil recovery factor from such a method is much lower than that the case of CO_2 huff-n-puff scenario. Altawati [58] conducted core displacement experiments in a huff-n-puff setting; he considered the effect of initial water saturation in their experiments. According to his experimental work, the presence of initial water saturation affects the oil recovery performance drastically; in a case of no initial water saturation, the amount of recovered oil was much higher than that the case with initial water saturation. Sheng and Chen [11] performed numerical simulations to compare the performance of the water and gas injection in a huff-n-puff mode. They concluded that the oil recovery factor of the CO_2 huff-n-puff method is 2–3 times higher than the case of water injection in the huff-n-puff setting. Some field tests are discussed below and the performances are reported in Table 9.1.

Table 9.1 Performance of Water Injection in Huff-n-puff Mode [1]

Field	Huff Time (Day)	Soak Time (Day)	Puff Time (Day)	Performance
Bakken, ND	30	15	90–120	No considerable increase in oil production
Parshall	30	10		No considerable increase in oil production
Parshall	At first, 439,000 barrels injected through the reservoir and then WAG applied			No considerable increase in oil production

WAG, water alternating gas.

9.3.2.3 Field Test of Water Injection

Water injection process was started in 2006 in Bakken and Lower Shaunavon formations. The injection patterns were such that horizontal injectors paralleled horizontal producers with their spacing of hundreds of meters [60]. Later a simulation study was conducted for a Lower Shaunavon of 1 injector and 18 producers. After 50 years, the recovery factor was predicted to be 5.1% [61].

Water was injected by Meridian Oil in the Bicentennial Field in McKenzie County in 1994. Approximately 13,200 barrels of freshwater were injected into a horizontal well in the Upper Bakken Shale for 50 days, then the soaking time was 60 days, after soaking period, oil production remained below the rates before water injection for the rest of the well's operational life. Another water injection pilot test was conducted in the Bakken formation located in the North Dakota; as shown in Fig. 9.8, the ratio of production wells to the injection ones is 4 to 1. The injection rate was about 1350 bbl/day for 8 months in the middle of 2012; However, no incremental oil was observed [40]. The failure of this waterflooding pilot seemed to be caused by low water sweep efficiency because much less water was produced than the injected (water lost). Therefore, this case may not be considered as general rule in designing water flooding scenario in tight and shale reservoirs [1].

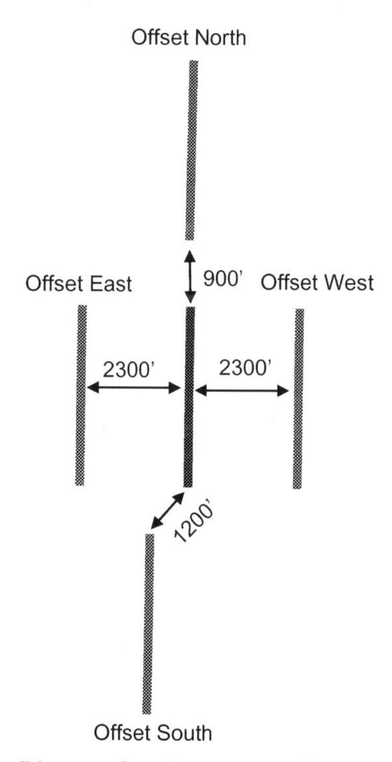

Figure 9.8 Schematic of the well layout of a pilot test in Bakken Formation [40].

The test was conducted in 2014. There were one injection well and several offset wells. In the first 3 months, the well injection rate was 1700 STB/day and later reduced to 1000 STB/day because of breakthrough at a close offset well [40]. The close offset well (about 880 ft away) had a huge increase in water production, but the oil rate did not increase during this time. The water broke after 1 week.

One water huff-n-puff pilot test was conducted in the North Dakota part of the Bakken Formation in 2012. The injection time was just over 1 month, the soaking time was 2 weeks, and the production time was 3–4 months. The injection rate was 1200 bbl/day. No water injectivity problem was observed, but little to no oil rate was increased [40].

According to the pilot test of produced water re-injection with huff-n-puff setting that has been done in the Parshall field, there is no significant improvement of the oil recovery factor after employing such a method. In this scenario that is performed on spring of 2012, the duration of water injection was about 30 days and 10 days as a soaking period; after that time the production was open to flow [1].

Table 9.2 presents several field tests that have been performed in shale oil reservoirs located in China. This table provides general information regarding the type of injection mode as well as rock and fluid characterization (some of them not reported). Also, the performance of each field test is reported in Table 9.2 [1].

Huff-n-puff water injection was also performed in the Chang 6 reservoir, Yanchang oil field near Qieli Village, Ordos Basin, China (huff-n-puff 4 in Table 9.2) [1].

Huff-n-puff water injection was also successfully tested in Toutai reservoir, Daqing. The permeability was 1.25 mD, and the huff-n-puff cycle was half to one year [1]. Huff-n-puff water injection was tested in 2007 in Well Niu 15-5 in Niuquanhu reservoir, Tuha Field. The permeability in this area was 0.42–7.84 mD. The soaking time was 108 days. Two cycles were performed with incremental oil of 1816 tons [1]. Huff-n-puff water injection was performed in Duha Field, Xingjiang, China (huff-n-puff 5 in Table 9.2). Water was injected in Well Ma-55 from July 18, 2014 to July 24, 2014. The injection rate was 285 m^3/day. The total injected water volume was 2000 m^3. The oil rate before water injection was 0.9 tons/day and the water cut was 16%. After the huff-n-puff water injection, the oil rate was 5 tons/day and the water rate was kept constant until the reporting date (August 2015) [1].

9.4 ENVIRONMENTAL ASPECTS OF SHALE OIL AND GAS PRODUCTION

Owing to the growing concern regarding climate change and energy demand, using natural gas as a clean-burning and affordable source of energy has been

Table 9.2 Performance of Water Injection in Reservoirs Located in China [1]

Injection	Field Name	Permeability (mD)	Oil Viscosity (cP)	Performance
Pulsed	An 83, Chang 7	0.17	1.01	When shut-in, p reduced sharply, fw not reduced
Asynchronous	An 83, Chang 7			Oil production increased
Huff-n-Puff 1	An 83, Chang 7			Oil production increased
Huff-n-Puff 2	An 83, Chang 7			Performance not as good as Huff-n-Puff 1
Huff-n-Puff 3	An 83, Chang 7			Six wells had one cycle and two wells had two cycles. Neighboring nonhuff-n-puff wells outperformed huff-n-puff wells. The second cycle performed not as good as the first cycle
Huff-n-Puff 4	Chang 6	0.54	4.67	Oil production rate increased. The scenario was performed for 7 days soaking and well spacing 300 m
Huff-n-Puff 5		0.1−1		Oil production increased from 0.9 to 5 t/day

p, pressure; *fw*, fractional flow to water.

attracting researchers' attention. One of the reliable natural gas resources is a shale gas reservoir. Wood et al. (2011) argue that development of shale gas production as the huge potential energy resource is a part of clean-burning fossil fuel production, especially, in many countries where other fossil fuel resources are becoming depleted [60,62]. Statistical forecasts reveal that the consumption of natural gas is anticipated to triple by 2035 and large shale gas reservoirs throughout the world have raised expectations for cheap energy and reliable energy supply [63]. Hydraulic fracturing is the key technology that has enabled shale gas development. Vidic et al. [64] argue that in hydraulic fracturing operations, a significant amount of fracturing fluid[1] [water and proppants[2]] is injected under high pressure into low-permeability shale formations to induce fracturing and improve the mobility of natural gas. Large-scale production of

[1] Fracturing fluid: "Fracturing fluid is a special type of fluid that is injected into a wellbore to induce fractures to facilitate access to the oil and gas resources" [65].

[2] Proppants: "Proppants are sand or other granular substances injected into the formation to hold or prop open the formation fractures created by hydraulic fracturing" [65].

shale gas has become economical through the application of hydraulic fracturing technologies [66]. Sovacool [67] argues that many people are opposed to shale gas because of various environmental and sustainability issues associated with shale gas production through the use of hydraulic fracturing technology. Mair et al. [68] maintain that in many circumstances, the environmental costs followed by the shale gas production dominates the economic benefits, comprising groundwater and drinking water contamination, greenhouse gas emissions, and earthquakes. As a result, shale gas has been banned in some European countries, notably, France and Bulgaria [63]. The main aim of this section is to discuss environmental issues following shale gas production using hydraulic fracturing method and to provide some recommendations for solving these issues or minimizing the consequences of these problems on the human life. From an environmental position, three main categories such as air emissions, water issues, and land problems should be taken into account for studying environmental impacts of shale gas production using the hydraulic fracturing method.

9.4.1 Air Emissions

A statement regarding the amount of shale gas contribution to global warming and climate change is an uncertain comment, which needs more research [69–72]. However, the most frequent recommendation is replacing shale gas to coal for power generation; this replacement could decrease greenhouse gases emissions [63]. An additional air emission issue in shale gas development is photochemical oxidants creation potential (POCP), which is also known as photochemical smog. Compared to other energy technologies, shale gas has a greater POCP: up to 45 times bigger than nuclear power, 26 times larger than offshore wind, and three times higher than solar PV [63]. Different air contaminants are produced during the shale gas production, comprising nitrogen oxides (NO_x), volatile organic compounds, alkenes, alkanes, and silica particles [63].

9.4.2 Impacts to Water

One of the main conflicts of interests in shale gas production is significant water consumption, which will increase stress on water supplies, mainly in semiarid and arid regions [73–76]. However, as explained in Laurenzi and Jersey [77], in comparison with conventional oil, oil sands, and coal, the water footprint of shale gas is lower [63]. As described before, fracturing fluid that is used for hydraulic fracturing process comprises water, sand, and different chemicals. These chemical additives that are utilized in fracturing fluid are mainly surfactants, polymers, and biocides (Biocides: "Additive applied to kill bacteria in the water" [63]). Moore et al. [78] maintain that using these chemical agents in fracturing fluid are a source of concern for the public owing to possible contamination of water sources [63]. Vengosh et al. [79] demonstrate three main ways for water pollution exposure including stray gas; spills, leaks, and illegal dumping/disposal; and accumulation in disposal locations. However, investigating the connection between shale gas production and contaminating water is inconclusive [63,80–84].

9.4.3 Impacts to Land

Productions from Shale gas reservoirs have various effects to land including land usage, terrestrial ecotoxicity [85–87], and earthquakes; however, the effect of shale gas production on earthquakes is an unclear question. Same as other oil and gas wells, whether conventional or unconventional, to initiate drilling of the well an area of land needs to be prepared to place the drilling rig and allow access to the drilling location [63]. A further land impact related to shale gas production is terrestrial ecotoxicity, mainly owing to the discarding of the drilling waste, which comprises toxic materials, for instance, barite. As Johri and Zoback [88] argue, tremor can be initiated by fractures created during hydraulic fracturing for shale gas production. However, in comparison with those related to other activities, for instance, coal mining and reservoir impoundment for hydroelectric projects, the number of tremors caused by shale gas production are much lower. Also, the magnitude of the tremor initiated by shale gas production is usually not felt or felt with little damage. As Davies et al. [89] have reported, shale gas production just caused three earthquakes in British Columbia (Canada), Lancashire (United Kingdom), and Oklahoma (United States). Compared to the tremors caused by other human activities three earthquakes is too small [63].

9.4.4 Recommendations

As Mair et al. [68] demonstrate, the best available technology (BAT) for defeating the obstacle of air emissions from shale gas production is the use of "green" well completion, not allowing gas to be vented, and detecting and fixing leaks in tools and pipes [63]. As Wang et al. [90] demonstrate, some strategies that can be used for eliminating the effect of water contamination are baseline data, continuous monitoring over the wells lifetime, and adaptive wastewater management. These strategies can decrease the probability of chemical exposure, while chemical tracking can be utilized to detect the source of pollution. As Rahm and Riha [91] propose, to reduce the water consumption during shale gas production, different plans can be used, which include regulation for water withdrawals, using brackish water instead of freshwater, and recycling/reuse of water [63]. As Manda et al. [92] recommend, one of the promising methods to reduce the land usage during shale gas production is using multiwell pads, which have a water usage per well two to four times lower than single-well pads [63].

Regardless of the uncertainties due to lack of data, some evidence associated with shale gas production and utilization are well established. For instance, water usage, wastewater disposal, the creation of POCP, land usage, and well leakages are the cause of many of the environmental issues; these problems can be solved by implementing of best practice and BAT [63]. Environmental implications of shale gas production from early stages should be evaluated on a life-cycle basis instead of focusing on a single activity [63]. Periodical environmental assessment in a long-term plan is also crucial to evaluate how shale gas can contribute to the environmental issues, i.e., climate change and global warming. Finally, various strategies over different time span could be proposed with and without shale gas production development to determine its impacts on the energy supply as well as environment.

REFERENCES

[1] J.J. Sheng, B.L. Herd, Critical review of field EOR projects in shale and tight reservoirs, J. Pet. Sci. Eng. 159 (2017) 654–665.
[2] A. Bahadori, Fluid Phase Behavior for Conventional and Unconventional Oil and Gas Reservoirs, Gulf Professional Publishing, Amsterdam, Netherlands, 2016.
[3] D. Gordon, Understanding Unconventional Oil, Carnegie Endowment for International Peace, Washington DC, United States, 2012.
[4] M. Ashraf, M. Satapathy, The global quest for light tight oil: myth or reality, Energy Perspect. (2013) 16–23.
[5] K. Brendow, Oil shale—a local asset under global constraint, Oil Shale 26 (3) (2009) 357–372. Available from: http://dx.doi.org/10.3176/oil.2009.3.02.
[6] U. EIA, Annual Energy Outlook 2010, US Energy Information Administration, Washington, DC, 2013, pp. 60–62.
[7] J.J. Sheng, Increase liquid oil production by huff-n-puff of produced gas in shale gas condensate reservoirs, J. Unconv. Oil Gas Res. 11 (2015) 19–26.
[8] Clark, A.J., 2009. Determination of recovery factor in the Bakken Formation, Mountrail County, ND. In: SPE Annual Technical Conference and Exhibition. Society of Petroleum Engineers.
[9] C. Jia, C. Zou, The evaluation standard, the main types, basic characteristics and resources prospect of tight oil in China, APS 33 (2012) 343–350.
[10] Z. Qingfan, Y. Guofeng, Definition and application of tight oil and shale oil terms, Oil Gas Geol. 33 (2012) 541–544.
[11] J.J. Sheng, K. Chen, Evaluation of the EOR potential of gas and water injection in shale oil reservoirs, J. Unconv. Oil Gas Res. 5 (2014) 1–9.
[12] Y. Yu, X. Meng, J.J. Sheng, Experimental and numerical evaluation of the potential of improving oil recovery from shale plugs by nitrogen gas flooding, J. Unconv. Oil Gas Res. 15 (2016) 56–65.
[13] Kovscek, A.R., Tang, G.-Q., Vega, B., 2008. Experimental investigation of oil recovery from siliceous shale by CO_2 injection. In: SPE Annual Technical Conference and Exhibition. Society of Petroleum Engineers.
[14] Vega, B., O'Brien, W.J., Kovscek, A.R., 2010. Experimental investigation of oil recovery from siliceous shale by miscible CO_2 injection. In: SPE Annual Technical Conference and Exhibition. Society of Petroleum Engineers.
[15] Wang, X., Luo, P., Er, V., Huang, S.-S.S., 2010. Assessment of CO_2 flooding potential for Bakken Formation, Saskatchewan. In: Canadian Unconventional Resources and International Petroleum Conference. Society of Petroleum Engineers.
[16] Joslin, K., Ghedan, S., Abraham, A., Pathak, V., 2017. EOR in tight reservoirs, technical and economical feasibility. In: SPE Unconventional Resources Conference. Society of Petroleum Engineers.
[17] Dong, C., Hoffman, B.T., 2013. Modeling gas injection into shale oil reservoirs in the Sanish Field, North Dakota. In: Unconventional Resources Technology Conference. Society of Exploration Geophysicists, American Association of Petroleum Geologists, Society of Petroleum Engineers, pp. 1824–1833.
[18] Wan, T., 2013. Evaluation of the EOR potential in shale oil reservoirs by cyclic gas injection.

[19] Gamadi, T.D., Sheng, J., Soliman, M., 2013. An experimental study of cyclic gas injection to improve shale oil recovery. In: SPE Annual Technical Conference and Exhibition. Society of Petroleum Engineers.
[20] Tovar, F.D., Eide, O., Graue, A., Schechter, D.S., 2014. Experimental investigation of enhanced recovery in unconventional liquid reservoirs using CO_2: a look ahead to the future of unconventional EOR. In: SPE Unconventional Resources Conference. Society of Petroleum Engineers.
[21] T. Wan, Y. Yu, J.J. Sheng, Experimental and numerical study of the EOR potential in liquid-rich shales by cyclic gas injection, J. Unconv. Oil Gas Res. 12 (2015) 56–67.
[22] Yu, Y., Sheng, J., Mody, F., 2015. Evaluation of cyclic gas injection EOR performance on shale core samples using X-ray CT scanner. In: AIChE 407411 Presented at the AIChE Annual Meeting, 8–13 November 2015, Salt Lake City, UT.
[23] Yu, Y., Sheng, J.J., 2015. An experimental investigation of the effect of pressure depletion rate on oil recovery from shale cores by cyclic N_2 injection. In: Unconventional Resources Technology Conference, 20–22 July 2015, San Antonio, TX. Society of Exploration Geophysicists, American Association of Petroleum Geologists, Society of Petroleum Engineers, pp. 548–557.
[24] L. Li, J.J. Sheng, Experimental study of core size effect on CH_4 huff-n-puff enhanced oil recovery in liquid-rich shale reservoirs, J. Nat. Gas Sci. Eng. 34 (2016) 1392–1402.
[25] Denoyelle, L., Lemonnier, P., 1987. Simulation of CO_2 huff 'n' puff using relative permeability hysteresis. In: SPE Annual Technical Conference and Exhibition. Society of Petroleum Engineers.
[26] Li, L., Sheng, J.J., Sheng, J., 2016. Optimization of huff-n-puff gas injection to enhance oil recovery in shale reservoirs. In: SPE Low Permeability Symposium. Society of Petroleum Engineers.
[27] X. Meng, J.J. Sheng, Optimization of huff-n-puff gas injection in a shale gas condensate reservoir, J. Unconv. Oil Gas Res. 16 (2016) 34–44.
[28] Gamadi, T., Sheng, J., Soliman, M., Menouar, H., Watson, M., Emadibaladehi, H., 2014. An experimental study of cyclic CO_2 injection to improve shale oil recovery. In: SPE Improved Oil Recovery Symposium. Society of Petroleum Engineers.
[29] Yu, Y., Li, L., Sheng, J.J., 2016. Further discuss the roles of soaking time and pressure depletion rate in gas Huff-n-Puff process in fractured liquid-rich shale reservoirs. In: SPE Annual Technical Conference and Exhibition. Society of Petroleum Engineers.
[30] Meng, X., Sheng, J.J., Yu, Y., 2015. Evaluation of enhanced condensate recovery potential in shale plays by huff-n-puff gas injection. In: SPE Eastern Regional Meeting. Society of Petroleum Engineers.
[31] Meng, X., Yu, Y., Sheng, J.J., 2015. An experimental study on huff-n-puff gas injection to enhance condensate recovery in shale gas reservoirs. In: Unconventional Resources Technology Conference, 20–22 July 2015, San Antonio, TX. Society of Exploration Geophysicists, American Association of Petroleum Geologists, Society of Petroleum Engineers, 853–863.
[32] Wan, T., Meng, X., Sheng, J.J., Watson, M., 2014. Compositional modeling of EOR process in stimulated shale oil reservoirs by cyclic gas injection. In: SPE Improved Oil Recovery Symposium. Society of Petroleum Engineers.
[33] Wan, T., Yang, Y., Sheng, J., 2014. Comparative study of enhanced oil recovery efficiency by CO_2 injection and CO_2 huff-n-puff in stimulated shale oil reservoirs. In: AICHE Annual Meeting, Atlanta, GA, USA.
[34] Y. Yu, L. Li, J.J. Sheng, A comparative experimental study of gas injection in shale plugs by flooding and huff-n-puff processes, J. Nat. Gas Sci. Eng. 38 (2017) 195–202.
[35] X. Meng, J.J. Sheng, Y. Yu, Experimental and numerical study of enhanced condensate recovery by gas injection in shale gas-condensate reservoirs, SPE Reserv. Eval. Eng. 20 (2017) 471–477.
[36] Shoaib, S., Hoffman, B.T., 2009. CO_2 flooding the Elm Coulee Field. In: SPE Rocky Mountain Petroleum Technology Conference. Society of Petroleum Engineers.
[37] Kurtoglu, B., Sorensen, J.A., Braunberger, J., Smith, S., Kazemi, H., 2013. Geologic characterization of a Bakken reservoir for potential CO_2 EOR. In: Unconventional Resources Technology Conference (URTEC).
[38] Schmidt, M., Sekar, B., 2014. Innovative unconventional^2 EOR-A light EOR an unconventional tertiary recovery approach to an unconventional Bakken reservoir in Southeast Saskatchewan. In: 21st World Petroleum Congress. World Petroleum Congress.

[39] D. Kohlruss, E. Nickel, Facies analysis of the Upper Devonian−Lower Mississippian Bakken Formation, southeastern Saskatchewan: summary of investigations 2009. Saskatchewan Geological Survey, Saskatchewan Ministry of Energy and Resources, Miscellaneous Report, 4, 2009.
[40] Todd, H.B., Evans, J.G., 2016. Improved oil recovery IOR pilot projects in the Bakken Formation. In: SPE Low Permeability Symposium. Society of Petroleum Engineers.
[41] Sorensen, J.A., Hamling, J.G., 2016. Historical Bakken test data provide critical insights on EOR in tight oil plays.
[42] H. Dehghanpour, Q. Lan, Y. Saeed, H. Fei, Z. Qi, Spontaneous imbibition of brine and oil in gas shales: effect of water adsorption and resulting microfractures, Energy Fuels 27 (2013) 3039−3049.
[43] Morsy, S., Sheng, J., Gomaa, A.M., Soliman, M., 2013. Potential of improved waterflooding in acid-hydraulically-fractured shale formations. In: SPE Annual Technical Conference and Exhibition. Society of Petroleum Engineers.
[44] Morsy, S.S., Sheng, J., Soliman, M., 2013. Improving hydraulic fracturing of shale formations by acidizing. In: SPE Eastern Regional Meeting. Society of Petroleum Engineers.
[45] Morsy, S., Sheng, J., Ezewu, R.O., 2013. Potential of waterflooding in shale formations. In: SPE Nigeria Annual International Conference and Exhibition. Society of Petroleum Engineers.
[46] Morsy, S., Sheng, J., 2014. Imbibition characteristics of the Barnett Shale formation. In: SPE Unconventional Resources Conference. Society of Petroleum Engineers.
[47] J. Behnsen, D. Faulkner, Water and argon permeability of phyllosilicate powders under medium to high pressure, J. Geophy. Res.: Solid Earth 116 (B12) (2011) 1−13.
[48] Q. Duan, X. Yang, Experimental studies on gas and water permeability of fault rocks from the rupture of the 2008 Wenchuan earthquake, China, Sci. China Earth Sci. 57 (2014) 2825−2834.
[49] D. Faulkner, E. Rutter, Comparisons of water and argon permeability in natural clay-bearing fault gouge under high pressure at 20°C, J. Geophy. Res.: Solid Earth 105 (2000) 16415−16426.
[50] Z. Chen, S. Narayan, Z. Yang, S. Rahman, An experimental investigation of hydraulic behaviour of fractures and joints in granitic rock, Int. J. Rock Mech. Min. Sci. 37 (2000) 1061−1071.
[51] Weng, X., Sesetty, V., Kresse, O., 2015. Investigation of shear-induced permeability in unconventional reservoirs. In: 49th US Rock Mechanics/Geomechanics Symposium, San Francisco, CA, USA.
[52] Zoback, M.D., Kohli, A., Das, I., Mcclure, M.W., 2012. The importance of slow slip on faults during hydraulic fracturing stimulation of shale gas reservoirs. In: SPE Americas Unconventional Resources Conference. Society of Petroleum Engineers.
[53] R. Wong, Swelling and softening behaviour of La Biche shale, Can. Geotech. J. 35 (1998) 206−221.
[54] A.-B. Talal, A novel experimental technique to monitor the time-dependent water and ions uptake when shale interacts with aqueous solutions, Rock Mech. Rock Eng. 46 (2013) 1145−1156.
[55] J. Cheng, Z. Wan, Y. Zhang, W. Li, S.S. Peng, P. Zhang, Experimental study on anisotropic strength and deformation behavior of a coal measure shale under room dried and water saturated conditions, Shock Vib. 2015 (2015) 1−13.
[56] Pedlow, J., Sharma, M., 2014. Changes in shale fracture conductivity due to interactions with water-based fluids. In: SPE Hydraulic Fracturing Technology Conference. Society of Petroleum Engineers.
[57] Jansen, T.A., Zhu, D., Hill, A.D., 2015. The effect of rock mechanical properties on fracture conductivity for shale formations. In: SPE Hydraulic Fracturing Technology Conference. Society of Petroleum Engineers.
[58] Yu, Y., Sheng, J.J., 2016. Experimental investigation of light oil recovery from fractured shale reservoirs by cyclic water injection. In: SPE Western Regional Meeting. Society of Petroleum Engineers.
[59] Altawati, F.S., 2016. An Experimental Study of the Effect of Water Saturation on Cyclic N_2 and CO_2 Injection in Shale Oil Reservoir, Master of Science Thesis, Texas Tech University.
[60] T. Wood, B. Milne, Waterflood Potential Could Unlock Billions of Barrels, Crescent Point Energy, 2011.

[61] Thomas, A., Kumar, A., Rodrigues, K., Sinclair, R.I., Lackie, C., Galipeault, A., et al., 2014. Understanding water flood response in tight oil formations: a case study of the Lower Shaunavon. In: SPE/CSUR Unconventional Resources Conference—Canada. Society of Petroleum Engineers.
[62] R.W. Howarth, R. Santoro, A. Ingraffea, Methane and the greenhouse-gas footprint of natural gas from shale formations, Clim. Change 106 (2011) 679.
[63] J. Cooper, L. Stamford, A. Azapagic, Shale gas: a review of the economic, environmental, and social sustainability, Energy Technol. 4 (2016) 772–792.
[64] R.D. Vidic, S.L. Brantley, J.M. Vandenbossche, D. Yoxtheimer, J.D. Abad, Impact of shale gas development on regional water quality, Science 340 (2013) 1235009.
[65] J.G. Speight, Handbook of Hydraulic Fracturing, John Wiley & Sons, Hoboken, New Jersey, 2016.
[66] X. Zhang, A.Y. Sun, I.J. Duncan, Shale gas wastewater management under uncertainty, J. Environ. Manag. 165 (2016) 188–198.
[67] B.K. Sovacool, Cornucopia or curse? Reviewing the costs and benefits of shale gas hydraulic fracturing (fracking), Renewable Sustainable Energy Rev. 37 (2014) 249–264.
[68] Mair, R., Bickle, M., Goodman, D., Koppelman, B., Roberts, J., Selley, R., et al., 2012. Shale Gas Extraction in the UK: A Review of Hydraulic Fracturing, The Royal Society and The Royal Academy of Engineering, London, United Kingdom.
[69] F. O'Sullivan, S. Paltsev, Shale gas production: potential versus actual greenhouse gas emissions, Environ. Res. Lett. 7 (2012) 044030.
[70] A.R. Brandt, G. Heath, E. Kort, F. O'sullivan, G. Pétron, S. Jordaan, et al., Methane leaks from North American natural gas systems, Science 343 (2014) 733–735.
[71] D.R. Caulton, P.B. Shepson, R.L. Santoro, J.P. Sparks, R.W. Howarth, A.R. Ingraffea, et al., Toward a better understanding and quantification of methane emissions from shale gas development, Proc. Natl. Acad. Sci. 111 (2014) 6237–6242.
[72] A. Karion, C. Sweeney, G. Pétron, G. Frost, R. Michael Hardesty, J. Kofler, et al., Methane emissions estimate from airborne measurements over a western United States natural gas field, Geophys. Res. Lett. 40 (2013) 4393–4397.
[73] E.M. Thurman, I. Ferrer, J. Blotevogel, T. Borch, Analysis of hydraulic fracturing flowback and produced waters using accurate mass: identification of ethoxylated surfactants, Anal. Chem. 86 (2014) 9653–9661.
[74] S. Gamper-Rabindran, Information collection, access, and dissemination to support evidence-based shale gas policies, Energy Technol. 2 (2014) 977–987.
[75] T. Colborn, C. Kwiatkowski, K. Schultz, M. Bachran, Natural gas operations from a public health perspective, Hum. Ecol. Risk Assess.: Int. J. 17 (2011) 1039–1056.
[76] B.C. Gordalla, U. Ewers, F.H. Frimmel, Hydraulic fracturing: a toxicological threat for groundwater and drinking-water? Environ. Earth Sci. 70 (2013) 3875–3893.
[77] I.J. Laurenzi, G.R. Jersey, Life cycle greenhouse gas emissions and freshwater consumption of Marcellus shale gas, Environ. Sci. Technol. 47 (2013) 4896–4903.
[78] V. Moore, A. Beresford, B. Gove, Hydraulic Fracturing for Shale Gas in the UK: Examining the Evidence for Potential Environmental Impacts, RSPB, Sandy, UK, 2014.
[79] A. Vengosh, R.B. Jackson, N. Warner, T.H. Darrah, A. Kondash, A critical review of the risks to water resources from unconventional shale gas development and hydraulic fracturing in the United States, Environ. Sci. Technol. 48 (2014) 8334–8348.
[80] J. Riedl, S. Rotter, S. Faetsch, M. Schmitt-Jansen, R. Altenburger, Proposal for applying a component-based mixture approach for ecotoxicological assessment of fracturing fluids, Environ. Earth Sci. 70 (2013) 3907–3920.
[81] E.W. Boyer, B.R. Swistock, J. Clark, M. Madden, D.E. Rizzo, The Impact of Marcellus Gas Drilling on Rural Drinking Water Supplies, Center for Rural Pennsylvania, United States, 2012.
[82] B.E. Fontenot, Z.L. Hildenbrand, D.D. Carlton Jr, J.L. Walton, K.A. Schug, Response to comment on "an evaluation of water quality in private drinking water wells near natural gas extraction sites in the Barnett Shale formation", Environ. Sci. Technol. 48 (2014) 3597–3599.

[83] B.E. Fontenot, L.R. Hunt, Z.L. Hildenbrand, D.D. Carlton Jr, H. Oka, J.L. Walton, et al., An evaluation of water quality in private drinking water wells near natural gas extraction sites in the Barnett Shale formation, Environ. Sci. Technol. 47 (2013) 10032–10040.
[84] T. McHugh, L. Molofsky, A. Daus, J. Connor, Comment on "an evaluation of water quality in private drinking water wells near natural gas extraction sites in the Barnett Shale formation", Environ. Sci. Technol. 48 (2014) 3595–3596.
[85] M.C. Brittingham, K.O. Maloney, A.M. Farag, D.D. Harper, Z.H. Bowen, Ecological risks of shale oil and gas development to wildlife, aquatic resources and their habitats, Environ. Sci. Technol. 48 (2014) 11034–11047.
[86] J.R. Barber, C.L. Burdett, S.E. Reed, K.A. Warner, C. Formichella, K.R. Crooks, et al., Anthropogenic noise exposure in protected natural areas: estimating the scale of ecological consequences, Landsc. Ecol. 26 (2011) 1281.
[87] J.R. Barber, K.R. Crooks, K.M. Fristrup, The costs of chronic noise exposure for terrestrial organisms, Trends Ecol. Evol. 25 (2010) 180–189.
[88] Johri, M., Zoback, M.D., 2013. The evolution of stimulated reservoir volume during hydraulic stimulation of shale gas formations. In: Unconventional Resources Technology Conference. Society of Exploration Geophysicists, American Association of Petroleum Geologists, Society of Petroleum Engineers, pp. 1661–1671.
[89] R. Davies, G. Foulger, A. Bindley, P. Styles, Induced seismicity and hydraulic fracturing for the recovery of hydrocarbons, Mar. Pet. Geol. 45 (2013) 171–185.
[90] Q. Wang, X. Chen, A.N. Jha, H. Rogers, Natural gas from shale formation—the evolution, evidences and challenges of shale gas revolution in United States, Renewable Sustainable Energy Rev. 30 (2014) 1–28.
[91] B.G. Rahm, S.J. Riha, Toward strategic management of shale gas development: regional, collective impacts on water resources, Environ. Sci. Policy 17 (2012) 12–23.
[92] A.K. Manda, J.L. Heath, W.A. Klein, M.T. Griffin, B.E. Montz, Evolution of multi-well pad development and influence of well pads on environmental violations and wastewater volumes in the Marcellus shale (USA), J. Environ. Manag. 142 (2014) 36–45.

CHAPTER TEN

Microbial Enhanced Oil Recovery: Microbiology and Fundamentals

Afshin Tatar
Islamic Azad University, Tehran, Iran

10.1 INTRODUCTION

The human population will increase to about 9.5 billion people by 2050 [1]. Per capita energy consumption is directly related to the standard of living, which is desired to be ever-increasing. This is indicative of the inevitable increase in the global energy demand [2]. The global demand for energy well result in 49% expansion by 2035 compared with 2007 [3]. Based on the US Energy Information Administration [4], the world energy consumption will increases from 575 quadrillion (575×10^{15}) Btu in 2015 to 663 quadrillion (663×10^{15}) Btu by 2030 and then to 736 (736×10^{15}) quadrillion Btu by 2040. Based on the same report [4], although consumption of nonfossil fuels is expected to grow faster than fossil fuels (2.3 and 1.5%/year for respective renewable and nuclear energy), fossil fuels will still account for 77% of energy use in 2040. Transportation sector is the main energy consumer. Contribution of nonpetroleum sources such as ethanol in spite of its increase will still supply less than the 10% of the demand by 2030 [5]. Crude oil will be the most probable energy source for transportation use in the close future [6].

In general, about 10% of the initial oil in place (IOIP) is recoverable through primary production [7] and secondary recovery can promote the production to one-thirds of the IOIP and still two-thirds is left on place [8]. In other words, a great volume of oil remains unrecoverable after the convention technologies reach their economic limit. For example the case for United States is 300 billion barrels [9]. Finding new oil fields cannot satisfy the increasing demand for the oil. Furthermore, new exploration of new oil resources is becoming increasingly limited [10]. The best solution is devising new methods to recover more oil from the existing oil fields [11–14]. In the United States, it is estimated that nearly 17.5 m^3 of oil has been lost because of abandonment or plugging of stripper (marginal) wells in a 10-year period between 1994 and 2003 [6]. In addition, increasing of the oil price and the future lack of access to the easy supplies will make the tertiary methods more viable.

Fundamentals of Enhanced Oil and Gas Recovery from Conventional and Unconventional Reservoirs.
DOI: https://doi.org/10.1016/B978-0-12-813027-8.00010-2

10.2 DEFINITION

Microbial enhanced oil recovery (MEOR) is one of the tertiary oil recovery categories. Any process utilizing microorganisms and/or their metabolic products including biosurfactants, biomass, biopolymers, bioacids, biosolvents, biogases, and also enzymes to improve the petroleum production from marginal or depleted reservoirs is referred to as MEOR [7,15–19]. This would improve the life of the oil wells. Microbes are single-celled organisms existing everywhere in the nature, including the hydrocarbon reservoirs [15,20] having a substantial impression on the reservoir geochemistry and behavior as well as oil mobilization [10,21]. The microbes utilized in MEOR are typically nonpathogenic hydrocarbon-utilizing microorganisms [22]. The bioproducts generated by the microorganisms amend the physical-chemical properties and consequently the oil – water – rock interaction to enhance oil recovery [22]. Microorganisms can also be employed to clean up the wellbore and remove the built-up hydrocarbons. Although this would promote the well injectivity as well as the flow out of the well, this is not categorized as MEOR. MEOR applies biotechnology to enhance the oil recovery. Despite other tertiary recovery methods, there has not been great attention regarding this method. However, numerous experimental studies has proved that certain microorganisms are capable to develop under the high pressure, temperature, and salinity condition of underground reservoir and produce metabolic products such as biosurfactants, biopolymers, alcohols, acids, and gases. The mentioned compounds can displace the trapped oil via several mechanisms, which will be discussed in the following sections. The combination of multiple mechanisms, which work simultaneously, makes this method highly effective.

MEOR may receive especial attention due to its cost advantage over the synthetic surfactants. Several tertiary oil recovery techniques such as steam flooding, polymer flooding, chemical surfactant flooding, and in-situ combustion have been found to be complex or induce heavy costs to be applicable for extensive utilization [23]. The main expenditures of Enhanced Oil Recovery (EOR) projects are related with production and transportation of EOR chemicals. In case of in situ production of EOR chemicals, at the oil droplet, the costs will significantly reduce [10]. MEOR does not require exceptional investments [23]. It is possible to use the existing waterflooding facilities and equipment with minor modification in this regard [24]. Although the microbes require plentiful nutrients to grow and develop, these nutrients are quite cheap, thus the MEOR technique relatively costs less [25–27]. Fig. 10.1 shows the incremental cost per barrel for various different EOR techniques. Microbial growth takes place at exponential rates [28,29]. This may make it possible to rapidly generate the desired bioproducts from cheap and renewable resources [6]. Moreover, the metabolic generated biochemicals are independent of the crude price, whereas most of

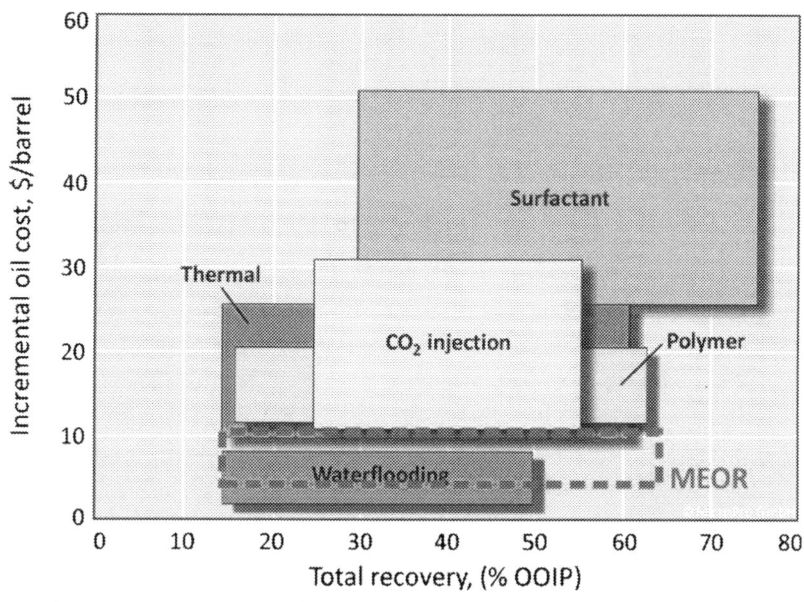

Figure 10.1 The incremental oil costs for different EOR methods [26].

other tertiary methods used petroleum-based chemicals [6,15]. In MEOR, several mechanisms works at the same time and this highly improves the effectiveness of this method [30]. This method is relatively environmental friendly [30]. The biochemicals generated by the microbes are largely biodegradable and nontoxic [24,31].

However, there are some challenges regarding employing MEOR. Unfortunately, considering the other tertiary techniques, MEOR is confined to laboratory investigations and the field trials are scattered and also small-scale [32]. Although there are many successful reported field trials in literature, in some cases there is still a place of skepticism about the validity of the results [32]. Moreover, the major uncertainty is about the process reliability and reproducibility [32]. Amongst the several reported successful MEOR field applications, still there are some cases reporting MEOR to be ineffective. Extreme caution should be considered in changing the microenvironment of the reservoir by introducing new compounds [31]. Accidental or uncontrolled development of some bacteria populations such as sulfate-reducing bacteria (SRB) can induce extra costs, damage the facilities, and bring some safety hazardous [15,16,33–36]. The injected oxygen for aerobic MEOR trials is a corrosive agent and can damage the facilities and downhole piping. Anaerobic MEOR needs a relatively huge amount of sugar as the nutrient, which makes the application of MEOR limited in offshore platforms due to logistical issues [24]. For cases dealing with cultivation of bacteria and producing the biochemicals on surface, the high costs of laboratory equipment, maintenance of the bioreactors, facilities, and purification practices compared with lower yield of the induced production are the main drawbacks [24,31]. In MEOR, the metabolically generated hydrogen sulfide induces souring and significant damages to equipment and piping due corrosion [19,24,37,38]. However, in contrast, some study has reported that MEOR reduced the reservoir souring [39]. Nitrate reducer microbe can contribute to reduce souring [24].

The inconsistency between the lab results and the field trials has been a primary reason for the MEOR not to become a popular technology [23]. The other reason preventing MEOR becoming a routine and accepted procedure for enhancing oil recovery may be the lack of scientific understandings and knowledge about the basis and details of the different MEOR approaches [32]. New advanced technologies will provide acceptable explanations for this fact and remove this barrier. Cheap oil supplies and low-price oil in the last decades in addition to inexpensive simple CO_2 injection EOR have hindered considerable investments on investigation of tertiary recovery methods including MEOR. Youssef et al. [6] specified 96% of all the studied MEOR projects have been successful. The ever growing field trials and laboratory experimental studies and issued patents regarding MEOR indicate the potential of this method to become an important and reliable tool in EOR.

10.3 RECOVERY EFFICIENCY

The recovery efficiency can be expressed as [40]:

$$E_r = E_d \times E_v \tag{10.1}$$

where E_r is the recovery efficiency. E_d denotes the microscopic oil displacement efficiency expressed as the fraction of the total volume of the oil displaced from a unit segment of rock and E_v is the volumetric or macroscopic sweep efficiency expressed as the fraction of the total reservoir that is contacted by the recovery fluid.

Trapping of oil in the porous media depends on fluid/rock interactions (reflected by wettability), fluid/fluid interaction (reflected by interfacial tension (IFT)), and pore structure [30]. The remaining oil in the reservoir is often located in difficultly accessible area such as small pores and dead-end pores and is trapped by capillary pressure [15,41–43]. Capillary forces in the porous media are governed by the combined effect of the IFTs between the rock and fluids, the pore size and geometry, and the wetting characteristics of the system [44]. Viscous forces denote the pressure gradients associated fluid flow within the porous media [30]. It is possible to show the effect of viscous and capillary forces on the trapping of oil within the porous media using a dimensionless number referred as the capillary number (N_C). This number is defined as the ratio of viscous forces to the capillary forces [6,15,45–48]:

$$N_C = \frac{\text{Viscous Forces}}{\text{Capillary Forces}} = \frac{\nu\mu}{\sigma\cos\theta} \tag{10.2}$$

where ν is the displacing fluid velocity, μ is the displacing fluid dynamic viscosity, σ is the oil-water IFT, and θ represents the contact angle. This number shows the relative importance of the viscous to the capillary forces. Higher values of this parameter denote lower residual oil saturation in the porous media and consequently higher oil recovery [46]. Capillary number is usually large for high-speed flows and low for low-speed flows. Typically, for fluid flow within the pores media in an oil reservoir capillary number is $\sim 10^{-6}$ and for flow in production pipelines is ~ 1 [49]. To enhance the microscopic oil displacement, the capillary number should be increased via either increasing the displacing fluid viscosity or decreasing the oil – water IFT. Chemicals such as surfactants decrease the IFT and polymers increase the water viscosity. Reed and Healy [50] specified that significant oil recovery demands an increase of 100–1000 folds in the capillary number. Microbially generated surfactants can be the suitable agent for this purpose [51–54], the detail of which will be debated in Section 10.1. Effect of the capillary number on the residual oil saturation has been investigated by several researchers [55–59]. The governing relationship can be schematically shown via a plot known as capillary desaturation curve. In this plot, the capillary number and residual oil saturation are on the x-, and y-axes, respectively. Typically, this plot shows a residual oil saturation plateau region in very low capillary numbers through approximately $N_c = 10^{-6}$ and after that the residual oil saturation drops with the increase in the capillary number [60]. The point at which the residual oil saturation starts to drop is called critical capillary number (N_{CC}). Parameters such as rock structure, wettability, fluid types, and also testing condition affect the N_{CC} magnitude [61]. For waterfloods, N_c is typically equal to 10^{-6} [62]. This value is generally considerably less than the N_{CC} and a moderate enhancement on N_C will not significantly decrease the residual oil saturation [63].

The other important parameter is the mobility ratio. In cases in which there are large variations between the viscosity of the displacing fluid and oil, the volumetric sweep efficiency will play the main role in the recovery process [40]. Moreover, in field implemented EOR process, the recovery efficiency is often dominated by the volumetric sweep efficiency [17,40]. In case of large differences between the viscosities, it is likely water moves more rapidly than oil and reaches the producing well sooner. The parameter denoting the relative mobility of the water and oil phases is the mobility ratio (M):

$$M = \frac{k_w\mu_o}{k_o\mu_w} \tag{10.3}$$

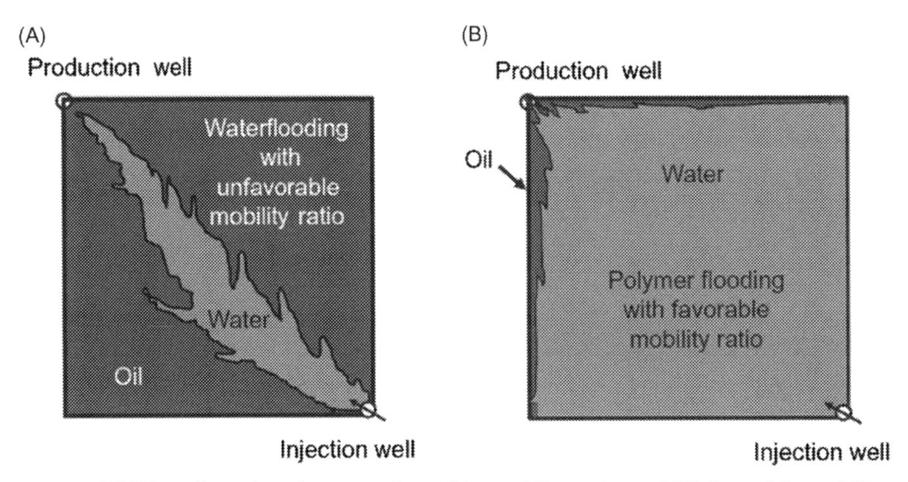

Figure 10.2 (A) Water fingering due to unfavorable mobility ratio and (B) favorable mobility ratio as a result of increasing water viscosity by polymer injection [64].

where M is the mobility ratio, k_w and k_o are water relative permeability in the waterflood zone and oil relative permeability in oil-saturated zone, respectively. μ_o and μ_w are the respective oil and water viscosities. The difference between the oil and water phases' mobilities is a factor leading to poor volumetric sweep efficiency [17]. The favorable condition in which a uniform oil displacement will occur is mobility ratios less than 1, while much greater values than unity are not favorable and will result in water fingering. Polymers such as xanthan gum can be used in waterfloods to increase the waterflood viscosity, which consequently decrease the mobility ratio and increase the volumetric sweep efficiency (see Fig. 10.2).

The other factor that can adversely affect the volumetric sweep efficiency is the differences between the permeability of different zones of the formation [40,65]. In general, most of the oil reservoirs are composed of different layers of rocks with different permeabilities thus, permeability variation is often the most important factor that control the sweep efficiency and consequently the overall recovery factor [17]. Employing biomass to plug the highly permeable channels and correct the permeability profile can be a suitable solution, which will be explained in details in Section 10.6.

The other influential factor would be the density difference between the displacing (water) and displaced (oil) fluids. Large differences may lead to gravitational segregation through which underriding or overriding of the displaced fluid would take place. As the result, the fluid would be bypassed at the bottom or top of the reservoir, which reduces the macroscopic efficiency [60].

10.4 HISTORY

The inherent risks associated with common tertiary recovery methods such as economic costs, toxicity, and damages to the environment propelled the attention toward finding new technics, which are economically feasible and environmentally friendly [31]. Bastin [67] was the first researcher who reported microbial life in the oil fields. The idea of incorporating biotechnology to enhance oil recovery was suggested for the first time by Beckman in 1926 [68]. He indicated that it is possible to incorporate microorganisms to free the oil droplets from porous media. Nothing notable was performed between 1926 and 1940. After two decades, ZoBell [69,70] was the pioneer researcher, who conducted tests to evaluate the bacterial release of oil from the oil bearing materials and received a patent [71] on his process. This patent was about injecting *Desulfovibrio hydrocarbonoclasticus* as well as nutrients into a well to improve the oil recovery. This researcher mentioned five different mechanisms in his patent. After this, broad experiments by Beck [72] based on the ZoBell culture failed to support it and yielded just inconsistent results. The author stated that ZoBell culture would be unserviceable in the field. ZoBell continued his studies in this field and could receive another patent [73] in 1953. In this patent, ZoBell employed the genus *Clostridium* along with other hydrogen-producing microorganisms. It is noteworthy to mention that both patents by ZoBell were based on laboratory results.

Another patent [74] was received by Updegraff and Wren in 1953 employing *Desulfovibrio* and possibly a symbiont bacterium. The researchers considered injection of molasses with the bacteria to promote the growth rate as they noticed that consumption of crude oil by the microbes is too slow. Four years later, Updegraff received another patent [75] proposing injection of a gas-producing facultative or obligate anaerobe along with a water-soluble carbohydrate such as sugar. This patent was not based on the field experiments too.

The Socony Mobil Research laboratory performed the first MEOR field trial in the Lisbon field, Union County, Arkansas in 1954 and reported marginal success due to the increase in the wells oil follow rates [16,31,76–88] and the analyses denoted the complexity of using microbes. However, Volk and Liu [32] mentioned that the pioneering field studies were performed in the United States in the 1930s and 1940s by Claude ZoBell and his group at the Scripps Institute of Oceanography in La Jolla, California. The starting point of MEOR field trials in former Soviet Union backs to the 1960s. The Soviet scientist [89] specified that there are some bacteria in the oil deposits, which can degrade the oil to gases such as CH_4, H_2, CO_2, and N_2 [16]. Encouraged by this, some Eastern European countries such as former Czechoslovakia, Poland, Hungary, and Romania performed extensive research activities and some field trials [31], the details of which are available in literature [90–103]. In Romania, several MEOR field tests were performed between 1971 and 1982, reporting successful results [31,104]. The field trials performed in the mentioned countries employed mixed anaerobic or facultative anaerobic bacteria, which were selected based on their capability to produce adequate volumes of biochemicals and also biomass [16]. Youssef et al. [6] mentioned the improvement in the technology of MEOR as using mixture cultures adapted to the nutrients and reservoir condition such as temperature and pressure and also incorporating larger volumes of nutrients [105–114]. By the end of the 20th century, MEOR was proved as a scientific and interdisciplinary approach for promoting oil recovery [16].

In 2003, Van Hamme et al. [19] stated that more than 400 MEOR field projects have been done so far just in the United States. Based on Khire and Khan [43,115], MEOR projects has been applied on over 400 wells in the same country. In addition, this recovery method has been tested on more than 1000 wells in numerous oil fields in China [32]. Based on Thomas [116], an estimated amount of 2.5 million oil barrels per day were produced in 2007 using EOR method of which the role of MEOR was negligible. On the other hand, based on a report by Chinese Ministry of Land and Resources (www.mlr.gov.cn), nearly 50 billion oil barrels in onshore Chinese oil fields have the potential to be treated by MEOR [32]. Youssef et al. [6] specified that the residual oil recovery in many MEOR field trials increased by 10%–340% for 2–8 years [78,92,99,101,106,108,117].

10.5 MICROBIAL ECOLOGY

Microbes are single-celled organisms existing everywhere in the nature, including the hydrocarbon reservoirs [20]. In other words, it is possible to find microbes everywhere in biosphere [118]. Several groups of microorganisms including bacterial and archaeal communities have been isolated from oil fields, which are different in physiological and metabolic abilities and phylogenetic affiliations [6]. Microbial life may extend the biosphere up to 4 km below the surface [118]. Single-celled bacteria presented in the oil reservoirs live in the aqueous phase and take in water through the cell membranes [63]. Accessible water is the most important resource for microorganisms to survive [63]. More than two-thirds of a bacterial cell is composed of water. Water provides the nutrients and various vitamins essential for cell construction. Moreover, the unwanted waste products are removed via hydration [63]. Many bacteria found in the oil reservoirs are heterotrophic, i.e., they are not capable to produce their own food and must gain energy and carbon via consuming organic substances [63]. As it was mentioned, the microorganisms are in aqueous phase, thus, the only way is to obtain the nutrient from sources, which are in contact with the aqueous phase. If the residual oil is the nutrient, the microorganism will preferentially thrive in the oil—water interface [63]. In case of injecting the nutrients such as molasses the microbial growth is boosted but discouraged on the oil—water interface.

Several parameters affect the subsurface microbial life including [119—121]:

1. Chemical factors such as nutrient composition, electrolyte composition, redox potential (activity of electrons (Eh)), and activity of hydrogen ions (pH);
2. Physical factors such as pressure, temperature, salinity, pore size, and pore geometry, porosity, permeability, and dissolved solids; and
3. Biological factors such as cytotoxity of the microbial metabolites and also specific type of microorganisms.

Studying the microbial life in deep biosphere, Jørgensen and Boetius [118] stated that temperature is the prevailing influential parameter. However, there are evidences of bacteria life at an extreme temperature of 120°C [122]. Youssef et al. [6] mentioned the three parameters of redox potential, temperature, and salinity as the most important influential parameters on the oil reservoir microbial communities. Redox activity is corresponding with availability of electron accepters and donors. Generally, oil reservoirs are associated with low redox potential [6]. The electron donors in the oil reservoirs are hydrogen, volatile fatty acids such as propionate, acetate, and benzoate [123], and inorganic electron donors and electron acceptors include sulfate and carbonate minerals in many reservoirs and also iron(III) in some reservoirs [6]. Presence of nitrate and oxygen as the other electron acceptors are not likely unless they are added via injection [6]. Van Hamme et al. [19] specified that considering the size and ability to grow under the harsh subsurface petroleum condition, only prokaryotes are the capable candidates for MEOR and yeasts, algae, molds, and protozoa are not suitable microorganisms. The microbes utilized in MEOR are typically nonpathogenic hydrocarbon-utilizing microorganisms and can be found naturally in the oil reservoirs [15,18,19,22,124]. These microbes may naturally inhabit the subsurface reservoirs or injected.

Aerobic condition is not common in the petroleum reservoirs. The aerobic microbes would have consumed the existed oxygen already. The other point is that ferrous iron and sulfur, which are commonplace in oil reservoirs, deplete the free oxygen too [22].

Microorganisms for MEOR can be classified based on the

1. Origin;
2. Metabolic processes; and
3. Action.

In the following sections, a brief explanation is provided.

10.5.1 Microorganisms Based on Origin

Based on the origin, microorganisms are categorized as:

1. Indigenous (autochthonous) and
2. Exogenous (allochthonous or foreign).

Indigenous microorganisms have been originally in place and are not artificially transferred or injected. Various studies have reported isolation of different microorganisms, however in some cases there is suspicion that the microbes may have an external source like injection of seawater [125]. Without direct injection of target microorganisms, waterflooding may introduce some microorganisms to the formation. Even reinjecting the produced water, which has been exposed to the surface condition, may introduce surface microorganisms to the reservoir [6]. In addition to inoculation with the surface microorganisms, it is likely that the injected water (as brine of seawater) change the reservoir geochemistry even permanently. For example, presence of oxygen or sulfate may change the structure of the pristine microbial consortia [6]. There is a common fact that MEOR is always applied on mature fields. A mature field is defined as a field that has passed its production peak and production is declining. It is unlikely that the microorganisms present in such fields are the real indigenous ones and represent the biome presented before start of oil production [22]. Even in fields with no history of waterflooding, still there is a possibility of introduction of new microorganism species during drilling or well equipment operations. Damaged tubings might be another potential source thorough fluid leaking. Petroleum industry frequently uses biocides to avoid bacterial induce problems such as souring and corrosion problems. Although the microbial colonies will grow and flourish again in the treated area, the microbial dynamic is not original anymore [22]. Stringent sampling methods should be employed to isolate the bacteria from the oil reservoirs. Some expounded and detailed procedures have been proposed [126–128]; however, they are too expensive to be frequently used for sampling the hydrocarbon reservoirs [6]. Considering the economic issues, Magot et al. [129] specified that wellhead sampling is the only way of collecting samples from petroleum reservoir, which should deal with several sources of contamination. In order to have the best judgment about the origin of the recovered microorganisms, Magot [130] proposed two key criteria:

1. Comparing the isolate's growth optima with the in situ conditions in the oil reservoir and
2. Comparing the global distribution of the strain's phylotype in oil reservoir samples from all over the world.

Youssef et al. [6] argued the first criterion by mentioning some counterexamples. The authors mentioned reported existence of thermophilic isolates with much lower temperature optima than their ecosystem [131], thermotolerant isolates with a low temperature optima from high-temperature environments [132], and halophilic and halotolerant microorganisms recovered from salt crystals with a relatively low salt tolerance [133–135]. They modified the criterion as considering range (minimum and maximum growth limits) or the ability to survive for prolonged periods of time at the in situ reservoir condition. However, they mentioned the difficulties associated with evaluating the growth limits of slow-growing recovered microorganisms due to the probability of resulting false negatives because of extended incubation period.

Several studies have reported microorganisms isolated from the petroleum reservoirs or effective ones in degrading the petroleum. Table 10.1 lists some of the mentioned microorganisms along with the corresponding taxonomy and isolation source. The taxonomic details are acquired from the SILVA [136].

10.5.2 Microorganisms Based on Action

Based on their actions, microorganisms for MEOR can be classified as the main categories of:

1. Methanogens;
2. Sulfate-reducing bacteria (SRB);
3. Fermentative microorganisms;
4. Nitrate-reducing bacteria (NRB); and
5. Iron-reducing bacteria (IRB);

The mentioned classes are explained briefly in following.

10.5.2.1 Methanogens

Methanogenesis is a microbial metabolism through which methane is generated by microorganisms called methanogens. This is a common process in the oil fields and the corresponding first report was prepared in the early 1950s [268,269] and then was further elucidated especially by Russian scientists [270–277]. Through methanogenesis, methanogens metabolize the substrates of hydrogen and CO_2, methylamines, acetate, and dimethysulfides and produce methane as a biogenic gas [6]. All methanogens belong to the domain Archaea, phylum Euryarchaeota [278]. These microorganisms are distributed in four classes and five orders of *Methanobacteriales*, *Methanococcales*, *Methanomicrobiales*, *Methanosarcinales*, and *Methanopyrales* [278]. The ecological niches of methanogens are widely distributed. All the mentioned orders have been isolated from oil reservoir except *Methanopyrales* [6]. Based on the utilized substrate by the isolates from the oil fields, the main categories of methanogens include [6]

1. Hydrogenotrophic methanogens [277,279–288];
2. Methylotrophic methanogens [148,289–293]; and
3. Aceticlastic methanogens [147,171,271,283,294–300].

Table 10.1 Microorganisms Isolated From the Petroleum Reservoirs or Effective Ones in Degrading the Petroleum

Index	Kingdom	Phylum	Class	Order	Family	Genus	Species	Isolation Source	Reference
1	Archaea	Bathyarchaeota					Uncultured *Thermoprotei archaeon*	Operating oil well NR-6 in the Niibori oil field, which is located in the northeast part of the main island of Japan (39° 43′ N, 139° 53′ E)	[137]
2	Archaea	Candidate division YNPFFA					Uncultured *Thermoprotei archaeon*	Operating oil well NR-6 in the Niibori oil field, which is located in the northeast part of the main island of Japan (39° 43′ N, 139° 53′ E)	[137]
3	Archaea	Euryarchaeota	Archaeoglobi	Archaeoglobales	Archaeoglobaceae		Uncultured *Archaeoglobi archaeon*	Operating oil well NR-6 in the Niibori oil field, which is located in the northeast part of the main island of Japan (39° 43′ N, 139° 53′ E)	[137]
4	Archaea	Euryarchaeota	Archaeoglobi	Archaeoglobales	Archaeoglobaceae	*Archaeoglobus*	*Archaeoglobus* sp. NS-tSRB-2	Ekofisk field in the Norwegian sector of the North Sea	[138]
5	Archaea	Euryarchaeota	Archaeoglobi	Archaeoglobales	Archaeoglobaceae	*Archaeoglobus*	Archaeon enrichment culture clone EA8.1	Produced water samples of a high temperature and fractured chalk reservoir, the Ekofisk oil field, in block 2/4 of the Norwegian sector of the North Sea about 320 km southwest of Stavanger.	[139]

6	Archaea	Euryarchaeota	Archaeoglobi	Archaeoglobales	Archaeoglobaceae	*Archaeoglobus*	Archaeon enrichment culture clone EA8.8	Produced water samples of a high temperature and fractured chalk reservoir, the Ekofisk oil field, in block 2/4 of the Norwegian sector of the North Sea about 320 km southwest of Stavanger.	[139]
7	Archaea	Euryarchaeota	Archaeoglobi	Archaeoglobales	Archaeoglobaceae	*Archaeoglobus*	Archaeon enrichment culture clone PW45.1	Produced water samples of a high temperature and fractured chalk reservoir, the Ekofisk oil field, in block 2/4 of the Norwegian sector of the North Sea about 320 km southwest of Stavanger	[139]
8	Archaea	Euryarchaeota	Halobacteria	Halobacteriales	Haloferacaceae	*Haloferax*	*Haloferax* sp. BO3	Five hypersaline locations; salt marshes in the Uyuni salt flats in Bolivia, crystallizer ponds in Chile and Cabo Rojo (Puerto Rico), and sabkhas (salt flats) in the Persian Gulf (Saudi Arabia) and the Dead Sea (Israel and Jordan)	[140]

(*continued*)

Table 10.1 (Continued)

Index	Kingdom	Phylum	Class	Order	Family	Genus	Species	Isolation Source	Reference
9	Archaea	Euryarchaeota	Methanobacteria	Methanobacteriales	Methanobacteriaceae		Uncultured archaeon	A working block of the waterflooded Gudao petroleum reservoir located in the Yellow River Delta, China	[141]
10	Archaea	Euryarchaeota	Methanobacteria	Methanobacteriales	Methanobacteriaceae		Uncultured *Methanothermobacter* sp.	The Kongdian bed of the Dagang oil field, Hebei Province, China	[142]
11	Archaea	Euryarchaeota	Methanobacteria	Methanobacteriales	Methanobacteriaceae	*Methanobacterium*	Uncultured bacterium	A high-temperature petroleum reservoir at an offshore oil field, China	[143]
12	Archaea	Euryarchaeota	Methanobacteria	Methanobacteriales	Methanobacteriaceae	*Methanobacterium*	Uncultured *Methanobacterium* sp.	Operating oil well NR-6 in the Niibori oil field, which is located in the northeast part of the main island of Japan (39° 43′ N, 139° 53′ E)	[137]
13	Archaea	Euryarchaeota	Methanobacteria	Methanobacteriales	Methanobacteriaceae	*Methanothermobacter*	Uncultured bacterium	A high-temperature petroleum reservoir at an offshore oil field, China	[143]
14	Archaea	Euryarchaeota	Methanobacteria	Methanobacteriales	Methanobacteriaceae	*Methanothermobacter*	Uncultured *Methanobacteria archaeron*	Operating oil well NR-6 in the Niibori oil field, which is located in the northeast part of the main island of Japan (39° 43′ N, 139° 53′ E)	[137]

15	Archaea	Euryarchaeota	Methanobacteria	Methanobacteriales	Methanobacteriaceae	*Methanothermobacter*	Uncultured *Methanothermobacter* sp.	Operating oil well NR-6 in the Niibori oil field, which is located in the northeast part of the main island of Japan (39° 43′ N, 139° 53′ E)	[137]
16	Archaea	Euryarchaeota	Methanobacteria	Methanobacteriales	Methanobacteriaceae	*Methanothermobacter*	Uncultured *Methanothermobacter* sp.	Two production wells (AR-80 and OR-79) in the Yabase oil field, a formation of tuffaceous sandstone of Miocene–Pliocene age, located around 1293–1436 m under the surface, with in situ temperature of 40–82°C and pressure of 5 MPa, Japan	[144]
17	Archaea	Euryarchaeota	Methanococci	Methanococcales	Methanococcaceae	*Methanococcus*	Uncultured bacterium	A high-temperature petroleum reservoir at an offshore oil field, China	[143]
18	Archaea	Euryarchaeota	Methanococci	Methanococcales	Methanococcaceae	*Methanothermococcus*	Archaeon enrichment culture clone EA29.6	Produced water samples of a high temperature and fractured chalk reservoir, the Ekofisk oil field, in block 2/4 of the Norwegian sector of the North Sea about 320 km southwest of Stavanger	[139]

(continued)

Table 10.1 (Continued)

Index	Kingdom	Phylum	Class	Order	Family	Genus	Species	Isolation Source	Reference
19	Archaea	Euryarchaeota	Methanomicrobia	D-C06			Uncultured *Methanosarcinales archaeon*	Operating oil well NR-6 in the Niibori oil field, which is located in the northeast part of the main island of Japan (39° 43′ N, 139° 53′ E)	[137]
20	Archaea	Euryarchaeota	Methanomicrobia	Methanomicrobiales	Methanomicrobiaceae	*Methanoculleus*	Uncultured *Methanoculleus* sp.	Operating oil well NR-6 in the Niibori oil field, which is located in the northeast part of the main island of Japan (39° 43′ N, 139° 53′ E)	[137]
21	Archaea	Euryarchaeota	Methanomicrobia	Methanomicrobiales	Methanomicrobiaceae	*Methanoculleus*	Uncultured *Methanoculleus* sp.	Two production wells (AR-80 and OR-79) in the Yabase oil field, a formation of tuffaceous sandstone of Miocene–Pliocene age, located around 1293–1436 m under the surface, with in situ temperature of 40–82°C and pressure of 5 MPa, Japan	[144]
22	Archaea	Euryarchaeota	Methanomicrobia	Methanomicrobiales	Methanomicrobiaceae	Uncultured	Uncultured *Methanoplanus* sp.	Operating oil well NR-6 in the Niibori oil field, which is located in the northeast part of the main island of Japan (39° 43′ N, 139° 53′ E)	[137]

23	Archaea	Euryarchaeota	Methanomicrobia	Methanomicrobiales	Methanomicrobiales incertae sedis	*Methanocalculus*	Uncultured *Methanocalculus* sp.	Operating oil well NR-6 in the Niibori oil field, which is located in the northeast part of the main island of Japan (39° 43′ N, 139° 53′ E)	[137]
24	Archaea	Euryarchaeota	Methanomicrobia	Methanosarcinales	Methanosaetaceae	*Methanosaeta*	Uncultured archaeon	Unknown	[145]
25	Archaea	Euryarchaeota	Methanomicrobia	Methanosarcinales	Methanosaetaceae	*Methanosaeta*	Uncultured *Methanosaeta* sp.	Production water sample from the mesothermic and highly degraded Schrader Bluff petroleum field in Alaska's North Slope region, USA	[146]
26	Archaea	Euryarchaeota	Methanomicrobia	Methanosarcinales	Methanosaetaceae	*Methanosaeta*	Uncultured *Methanosaeta* sp.	Operating oil well NR-6 in the Niibori oil field, which is located in the northeast part of the main island of Japan (39° 43′ N, 139° 53′ E)	[137]
27	Archaea	Euryarchaeota	Methanomicrobia	Methanosarcinales	Methanosarcinaceae	*Methanolobus*	Archaeon enrichment culture clone EA17.1	Produced water samples of a high temperature and fractured chalk reservoir, the Ekofisk oil field, in block 2/4 of the Norwegian sector of the North Sea about 320 km southwest of Stavanger	[139]

(*continued*)

Table 10.1 (Continued)

Index	Kingdom	Phylum	Class	Order	Family	Genus	Species	Isolation Source	Reference
28	Archaea	Euryarchaeota	Methanomicrobia	Methanosarcinales	Methanosarcinaceae	*Methanolobus*	Uncultured archaeon	A low-temperature and low-salinity petroleum reservoir in Canada	[147]
29	Archaea	Euryarchaeota	Methanomicrobia	Methanosarcinales	Methanosarcinaceae	*Methanomethylovorans*	Uncultured *Methanomethylovorans* sp.	Operating oil well NR-6 in the Niibori oil field, which is located in the northeast part of the main island of Japan (39° 43′ N, 139° 53′ E)	[137]
30	Archaea	Euryarchaeota	Methanomicrobia	Methanosarcinales	Methermicoccaceae	*Methermicoccus*	Methermicoccus shengliensis	Oil-production water of Shengli oil field, China	[148]
31	Archaea	Euryarchaeota	Thermococci	Thermococcales	Thermococcaceae	*Thermococcus*	Archaeon enrichment culture clone EA3.5	Produced water samples of a high temperature and fractured chalk reservoir, the Ekofisk oil field, in block 2/4 of the Norwegian sector of the North Sea about 320 km southwest of Stavanger	[139]
32	Archaea	Euryarchaeota	Thermococci	Thermococcales	Thermococcaceae	*Thermococcus*	*Thermococcus alcaliphilus*	Samotlor oil reservoir, Nizhnevartovsk, Western Siberia, Russia	[149]
33	Archaea	Euryarchaeota	Thermococci	Thermococcales	Thermococcaceae	*Thermococcus*	*Thermococcus peptonophilus*	Samotlor oil reservoir, Nizhnevartovsk, Western Siberia, Russia	[149]
34	Archaea	Euryarchaeota	Thermococci	Thermococcales	Thermococcaceae	*Thermococcus*	*Thermococcus sibiricus*	Samotlor oil reservoir, Nizhnevartovsk, Western Siberia, Russia	[149]

35	Archaea	Euryarchaeota	Thermococci	Thermococcales	Thermococcaceae	*Thermococcus*	*Thermococcus siculi*	Samotlor oil reservoir, Nizhnevartovsk, Western Siberia, Russia	[149]
36	Archaea	Euryarchaeota	Thermococci	Thermococcales	Thermococcaceae	*Thermococcus*	*Thermococcus* sp. CKU-1	Kubiki oil reservoir located near the coast of the Sea of Japan in Niigata prefecture, Japan	[150]
37	Archaea	Euryarchaeota	Thermococci	Thermococcales	Thermococcaceae	*Thermococcus*	*Thermococcus* sp. CKU-199	Kubiki oil reservoir located near the coast of the Sea of Japan in Niigata prefecture, Japan	[150]
38	Archaea	Euryarchaeota	Thermococci	Thermococcales	Thermococcaceae	*Thermococcus*	Uncultured *Thermococcus* sp.	Operating oil well NR-6 in the Niibori oil field, which is located in the northeast part of the main island of Japan (39° 43′ N, 139° 53′ E)	[137]
39	Archaea	Euryarchaeota	Thermoplasmata	Thermoplasmatales	Terrestrial Miscellaneous Gp(TMEG)		Uncultured *Thermoplasmata archaeon*	Operating oil well NR-6 in the Niibori oil field, which is located in the northeast part of the main island of Japan (39° 43′ N, 139° 53′ E)	[137]
40	Archaea	Euryarchaeota	Thermoplasmata	Thermoplasmatales	Thermoplasmatales incertae sedis	Uncultured	Uncultured *Thermoplasmata archaeon*	Operating oil well NR-6 in the Niibori oil field, which is located in the northeast part of the main island of Japan (39° 43′ N, 139° 53′ E)	[137]

(continued)

Table 10.1 (Continued)

Index	Kingdom	Phylum	Class	Order	Family	Genus	Species	Isolation Source	Reference
41	Archaea	Thaumarchaeota	Marine Group I	Unknown Order	Unknown Family	Candidatus Nitrosopumilus	Archaeon enrichment culture clone EA3.3	Produced water samples of a high temperature and fractured chalk reservoir, the Ekofisk oil field, in block 2/4 of the Norwegian sector of the North Sea about 320 km southwest of Stavanger	[139]
42	Bacteria	Actinobacteria	Actinobacteria	Micrococcales	Cellulomonadaceae	*Cellulomonas*	*Cellulomonas* sp. MIXRI54	The petroleum oil-contaminated soil (17.2 g total hydrocarbon kg^{-1} soil) from a landfill used for deposition of crude oil-contaminated soil from oil pumping sites in Zistersdorf, Austria	[151]
43	Bacteria	AC1					Bacterium enrichment culture clone B31149	A disposal field that treats mixtures of crude oil-contaminated soil and oily sludge in the Shengli oil field, China	[152]
44	Bacteria	AC1					Bacterium enrichment culture clone B312151	A disposal field that treats mixtures of crude oil-contaminated soil and oily sludge in the Shengli oil field, China	[152]

45	Bacteria	Acetothermia					Uncultured *Acetothermia bacterium*	Operating oil well NR-6 in the Niibori oil field, which is located in the northeast part of the main island of Japan (39° 43′ N, 139° 53′ E)	[137]
46	Bacteria	Acidobacteria	Subgroup 9				Uncultured bacterium	The Ban 876 Gas and Oil Field within the DaGang Area (39°32′N, 117°38′E), TianJin, China	[153]
47	Bacteria	Actinobacteria	Acidimicrobiia	Acidimicrobiales	OM1 clade		Uncultured bacterium	The Ban 876 Gas and Oil Field within the DaGang Area (39°32′N, 117°38′E), TianJin, China	[153]
48	Bacteria	Actinobacteria	Actinobacteria	Corynebacteriales	Dietziaceae	*Dietzia*	*Dietzia psychralcaliphila*	Water (6°C, pH 7) obtained from a drain pool of a fish-egg-processing plant	[154]
49	Bacteria	Actinobacteria	Actinobacteria	Corynebacteriales	Dietziaceae	*Dietzia*	*Dietzia* sp. DQ12-45-1b	An oil production water sample collected from a well-head, in a deep subterranean oil-reservoir in Daqing Oil Field, China	[155]
50	Bacteria	Actinobacteria	Actinobacteria	Corynebacteriales	Dietziaceae	*Dietzia*	*Dietzia* sp. SG-3	Sagara oil reservoir, Shizuoka Prefecture, Japan	[156]

(*continued*)

Table 10.1 (Continued)

Index	Kingdom	Phylum	Class	Order	Family	Genus	Species	Isolation Source	Reference
51	Bacteria	Actinobacteria	Actinobacteria	Corynebacteriales	Mycobacteriaceae	*Mycobacterium*	*Mycobacterium aromaticivorans*	Hawaiian soils, Hawaii, USA	[157]
52	Bacteria	Actinobacteria	Actinobacteria	Corynebacteriales	Mycobacteriaceae	*Mycobacterium*	*Mycobacterium crocinum*	Hawaiian soils, Hawaii, USA	[157]
53	Bacteria	Actinobacteria	Actinobacteria	Corynebacteriales	Mycobacteriaceae	*Mycobacterium*	*Mycobacterium frederiksbergense*	Coal tar-contaminated soil on the site of a former gas works at Frederiksberg, Denmark	[158]
54	Bacteria	Actinobacteria	Actinobacteria	Corynebacteriales	Mycobacteriaceae	*Mycobacterium*	*Mycobacterium gilvum*	Five different oil- or h contaminated soil sites, Hamburg, Germany	[159]
55	Bacteria	Actinobacteria	Actinobacteria	Corynebacteriales	Mycobacteriaceae	*Mycobacterium*	*Mycobacterium gilvum*	Hawaiian soils, Hawaii, USA	[157]
56	Bacteria	Actinobacteria	Actinobacteria	Corynebacteriales	Mycobacteriaceae	*Mycobacterium*	*Mycobacterium pallens*	Hawaiian soils, Hawaii, USA	[157]
57	Bacteria	Actinobacteria	Actinobacteria	Corynebacteriales	Mycobacteriaceae	*Mycobacterium*	*Mycobacterium pyrenivorans*	Polycyclic aromatic hydrocarbons (PAH) contaminated soil from a former coking plant at Übach – Palenberg, Germany	[160]
58	Bacteria	Actinobacteria	Actinobacteria	Corynebacteriales	Mycobacteriaceae	*Mycobacterium*	*Mycobacterium rufum*	Hawaiian soils, Hawaii, USA	[157]
59	Bacteria	Actinobacteria	Actinobacteria	Corynebacteriales	Mycobacteriaceae	*Mycobacterium*	*Mycobacterium rutilum*	Hawaiian soils, Hawaii, USA	[157]
60	Bacteria	Actinobacteria	Actinobacteria	Corynebacteriales	Mycobacteriaceae	*Mycobacterium*	Uncultured bacterium	A continental high-temperature, waterflooded petroleum reservoir in the J-12 Unit at Huabei Oil field, Hebei Province, China	[161]

61	Bacteria	Actinobacteria	Actinobacteria	Corynebacteriales	Nocardiaceae	*Gordonia*	*Gordonia paraffinivorans*	An oil-producing well of Daqing Oil Field	[162]
62	Bacteria	Actinobacteria	Actinobacteria	Corynebacteriales	Nocardiaceae	*Rhodococcus*	*Rhodococcus cerastii*	Germany	[163]
63	Bacteria	Actinobacteria	Actinobacteria	Corynebacteriales	Nocardiaceae	*Rhodococcus*	*Rhodococcus* sp. ITRH42	The petroleum oil-contaminated soil (17.2 g total hydrocarbon kg^{-1} soil) from a landfill used for deposition of crude oil-contaminated soil from oil pumping sites in Zistersdorf, Austria	[151]
64	Bacteria	Actinobacteria	Actinobacteria	Micrococcales	Brevibacteriaceae	*Brevibacterium*	Uncultured bacterium	The Ban 876 Gas and Oil Field within the DaGang Area (39°32′N, 117°38′E), TianJin, China	[153]
65	Bacteria	Actinobacteria	Actinobacteria	Micrococcales	Intrasporangiaceae	*Janibacter*	*Janibacter terrae*	Hawaiian soils, Hawaii, USA	[157]
66	Bacteria	Actinobacteria	Actinobacteria	Micrococcales	Microbacteriaceae	*Leifsonia*	Naphthalene-utilizing bacterium IS1	The microbial community in soil surrounding an outdoor coal storage pile in northern Indiana, USA	[164]
67	Bacteria	Actinobacteria	Actinobacteria	Micrococcales	Microbacteriaceae	*Microbacterium*	Actinobacterium MIXRI55	The petroleum oil-contaminated soil (17.2 g total hydrocarbon kg^{-1} soil) from a landfill used for deposition of crude oil-contaminated soil from oil pumping sites in Zistersdorf, Austria	[151]

(continued)

Table 10.1 (Continued)

Index	Kingdom	Phylum	Class	Order	Family	Genus	Species	Isolation Source	Reference
68	Bacteria	Actinobacteria	Actinobacteria	Micrococcales	Microbacteriaceae	*Microbacterium*	*Bacillus subtilis*	Hawaiian soils, Hawaii, USA	[157]
69	Bacteria	Actinobacteria	Actinobacteria	Micrococcales	Microbacteriaceae	*Microbacterium*	*Microbacterium hydrocarbonoxydans*	Germany	[163]
70	Bacteria	Actinobacteria	Actinobacteria	Micrococcales	Microbacteriaceae	*Microbacterium*	*Microbacterium oleivorans*	Germany	[163]
71	Bacteria	Actinobacteria	Actinobacteria	Micrococcales	Microbacteriaceae	*Microbacterium*	*Microbacterium oxydans*	Hawaiian soils, Hawaii, USA	[157]
72	Bacteria	Actinobacteria	Actinobacteria	Micrococcales	Microbacteriaceae	*Microbacterium*	*Microbacterium* sp. ITRH47	The petroleum oil-contaminated soil (17.2 g total hydrocarbon kg^{-1} soil) from a landfill used for deposition of crude oil-contaminated soil from oil pumping sites in Zistersdorf, Austria	[151]
73	Bacteria	Actinobacteria	Actinobacteria	Micrococcales	Micrococcaceae	*Arthrobacter*	*Arthrobacter* sp. ITRH48	The petroleum oil-contaminated soil (17.2 g total hydrocarbon kg^{-1} soil) from a landfill used for deposition of crude oil-contaminated soil from oil pumping sites in Zistersdorf, Austria	[151]
74	Bacteria	Actinobacteria	Actinobacteria	Micrococcales	Micrococcaceae	*Arthrobacter*	Naphthalene-utilizing bacterium IS13	The microbial community in soil surrounding an outdoor coal storage pile in northern Indiana, USA	[164]

75	Bacteria	Actinobacteria	Actinobacteria	Micrococcales	Micrococcaceae	*Arthrobacter*	Naphthalene-utilizing bacterium IS4	The microbial community in soil surrounding an outdoor coal storage pile in northern Indiana, USA	[164]
76	Bacteria	Actinobacteria	Actinobacteria	Micrococcales	Micrococcaceae	*Arthrobacter*	Naphthalene-utilizing bacterium IS5	The microbial community in soil surrounding an outdoor coal storage pile in northern Indiana, USA	[164]
77	Bacteria	Actinobacteria	Actinobacteria	Micrococcales	Micrococcaceae	*Micrococcus*	Bacterium enrichment culture clone 57.8B	Produced water samples of a high temperature and fractured chalk reservoir, the Ekofisk oil field, in block 2/4 of the Norwegian sector of the North Sea about 320 km southwest of Stavanger	[139]
78	Bacteria	Actinobacteria	Actinobacteria	Micrococcales	Micrococcaceae	*Micrococcus*	*Micrococcus luteus*	Hawaiian soils, Hawaii, USA	[157]
79	Bacteria	Actinobacteria	Actinobacteria	Micrococcales	Micrococcaceae	*Micrococcus*	*Micrococcus* sp. BTSI50	The petroleum oil-contaminated soil (17.2 g total hydrocarbon kg^{-1} soil) from a landfill used for deposition of crude oil-contaminated soil from oil pumping sites in Zistersdorf, Austria	[151]

(*continued*)

Table 10.1 (Continued)

Index	Kingdom	Phylum	Class	Order	Family	Genus	Species	Isolation Source	Reference
80	Bacteria	Actinobacteria	Actinobacteria	Micrococcales	Micrococcaceae	*Pseudarthrobacter*	*Arthrobacter* sp. ITRH49	The petroleum oil-contaminated soil (17.2 g total hydrocarbon kg^{-1} soil) from a landfill used for deposition of crude oil-contaminated soil from oil pumping sites in Zistersdorf, Austria	[151]
81	Bacteria	Actinobacteria	Actinobacteria	Micrococcales	Micrococcaceae	*Pseudarthrobacter*	*Arthrobacter* sp. P1-1	Hawaiian soils, Hawaii, USA	[157]
82	Bacteria	Actinobacteria	Actinobacteria	Micrococcales	Micrococcaceae	*Pseudarthrobacter*	*Pseudarthrobacter oxydans*	Germany	[163]
83	Bacteria	Actinobacteria	Actinobacteria	Micrococcales	Sanguibacteraceae	*Sanguibacter*	*Sanguibacter* sp. SG-4	Sagara oil reservoir, Shizuoka Prefecture, Japan	[156]
84	Bacteria	Actinobacteria	Actinobacteria	Propionibacteriales	Nocardioidaceae	*Nocardioides*	*Nocardioides aromaticivorans*	Contaminated surface water and sediments, Hikichi river, Kanagawa, Japan	[165]
85	Bacteria	Actinobacteria	Actinobacteria	Propionibacteriales	Nocardioidaceae	*Nocardioides*	*Nocardioides oleivorans*	Crude oil sample 19 from the oil field Oerrel of the Gifhorn Trough, Germany	[166]
86	Bacteria	Actinobacteria	Actinobacteria	Propionibacteriales	Propionibacteriaceae	*Propionicicella*	*Propionicimonas* sp. F6	A low-temperature and low-salinity petroleum reservoir in Canada	[147]
87	Bacteria	Actinobacteria	Actinobacteria	Propionibacteriales	Propionibacteriaceae	*Tessaracoccus*	*Tessaracoccus oleiagri*	A crude oil-contaminated saline soil of Shengli Oil field, China	[167]

88	Bacteria	Actinobacteria	Actinobacteria	Streptomycetales	Streptomycetaceae	*Streptomyces*	Actinomycetales bacterium ITRH51	The petroleum oil-contaminated soil (17.2 g total hydrocarbon kg^{-1} soil) from a landfill used for deposition of crude oil-contaminated soil from oil pumping sites in Zistersdorf, Austria	[151]
89	Bacteria	Actinobacteria	Actinobacteria	Streptomycetales	Streptomycetaceae	*Streptomyces*	*Streptomyces* sp. ERI-CPDA-1	Oil spilled areas in petrol bunk in Chetpet, Chennai, Tamil Nadu, India	[168]
90	Bacteria	Actinobacteria	Actinobacteria	Streptomycetales	Streptomycetaceae	*Streptomyces*	Uncultured bacterium	Daqing Oil Field, China	[169]
91	Bacteria	Actinobacteria	Coriobacteriia	Coriobacteriales	Coriobacteriaceae	Uncultured	Coriobacteriaceae bacterium enrichment culture clone B3113	A disposal field that treats mixtures of crude oil-contaminated soil and oily sludge in the Shengli oil field, China	[152]
92	Bacteria	Actinobacteria	Coriobacteriia	Coriobacteriales	Coriobacteriaceae	Uncultured	Uncultured actinobacterium	Operating oil well NR-6 in the Niibori oilfield, which is located in the northeast part of the main island of Japan (39° 43′ N, 139° 53′ E)	[137]
93	Bacteria	Actinobacteria	Coriobacteriia	Coriobacteriales	Coriobacteriaceae	Uncultured	Uncultured actinobacterium	Production water sample from the mesothermic and highly degraded Schrader Bluff petroleum field in Alaska's North Slope region, USA	[146]

(continued)

Table 10.1 (Continued)

Index	Kingdom	Phylum	Class	Order	Family	Genus	Species	Isolation Source	Reference
94	Bacteria	Actinobacteria	Thermoleophilia	Solirubrobacterales	TM146		Uncultured bacterium	The Ban 876 Gas and Oil Field within the DaGang Area (39°32′N, 117°38′E), TianJin, China	[153]
95	Bacteria	Aminicenantes					Uncultured bacterium	Daqing Oil Field, China	[169]
96	Bacteria	Atribacteria					Candidate division JS1 bacterium enrichment culture clone B3111	A disposal field that treats mixtures of crude oil-contaminated soil and oily sludge in the Shengli oil field, China	[152]
97	Bacteria	Atribacteria					Candidate division JS1 bacterium enrichment culture clone B31137	A disposal field that treats mixtures of crude oil-contaminated soil and oily sludge in the Shengli oil field, China	[152]
98	Bacteria	Atribacteria					Candidate division JS1 bacterium enrichment culture clone B31141	A disposal field that treats mixtures of crude oil-contaminated soil and oily sludge in the Shengli oil field, China	[152]
99	Bacteria	Atribacteria					Candidate division JS1 bacterium enrichment culture clone B31147	A disposal field that treats mixtures of crude oil-contaminated soil and oily sludge in the Shengli oil field, China	[152]
100	Bacteria	Atribacteria					Candidate division JS1 bacterium enrichment culture clone B31158	A disposal field that treats mixtures of crude oil-contaminated soil and oily sludge in the Shengli oil field, China	[152]

101	Bacteria	Atribacteria	Candidate division JS1 bacterium enrichment culture clone B31162	A disposal field that treats mixtures of crude oil-contaminated soil and oily sludge in the Shengli oil field, China	[152]
102	Bacteria	Atribacteria	Candidate division JS1 bacterium enrichment culture clone B31164	A disposal field that treats mixtures of crude oil-contaminated soil and oily sludge in the Shengli oil field, China	[152]
103	Bacteria	Atribacteria	Candidate division JS1 bacterium enrichment culture clone B312103	A disposal field that treats mixtures of crude oil-contaminated soil and oily sludge in the Shengli oil field, China	[152]
104	Bacteria	Atribacteria	Candidate division JS1 bacterium enrichment culture clone B312119	A disposal field that treats mixtures of crude oil-contaminated soil and oily sludge in the Shengli oil field, China	[152]
105	Bacteria	Atribacteria	Candidate division JS1 bacterium enrichment culture clone B312128	A disposal field that treats mixtures of crude oil-contaminated soil and oily sludge in the Shengli oil field, China	[152]
106	Bacteria	Atribacteria	Candidate division JS1 bacterium enrichment culture clone B312155	A disposal field that treats mixtures of crude oil-contaminated soil and oily sludge in the Shengli oil field, China	[152]

(continued)

Table 10.1 (Continued)

Index	Kingdom	Phylum	Class	Order	Family	Genus	Species	Isolation Source	Reference
107	Bacteria	Atribacteria					Candidate division JS1 bacterium enrichment culture clone B312156	A disposal field that treats mixtures of crude oil-contaminated soil and oily sludge in the Shengli oil field, China	[152]
108	Bacteria	Atribacteria					Candidate division JS1 bacterium enrichment culture clone B312163	A disposal field that treats mixtures of crude oil-contaminated soil and oily sludge in the Shengli oil field, China	[152]
109	Bacteria	Atribacteria					Candidate division JS1 bacterium enrichment culture clone B31283	A disposal field that treats mixtures of crude oil-contaminated soil and oily sludge in the Shengli oil field, China	[152]
110	Bacteria	Atribacteria					Candidate division JS1 bacterium enrichment culture clone B31297	A disposal field that treats mixtures of crude oil-contaminated soil and oily sludge in the Shengli oil field, China	[152]
111	Bacteria	Atribacteria					Uncultured bacterium	Daqing Oil Field, China	[169]
112	Bacteria	Atribacteria	Atribacteria incertae sedis	Unknown Order	Unknown Family	Candidatus Caldatribacterium	Bacterium enrichment culture clone B31288	A disposal field that treats mixtures of crude oil-contaminated soil and oily sludge in the Shengli oil field, China	[152]

113	Bacteria	Atribacteria	Atribacteria incertae sedis	Unknown Order	Unknown Family	Candidatus Caldatribacterium	Uncultured Candidatus Atribacteria bacterium	Operating oil well NR-6 in the Niibori oil field, which is located in the northeast part of the main island of Japan (39° 43′ N, 139° 53′ E)	[137]
114	Bacteria	Bacteroidetes	Bacteroidetes incertae sedis	Order II	Rhodothermaceae	Uncultured	Uncultured bacterium	The Ban 876 Gas and Oil Field within the DaGang Area (39°32′N, 117°38′E), TianJin, China	[153]
115	Bacteria	Bacteroidetes	Bacteroidetes incertae sedis	Order II	Rhodothermaceae	Uncultured	Uncultured bacterium	The Ban 876 Gas and Oil Field within the DaGang Area (39°32′N, 117°38′E), TianJin, China	[153]
116	Bacteria	Bacteroidetes	Bacteroidetes incertae sedis	Order III	Uncultured		Uncultured bacterium	The Ban 876 Gas and Oil Field within the DaGang Area (39°32′N, 117°38′E), TianJin, China	[153]
117	Bacteria	Bacteroidetes	Bacteroidetes incertae sedis	Order III	Unknown Family	Uncultured	Uncultured bacterium	The Ban 876 Gas and Oil Field within the DaGang Area (39°32′N, 117°38′E), TianJin, China	[153]
118	Bacteria	Bacteroidetes	Bacteroidetes VC2.1 Bac22				Unidentified	Unknown	[170]
119	Bacteria	Bacteroidetes	Bacteroidia	Bacteroidales	Bacteroidaceae	*Bacteroides*	Uncultured bacterium	A low-temperature and low-salinity petroleum reservoir in Canada	[147]

(continued)

Table 10.1 (Continued)

Index	Kingdom	Phylum	Class	Order	Family	Genus	Species	Isolation Source	Reference
120	Bacteria	Bacteroidetes	Bacteroidia	Bacteroidales	Dysgonomonadaceae	*Petrimonas*	*Petrimonas sulfuriphila*	The Pelican Lake oil field, which is located in the Western Canadian Sedimentary Basin, Canada	[171]
121	Bacteria	Bacteroidetes	Bacteroidia	Bacteroidales	Marinilabiaceae	*Anaerophaga*	Uncultured *Anaerophaga* sp.	A high-temperature oil-bearing formation in the North Sea	[172]
122	Bacteria	Bacteroidetes	Bacteroidia	Bacteroidales	Marinilabiaceae	*Marinifilum*	Bacterium enrichment culture clone EB24.11	Produced water samples of a high temperature and fractured chalk reservoir, the Ekofisk oil field, in block 2/4 of the Norwegian sector of the North Sea about 320 km southwest of Stavanger	[139]
123	Bacteria	Bacteroidetes	Bacteroidia	Bacteroidales	Porphyromonadaceae	*Petrimonas*	*Petrimonas sulfuriphila*	A low-temperature and low-salinity petroleum reservoir in Canada	[147]
124	Bacteria	Bacteroidetes	Bacteroidia	Bacteroidales	Porphyromonadaceae	*Proteiniphilum*	Porphyromonadaceae bacterium enrichment culture clone B31181	A disposal field that treats mixtures of crude oil-contaminated soil and oily sludge in the Shengli oil field, China	[152]
125	Bacteria	Bacteroidetes	Bacteroidia	Bacteroidales	Porphyromonadaceae	*Proteiniphilum*	Porphyromonadaceae bacterium enrichment culture clone B312134	A disposal field that treats mixtures of crude oil-contaminated soil and oily sludge in the Shengli oil field, China	[152]

126	Bacteria	Bacteroidetes	Bacteroidia	Bacteroidales	Porphyromonadaceae	*Proteiniphilum*	Uncultured bacterium	Daqing Oil Field, China	[169]
127	Bacteria	Bacteroidetes	Bacteroidia	Bacteroidales	Porphyromonadaceae	*Proteiniphilum*	Uncultured *Bacteroidetes bacterium*	Operating oil well NR-6 in the Niibori oil field, which is located in the northeast part of the main island of Japan (39° 43′ N, 139° 53′ E)	[137]
128	Bacteria	Bacteroidetes	Bacteroidia	Bacteroidales	Porphyromonadaceae	Uncultured	Uncultured bacterium	A crude oil-producing well in the Shengli oil field, Shandong, China	[173]
129	Bacteria	Bacteroidetes	Bacteroidia	Bacteroidales	Porphyromonadaceae	Uncultured	Uncultured bacterium	Daqing Oil Field, China	[169]
130	Bacteria	Bacteroidetes	Bacteroidia	Bacteroidales	Prolixibacteraceae	*Prolixibacter*	Bacterium enrichment culture clone EB35.8	Produced water samples of a high temperature and fractured chalk reservoir, the Ekofisk oil field, in block 2/4 of the Norwegian sector of the North Sea about 320 km southwest of Stavanger	[139]
131	Bacteria	Bacteroidetes	Bacteroidia	Cytophagales	Cyclobacteriaceae	*Cyclobacterium*	*Cyclobacterium lianum*	Sediment from the Xijiang oil field in the South China Sea, China	[174]
132	Bacteria	Bacteroidetes	Bacteroidia	Flavobacteriales	Flavobacteriaceae	*Arenibacter*	*Arenibacter algicola*	Nonaxenic laboratory culture of the marine diatom Skeletonema costatum CCAP1077/1C, North Sea	[175]

(continued)

Table 10.1 (Continued)

Index	Kingdom	Phylum	Class	Order	Family	Genus	Species	Isolation Source	Reference
133	Bacteria	Bacteroidetes	Bacteroidia	Flavobacteriales	Flavobacteriaceae	*Flavobacterium*	Bacterium ITRI59	The petroleum oil-contaminated soil (17.2 g total hydrocarbon kg^{-1} soil) from a landfill used for deposition of crude oil-contaminated soil from oil pumping sites in Zistersdorf, Austria	[151]
134	Bacteria	Bacteroidetes	Bacteroidia	Flavobacteriales	Weeksellaceae	*Chryseobacterium*	*Chryseobacterium* sp. ITRH57	The petroleum oil-contaminated soil (17.2 g total hydrocarbon kg^{-1} soil) from a landfill used for deposition of crude oil-contaminated soil from oil pumping sites in Zistersdorf, Austria	[151]
135	Bacteria	Bacteroidetes	Bacteroidia	Sphingobacteriales	Lentimicrobiaceae		Uncultured bacterium	Unknown	[145]
136	Bacteria	Bacteroidetes	Bacteroidia	Sphingobacteriales	Sphingobacteriaceae	*Olivibacter*	*Olivibacter oleidegradans*	An on-site, ex situ groundwater-cleaning biofilter facility on a former airbase, Hungary	[176]
137	Bacteria	Bacteroidetes	Flavobacteriia	Flavobacteriales	Flavobacteriaceae	*Actibacter*	Uncultured bacterium	The Ban 876 Gas and Oil Field within the DaGang Area (39°32′N, 117°38′E), TianJin, China	[153]

138	Bacteria	Bacteroidetes	Flavobacteriia	Flavobacteriales	Flavobacteriaceae	*Chryseobacterium*	Bacterium enrichment culture clone PW.25.5B	Produced water samples of a high temperature and fractured chalk reservoir, the Ekofisk oil field, in block 2/4 of the Norwegian sector of the North Sea about 320 km southwest of Stavanger	[139]
139	Bacteria	Bacteroidetes	Flavobacteriia	Flavobacteriales	Flavobacteriaceae	*Flavobacterium*	Bacterium enrichment culture clone PW25.9B	Produced water samples of a high temperature and fractured chalk reservoir, the Ekofisk oil field, in block 2/4 of the Norwegian sector of the North Sea about 320 km southwest of Stavanger	[139]
140	Bacteria	Bacteroidetes	Sphingobacteriia	Sphingobacteriales	Lentimicrobiaceae		Bacterium enrichment culture clone B312120	A disposal field that treats mixtures of crude oil-contaminated soil and oily sludge in the Shengli oil field, China	[152]
141	Bacteria	Bacteroidetes	Sphingobacteriia	Sphingobacteriales	Lentimicrobiaceae		Uncultured bacterium	Daqing Oil Field, China	[169]
142	Bacteria	Bacteroidetes	Sphingobacteriia	Sphingobacteriales	Lentimicrobiaceae	*Lentimicrobium*	Uncultured bacterium	A low-temperature and low-salinity petroleum reservoir in Canada	[147]

(*continued*)

Table 10.1 (Continued)

Index	Kingdom	Phylum	Class	Order	Family	Genus	Species	Isolation Source	Reference
143	Bacteria	Caldiserica	Caldisericia	Caldisericales	TTA-B1		Uncultured *Caldiserica bacterium*	Operating oil well NR-6 in the Niibori oil field, which is located in the northeast part of the main island of Japan (39° 43′ N, 139° 53′ E)	[137]
144	Bacteria	Caldiserica	Caldisericia	Caldisericales	TTA-B15		Uncultured bacterium	Daqing Oil Field, China	[169]
145	Bacteria	Caldiserica	Caldisericia	Caldisericales	WCHB1-02		Caldiserica bacterium enrichment culture clone B31168	A disposal field that treats mixtures of crude oil-contaminated soil and oily sludge in the Shengli oil field, China	[152]
146	Bacteria	Caldiserica	Caldisericia	Caldisericales	WCHB1-02		Caldiserica bacterium enrichment culture clone B31178	A disposal field that treats mixtures of crude oil-contaminated soil and oily sludge in the Shengli oil field, China	[152]
147	Bacteria	Caldiserica	Caldisericia	Caldisericales	WCHB1-02		Uncultured *Caldiserica bacterium*	Operating oil well NR-6 in the Niibori oil field, which is located in the northeast part of the main island of Japan (39° 43′ N, 139° 53′ E)	[137]
148	Bacteria	Chloroflexi	Anaerolineae	Anaerolineales	Anaerolineaceae	*Leptolinea*	Chloroflexi bacterium enrichment culture clone B312100	A disposal field that treats mixtures of crude oil-contaminated soil and oily sludge in the Shengli oil field, China	[152]

149	Bacteria	Chloroflexi	Anaerolineae	Anaerolineales	Anaerolineaceae	*Longilinea*	Chloroflexi bacterium enrichment culture clone B312136	A disposal field that treats mixtures of crude oil-contaminated soil and oily sludge in the Shengli oil field, China	[152]
150	Bacteria	Chloroflexi	Anaerolineae	Anaerolineales	Anaerolineaceae	*Pelolinea*	Chloroflexi bacterium enrichment culture clone B312139	A disposal field that treats mixtures of crude oil-contaminated soil and oily sludge in the Shengli oil field, China	[152]
151	Bacteria	Chloroflexi	Anaerolineae	Anaerolineales	Anaerolineaceae	Uncultured	Chloroflexi bacterium enrichment culture clone B31110	A disposal field that treats mixtures of crude oil-contaminated soil and oily sludge in the Shengli oil field, China	[152]
152	Bacteria	Chloroflexi	Anaerolineae	Anaerolineales	Anaerolineaceae	Uncultured	Chloroflexi bacterium enrichment culture clone B31112	A disposal field that treats mixtures of crude oil-contaminated soil and oily sludge in the Shengli oil field, China	[152]
153	Bacteria	Chloroflexi	Anaerolineae	Anaerolineales	Anaerolineaceae	Uncultured	Chloroflexi bacterium enrichment culture clone B31113	A disposal field that treats mixtures of crude oil-contaminated soil and oily sludge in the Shengli oil field, China	[152]
154	Bacteria	Chloroflexi	Anaerolineae	Anaerolineales	Anaerolineaceae	Uncultured	Chloroflexi bacterium enrichment culture clone B31117	A disposal field that treats mixtures of crude oil-contaminated soil and oily sludge in the Shengli oil field, China	[152]

(*continued*)

Table 10.1 (Continued)

Index	Kingdom	Phylum	Class	Order	Family	Genus	Species	Isolation Source	Reference
155	Bacteria	Chloroflexi	Anaerolineae	Anaerolineales	Anaerolineaceae	Uncultured	Chloroflexi bacterium enrichment culture clone B31120	A disposal field that treats mixtures of crude oil-contaminated soil and oily sludge in the Shengli oil field, China	[152]
156	Bacteria	Chloroflexi	Anaerolineae	Anaerolineales	Anaerolineaceae	Uncultured	Chloroflexi bacterium enrichment culture clone B31122	A disposal field that treats mixtures of crude oil-contaminated soil and oily sludge in the Shengli oil field, China	[152]
157	Bacteria	Chloroflexi	Anaerolineae	Anaerolineales	Anaerolineaceae	Uncultured	Chloroflexi bacterium enrichment culture clone B31126	A disposal field that treats mixtures of crude oil-contaminated soil and oily sludge in the Shengli oil field, China	[152]
158	Bacteria	Chloroflexi	Anaerolineae	Anaerolineales	Anaerolineaceae	Uncultured	Chloroflexi bacterium enrichment culture clone B31128	A disposal field that treats mixtures of crude oil-contaminated soil and oily sludge in the Shengli oil field, China	[152]
159	Bacteria	Chloroflexi	Anaerolineae	Anaerolineales	Anaerolineaceae	Uncultured	Chloroflexi bacterium enrichment culture clone B31129	A disposal field that treats mixtures of crude oil-contaminated soil and oily sludge in the Shengli oil field, China	[152]
160	Bacteria	Chloroflexi	Anaerolineae	Anaerolineales	Anaerolineaceae	Uncultured	Chloroflexi bacterium enrichment culture clone B31132	A disposal field that treats mixtures of crude oil-contaminated soil and oily sludge in the Shengli oil field, China	[152]

161	Bacteria	Chloroflexi	Anaerolineae	Anaerolineales	Anaerolineaceae	Uncultured	Chloroflexi bacterium enrichment culture clone B31133	A disposal field that treats mixtures of crude oil-contaminated soil and oily sludge in the Shengli oil field, China	[152]
162	Bacteria	Chloroflexi	Anaerolineae	Anaerolineales	Anaerolineaceae	Uncultured	Chloroflexi bacterium enrichment culture clone B31134	A disposal field that treats mixtures of crude oil-contaminated soil and oily sludge in the Shengli oil field, China	[152]
163	Bacteria	Chloroflexi	Anaerolineae	Anaerolineales	Anaerolineaceae	Uncultured	Chloroflexi bacterium enrichment culture clone B31138	A disposal field that treats mixtures of crude oil-contaminated soil and oily sludge in the Shengli oil field, China	[152]
164	Bacteria	Chloroflexi	Anaerolineae	Anaerolineales	Anaerolineaceae	Uncultured	Chloroflexi bacterium enrichment culture clone B31146	A disposal field that treats mixtures of crude oil-contaminated soil and oily sludge in the Shengli oil field, China	[152]
165	Bacteria	Chloroflexi	Anaerolineae	Anaerolineales	Anaerolineaceae	Uncultured	Chloroflexi bacterium enrichment culture clone B31153	A disposal field that treats mixtures of crude oil-contaminated soil and oily sludge in the Shengli oil field, China	[152]
166	Bacteria	Chloroflexi	Anaerolineae	Anaerolineales	Anaerolineaceae	Uncultured	Chloroflexi bacterium enrichment culture clone B31165	A disposal field that treats mixtures of crude oil-contaminated soil and oily sludge in the Shengli oil field, China	[152]

(continued)

Table 10.1 (Continued)

Index	Kingdom	Phylum	Class	Order	Family	Genus	Species	Isolation Source	Reference
167	Bacteria	Chloroflexi	Anaerolineae	Anaerolineales	Anaerolineaceae	Uncultured	Chloroflexi bacterium enrichment culture clone B312104	A disposal field that treats mixtures of crude oil-contaminated soil and oily sludge in the Shengli oil field, China	[152]
168	Bacteria	Chloroflexi	Anaerolineae	Anaerolineales	Anaerolineaceae	Uncultured	Chloroflexi bacterium enrichment culture clone B312106	A disposal field that treats mixtures of crude oil-contaminated soil and oily sludge in the Shengli oil field, China	[152]
169	Bacteria	Chloroflexi	Anaerolineae	Anaerolineales	Anaerolineaceae	Uncultured	Chloroflexi bacterium enrichment culture clone B312107	A disposal field that treats mixtures of crude oil-contaminated soil and oily sludge in the Shengli oil field, China	[152]
170	Bacteria	Chloroflexi	Anaerolineae	Anaerolineales	Anaerolineaceae	Uncultured	Chloroflexi bacterium enrichment culture clone B312117	A disposal field that treats mixtures of crude oil-contaminated soil and oily sludge in the Shengli oil field, China	[152]
171	Bacteria	Chloroflexi	Anaerolineae	Anaerolineales	Anaerolineaceae	Uncultured	Chloroflexi bacterium enrichment culture clone B312124	A disposal field that treats mixtures of crude oil-contaminated soil and oily sludge in the Shengli oil field, China	[152]
172	Bacteria	Chloroflexi	Anaerolineae	Anaerolineales	Anaerolineaceae	Uncultured	Chloroflexi bacterium enrichment culture clone B312132	A disposal field that treats mixtures of crude oil-contaminated soil and oily sludge in the Shengli oil field, China	[152]

173	Bacteria	Chloroflexi	Anaerolineae	Anaerolineales	Anaerolineaceae	Uncultured	Chloroflexi bacterium enrichment culture clone B312142	A disposal field that treats mixtures of crude oil-contaminated soil and oily sludge in the Shengli oil field, China	[152]
174	Bacteria	Chloroflexi	Anaerolineae	Anaerolineales	Anaerolineaceae	Uncultured	Chloroflexi bacterium enrichment culture clone B312144	A disposal field that treats mixtures of crude oil-contaminated soil and oily sludge in the Shengli oil field, China	[152]
175	Bacteria	Chloroflexi	Anaerolineae	Anaerolineales	Anaerolineaceae	Uncultured	Chloroflexi bacterium enrichment culture clone B312145	A disposal field that treats mixtures of crude oil-contaminated soil and oily sludge in the Shengli oil field, China	[152]
176	Bacteria	Chloroflexi	Anaerolineae	Anaerolineales	Anaerolineaceae	Uncultured	Chloroflexi bacterium enrichment culture clone B312146	A disposal field that treats mixtures of crude oil-contaminated soil and oily sludge in the Shengli oil field, China	[152]
177	Bacteria	Chloroflexi	Anaerolineae	Anaerolineales	Anaerolineaceae	Uncultured	Chloroflexi bacterium enrichment culture clone B312149	A disposal field that treats mixtures of crude oil-contaminated soil and oily sludge in the Shengli oil field, China	[152]
178	Bacteria	Chloroflexi	Anaerolineae	Anaerolineales	Anaerolineaceae	Uncultured	Chloroflexi bacterium enrichment culture clone B312150	A disposal field that treats mixtures of crude oil-contaminated soil and oily sludge in the Shengli oil field, China	[152]

(*continued*)

Table 10.1 (Continued)

Index	Kingdom	Phylum	Class	Order	Family	Genus	Species	Isolation Source	Reference
179	Bacteria	Chloroflexi	Anaerolineae	Anaerolineales	Anaerolineaceae	Uncultured	Chloroflexi bacterium enrichment culture clone B312157	A disposal field that treats mixtures of crude oil-contaminated soil and oily sludge in the Shengli oil field, China	[152]
180	Bacteria	Chloroflexi	Anaerolineae	Anaerolineales	Anaerolineaceae	Uncultured	Chloroflexi bacterium enrichment culture clone B312159	A disposal field that treats mixtures of crude oil-contaminated soil and oily sludge in the Shengli oil field, China	[152]
181	Bacteria	Chloroflexi	Anaerolineae	Anaerolineales	Anaerolineaceae	Uncultured	Chloroflexi bacterium enrichment culture clone B31293	A disposal field that treats mixtures of crude oil-contaminated soil and oily sludge in the Shengli oil field, China	[152]
182	Bacteria	Chloroflexi	Anaerolineae	Anaerolineales	Anaerolineaceae	Uncultured	Uncultured bacterium	The Ban 876 Gas and Oil Field within the DaGang Area (39°32′N, 117°38′E), TianJin, China	[153]
183	Bacteria	Chloroflexi	Anaerolineae	Anaerolineales	Anaerolineaceae	Uncultured	Uncultured *Chloroflexi bacterium*	Operating oil well NR-6 in the Niibori oil field, which is located in the northeast part of the main island of Japan (39° 43′ N, 139° 53′ E)	[137]
184	Bacteria	Chloroflexi	Ardenticatenia	Uncultured			Uncultured bacterium	The Ban 876 Gas and Oil Field within the DaGang Area (39°32′N, 117°38′E), TianJin, China	[153]

185	Bacteria	Chloroflexi	SAR202 clade	Uncultured bacterium	The Ban 876 Gas and Oil Field within the DaGang Area (39°32′N, 117°38′E), TianJin, China	[153]
186	Bacteria	Cloacimonetes	LNR A2-18	Bacterium enrichment culture clone B312121	A disposal field that treats mixtures of crude oil-contaminated soil and oily sludge in the Shengli oil field, China	[152]
187	Bacteria	Cloacimonetes	MSBL2	Uncultured *Spirochaetes bacterium*	Operating oil well NR-6 in the Niibori oil field, which is located in the northeast part of the main island of Japan (39° 43′ N, 139° 53′ E)	[137]
188	Bacteria	Cloacimonetes	W27	Bacterium enrichment culture clone B31160	A disposal field that treats mixtures of crude oil-contaminated soil and oily sludge in the Shengli oil field, China	[152]
189	Bacteria	Cloacimonetes	W27	Bacterium enrichment culture clone B312105	A disposal field that treats mixtures of crude oil-contaminated soil and oily sludge in the Shengli oil field, China	[152]
190	Bacteria	Cloacimonetes	W27	Bacterium enrichment culture clone B312129	A disposal field that treats mixtures of crude oil-contaminated soil and oily sludge in the Shengli oil field, China	[152]

(*continued*)

Table 10.1 (Continued)

Index	Kingdom	Phylum	Class	Order	Family	Genus	Species	Isolation Source	Reference
191	Bacteria	Cloacimonetes	W27				Bacterium enrichment culture clone B312131	A disposal field that treats mixtures of crude oil-contaminated soil and oily sludge in the Shengli oil field, China	[152]
192	Bacteria	Cloacimonetes	W27				Bacterium enrichment culture clone B312141	A disposal field that treats mixtures of crude oil-contaminated soil and oily sludge in the Shengli oil field, China	[152]
193	Bacteria	Cloacimonetes	W27				Bacterium enrichment culture clone B312152	A disposal field that treats mixtures of crude oil-contaminated soil and oily sludge in the Shengli oil field, China	[152]
194	Bacteria	Cloacimonetes	W27				Bacterium enrichment culture clone B312153	A disposal field that treats mixtures of crude oil-contaminated soil and oily sludge in the Shengli oil field, China	[152]
195	Bacteria	Cloacimonetes	W27				Bacterium enrichment culture clone B312158	A disposal field that treats mixtures of crude oil-contaminated soil and oily sludge in the Shengli oil field, China	[152]
196	Bacteria	Cloacimonetes	W27				Bacterium enrichment culture clone B312162	A disposal field that treats mixtures of crude oil-contaminated soil and oily sludge in the Shengli oil field, China	[152]

197	Bacteria	Cyanobacteria	Cyanobacteria	Subsection I	Family I	*Prochlorococcus*	Uncultured bacterium	The Ban 876 Gas and Oil Field within the DaGang Area (39°32′N, 117°38′E), TianJin, China	[153]
198	Bacteria	Deferribacteres	Deferribacteres	Deferribacterales	Deferribacteraceae	*Deferribacter*	*Deferribacter thermophilus*	Produced formation water collected from well AOl(07) in the Beatrice oil field, Scotland	[177]
199	Bacteria	Deferribacteres	Deferribacteres	Deferribacterales	Deferribacteraceae	Uncultured	Unidentified	Unknown	[170]
200	Bacteria	Deinococcus-Thermus	Deinococci	Deinococcales	Deinococcaceae	*Deinococcus*	Bacterium enrichment culture clone PW71.11B	Produced water samples of a high temperature and fractured chalk reservoir, the Ekofisk oil field, in block 2/4 of the Norwegian sector of the North Sea about 320 km southwest of Stavanger	[139]
201	Bacteria	Deinococcus-Thermus	Deinococci	KD3-62			Uncultured bacterium	The Ban 876 Gas and Oil Field within the DaGang Area (39°32′N, 117°38′E), TianJin, China	[153]
202	Bacteria	Elusimicrobia	Elusimicrobia	Lineage I	Unknown Family	Candidatus Endomicrobium	Uncultured bacterium	Daqing Oil Field, China	[169]
203	Bacteria	Elusimicrobia	Elusimicrobia	Lineage IV			Uncultured bacterium	Daqing Oil Field, China	[169]
204	Bacteria	Epsilonbacteraeota	Campylobacteria	Campylobacterales	Arcobacteraceae	*Arcobacter*	Uncultured bacterium	Unknown	[145]

(continued)

Table 10.1 (Continued)

Index	Kingdom	Phylum	Class	Order	Family	Genus	Species	Isolation Source	Reference
205	Bacteria	Fibrobacteres	Fibrobacteria	Fibrobacterales	FD035		Bacterium enrichment culture clone B31144	A disposal field that treats mixtures of crude oil-contaminated soil and oily sludge in the Shengli oil field, China	[152]
206	Bacteria	Fibrobacteres	Fibrobacteria	Fibrobacterales	FD035		Bacterium enrichment culture clone B31180	A disposal field that treats mixtures of crude oil-contaminated soil and oily sludge in the Shengli oil field, China	[152]
207	Bacteria	Fibrobacteres	Fibrobacteria	Fibrobacterales	FD035		Bacterium enrichment culture clone B312118	A disposal field that treats mixtures of crude oil-contaminated soil and oily sludge in the Shengli oil field, China	[152]
208	Bacteria	Fibrobacteres	Fibrobacteria	Fibrobacterales	FD035		Bacterium enrichment culture clone B312126	A disposal field that treats mixtures of crude oil-contaminated soil and oily sludge in the Shengli oil field, China	[152]
209	Bacteria	Fibrobacteres	Fibrobacteria	Fibrobacterales	FD035		Bacterium enrichment culture clone B312135	A disposal field that treats mixtures of crude oil-contaminated soil and oily sludge in the Shengli oil field, China	[152]
210	Bacteria	Fibrobacteres	Fibrobacteria	Fibrobacterales	FD035		Bacterium enrichment culture clone B312147	A disposal field that treats mixtures of crude oil-contaminated soil and oily sludge in the Shengli oil field, China	[152]

211	Bacteria	Fibrobacteres	Fibrobacteria	Fibrobacterales	FD035		Bacterium enrichment culture clone B312161	A disposal field that treats mixtures of crude oil-contaminated soil and oily sludge in the Shengli oil field, China	[152]
212	Bacteria	Fibrobacteres	Fibrobacteria	Fibrobacterales	FD035		Bacterium enrichment culture clone B31290	A disposal field that treats mixtures of crude oil-contaminated soil and oily sludge in the Shengli oil field, China	[152]
213	Bacteria	Fibrobacteres	Fibrobacteria	Fibrobacterales	FD035		Bacterium enrichment culture clone B31299	A disposal field that treats mixtures of crude oil-contaminated soil and oily sludge in the Shengli oil field, China	[152]
214	Bacteria	Fibrobacteres	Fibrobacteria	Fibrobacterales	MAT-CR-H6-H10		Uncultured bacterium	The Ban 876 Gas and Oil Field within the DaGang Area (39°32′N, 117°38′E), TianJin, China	[153]
215	Bacteria	Firmicutes	Bacilli	Bacillales	Bacillaceae	*Aeribacillus*	*Aeribacillus pallidus*	Production water (an oil/water mixture) of the oil fields TPS "Thyna Petroleum Services," Tunisia	[178]
216	Bacteria	Firmicutes	Bacilli	Bacillales	Bacillaceae	*Aeribacillus*	*Aeribacillus pallidus*	Yumen Oil field, China	[179]
217	Bacteria	Firmicutes	Bacilli	Bacillales	Bacillaceae	*Bacillus*	*Bacillus cereus*	A virgin field located in the Atlantic Ocean, Rio de Janeiro, Brazil	[180]

(continued)

Table 10.1 (Continued)

Index	Kingdom	Phylum	Class	Order	Family	Genus	Species	Isolation Source	Reference
218	Bacteria	Firmicutes	Bacilli	Bacillales	Bacillaceae	*Bacillus*	*Bacillus cereus*	A petroleum reservoir in the Daqing Oil Field, China	[181]
219	Bacteria	Firmicutes	Bacilli	Bacillales	Bacillaceae	*Bacillus*	*Bacillus galliciensis*	Germany	[163]
220	Bacteria	Firmicutes	Bacilli	Bacillales	Bacillaceae	*Bacillus*	*Bacillus licheniformis*	Production water (an oil/water mixture) of the oil fields TPS "Thyna Petroleum Services," Tunisia	[178]
221	Bacteria	Firmicutes	Bacilli	Bacillales	Bacillaceae	*Bacillus*	*Bacillus licheniformis*	A virgin field located in the Atlantic Ocean, Rio de Janeiro, Brazil	[180]
222	Bacteria	Firmicutes	Bacilli	Bacillales	Bacillaceae	*Bacillus*	*Bacillus licheniformis*	A petroleum reservoir in the Daqing Oil Field, China	[181]
223	Bacteria	Firmicutes	Bacilli	Bacillales	Bacillaceae	*Bacillus*	*Bacillus licheniformis*	Ahvaz and Masjid Suleiman oil fields, Khuzestan, Iran	[182]
224	Bacteria	Firmicutes	Bacilli	Bacillales	Bacillaceae	*Bacillus*	*Bacillus licheniformis*	Ahvaz and Masjid Suleiman oil fields, Khuzestan, Iran	[182]
225	Bacteria	Firmicutes	Bacilli	Bacillales	Bacillaceae	*Bacillus*	*Bacillus psychrosaccharolyticus*	Germany	[163]
226	Bacteria	Firmicutes	Bacilli	Bacillales	Bacillaceae	*Bacillus*	*Bacillus* sp. BTSI34	The petroleum oil-contaminated soil (17.2 g total hydrocarbon kg^{-1} soil) from a landfill used for deposition of crude oil-contaminated soil from oil pumping sites in Zistersdorf, Austria	[151]

227	Bacteria	Firmicutes	Bacilli	Bacillales	Bacillaceae	*Bacillus*	*Bacillus* sp. ITRI46	The petroleum oil-contaminated soil (17.2 g total hydrocarbon kg^{-1} soil) from a landfill used for deposition of crude oil-contaminated soil from oil pumping sites in Zistersdorf, Austria	[151]
228	Bacteria	Firmicutes	Bacilli	Bacillales	Bacillaceae	*Bacillus*	*Bacillus* sp. T4.3	A virgin field located in the Atlantic Ocean, Rio de Janeiro, Brazil	[180]
229	Bacteria	Firmicutes	Bacilli	Bacillales	Bacillaceae	*Bacillus*	Bacterium enrichment culture clone DT1-8	Daqing Oil Field, China	[169]
230	Bacteria	Firmicutes	Bacilli	Bacillales	Bacillaceae	*Bacillus*	Bacterium enrichment culture clone DT2-1	Daqing Oil Field, China	[169]
231	Bacteria	Firmicutes	Bacilli	Bacillales	Bacillaceae	*Bacillus*	Bacterium enrichment culture clone DT2-39	Daqing Oil Field, China	[169]
232	Bacteria	Firmicutes	Bacilli	Bacillales	Bacillaceae	*Bacillus*	Uncultured bacterium	A high-temperature petroleum reservoir at an offshore oil field, China	[143]
233	Bacteria	Firmicutes	Bacilli	Bacillales	Bacillaceae	*Geobacillus*	*Geobacillus jurassicus*	The formation water of the Dagang oil field (the Kongdian area), located in the Hebei Province, China	[183]
234	Bacteria	Firmicutes	Bacilli	Bacillales	Bacillaceae	*Geobacillus*	*Geobacillus jurassicus*	The formation water of the Dagang oil field (the Kongdian area), located in the Hebei Province of China	[183]

(*continued*)

Table 10.1 (Continued)

Index	Kingdom	Phylum	Class	Order	Family	Genus	Species	Isolation Source	Reference
235	Bacteria	Firmicutes	Bacilli	Bacillales	Bacillaceae	*Geobacillus*	*Geobacillus lituanicus*	Crude oil of the oil field Girkaliai, Lithuania	[184]
236	Bacteria	Firmicutes	Bacilli	Bacillales	Bacillaceae	*Geobacillus*	*Geobacillus* sp. SH-1	A deep oil well in Shengli Oilfield, China	[185]
237	Bacteria	Firmicutes	Bacilli	Bacillales	Bacillaceae	*Geobacillus*	*Geobacillus stearothermophilus*	The formation water of the Dagang oil field (the Kongdian area), located in the Hebei Province, China	[183]
238	Bacteria	Firmicutes	Bacilli	Bacillales	Bacillaceae	*Geobacillus*	*Geobacillus stearothermophilus*	The formation water of the Dagang oil field (the Kongdian area), located in the Hebei Province of China	[183]
239	Bacteria	Firmicutes	Bacilli	Bacillales	Bacillaceae	*Geobacillus*	*Geobacillus subterraneus* subsp. subterraneus	Samotlor oil field, Western Siberia, Russia; Liaohe oil field, China	[186]
240	Bacteria	Firmicutes	Bacilli	Bacillales	Bacillaceae	*Geobacillus*	Geobacillus thermodenitrificans NG80-2	A deep oil reservoir in Northern China	[187]
241	Bacteria	Firmicutes	Bacilli	Bacillales	Bacillaceae	*Geobacillus*	*Geobacillus thermoleovorans*	Deep subterranean petroleum reservoirs in the Minamiaga (Niigata) oil field, AA-5 and the Yabase (Akita) oil field, S-114, Japan	[188]
242	Bacteria	Firmicutes	Bacilli	Bacillales	Bacillaceae	*Geobacillus*	*Geobacillus uzenensis*	Uzen oilfeld, Kazakhstan	[186]
243	Bacteria	Firmicutes	Bacilli	Bacillales	Bacillaceae	*Geobacillus*	Uncultured bacterium	A high-temperature petroleum reservoir at an offshore oil field, China	[143]

244	Bacteria	Firmicutes	Bacilli	Bacillales	Bacillaceae	*Oceanobacillus*	*Oceanobacillus massiliensis*	Germany	[163]
245	Bacteria	Firmicutes	Bacilli	Bacillales	Bacillaceae	*Salibacterium*	*Bacillus* sp. B21(2010)	Four different oil fields located in the southern Algerian Sahara	[189]
246	Bacteria	Firmicutes	Bacilli	Bacillales	Bacillaceae	*Ureibacillus*	Uncultured bacterium	The Ban 876 Gas and Oil Field within the DaGang Area (39°32′N, 117°38′E), TianJin, China	[153]
247	Bacteria	Firmicutes	Bacilli	Bacillales	Family XII	*Exiguobacterium*	*Exiguobacterium mexicanum*	Germany	[163]
248	Bacteria	Firmicutes	Bacilli	Bacillales	Paenibacillaceae	*Brevibacillus*	Bacterium enrichment culture clone DT1-3	Daqing Oil Field, China	[169]
249	Bacteria	Firmicutes	Bacilli	Bacillales	Paenibacillaceae	*Brevibacillus*	Bacterium enrichment culture clone DT3-1	Daqing Oil Field, China	[169]
250	Bacteria	Firmicutes	Bacilli	Bacillales	Paenibacillaceae	*Brevibacillus*	Bacterium enrichment culture clone DT3-10	Daqing Oil Field, China	[169]
251	Bacteria	Firmicutes	Bacilli	Bacillales	Paenibacillaceae	*Brevibacillus*	Bacterium enrichment culture clone DT3-5	Daqing Oil Field, China	[169]
252	Bacteria	Firmicutes	Bacilli	Bacillales	Paenibacillaceae	*Brevibacillus*	*Brevibacillus thermoruber*	Production water (an oil/water mixture) of the oil fields TPS "Thyna Petroleum Services," Tunisia	[178]
253	Bacteria	Firmicutes	Bacilli	Bacillales	Paenibacillaceae	*Cohnella*	*Paenibacillus* sp. czh-CC13	Hawaiian soils, Hawaii, USA	[157]

(*continued*)

Table 10.1 (Continued)

Index	Kingdom	Phylum	Class	Order	Family	Genus	Species	Isolation Source	Reference
254	Bacteria	Firmicutes	Bacilli	Bacillales	Paenibacillaceae	*Paenibacillus*	*Paenibacillus* sp. MIXRH44	The petroleum oil-contaminated soil (17.2 g total hydrocarbon kg^{-1} soil) from a landfill used for deposition of crude oil-contaminated soil from oil pumping sites in Zistersdorf, Austria	[151]
255	Bacteria	Firmicutes	Bacilli	Bacillales	Planococcaceae	*Lysinibacillus*	*Lysinibacillus* sp. C250R	Production water (an oil/water mixture) of the oil fields TPS "Thyna Petroleum Services," Tunisia	[178]
256	Bacteria	Firmicutes	Bacilli	Bacillales	Planococcaceae	*Planococcus*	*Planococcus* sp. ZD22	The Daqing Oil Field, located between the Songhua river and Nen River in Heilongjiang Province, China	[190]
257	Bacteria	Firmicutes	Bacilli	Bacillales	Planococcaceae	*Planomicrobium*	*Planomicrobium alkanoclasticum*	A fine sandy sediment (80% of sediment particles in the range 125–180 μm) in the intertidal zone of Stert Flats, Bridgewater Bay, Somerset, UK	[191]

258	Bacteria	Firmicutes	Bacilli	Bacillales	Planococcaceae	*Planomicrobium*	*Planomicrobium chinense*	Diesel contaminated sites from different petrol filling stations at Bilaspur, Chhattisgarh, India	[192]
259	Bacteria	Firmicutes	Bacilli	Lactobacillales	Carnobacteriaceae	*Marinilactibacillus*	Bacterium enrichment culture clone PW32.7	Produced water samples of a high temperature and fractured chalk reservoir, the Ekofisk oil field, in block 2/4 of the Norwegian sector of the North Sea about 320 km southwest of Stavanger	[139]
260	Bacteria	Firmicutes	Bacilli	Lactobacillales	Carnobacteriaceae	*Trichococcus*	*Bacterium* sp. A15	A low-temperature and low-salinity petroleum reservoir in Canada	[147]
261	Bacteria	Firmicutes	Clostridia	Clostridiales	Caldicoprobacteraceae	*Caldicoprobacter*	Uncultured *Firmicutes bacterium*	Operating oil well NR-6 in the Niibori oil field, which is located in the northeast part of the main island of Japan (39° 43′ N, 139° 53′ E)	[137]
262	Bacteria	Firmicutes	Clostridia	Clostridiales	Clostridiaceae 1	*Proteiniclasticum*	Uncultured *Clostridium* sp.	Operating oil well NR-6 in the Niibori oil field, which is located in the northeast part of the main island of Japan (39° 43′ N, 139° 53′ E)	[137]

(continued)

Table 10.1 (Continued)

Index	Kingdom	Phylum	Class	Order	Family	Genus	Species	Isolation Source	Reference
263	Bacteria	Firmicutes	Clostridia	Clostridiales	Clostridiaceae 1	*Proteiniclasticum*	Unidentified	Unknown	[170]
264	Bacteria	Firmicutes	Clostridia	Clostridiales	Clostridiaceae 1	Uncultured	Uncultured *Firmicutes bacterium*	Operating oil well NR-6 in the Niibori oil field, which is located in the northeast part of the main island of Japan (39° 43′ N, 139° 53′ E)	[137]
265	Bacteria	Firmicutes	Clostridia	Clostridiales	Clostridiaceae 1	Uncultured	Uncultured *Moorella group bacterium*	Two production wells (AR-80 and OR-79) in the Yabase oil field, a formation of tuffaceous sandstone of Miocene–Pliocene age, located around 1293–1436 m under the surface, with in situ temperature of 40–82°C and pressure of 5 MPa, Japan	[144]
266	Bacteria	Firmicutes	Clostridia	Clostridiales	Clostridiaceae 1	Uncultured	Unidentified	Unknown	[170]
267	Bacteria	Firmicutes	Clostridia	Clostridiales	Clostridiaceae 4	*Caminicella*	Bacterium enrichment culture clone PW10.10B	Produced water samples of a high temperature and fractured chalk reservoir, the Ekofisk oil field, in block 2/4 of the Norwegian sector of the North Sea about 320 km southwest of Stavanger	[139]

268	Bacteria	Firmicutes	Clostridia	Clostridiales	Clostridiaceae 4	*Caminicella*	Bacterium enrichment culture clone PW32.3B	Produced water samples of a high temperature and fractured chalk reservoir, the Ekofisk oil field, in block 2/4 of the Norwegian sector of the North Sea about 320 km southwest of Stavanger	[139]
269	Bacteria	Firmicutes	Clostridia	Clostridiales	Clostridiaceae 4	Clostridium sensu stricto	Bacterium enrichment culture clone EB35.5	Produced water samples of a high temperature and fractured chalk reservoir, the Ekofisk oil field, in block 2/4 of the Norwegian sector of the North Sea about 320 km southwest of Stavanger	[139]
270	Bacteria	Firmicutes	Clostridia	Clostridiales	Clostridiales incertae sedis	*Dethiosulfatibacter*	Uncultured *Firmicutes bacterium*	Operating oil well NR-6 in the Niibori oil field, which is located in the northeast part of the main island of Japan (39° 43′ N, 139° 53′ E)	[137]
271	Bacteria	Firmicutes	Clostridia	Clostridiales	Defluviitaleaceae	Defluviitaleaceae UCG-011	Uncultured *Clostridiaceae bacterium*	A high-temperature oil-bearing formation in the North Sea	[172]
272	Bacteria	Firmicutes	Clostridia	Clostridiales	Defluviitaleaceae	Defluviitaleaceae UCG-011	Uncultured *Firmicutes bacterium*	Operating oil well NR-6 in the Niibori oil field, which is located in the northeast part of the main island of Japan (39° 43′ N, 139° 53′ E)	[137]

(continued)

Table 10.1 (Continued)

Index	Kingdom	Phylum	Class	Order	Family	Genus	Species	Isolation Source	Reference
273	Bacteria	Firmicutes	Clostridia	Clostridiales	Eubacteriaceae	*Acetobacterium*	*Acetobacterium* sp. Ha4	A low-temperature and low-salinity petroleum reservoir in Canada	[147]
274	Bacteria	Firmicutes	Clostridia	Clostridiales	Eubacteriaceae	*Acetobacterium*	Uncultured *Acetobacterium* sp.	Operating oil well NR-6 in the Niibori oil field, which is located in the northeast part of the main island of Japan (39° 43′ N, 139° 53′ E)	[137]
275	Bacteria	Firmicutes	Clostridia	Clostridiales	Eubacteriaceae	*Acetobacterium*	Uncultured bacterium	A low-temperature and low-salinity petroleum reservoir in Canada	[147]
276	Bacteria	Firmicutes	Clostridia	Clostridiales	Eubacteriaceae	*Acetobacterium*	Uncultured *Dehalobacter* sp.	Operating oil well NR-6 in the Niibori oil field, which is located in the northeast part of the main island of Japan (39° 43′ N, 139° 53′ E)	[137]
277	Bacteria	Firmicutes	Clostridia	Clostridiales	Eubacteriaceae	*Alkalibacter*	Uncultured *Alkalibacter* sp.	Operating oil well NR-6 in the Niibori oil field, which is located in the northeast part of the main island of Japan (39° 43′ N, 139° 53′ E)	[137]
278	Bacteria	Firmicutes	Clostridia	Clostridiales	Eubacteriaceae	*Garciella*	*Garciaella* sp. "TERI MEOR 02"	Sea buried oil pipeline known as Mumbai Uran trunk line (MUT) located on western coast of India, India	[193]

279	Bacteria	Firmicutes	Clostridia	Clostridiales	Family XI	*Soehngenia*	*Soehngenia* sp. enrichment culture clone B31151	A disposal field that treats mixtures of crude oil-contaminated soil and oily sludge in the Shengli oil field, China	[152]
280	Bacteria	Firmicutes	Clostridia	Clostridiales	Family XI	*Soehngenia*	*Soehngenia* sp. enrichment culture clone B31154	A disposal field that treats mixtures of crude oil-contaminated soil and oily sludge in the Shengli oil field, China	[152]
281	Bacteria	Firmicutes	Clostridia	Clostridiales	Family XI	*Soehngenia*	*Soehngenia* sp. enrichment culture clone B31161	A disposal field that treats mixtures of crude oil-contaminated soil and oily sludge in the Shengli oil field, China	[152]
282	Bacteria	Firmicutes	Clostridia	Clostridiales	Family XI	*Soehngenia*	*Soehngenia* sp. enrichment culture clone B3118	A disposal field that treats mixtures of crude oil-contaminated soil and oily sludge in the Shengli oil field, China	[152]
283	Bacteria	Firmicutes	Clostridia	Clostridiales	Family XI	*Soehngenia*	*Soehngenia* sp. enrichment culture clone B312143	A disposal field that treats mixtures of crude oil-contaminated soil and oily sludge in the Shengli oil field, China	[152]
284	Bacteria	Firmicutes	Clostridia	Clostridiales	Family XI	*Soehngenia*	*Soehngenia* sp. enrichment culture clone B31284	A disposal field that treats mixtures of crude oil-contaminated soil and oily sludge in the Shengli oil field, China	[152]

(continued)

Table 10.1 (Continued)

Index	Kingdom	Phylum	Class	Order	Family	Genus	Species	Isolation Source	Reference
285	Bacteria	Firmicutes	Clostridia	Clostridiales	Family XI	*Soehngenia*	Uncultured *Soehngenia* sp.	Two production wells (AR-80 and OR-79) in the Yabase oil field, a formation of tuffaceous sandstone of Miocene–Pliocene age, located around 1293–1436 m under the surface, with in situ temperature of 40–82°C and pressure of 5 MPa, Japan	[144]
286	Bacteria	Firmicutes	Clostridia	Clostridiales	Family XI	*Soehngenia*	Unidentified	Unknown	[170]
287	Bacteria	Firmicutes	Clostridia	Clostridiales	Family XI	Uncultured	*Soehngenia* sp. enrichment culture clone B31136	A disposal field that treats mixtures of crude oil-contaminated soil and oily sludge in the Shengli oil field, China	[152]
288	Bacteria	Firmicutes	Clostridia	Clostridiales	Family XI	Uncultured	*Soehngenia* sp. enrichment culture clone B31145	A disposal field that treats mixtures of crude oil-contaminated soil and oily sludge in the Shengli oil field, China	[152]
289	Bacteria	Firmicutes	Clostridia	Clostridiales	Family XI	Uncultured	*Soehngenia* sp. enrichment culture clone B31170	A disposal field that treats mixtures of crude oil-contaminated soil and oily sludge in the Shengli oil field, China	[152]

290	Bacteria	Firmicutes	Clostridia	Clostridiales	Family XI	Uncultured	*Soehngenia* sp. enrichment culture clone B312123	A disposal field that treats mixtures of crude oil-contaminated soil and oily sludge in the Shengli oil field, China	[155]
291	Bacteria	Firmicutes	Clostridia	Clostridiales	Family XI	Uncultured	*Soehngenia* sp. enrichment culture clone B312138	A disposal field that treats mixtures of crude oil-contaminated soil and oily sludge in the Shengli oil field, China	[152]
292	Bacteria	Firmicutes	Clostridia	Clostridiales	Family XI	Uncultured	*Soehngenia* sp. enrichment culture clone B312140	A disposal field that treats mixtures of crude oil-contaminated soil and oily sludge in the Shengli oil field, China	[152]
293	Bacteria	Firmicutes	Clostridia	Clostridiales	Family XI	Uncultured	*Soehngenia* sp. enrichment culture clone B31289	A disposal field that treats mixtures of crude oil-contaminated soil and oily sludge in the Shengli oil field, China	[152]
294	Bacteria	Firmicutes	Clostridia	Clostridiales	Family XI	Uncultured	Uncultured *Firmicutes bacterium*	Operating oil well NR-6 in the Niibori oil field, which is located in the northeast part of the main island of Japan (39° 43′ N, 139° 53′ E)	[137]
295	Bacteria	Firmicutes	Clostridia	Clostridiales	Family XI	Uncultured	Unidentified	Unknown	[170]

(continued)

Table 10.1 (Continued)

Index	Kingdom	Phylum	Class	Order	Family	Genus	Species	Isolation Source	Reference
296	Bacteria	Firmicutes	Clostridia	Clostridiales	Family XII	*Fusibacter*	*Fusibacter paucivorans*	A reservoir water sample from an offshore oil-producing well (Emeraude oil field), Congo	[194]
297	Bacteria	Firmicutes	Clostridia	Clostridiales	Family XII	*Fusibacter*	Uncultured *Firmicutes bacterium*	The Statfjord Weld located in the Tampen Spur area in the Norwegian sector of the North Sea	[195]
298	Bacteria	Firmicutes	Clostridia	Clostridiales	Family XII	*Fusibacter*	Uncultured *Fusibacter* sp.	Operating oil well NR-6 in the Niibori oil field, which is located in the northeast part of the main island of Japan (39° 43′ N, 139° 53′ E)	[137]
299	Bacteria	Firmicutes	Clostridia	Clostridiales	Family XIII	*Anaerovorax*	Unidentified	Unknown	[170]
300	Bacteria	Firmicutes	Clostridia	Clostridiales	Family XVIII	Uncultured	Uncultured bacterium	The Ban 876 Gas and Oil Field within the DaGang Area (39°32′N, 117°38′E), TianJin, China	[153]
301	Bacteria	Firmicutes	Clostridia	Clostridiales	Lachnospiraceae	Uncultured	Uncultured bacterium	The Statfjord Weld located in the Tampen Spur area in the Norwegian sector of the North Sea	[195]
302	Bacteria	Firmicutes	Clostridia	Clostridiales	Lachnospiraceae	Uncultured	Uncultured *Firmicutes bacterium*	Operating oil well NR-6 in the Niibori oil field, which is located in the northeast part of the main island of Japan (39° 43′ N, 139° 53′ E)	[137]

303	Bacteria	Firmicutes	Clostridia	Clostridiales	Peptococcaceae	*Desulfitibacter*	Uncultured *Firmicutes bacterium*	Operating oil well NR-6 in the Niibori oil field, which is located in the northeast part of the main island of Japan (39° 43′ N, 139° 53′ E)	[137]
304	Bacteria	Firmicutes	Clostridia	Clostridiales	Peptococcaceae	*Desulfitobacterium*	*Desulfitobacterium aromaticivorans* UKTL	Soil of a former coal-gasification site in Gliwice, Poland	[196]
305	Bacteria	Firmicutes	Clostridia	Clostridiales	Peptococcaceae	*Desulfosporosinus*	Desulfosporosinus youngiae DSM 17734	A constructed treatment Wetland receiving acid mine drainage	[197]
306	Bacteria	Firmicutes	Clostridia	Clostridiales	Peptococcaceae	*Desulfotomaculum*	*Desulfotomaculum* sp.	A wellhead on the Statfjord A platform in the Norwegian sector of the North Sea, Norway	[198]
307	Bacteria	Firmicutes	Clostridia	Clostridiales	Peptococcaceae	*Desulfotomaculum*	*Desulfotomaculum thermocisternum*	A wellhead on the Statfjord A platform in the Norwegian sector of the North Sea, Norway	[198]
308	Bacteria	Firmicutes	Clostridia	Clostridiales	Peptococcaceae	*Desulfurispora*	*Desulfotomaculum* sp. Ox39	Sediment of a drilling core taken at a former gasworks plant near Stuttgart, Germany	[199]
309	Bacteria	Firmicutes	Clostridia	Clostridiales	Peptococcaceae	Uncultured	Uncultured *Firmicutes bacterium*	Operating oil well NR-6 in the Niibori oil field, which is located in the northeast part of the main island of Japan (39° 43′ N, 139° 53′ E)	[137]

(continued)

Table 10.1 (Continued)

Index	Kingdom	Phylum	Class	Order	Family	Genus	Species	Isolation Source	Reference
310	Bacteria	Firmicutes	Clostridia	Clostridiales	Ruminococcaceae	*Ruminiclostridium*	Uncultured bacterium	The Ban 876 Gas and Oil Field within the DaGang Area (39°32′N, 117°38′E), TianJin, China	[153]
311	Bacteria	Firmicutes	Clostridia	Clostridiales	Syntrophomonadaceae		Uncultured *Firmicutes bacterium*	Operating oil well NR-6 in the Niibori oil field, which is located in the northeast part of the main island of Japan (39° 43′ N, 139° 53′ E)	[137]
312	Bacteria	Firmicutes	Clostridia	Clostridiales	Syntrophomonadaceae	*Thermosyntropha*	Uncultured *Syntrophomonadaceae bacterium*	Two production wells (AR-80 and OR-79) in the Yabase oil field, a formation of tuffaceous sandstone of Miocene–Pliocene age, located around 1293–1436 m under the surface, with in situ temperature of 40–82°C and pressure of 5 MPa, Japan	[144]

313	Bacteria	Firmicutes	Clostridia	Clostridiales	TTA-B61		*Uncultured Firmicutes bacterium*	Two production wells (AR-80 and OR-79) in the Yabase oil field, a formation of tuffaceous sandstone of Miocene–Pliocene age, located around 1293–1436 m under the surface, with in situ temperature of 40–82°C and pressure of 5 MPa, Japan	[144]
314	Bacteria	Firmicutes	Clostridia	Halanaerobiales	Halanaerobiaceae	*Halocella*	Uncultured bacterium	The Ban 876 Gas and Oil Field within the DaGang Area (39°32′N, 117°38′E), TianJin, China	[153]
315	Bacteria	Firmicutes	Clostridia	Halanaerobiales	Halobacteroidaceae	*Orenia*	Uncultured *Halobacteroidaceae bacterium*	A high-temperature oil-bearing formation in the North Sea	[172]
316	Bacteria	Firmicutes	Clostridia	Halanaerobiales	ODP1230B8.23		Firmicutes bacterium enrichment culture clone B31179	A disposal field that treats mixtures of crude oil-contaminated soil and oily sludge in the Shengli oil field, China	[152]
317	Bacteria	Firmicutes	Clostridia	NRB23			Uncultured *Firmicutes bacterium*	Operating oil well NR-6 in the Niibori oil field, which is located in the northeast part of the main island of Japan (39° 43′ N, 139° 53′ E)	[137]

(continued)

Table 10.1 (Continued)

Index	Kingdom	Phylum	Class	Order	Family	Genus	Species	Isolation Source	Reference
318	Bacteria	Firmicutes	Clostridia	Thermoanaerobacterales	Family III	*Tepidanaerobacter*	Uncultured *Tepidanaerobacter* sp.	Two production wells (AR-80 and OR-79) in the Yabase oil field, a formation of tuffaceous sandstone of Miocene–Pliocene age, located around 1293–1436 m under the surface, with in situ temperature of 40–82°C and pressure of 5 MPa, Japan	[144]
319	Bacteria	Firmicutes	Clostridia	Thermoanaerobacterales	Family III	*Thermovenabulum*	Uncultured bacterium	A high-temperature petroleum reservoir at an offshore oil field, China	[143]
320	Bacteria	Firmicutes	Clostridia	Thermoanaerobacterales	Family III	Uncultured	Uncultured bacterium	A continental high-temperature, waterflooded petroleum reservoir in the J-12 Unit at Huabei Oil field, Hebei Province, China	[161]
321	Bacteria	Firmicutes	Clostridia	Thermoanaerobacterales	Family IV	*Mahella*	Mahella australiensis	The Riverslea oil field in the Bowen-Surat Basin of Queensland, Australia	[200]

322	Bacteria	Firmicutes	Clostridia	Thermoanaerobacterales	Thermoanaerobacteraceae		Bacterium enrichment culture clone PW54.9B	Produced water samples of a high temperature and fractured chalk reservoir, the Ekofisk oil field, in block 2/4 of the Norwegian sector of the North Sea about 320 km southwest of Stavanger	[139]
323	Bacteria	Firmicutes	Clostridia	Thermoanaerobacterales	Thermoanaerobacteraceae		Uncultured *Firmicutes bacterium*	Two injection wells and one production water tank on two different platforms, Statfjord A and Statfjord C, in the Tampen Spur area in the northern part of the North Sea	[201]
324	Bacteria	Firmicutes	Clostridia	Thermoanaerobacterales	Thermoanaerobacteraceae	*Caldanaerobacter*	*Caldanaerobacter subterraneus* subsp. subterraneus	Lacq Superieur Oil Field located in south-west France	[202]
325	Bacteria	Firmicutes	Clostridia	Thermoanaerobacterales	Thermoanaerobacteraceae	*Caldanaerobacter*	Uncultured bacterium	A high-temperature petroleum reservoir at an offshore oil field, China	[143]
326	Bacteria	Firmicutes	Clostridia	Thermoanaerobacterales	Thermoanaerobacteraceae	*Caldanaerobacter*	Uncultured bacterium	A continental high-temperature, waterflooded petroleum reservoir in the J-12 Unit at Huabei Oil field, Hebei Province, China	[161]

(continued)

Table 10.1 (Continued)

Index	Kingdom	Phylum	Class	Order	Family	Genus	Species	Isolation Source	Reference
327	Bacteria	Firmicutes	Clostridia	Thermoanaerobacterales	Thermoanaerobacteraceae	*Gelria*	Uncultured *Firmicutes bacterium*	Operating oil well NR-6 in the Niibori oil field, which is located in the northeast part of the main island of Japan (39° 43′ N, 139° 53′ E)	[137]
328	Bacteria	Firmicutes	Clostridia	Thermoanaerobacterales	Thermoanaerobacteraceae	*Gelria*	Uncultured *Firmicutes bacterium*	Two production wells (AR-80 and OR-79) in the Yabase oil field, a formation of tuffaceous sandstone of Miocene–Pliocene age, located around 1293–1436 m under the surface, with in situ temperature of 40–82°C and pressure of 5 MPa, Japan	[144]
329	Bacteria	Firmicutes	Clostridia	Thermoanaerobacterales	Thermoanaerobacteraceae	*Thermacetogenium*	Uncultured *Thermacetogenium* sp.	Two production wells (AR-80 and OR-79) in the Yabase oil field, a formation of tuffaceous sandstone of Miocene–Pliocene age, located around 1293–1436 m under the surface, with in situ temperature of 40–82°C and pressure of 5 MPa, Japan	[144]

330	Bacteria	Firmicutes	Clostridia	Thermoanaerobacterales	Thermoanaerobacteraceae	*Thermoanaerobacter*	Thermoanaerobacter brockii subsp. lactiethylicus	A French oil field, France	[203]
331	Bacteria	Firmicutes	Clostridia	Thermoanaerobacterales	Thermodesulfobiaceae	*Coprothermobacter*	Thermodesulfobiaceae bacterium enrichment culture clone B312109	A disposal field that treats mixtures of crude oil-contaminated soil and oily sludge in the Shengli oil field, China	[152]
332	Bacteria	Firmicutes	Clostridia	Thermoanaerobacterales	Thermodesulfobiaceae	*Coprothermobacter*	Uncultured bacterium	Daqing Oil Field, China	[169]
333	Bacteria	Firmicutes	Clostridia	Thermoanaerobacterales	Thermodesulfobiaceae	*Coprothermobacter*	Uncultured *Coprothermobacter* sp.	Two production wells (AR-80 and OR-79) in the Yabase oil field, a formation of tuffaceous sandstone of Miocene–Pliocene age, located around 1293–1436 m under the surface, with in situ temperature of 40–82°C and pressure of 5 MPa, Japan	[144]
334	Bacteria	Gemmatimonadetes	BD2-11 terrestrial group				Uncultured bacterium	The Ban 876 Gas and Oil Field within the DaGang Area (39°32′N, 117°38′E), TianJin, China	[153]
335	Bacteria	Gemmatimonadetes	Gemmatimonadetes	Gemmatimonadales	Gemmatimonadaceae	Uncultured	Uncultured bacterium	The Ban 876 Gas and Oil Field within the DaGang Area (39°32′N, 117°38′E), TianJin, China	[153]

(*continued*)

Table 10.1 (Continued)

Index	Kingdom	Phylum	Class	Order	Family	Genus	Species	Isolation Source	Reference
336	Bacteria	Gemmatimonadetes	PAUC43f marine benthic group				Uncultured bacterium	The Ban 876 Gas and Oil Field within the DaGang Area (39°32′N, 117°38′E), TianJin, China	[153]
337	Bacteria	Halanaerobiaeota	Halanaerobiia	Halanaerobiales	Halanaerobiaceae	*Halanaerobium*	Halanaerobium congolense	An offshore Congolese oil field, Congo	[204]
338	Bacteria	Ignavibacteriae	Ignavibacteria	Ignavibacteriales	Ignavibacteriaceae	*Ignavibacterium*	Uncultured bacterium	Daqing Oil Field, China	[169]
339	Bacteria	Ignavibacteriae	Ignavibacteria	Ignavibacteriales	Ignavibacteriaceae	*Ignavibacterium*	Uncultured bacterium	Daqing Oil Field, China	[169]
340	Bacteria	Marinimicrobia (SAR406 clade)					Uncultured bacterium	Daqing Oil Field, China	[169]
341	Bacteria	Marinimicrobia (SAR406 clade)					*Wolinella* sp. enrichment culture clone B31166	A disposal field that treats mixtures of crude oil-contaminated soil and oily sludge in the Shengli oil field, China	[152]
342	Bacteria	Microgenomates	Candidatus Woesebacteria				Proteobacterium enrichment culture clone B31148	A disposal field that treats mixtures of crude oil-contaminated soil and oily sludge in the Shengli oil field, China	[152]
343	Bacteria	Nitrospirae	Nitrospira	Nitrospirales	Nitrospiraceae	*Thermodesulfovibrio*	Uncultured bacterium	A high-temperature petroleum reservoir at an offshore oil field, China	[143]
344	Bacteria	Nitrospirae	Nitrospira	Nitrospirales	Nitrospiraceae	*Thermodesulfovibrio*	Uncultured bacterium	A continental high-temperature, waterflooded petroleum reservoir in the J-12 Unit at Huabei Oil field, Hebei Province, China	[161]

345	Bacteria	Omnitrophica					Uncultured bacterium	The Ban 876 Gas and Oil Field within the DaGang Area (39°32′N, 117°38′E), TianJin, China	[153]
346	Bacteria	Parcubacteria					Uncultured bacterium	The Ban 876 Gas and Oil Field within the DaGang Area (39°32′N, 117°38′E), TianJin, China	[153]
347	Bacteria	Parcubacteria					Uncultured bacterium	Daqing Oil Field, China	[169]
348	Bacteria	Parcubacteria	Candidatus Campbellbacteria				Uncultured bacterium	The Statfjord Weld located in the Tampen Spur area in the Norwegian sector of the North Sea	[195]
349	Bacteria	Parcubacteria	Candidatus Falkowbacteria				Bacterium enrichment culture clone B31152	A disposal field that treats mixtures of crude oil-contaminated soil and oily sludge in the Shengli oil field, China	[152]
350	Bacteria	Parcubacteria	Candidatus Falkowbacteria				Bacterium enrichment culture clone B312133	A disposal field that treats mixtures of crude oil-contaminated soil and oily sludge in the Shengli oil field, China	[152]
351	Bacteria	Planctomycetes	Phycisphaerae	MSBL9			Uncultured bacterium	The Ban 876 Gas and Oil Field within the DaGang Area (39°32′N, 117°38′E), TianJin, China	[153]

(continued)

Table 10.1 (Continued)

Index	Kingdom	Phylum	Class	Order	Family	Genus	Species	Isolation Source	Reference
352	Bacteria	Planctomycetes	Planctomycetacia	Brocadiales	Brocadiaceae	Candidatus Brocadia	Uncultured anaerobic ammonium-oxidizing bacterium	The Enermark Medicine Hat Glauconitic C field (the Enermark field) in southeastern Alberta, Canada	[205]
353	Bacteria	Proteobacteria	Alphaproteobacteria	Acetobacterales	Acetobacteraceae	*Acidocella*	*Acidocella* sp. IS10	The microbial community in soil surrounding an outdoor coal storage pile in northern Indiana, USA	[164]
354	Bacteria	Proteobacteria	Alphaproteobacteria	Caulobacterales	Caulobacteraceae	*Brevundimonas*	*Brevundimonas bullata*	Germany	[163]
355	Bacteria	Proteobacteria	Alphaproteobacteria	Caulobacterales	Caulobacteraceae	*Brevundimonas*	Uncultured bacterium	Daqing Oil Field, China	[169]
356	Bacteria	Proteobacteria	Alphaproteobacteria	Caulobacterales	Hyphomonadaceae	*Hyphomonas*	Uncultured bacterium	Daqing Oil Field, China	[169]
357	Bacteria	Proteobacteria	Alphaproteobacteria	Caulobacterales	Hyphomonadaceae	*Woodsholea*	Uncultured bacterium	Daqing Oil Field, China	[169]
358	Bacteria	Proteobacteria	Alphaproteobacteria	Rhizobiales	Aurantimonadaceae		Uncultured bacterium	Daqing Oil Field, China	[169]
359	Bacteria	Proteobacteria	Alphaproteobacteria	Rhizobiales	Brucellaceae	*Ochrobactrum*	Unidentified	Unknown	[170]
360	Bacteria	Proteobacteria	Alphaproteobacteria	Rhizobiales	Hyphomicrobiaceae	*Devosia*	Uncultured bacterium	Daqing Oil Field, China	[169]
361	Bacteria	Proteobacteria	Alphaproteobacteria	Rhizobiales	Methylobacteriaceae	*Methylobacterium*	Bacterium enrichment culture clone PW71.2B	Produced water samples of a high temperature and fractured chalk reservoir, the Ekofisk oil field, in block 2/4 of the Norwegian sector of the North Sea about 320 km southwest of Stavanger	[139]
362	Bacteria	Proteobacteria	Alphaproteobacteria	Rhizobiales	Phyllobacteriaceae	*Mesorhizobium*	Unidentified	Unknown	[170]
363	Bacteria	Proteobacteria	Alphaproteobacteria	Rhizobiales	Rhizobiaceae	Allorhizobium-Neorhizobium-Pararhizobium-Rhizobium	Agrobacterium radiobacter	Hawaiian soils, Hawaii, USA	[157]

364	Bacteria	Proteobacteria	Alphaproteobacteria	Rhizobiales	Rhizobiaceae	Allorhizobium-Neorhizobium-Pararhizobium-Rhizobium	*Rhizobium* sp. ITR H_2	The petroleum oil-contaminated soil (17.2 g total hydrocarbon kg^{-1} soil) from a landfill used for deposition of crude oil-contaminated soil from oil pumping sites in Zistersdorf, Austria	[151]
365	Bacteria	Proteobacteria	Alphaproteobacteria	Rhizobiales	Rhizobiaceae	*Aurantimonas*	Aurantimonas coralicida	Germany	[163]
366	Bacteria	Proteobacteria	Alphaproteobacteria	Rhizobiales	Rhizobiaceae	*Ensifer*	*Sinorhizobium* sp. C4-2005	Hawaiian soils, Hawaii, USA	[157]
367	Bacteria	Proteobacteria	Alphaproteobacteria	Rhizobiales	Rhizobiaceae	*Mesorhizobium*	Phyllobacterium myrsinacearum	Hawaiian soils, Hawaii, USA	[157]
368	Bacteria	Proteobacteria	Alphaproteobacteria	Rhizobiales	Rhizobiaceae	*Neorhizobium*	Uncultured bacterium	Daqing Oil Field, China	[169]
369	Bacteria	Proteobacteria	Alphaproteobacteria	Rhizobiales	Rhizobiaceae	*Nitratireductor*	*Nitratireductor shengliensis*	An oil-polluted saline soil in Shengli Oil field, Eastern China	[206]
370	Bacteria	Proteobacteria	Alphaproteobacteria	Rhizobiales	Rhizobiaceae	*Ochrobactrum*	*Ochrobactrum anthropi*	Hawaiian soils, Hawaii, USA	[157]
371	Bacteria	Proteobacteria	Alphaproteobacteria	Rhizobiales	Rhizobiaceae	*Ochrobactrum*	*Ochrobactrum* sp. ITRH1	The petroleum oil-contaminated soil (17.2 g total hydrocarbon kg^{-1} soil) from a landfill used for deposition of crude oil-contaminated soil from oil pumping sites in Zistersdorf, Austria	[151]
372	Bacteria	Proteobacteria	Alphaproteobacteria	Rhizobiales	Rhizobiaceae	*Pseudorhizobium*	*Rhizobium selenitireducens*	Germany	[163]

(*continued*)

Table 10.1 (Continued)

Index	Kingdom	Phylum	Class	Order	Family	Genus	Species	Isolation Source	Reference
373	Bacteria	Proteobacteria	Alphaproteobacteria	Rhizobiales	Rhizobiaceae	*Rhizobium*	Bacterium enrichment culture clone PW30.6B	Produced water samples of a high temperature and fractured chalk reservoir, the Ekofisk oil field, in block 2/4 of the Norwegian sector of the North Sea about 320 km southwest of Stavanger	[139]
374	Bacteria	Proteobacteria	Alphaproteobacteria	Rhizobiales	Rhizobiaceae	*Rhizobium*	Uncultured bacterium	Daqing Oil Field, China	[169]
375	Bacteria	Proteobacteria	Alphaproteobacteria	Rhizobiales	Rhizobiaceae	*Rhizobium*	Uncultured bacterium	A high-temperature petroleum reservoir at an offshore oil field, China	[143]
376	Bacteria	Proteobacteria	Alphaproteobacteria	Rhodobacterales	Rhodobacteraceae	*Celeribacter*	Bacterium enrichment culture clone EB27.11	Produced water samples of a high temperature and fractured chalk reservoir, the Ekofisk oil field, in block 2/4 of the Norwegian sector of the North Sea about 320 km southwest of Stavanger	[139]

377	Bacteria	Proteobacteria	Alphaproteobacteria	Rhodobacterales	Rhodobacteraceae	*Paracoccus*	*Paracoccus carotinifaciens*	Germany	[163]
378	Bacteria	Proteobacteria	Alphaproteobacteria	Rhodobacterales	Rhodobacteraceae	*Planktotalea*	Bacterium enrichment culture clone EB39.6	Produced water samples of a high temperature and fractured chalk reservoir, the Ekofisk oil field, in block 2/4 of the Norwegian sector of the North Sea about 320 km southwest of Stavanger	[139]
379	Bacteria	Proteobacteria	Alphaproteobacteria	Rhodobacterales	Rhodobacteraceae	*Polymorphum*	*Polymorphum gilvum*	A crude oil contaminated saline soil in Shengli Oil field, China	[207]
380	Bacteria	Proteobacteria	Alphaproteobacteria	Rhodobacterales	Rhodobacteraceae	*Polymorphum*	Polymorphum gilvum SL003B-26A1	A crude oil contaminated saline soil in Shengli Oil field, China.	[207]
381	Bacteria	Proteobacteria	Alphaproteobacteria	Rhodobacterales	Rhodobacteraceae	*Rubrimonas*	*Rubrimonas shengliensis*	A crude oil contaminated saline soil in Shengli Oil field, China	[207]
382	Bacteria	Proteobacteria	Alphaproteobacteria	Rhodobacterales	Rhodobacteraceae	*Sulfitobacter*	*Sulfitobacter dubius*	Germany	[163]
383	Bacteria	Proteobacteria	Alphaproteobacteria	Rhodobacterales	Rhodobacteraceae	*Sulfitobacter*	*Sulfitobacter pontiacus*	Germany	[163]
384	Bacteria	Proteobacteria	Alphaproteobacteria	Rhodobacterales	Rhodobacteraceae	*Tropicibacter*	*Tropicibacter naphthalenivorans*	Seawater obtained from Semarang Port, Indonesia	[208]
385	Bacteria	Proteobacteria	Alphaproteobacteria	Rhodobacterales	Rhodobacteraceae	*Tropicimonas*	*Tropicimonas isoalkanivorans*	Seawater obtained from Semarang Port, Indonesia	[209]
386	Bacteria	Proteobacteria	Alphaproteobacteria	Rhodobacterales	Rhodobacteraceae	Uncultured	Uncultured alpha proteobacterium	The Statfjord Weld located in the Tampen Spur area in the Norwegian sector of the North Sea	[195]

(continued)

Table 10.1 (Continued)

Index	Kingdom	Phylum	Class	Order	Family	Genus	Species	Isolation Source	Reference
387	Bacteria	Proteobacteria	Alphaproteobacteria	Rhodobacterales	Rhodobacteraceae	Uncultured	Uncultured bacterium	A continental high-temperature, waterflooded petroleum reservoir in the J-12 Unit at Huabei Oil field, Hebei Province, China	[161]
388	Bacteria	Proteobacteria	Alphaproteobacteria	Rhodobacterales	Rhodobacteraceae	*Wenxinia*	*Wenxinia marina*	Sediment of the Xijiang oil field in the South China Sea near Fujian Province, China	[210]
389	Bacteria	Proteobacteria	Alphaproteobacteria	Rhodospirillales	Rhodospirillaceae	*Defluviicoccus*	Uncultured bacterium	The Ban 876 Gas and Oil Field within the DaGang Area (39°32′N, 117°38′E), TianJin, China	[153]
390	Bacteria	Proteobacteria	Alphaproteobacteria	Rhodospirillales	Rhodospirillaceae	*Magnetospirillum*	Uncultured bacterium	The Enermark Medicine Hat Glauconitic C field (the Enermark field) in southeastern Alberta, Canada	[205]
391	Bacteria	Proteobacteria	Alphaproteobacteria	Rhodospirillales	Rhodospirillaceae	*Terasakiella*	Uncultured alpha proteobacterium	The Statfjord Weld located in the Tampen Spur area in the Norwegian sector of the North Sea	[195]
392	Bacteria	Proteobacteria	Alphaproteobacteria	Rhodospirillales	Rhodospirillaceae	*Thalassospira*	Bacterium enrichment culture clone EB25.2	Produced water samples of a high temperature and fractured chalk reservoir, the Ekofisk oil field, in block 2/4 of the Norwegian sector of the North Sea about 320 km southwest of Stavanger.	[139]

393	Bacteria	Proteobacteria	Alphaproteobacteria	Rhodospirillales	Rhodospirillaceae	Uncultured	Uncultured alpha proteobacterium	Two injection wells and one production water tank on two different platforms, Statfjord A and Statfjord C, in the Tampen Spur area in the northern part of the North Sea	[201]
394	Bacteria	Proteobacteria	Alphaproteobacteria	Rhodospirillales	Thalassospiraceae	*Thalassospira*	*Thalassospira tepidiphila*	Petroleum-contaminated seawater	[211]
395	Bacteria	Proteobacteria	Alphaproteobacteria	Rhodospirillales	Thalassospiraceae	*Thalassospira*	*Thalassospira xianhensis*	Oil-polluted saline soil in Xianhe, Shandong Province, China	[212]
396	Bacteria	Proteobacteria	Alphaproteobacteria	Sneathiellales	Sneathiellaceae	*Sneathiella*	Uncultured alpha proteobacterium	The Statfjord Weld located in the Tampen Spur area in the Norwegian sector of the North Sea	[195]
397	Bacteria	Proteobacteria	Alphaproteobacteria	Sphingomonadales	Sphingomonadaceae	*Blastomonas*	*Blastobacter* sp. "SMCC B0477"	Deep saturated Atlantic coastal plain sediments	[213]
398	Bacteria	Proteobacteria	Alphaproteobacteria	Sphingomonadales	Sphingomonadaceae	*Erythrobacter*	*Erythrobacter citreus*	Germany	[163]
399	Bacteria	Proteobacteria	Alphaproteobacteria	Sphingomonadales	Sphingomonadaceae	*Erythrobacter*	*Lutibacterium anuloederans*	Burrow wall sediments of benthic macrofauna (mollusc) at the intertidal zone of Lowes Cove, Maine, USA	[214]
400	Bacteria	Proteobacteria	Alphaproteobacteria	Sphingomonadales	Sphingomonadaceae	*Novosphingobium*	*Novosphingobium aromaticivorans*	Deep saturated Atlantic coastal plain sediments	[213]
401	Bacteria	Proteobacteria	Alphaproteobacteria	Sphingomonadales	Sphingomonadaceae	*Novosphingobium*	*Novosphingobium aromaticivorans*	Deep saturated Atlantic coastal plain sediments	[213]

(continued)

Table 10.1 (Continued)

Index	Kingdom	Phylum	Class	Order	Family	Genus	Species	Isolation Source	Reference
402	Bacteria	Proteobacteria	Alphaproteobacteria	Sphingomonadales	Sphingomonadaceae	*Novosphingobium*	*Novosphingobium indicum*	Deep seawater (4,546 m below the surface) on the Southwest Indian Ridge, Indian Ocean	[215]
403	Bacteria	Proteobacteria	Alphaproteobacteria	Sphingomonadales	Sphingomonadaceae	*Novosphingobium*	*Novosphingobium naphthalenivorans*	Contaminated farmland soil and sediments Japan	[216]
404	Bacteria	Proteobacteria	Alphaproteobacteria	Sphingomonadales	Sphingomonadaceae	*Novosphingobium*	*Novosphingobium pentaromativorans*	Estuarine sediment at Ulsan Bay, Republic of Korea	[217]
405	Bacteria	Proteobacteria	Alphaproteobacteria	Sphingomonadales	Sphingomonadaceae	*Novosphingobium*	*Novosphingobium stygium*	Deep saturated Atlantic coastal plain sediments	[213]
406	Bacteria	Proteobacteria	Alphaproteobacteria	Sphingomonadales	Sphingomonadaceae	*Novosphingobium*	*Novosphingobium subterraneum*	Deep saturated Atlantic coastal plain sediments	[213]
407	Bacteria	Proteobacteria	Alphaproteobacteria	Sphingomonadales	Sphingomonadaceae	*Sphingobium*	Bacterium enrichment culture clone PW45.4B	Produced water samples of a high temperature and fractured chalk reservoir, the Ekofisk oil field, in block 2/4 of the Norwegian sector of the North Sea about 320 km southwest of Stavanger	[139]
408	Bacteria	Proteobacteria	Alphaproteobacteria	Sphingomonadales	Sphingomonadaceae	*Sphingobium*	*Sphingomonas* sp. BA2	Five different oil- or PAH contaminated soil sites, Hamburg, Germany	[159]

409	Bacteria	Proteobacteria	Alphaproteobacteria	Sphingomonadales	Sphingomonadaceae	*Sphingobium*	Uncultured bacterium	Daqing Oil Field, China	[169]
410	Bacteria	Proteobacteria	Alphaproteobacteria	Sphingomonadales	Sphingomonadaceae	*Sphingopyxis*	*Sphingopyxis* sp. ITRI4	The petroleum oil-contaminated soil (17.2 g total hydrocarbon kg^{-1} soil) from a landfill used for deposition of crude oil-contaminated soil from oil pumping sites in Zistersdorf, Austria	[151]
411	Bacteria	Proteobacteria	Alphaproteobacteria	Sphingomonadales	Sphingomonadaceae	*Sphingopyxis*	Uncultured bacterium	Daqing Oil Field, China	[169]
412	Bacteria	Proteobacteria	Alphaproteobacteria	Sphingomonadales	Sphingomonadaceae	*Sphingorhabdus*	Bacterium enrichment culture clone EB27.2	Produced water samples of a high temperature and fractured chalk reservoir, the Ekofisk oil field, in block 2/4 of the Norwegian sector of the North Sea about 320 km southwest of Stavanger	[139]
413	Bacteria	Proteobacteria	Betaproteobacteria	Burkholderiales	Comamonadaceae	*Comamonas*	Uncultured bacterium	A high-temperature petroleum reservoir at an offshore oil field, China	[143]
414	Bacteria	Proteobacteria	Betaproteobacteria	Hydrogenophilales	Hydrogenophilaceae	*Tepidiphilus*	Uncultured bacterium	Daqing Oil Field, China	[169]
415	Bacteria	Proteobacteria	Betaproteobacteria	Methylophilales	Methylophilaceae	*Methylophilus*	Methylophilaceae bacterium enrichment culture clone B31285	A disposal field that treats mixtures of crude oil-contaminated soil and oily sludge in the Shengli oil field, China	[152]

(*continued*)

Table 10.1 (Continued)

Index	Kingdom	Phylum	Class	Order	Family	Genus	Species	Isolation Source	Reference
416	Bacteria	Proteobacteria	Betaproteobacteria	Rhodocyclales	Rhodocyclaceae	*Azoarcus*	Unidentified	Unknown	[170]
417	Bacteria	Proteobacteria	Betaproteobacteria	Rhodocyclales	Rhodocyclaceae	*Thauera*	*Thauera* sp. Al7	A low-temperature and low-salinity petroleum reservoir in Canada	[147]
418	Bacteria	Proteobacteria	Betaproteobacteria	Rhodocyclales	Rhodocyclaceae	*Thauera*	Uncultured bacterium	Daqing Oil Field, China	[169]
419	Bacteria	Proteobacteria	Betaproteobacteria	Rhodocyclales	Rhodocyclaceae	*Thauera*	Unidentified	Unknown	[170]
420	Bacteria	Proteobacteria	Betaproteobacteria	Rhodocyclales	Rhodocyclaceae	Uncultured	Unidentified	Unknown	[170]
421	Bacteria	Proteobacteria	Deltaproteobacteria	Deltaproteobacteria incertae sedis	Syntrophorhabdaceae	*Syntrophorhabdus*	Bacterium enrichment culture clone B31175	A disposal field that treats mixtures of crude oil-contaminated soil and oily sludge in the Shengli oil field, China	[152]
422	Bacteria	Proteobacteria	Deltaproteobacteria	Desulfarculales	Desulfarculaceae	*Desulfatiglans*	Sulfate-reducing bacterium mXyS1	A previously described mesophilic enrichment culture growing anaerobically with crude oil and sulfate in seawater medium. The enrichment culture originated from the water phase of a North Sea oil tank in Wilhelmshaven, Germany	[218]
423	Bacteria	Proteobacteria	Deltaproteobacteria	Desulfarculales	Desulfarculaceae	*Desulfatiglans*	Uncultured bacterium	The Ban 876 Gas and Oil Field within the DaGang Area (39°32′N, 117°38′E), TianJin, China	[153]

424	Bacteria	Proteobacteria	Deltaproteobacteria	Desulfobacterales	Desulfobacteraceae	*Desulfatibacillum*	*Desulfatibacillum aliphaticivorans*	Marine sediment of Canal Vieil cove, polluted by petroleum refinery spills over a period of 15 years, Lavera, Gulf of Fos, France	[219]
425	Bacteria	Proteobacteria	Deltaproteobacteria	Desulfobacterales	Desulfobacteraceae	*Desulfatibacillum*	*Desulfatibacillum alkenivorans*	Oil-polluted sediments, Fos Harbour, France	[220]
426	Bacteria	Proteobacteria	Deltaproteobacteria	Desulfobacterales	Desulfobacteraceae	*Desulfatiferula*	*Desulfatiferula olefinivorans*	Brackish sediment of a wastewater decantation facility of an oil refinery, Berre lagoon, France	[221]
427	Bacteria	Proteobacteria	Deltaproteobacteria	Desulfobacterales	Desulfobacteraceae	*Desulfobacter*	*Desulfobacter vibrioformis*	A previously described mesophilic enrichment culture growing anaerobically with crude oil and sulfate in seawater medium. The enrichment culture originated from the water phase of a North Sea oil tank in Wilhelmshaven, Germany	[222]
428	Bacteria	Proteobacteria	Deltaproteobacteria	Desulfobacterales	Desulfobacteraceae	*Desulfobacula*	*Desulfobacula toluolica*	Anoxic, sulfide-rich marine sediment samples from Eel Pond, a seawater pond in Woods Hole, Massachusetts, USA	[223]

(*continued*)

Table 10.1 (Continued)

Index	Kingdom	Phylum	Class	Order	Family	Genus	Species	Isolation Source	Reference
429	Bacteria	Proteobacteria	Deltaproteobacteria	Desulfobacterales	Desulfobacteraceae	*Desulfosarcina*	*Desulfosarcina ovata*	A previously described mesophilic enrichment culture growing anaerobically with crude oil and sulfate in seawater medium. The enrichment culture originated from the water phase of a North Sea oil tank in Wilhelmshaven, Germany	[218]
430	Bacteria	Proteobacteria	Deltaproteobacteria	Desulfobacterales	Desulfobacteraceae	*Desulfotignum*	*Desulfotignum toluenicum*	An oil-reservoir model column	[224]
431	Bacteria	Proteobacteria	Deltaproteobacteria	Desulfobacterales	Desulfobacteraceae	*Desulfotignum*	*Desulfotignum toluenicum*	An oil-reservoir model column	[224]
432	Bacteria	Proteobacteria	Deltaproteobacteria	Desulfobacterales	Desulfobacteraceae	SEEP-SRB1	Desulfobacteraceae bacterium enrichment culture clone B31150	A disposal field that treats mixtures of crude oil-contaminated soil and oily sludge in the Shengli oil field, China	[152]
433	Bacteria	Proteobacteria	Deltaproteobacteria	Desulfobacterales	Desulfobacteraceae	SEEP-SRB1	Desulfobacteraceae bacterium enrichment culture clone B31157	A disposal field that treats mixtures of crude oil-contaminated soil and oily sludge in the Shengli oil field, China	[152]
434	Bacteria	Proteobacteria	Deltaproteobacteria	Desulfobacterales	Desulfobacteraceae	SEEP-SRB1	Desulfobacteraceae bacterium enrichment culture clone B312115	A disposal field that treats mixtures of crude oil-contaminated soil and oily sludge in the Shengli oil field, China	[152]

435	Bacteria	Proteobacteria	Deltaproteobacteria	Desulfobacterales	Desulfobacteraceae	Sva0081	Uncultured bacterium	The Ban 876 Gas and Oil Field within the DaGang Area (39°32′N, 117°38′E), TianJin, China	[153]
436	Bacteria	Proteobacteria	Deltaproteobacteria	Desulfobacterales	Desulfobacteraceae	Uncultured	Desulfobacteraceae bacterium enrichment culture clone B31172	A disposal field that treats mixtures of crude oil-contaminated soil and oily sludge in the Shengli oil field, China	[152]
437	Bacteria	Proteobacteria	Deltaproteobacteria	Desulfobacterales	Desulfobacteraceae	Uncultured	Desulfobacteraceae bacterium enrichment culture clone B31294	A disposal field that treats mixtures of crude oil-contaminated soil and oily sludge in the Shengli oil field, China	[152]
438	Bacteria	Proteobacteria	Deltaproteobacteria	Desulfobacterales	Desulfobulbaceae	*Desulfobulbus*	*Desulfobulbus rhabdoformis*	A water–oil separation system on the deck of the Statijord A field platform in the Norwegian sector of the North Sea, Norway.	[225]
439	Bacteria	Proteobacteria	Deltaproteobacteria	Desulfobacterales	Desulfobulbaceae	*Desulfobulbus*	Uncultured bacterium	Unknown	[145]
440	Bacteria	Proteobacteria	Deltaproteobacteria	Desulfobacterales	Desulfobulbaceae	MSBL7	Uncultured bacterium	The Ban 876 Gas and Oil Field within the DaGang Area (39°32′N, 117°38′E), TianJin, China	[153]

(continued)

Table 10.1 (Continued)

Index	Kingdom	Phylum	Class	Order	Family	Genus	Species	Isolation Source	Reference
441	Bacteria	Proteobacteria	Deltaproteobacteria	Desulfovibrionales	Desulfohalobiaceae	*Desulfohalobium*	*Desulfohalobium retbaense*	A water sample taken from an oil pipeline linking offshore production platforms to onshore treatment facilities, Africa	[226]
442	Bacteria	Proteobacteria	Deltaproteobacteria	Desulfovibrionales	Desulfohalobiaceae	*Desulfonauticus*	Uncultured delta proteobacterium	Two injection wells and one production water tank on two different platforms, Statfjord A and Statfjord C, in the Tampen Spur area in the northern part of the North Sea	[201]
443	Bacteria	Proteobacteria	Deltaproteobacteria	Desulfovibrionales	Desulfohalobiaceae	*Desulfothermus*	*Desulfothermus naphthae*	Guaymas Basin sediment and the water phase of a North Sea oil tank at Wilhelmshaven, Lower Saxony, Germany	[227]
444	Bacteria	Proteobacteria	Deltaproteobacteria	Desulfovibrionales	Desulfomicrobiaceae		Uncultured *Desulfocaldus* sp.	Production water sample from the mesothermic and highly degraded Schrader Bluff petroleum field in Alaska's North Slope region, USA	[146]
445	Bacteria	Proteobacteria	Deltaproteobacteria	Desulfovibrionales	Desulfomicrobiaceae	*Desulfomicrobium*	*Desulfomicrobium* sp. Bsl6	A low-temperature and low-salinity petroleum reservoir in Canada	[147]

446	Bacteria	Proteobacteria	Deltaproteobacteria	Desulfovibrionales	Desulfomicrobiaceae	*Desulfoplanes*	Uncultured *Desulfomicrobium* sp.	A high-temperature oil-bearing formation in the North Sea	[172]
447	Bacteria	Proteobacteria	Deltaproteobacteria	Desulfovibrionales	Desulfovibrionaceae	*Desulfovibrio*	*Desulfovibrio bastinii*	A pipeline of Emeraude Oil field, Congo	[228]
448	Bacteria	Proteobacteria	Deltaproteobacteria	Desulfovibrionales	Desulfovibrionaceae	*Desulfovibrio*	*Desulfovibrio gabonensis*	A water sample taken from an oil pipeline linking offshore production platforms to onshore treatment facilities, Africa	[226]
449	Bacteria	Proteobacteria	Deltaproteobacteria	Desulfovibrionales	Desulfovibrionaceae	*Desulfovibrio*	*Desulfovibrio gracilis*	A pipeline of Emeraude Oil field, Congo	[228]
450	Bacteria	Proteobacteria	Deltaproteobacteria	Desulfovibrionales	Desulfovibrionaceae	*Desulfovibrio*	*Desulfovibrio halophilus*	A water sample taken from an oil pipeline linking offshore production platforms to onshore treatment facilities, Africa	[226]
451	Bacteria	Proteobacteria	Deltaproteobacteria	Desulfovibrionales	Desulfovibrionaceae	*Desulfovibrio*	*Desulfovibrio longus*	A pipeline of Emeraude Oil field, Congo	[228]
452	Bacteria	Proteobacteria	Deltaproteobacteria	Desulfovibrionales	Desulfovibrionaceae	*Desulfovibrio*	*Desulfovibrio* sp. Bsl2	A low-temperature and low-salinity petroleum reservoir in Canada	[147]
453	Bacteria	Proteobacteria	Deltaproteobacteria	Desulfovibrionales	Desulfovibrionaceae	*Desulfovibrio*	Uncultured bacterium	The Statfjord Weld located in the Tampen Spur area in the Norwegian sector of the North Sea	[195]

(*continued*)

Table 10.1 (Continued)

Index	Kingdom	Phylum	Class	Order	Family	Genus	Species	Isolation Source	Reference
454	Bacteria	Proteobacteria	Deltaproteobacteria	Desulfurellales	Desulfurellaceae	H16	Uncultured bacterium	The Ban 876 Gas and Oil Field within the DaGang Area (39°32′N, 117°38′E), TianJin, China	[153]
455	Bacteria	Proteobacteria	Deltaproteobacteria	Desulfuromonadales	Desulfuromonadaceae	*Desulfuromonas*	Desulfuromonadaceae bacterium enrichment culture clone B31212	A disposal field that treats mixtures of crude oil-contaminated soil and oily sludge in the Shengli oil field, China	[152]
456	Bacteria	Proteobacteria	Deltaproteobacteria	Desulfuromonadales	Desulfuromonadaceae	*Pelobacter*	Uncultured *Pelobacter* sp.	A high-temperature oil-bearing formation in the North Sea	[172]
457	Bacteria	Proteobacteria	Deltaproteobacteria	Desulfuromonadales	Geobacteraceae	*Geoalkalibacter*	*Geoalkalibacter subterraneus*	Well 41-21B in the Red Wash oil field. Red Wash is an on-shore oil field located in Utah, USA	[229]
458	Bacteria	Proteobacteria	Deltaproteobacteria	Desulfuromonadales	Geobacteraceae	*Geobacter*	*Geobacter grbiciae*	A freshwater aquatic sediment collected from the estuary of the Potomac River in Virginia, USA	[230]
459	Bacteria	Proteobacteria	Deltaproteobacteria	Desulfuromonadales	Geobacteraceae	*Geobacter*	*Geobacter metallireducens*	A freshwater site in the Potomac River, Maryland, USA	[231]
460	Bacteria	Proteobacteria	Deltaproteobacteria	Desulfuromonadales	Geobacteraceae	*Geobacter*	*Geobacter toluenoxydans*	Well sediment from a tar-oil-contaminated site near Stuttgart, Germany	[196]

461	Bacteria	Proteobacteria	Deltaproteobacteria	Syntrophobacterales	Syntrophaceae	*Smithella*	*Smithella* sp. enrichment culture clone B312125	A disposal field that treats mixtures of crude oil-contaminated soil and oily sludge in the Shengli oil field, China	[152]
462	Bacteria	Proteobacteria	Deltaproteobacteria	Syntrophobacterales	Syntrophaceae	*Smithella*	Uncultured bacterium	Daqing Oil Field, China	[169]
463	Bacteria	Proteobacteria	Deltaproteobacteria	Syntrophobacterales	Syntrophobacteraceae		Uncultured delta proteobacterium	Two injection wells and one production water tank on two different platforms, Statfjord A and Statfjord C, in the Tampen Spur area in the northern part of the North Sea	[201]
464	Bacteria	Proteobacteria	Deltaproteobacteria	Syntrophobacterales	Syntrophobacteraceae	*Desulfacinum*	Uncultured *Desulfacinum* sp.	Operating oil well NR-6 in the Niibori oil field, which is located in the northeast part of the main island of Japan (39° 43′ N, 139° 53′ E)	[137]
465	Bacteria	Proteobacteria	Deltaproteobacteria	Syntrophobacterales	Syntrophobacteraceae	*Desulfoglaeba*	*Desulfoglaeba alkanexedens*	An oil—water separation tank in the Bebee-Konawa oil field, Oklahoma, USA.	[232]
466	Bacteria	Proteobacteria	Deltaproteobacteria	Syntrophobacterales	Syntrophobacteraceae	*Desulfoglaeba*	*Desulfoglaeba* sp. Lake	An oil—water separation tank in the Bebee-Konawa oil field, Oklahoma, USA.	[232]

(continued)

Table 10.1 (Continued)

Index	Kingdom	Phylum	Class	Order	Family	Genus	Species	Isolation Source	Reference
467	Bacteria	Proteobacteria	Deltaproteobacteria	Syntrophobacterales	Syntrophobacteraceae	*Thermodesulforhabdus*	*Thermodesulforhabdus norvegica*	North Sea oil field water from a Norwegian oil platform, Norway	[233]
468	Bacteria	Proteobacteria	Deltaproteobacteria	Syntrophobacterales	Syntrophobacteraceae	*Thermodesulforhabdus*	*Thermodesulforhabdus* sp. NS-tSRB-1	Ekofisk field in the Norwegian sector of the North Sea	[138]
469	Bacteria	Proteobacteria	Epsilonproteobacteria	Campylobacterales	Campylobacteraceae	*Arcobacter*	Bacterium enrichment culture clone EB24.3	Produced water samples of a high temperature and fractured chalk reservoir, the Ekofisk oil field, in block 2/4 of the Norwegian sector of the North Sea about 320 km southwest of Stavanger	[139]
470	Bacteria	Proteobacteria	Epsilonproteobacteria	Campylobacterales	Campylobacteraceae	*Arcobacter*	Uncultured *Epsilonproteobacteria bacterium*	The Statfjord Weld located in the Tampen Spur area in the Norwegian sector of the North Sea	[195]
471	Bacteria	Proteobacteria	Epsilonproteobacteria	Campylobacterales	Campylobacteraceae	*Sulfurospirillum*	Uncultured bacterium	A low-temperature and low-salinity petroleum reservoir in Canada	[147]
472	Bacteria	Proteobacteria	Epsilonproteobacteria	Campylobacterales	Campylobacteraceae	*Sulfurospirillum*	Uncultured bacterium	A continental high-temperature, waterflooded petroleum reservoir in the J-12 Unit at Huabei Oil field, Hebei Province, China	[161]

473	Bacteria	Proteobacteria	Epsilonproteobacteria	Campylobacterales	Helicobacteraceae	*Sulfuricurvum*	Uncultured bacterium	Daqing Oil Field, China	[169]
474	Bacteria	Proteobacteria	Epsilonproteobacteria	Campylobacterales	Helicobacteraceae	*Sulfurimonas*	Uncultured *Epsilonproteobacteria bacterium*	Two injection wells and one production water tank on two different platforms, Statfjord A and Statfjord C, in the Tampen Spur area in the northern part of the North Sea	[201]
475	Bacteria	Proteobacteria	Epsilonproteobacteria	Campylobacterales	Helicobacteraceae	*Sulfurovum*	Uncultured bacterium	The Ban 876 Gas and Oil Field within the DaGang Area (39°32′N, 117°38′E), TianJin, China	[153]
476	Bacteria	Proteobacteria	Epsilonproteobacteria	Nautiliales	Nautiliaceae	*Nitratifractor*	Uncultured *Epsilonproteobacteria bacterium*	Two injection wells and one production water tank on two different platforms, Statfjord A and Statfjord C, in the Tampen Spur area in the northern part of the North Sea	[201]
477	Bacteria	Proteobacteria	Gammaproteobacteria	Aeromonadales	Aeromonadaceae	*Aeromonas*	*Aeromonas* sp. MIXRI63	The petroleum oil-contaminated soil (17.2 g total hydrocarbon kg^{-1} soil) from a landfill used for deposition of crude oil-contaminated soil from oil pumping sites in Zistersdorf, Austria	[151]

(continued)

Table 10.1 (Continued)

Index	Kingdom	Phylum	Class	Order	Family	Genus	Species	Isolation Source	Reference
478	Bacteria	Proteobacteria	Gammaproteobacteria	Alteromonadales	Colwelliaceae	*Thalassotalea*	Uncultured bacterium	The Statfjord Weld located in the Tampen Spur area in the Norwegian sector of the North Sea	[195]
479	Bacteria	Proteobacteria	Gammaproteobacteria	Alteromonadales	Gallaecimonadaceae	*Gallaecimonas*	*Gallaecimonas pentaromativorans*	Isolated from intertidal sediment of Corcubion Ria in Cee, A Coruña, Spain	[234]
480	Bacteria	Proteobacteria	Gammaproteobacteria	Alteromonadales	Marinobacteraceae	*Marinobacter*	*Marinobacter hydrocarbonoclasticus*	Sediments collected in the Gulf of Fos, at the mouth of a petroleum refinery outlet chronically polluted by hydrocarbons, Mediterranean coast, 50 km north of Marseille, France	[235]
481	Bacteria	Proteobacteria	Gammaproteobacteria	Alteromonadales	Pseudoalteromonadaceae	*Pseudoalteromonas*	*Pseudoalteromonas agarivorans*	Germany	[163]
482	Bacteria	Proteobacteria	Gammaproteobacteria	Alteromonadales	Pseudoalteromonadaceae	*Pseudoalteromonas*	*Pseudoalteromonas fuliginea*	Germany	[163]
483	Bacteria	Proteobacteria	Gammaproteobacteria	Alteromonadales	Pseudoalteromonadaceae	*Pseudoalteromonas*	*Pseudoalteromonas haloplanktis*	Sediments collected in the Gulf of Fos, at the mouth of a petroleum refinery outlet chronically polluted by hydrocarbons, Mediterranean coast, 50 km north of Marseille, France	[235]

484	Bacteria	Proteobacteria	Gammaproteobacteria	Alteromonadales	Pseudoalteromonadaceae	*Pseudoalteromonas*	*Pseudoalteromonas haloplanktis*	Germany	[163]
485	Bacteria	Proteobacteria	Gammaproteobacteria	Alteromonadales	Pseudoalteromonadaceae	*Pseudoalteromonas*	*Pseudoalteromonas translucida*	Germany	[163]
486	Bacteria	Proteobacteria	Gammaproteobacteria	Alteromonadales	Shewanellaceae	*Shewanella*	*Shewanella arctica* Kim et al. 2012	Germany	[163]
487	Bacteria	Proteobacteria	Gammaproteobacteria	Alteromonadales	Shewanellaceae	*Shewanella*	*Shewanella basaltis*	Germany	[163]
488	Bacteria	Proteobacteria	Gammaproteobacteria	Alteromonadales	Shewanellaceae	*Shewanella*	*Shewanella putrefacien*	Germany	[163]
489	Bacteria	Proteobacteria	Gammaproteobacteria	Alteromonadales	Shewanellaceae	*Shewanella*	*Shewanella vesiculosa*	Germany	[163]
490	Bacteria	Proteobacteria	Gammaproteobacteria	Alteromonadales	Shewanellaceae	*Shewanella*	Uncultured bacterium	A high-temperature petroleum reservoir at an offshore oil field, China	[143]
491	Bacteria	Proteobacteria	Gammaproteobacteria	Alteromonadales	Shewanellaceae	*Shewanella*	Uncultured *Shewanella* sp.	Operating oil well NR-6 in the Niibori oil field, which is located in the northeast part of the main island of Japan (39° 43′ N, 139° 53′ E)	[137]
492	Bacteria	Proteobacteria	Gammaproteobacteria	Alteromonadales	Shewanellaceae	*Shewanella*	Uncultured *Shewanellaceae bacterium*	Production water sample from the mesothermic and highly degraded Schrader Bluff petroleum field in Alaska's North Slope region, USA	[146]
493	Bacteria	Proteobacteria	Gammaproteobacteria	Betaproteobacteriales	Burkholderiaceae		Alcaligenaceae bacterium BTRH65	The petroleum oil-contaminated soil (17.2 g total hydrocarbon kg^{-1} soil) from a landfill used for deposition of crude oil-contaminated soil from oil pumping sites in Zistersdorf, Austria	[151]

(continued)

Table 10.1 (Continued)

Index	Kingdom	Phylum	Class	Order	Family	Genus	Species	Isolation Source	Reference
494	Bacteria	Proteobacteria	Gammaproteobacteria	Betaproteobacteriales	Burkholderiaceae	*Achromobacter*	*Achromobacter* sp. C350R	Production water (an oil/water mixture) of the oil fields TPS "Thyna Petroleum Services," Tunisia	[178]
495	Bacteria	Proteobacteria	Gammaproteobacteria	Betaproteobacteriales	Burkholderiaceae	*Achromobacter*	*Achromobacter xylosoxidans* subsp. xylosoxidans	Hawaiian soils, Hawaii, USA	[157]
496	Bacteria	Proteobacteria	Gammaproteobacteria	Betaproteobacteriales	Burkholderiaceae	*Achromobacter*	*Alcaligenaceae bacterium* BTRH5	The petroleum oil-contaminated soil (17.2 g total hydrocarbon kg^{-1} soil) from a landfill used for deposition of crude oil-contaminated soil from oil pumping sites in Zistersdorf, Austria	[151]
497	Bacteria	Proteobacteria	Gammaproteobacteria	Betaproteobacteriales	Burkholderiaceae	*Achromobacter*	Burkholderiales bacterium ITSI70	The petroleum oil-contaminated soil (17.2 g total hydrocarbon kg^{-1} soil) from a landfill used for deposition of crude oil-contaminated soil from oil pumping sites in Zistersdorf, Austria	[151]
498	Bacteria	Proteobacteria	Gammaproteobacteria	Betaproteobacteriales	Burkholderiaceae	Burkholderia-Caballeronia-Paraburkholderia	*Burkholderia* sp. C3	Hawaiian soils, Hawaii, USA	[157]
499	Bacteria	Proteobacteria	Gammaproteobacteria	Betaproteobacteriales	Hydrogenophilaceae	*Tepidiphilus*	*Tepidiphilus succinatimandens*	The Riverslea oil field in the Bowen–Surat basin, Queensland, Australia	[236]

500	Bacteria	Proteobacteria	Gammaproteobacteria	Betaproteobacteriales	Rhodocyclaceae	*Azoarcus*	Aromatoleum aromaticum EbN1	A homogenized mixture of mud samples from ditches and the Weser river in Bremen, Germany	[237]
501	Bacteria	Proteobacteria	Gammaproteobacteria	Betaproteobacteriales	Rhodocyclaceae	*Azoarcus*	*Azoarcus* sp. PbN1	A homogenized mixture of mud samples from ditches and the Weser river in Bremen, Germany	[237]
502	Bacteria	Proteobacteria	Gammaproteobacteria	Betaproteobacteriales	Rhodocyclaceae	*Azoarcus*	*Azoarcus toluvorans*	An aquifer at Moffett Field, California, USA	[238]
503	Bacteria	Proteobacteria	Gammaproteobacteria	Betaproteobacteriales	Rhodocyclaceae	*Azoarcus*	Beta proteobacterium pCyN1	A previously described mesophilic enrichment culture growing anaerobically with crude oil and sulfate in seawater medium. The enrichment culture originated from the water phase of a North Sea oil tank in Wilhelmshaven, Germany	[218]
504	Bacteria	Proteobacteria	Gammaproteobacteria	Betaproteobacteriales	Rhodocyclaceae	*Dechloromonas*	*Dechloromonas* sp. JJ	Sediments collected from the Potomac River, Maryland, USA	[239]
505	Bacteria	Proteobacteria	Gammaproteobacteria	Betaproteobacteriales	Rhodocyclaceae	*Ferribacterium*	Dechloromonas aromatica RCB	Sediments collected from the Potomac River, Maryland, USA	[239]

(*continued*)

Table 10.1 (Continued)

Index	Kingdom	Phylum	Class	Order	Family	Genus	Species	Isolation Source	Reference
506	Bacteria	Proteobacteria	Gammaproteobacteria	Betaproteobacteriales	Rhodocyclaceae	*Georgfuchsia*	*Georgfuchsia toluolica*	An iron-reducing aquifer polluted by BTEX-containing Banisveld landfill leachate, near Boxtel, Netherlands	[240]
507	Bacteria	Proteobacteria	Gammaproteobacteria	Betaproteobacteriales	Rhodocyclaceae	*Thauera*	*Azoarcus* sp. mXyN1	A homogenized mixture of mud samples from ditches and the Weser river in Bremen, Germany	[237]
508	Bacteria	Proteobacteria	Gammaproteobacteria	Betaproteobacteriales	Rhodocyclaceae	*Thauera*	Beta proteobacterium pCy N_2	A previously described mesophilic enrichment culture growing anaerobically with crude oil and sulfate in seawater medium. The enrichment culture originated from the water phase of a North Sea oil tank in Wilhelmshaven, Germany	[218]
509	Bacteria	Proteobacteria	Gammaproteobacteria	Betaproteobacteriales	Rhodocyclaceae	*Thauera*	*Thauera aromatica*		[241]
510	Bacteria	Proteobacteria	Gammaproteobacteria	Betaproteobacteriales	Rhodocyclaceae	*Thauera*	*Thauera* sp. DNT-1	A wastewater treatment plant	[242]
511	Bacteria	Proteobacteria	Gammaproteobacteria	Cellvibrionales	Porticoccaceae	*Porticoccus*	*Porticoccus hydrocarbonoclasticus*	A nonaxenic laboratory culture of the marine dinoflagellate Lingulodinium polyedrum CCAP1121/2	[243]

512	Bacteria	Proteobacteria	Gammaproteobacteria	Chromatiales	Ectothiorhodospiraceae	*Thioalkalispira*	Uncultured bacterium	The Ban 876 Gas and Oil Field within the DaGang Area (39°32′N, 117°38′E), TianJin, China	[153]
513	Bacteria	Proteobacteria	Gammaproteobacteria	Enterobacteriales	Enterobacteriaceae		Enterobacteriaceae bacterium ITSI61	The petroleum oil-contaminated soil (17.2 g total hydrocarbon kg^{-1} soil) from a landfill used for deposition of crude oil-contaminated soil from oil pumping sites in Zistersdorf, Austria	[151]
514	Bacteria	Proteobacteria	Gammaproteobacteria	Enterobacteriales	Enterobacteriaceae		*Pantoea* sp. BTRH79	The petroleum oil-contaminated soil (17.2 g total hydrocarbon kg^{-1} soil) from a landfill used for deposition of crude oil-contaminated soil from oil pumping sites in Zistersdorf, Austria	[151]
515	Bacteria	Proteobacteria	Gammaproteobacteria	Enterobacteriales	Enterobacteriaceae		*Pantoea* sp. MIXSI9	The petroleum oil-contaminated soil (17.2 g total hydrocarbon kg^{-1} soil) from a landfill used for deposition of crude oil-contaminated soil from oil pumping sites in Zistersdorf, Austria	[151]

(*continued*)

Table 10.1 (Continued)

Index	Kingdom	Phylum	Class	Order	Family	Genus	Species	Isolation Source	Reference
516	Bacteria	Proteobacteria	Gammaproteobacteria	Enterobacteriales	Enterobacteriaceae	*Citrobacter*	Uncultured bacterium	A high-temperature petroleum reservoir at an offshore oil field, China	[143]
517	Bacteria	Proteobacteria	Gammaproteobacteria	Enterobacteriales	Enterobacteriaceae	*Enterobacter*	Enterobacteriaceae bacterium BTR $H_2$8	The petroleum oil-contaminated soil (17.2 g total hydrocarbon kg^{-1} soil) from a landfill used for deposition of crude oil-contaminated soil from oil pumping sites in Zistersdorf, Austria	[151]
518	Bacteria	Proteobacteria	Gammaproteobacteria	Enterobacteriales	Enterobacteriaceae	*Enterobacter*	*Leclercia adecarboxylata*	Subsurface soil collected from an oily sludge storage pit at the Digboi oil refinery, India	[244]
519	Bacteria	Proteobacteria	Gammaproteobacteria	Enterobacteriales	Enterobacteriaceae	*Klebsiella*	Enterobacteriaceae bacterium BTRH72	The petroleum oil-contaminated soil (17.2 g total hydrocarbon kg^{-1} soil) from a landfill used for deposition of crude oil-contaminated soil from oil pumping sites in Zistersdorf, Austria	[151]
520	Bacteria	Proteobacteria	Gammaproteobacteria	Enterobacteriales	Enterobacteriaceae	*Klebsiella*	*Klebsiella oxytoca*	An offshore "Sercina" oil field, located near the Kerkennah island, Tunisia	[245]

521	Bacteria	Proteobacteria	Gammaproteobacteria	Enterobacteriales	Enterobacteriaceae	*Kosakonia*	*Enterobacter* sp. ITSI60	The petroleum oil-contaminated soil (17.2 g total hydrocarbon kg^{-1} soil) from a landfill used for deposition of crude oil-contaminated soil from oil pumping sites in Zistersdorf, Austria	[151]
522	Bacteria	Proteobacteria	Gammaproteobacteria	Enterobacteriales	Enterobacteriaceae	*Pantoea*	*Pantoea* sp. ITSI8	The petroleum oil-contaminated soil (17.2 g total hydrocarbon kg^{-1} soil) from a landfill used for deposition of crude oil-contaminated soil from oil pumping sites in Zistersdorf, Austria	[151]
523	Bacteria	Proteobacteria	Gammaproteobacteria	Enterobacteriales	Enterobacteriaceae	*Serratia*	Uncultured bacterium	A continental high-temperature, waterflooded petroleum reservoir in the J-12 Unit at Huabei Oil field, Hebei Province, China	[161]
524	Bacteria	Proteobacteria	Gammaproteobacteria	Enterobacteriales	Enterobacteriaceae	*Siccibacter*	Enterobacteriaceae bacterium MIXRH30	The petroleum oil-contaminated soil (17.2 g total hydrocarbon kg^{-1} soil) from a landfill used for deposition of crude oil-contaminated soil from oil pumping sites in Zistersdorf, Austria	[151]

(continued)

Table 10.1 (Continued)

Index	Kingdom	Phylum	Class	Order	Family	Genus	Species	Isolation Source	Reference
525	Bacteria	Proteobacteria	Gammaproteobacteria	Methylococcales	Cycloclasticaceae	*Cycloclasticus*	*Cycloclasticus pugetii*	The surface sediments of Sinclair Inlet, Puget Sound, Bremerton, Washington, USA	[246]
526	Bacteria	Proteobacteria	Gammaproteobacteria	Methylococcales	Cycloclasticaceae	*Cycloclasticus*	*Cycloclasticus* sp. N3-PA321	PAH contaminated marine sediment Eagle Harbor Puget Sound, Washington, USA	[247]
527	Bacteria	Proteobacteria	Gammaproteobacteria	Methylococcales	Cycloclasticaceae	*Cycloclasticus*	*Cycloclasticus spirillensus*	Burrow wall sediments of benthic macrofauna (mollusc) at the intertidal zone of Lowes Cove, Maine, USA	[214]
528	Bacteria	Proteobacteria	Gammaproteobacteria	Oceanospirillales	Alcanivoracaceae	*Alcanivorax*	*Alcanivorax dieselolei*	Strains: oil contaminated surface water Bohai Sea near Shengli oil field, China/deep sea sediments from east Pacific Ocean (Pacific nodule region)	[248]
529	Bacteria	Proteobacteria	Gammaproteobacteria	Oceanospirillales	Alcanivoracaceae	*Alcanivorax*	*Alcanivorax jadensis*	An aerobic continuous culture with a suspension containing sediment of the intertidal zone of the North Sea coast (Jadebusen), Germany	[249]

530	Bacteria	Proteobacteria	Gammaproteobacteria	Oceanospirillales	Halomonadaceae	*Cobetia*	*Cobetia crustatorum*	Germany	[163]
531	Bacteria	Proteobacteria	Gammaproteobacteria	Oceanospirillales	Halomonadaceae	*Halomonas*	*Halomonas daqingensis*	Soil sample contaminated with crude oil from the Daqing Oil Field in Heilongjiang Province, northeastern China	[250]
532	Bacteria	Proteobacteria	Gammaproteobacteria	Oceanospirillales	Halomonadaceae	*Halomonas*	*Halomonas elongata*	Sediments collected in the Gulf of Fos, at the mouth of a petroleum refinery outlet chronically polluted by hydrocarbons, Mediterranean coast, 50 km north of Marseille, France	[235]
533	Bacteria	Proteobacteria	Gammaproteobacteria	Oceanospirillales	Halomonadaceae	*Halomonas*	*Halomonas titanicae*	Germany	[163]
534	Bacteria	Proteobacteria	Gammaproteobacteria	Oceanospirillales	Halomonadaceae	*Modicisalibacter*	*Modicisalibacter tunisiensis*	A sample of oil field-water injection collected in the Sidi Litayem area near Sfax, Tunisia	[251]
535	Bacteria	Proteobacteria	Gammaproteobacteria	Oceanospirillales	Marinomonadaceae	*Marinomonas*	*Marinomonas vaga*	Sediments collected in the Gulf of Fos, at the mouth of a petroleum refinery outlet chronically polluted by hydrocarbons, Mediterranean coast, 50 km north of Marseille, France	[235]

(*continued*)

Table 10.1 (Continued)

Index	Kingdom	Phylum	Class	Order	Family	Genus	Species	Isolation Source	Reference
536	Bacteria	Proteobacteria	Gammaproteobacteria	Oceanospirillales	Nitrincolaceae	*Neptunomonas*	*Neptunomonas naphthovorans*	Creosote contaminated sediment, Eagle Harbor Puget Sound, Washington, USA	[252]
537	Bacteria	Proteobacteria	Gammaproteobacteria	Oceanospirillales	Oceanospirillaceae	*Marinobacterium*	Bacterium enrichment culture clone EB1.12	Produced water samples of a high temperature and fractured chalk reservoir, the Ekofisk oil field, in block 2/4 of the Norwegian sector of the North Sea about 320 km southwest of Stavanger	[139]
538	Bacteria	Proteobacteria	Gammaproteobacteria	Oceanospirillales	Oceanospirillaceae	*Marinobacterium*	Uncultured *Marinobacterium* sp.	Operating oil well NR-6 in the Niibori oil field, which is located in the northeast part of the main island of Japan (39° 43′ N, 139° 53′ E)	[137]
539	Bacteria	Proteobacteria	Gammaproteobacteria	Oceanospirillales	Oceanospirillaceae	*Thalassolituus*	Uncultured *Thalassolituus* sp.	Operating oil well NR-6 in the Niibori oil field, which is located in the northeast part of the main island of Japan (39° 43′ N, 139° 53′ E)	[137]
540	Bacteria	Proteobacteria	Gammaproteobacteria	Oceanospirillales	OM182 clade		Uncultured bacterium	The Ban 876 Gas and Oil Field within the DaGang Area (39°32′N, 117°38′E), TianJin, China	[153]

541	Bacteria	Proteobacteria	Gammaproteobacteria	Oceanospirillales	Saccharospirillaceae	*Oleibacter*	*Oleibacter marinus*	Seawater collected at Pari Island (5.86u S 106.62u E) located near Jakarta, Indonesia	[253]
542	Bacteria	Proteobacteria	Gammaproteobacteria	Oceanospirillales	Saccharospirillaceae	*Oleispira*	*Oleispira antarctica*	Superficial seawater samples collected in the inlet Rod Bay, Ross Sea, Antarctica	[254]
543	Bacteria	Proteobacteria	Gammaproteobacteria	Oceanospirillales	Saccharospirillaceae	*Thalassolituus*	*Thalassolituus oleivorans*	Seawater/sediment samples that were collected in the harbor of Milazzo, Sicily, Italy	[255]
544	Bacteria	Proteobacteria	Gammaproteobacteria	Pseudomonadales	Moraxellaceae	*Acinetobacter*	Uncultured bacterium	Daqing Oil Field, China	[169]
545	Bacteria	Proteobacteria	Gammaproteobacteria	Pseudomonadales	Moraxellaceae	*Acinetobacter*	Uncultured bacterium	A high-temperature petroleum reservoir at an offshore oil field, China	[143]
546	Bacteria	Proteobacteria	Gammaproteobacteria	Pseudomonadales	Moraxellaceae	*Acinetobacter*	Uncultured bacterium	A continental high-temperature, waterflooded petroleum reservoir in the J-12 Unit at Huabei Oil field, Hebei Province, China	[161]
547	Bacteria	Proteobacteria	Gammaproteobacteria	Pseudomonadales	Moraxellaceae	*Alkanindiges*	*Alkanindiges illinoisensis*	Chronically crude oil-contaminated soil from an oil field in southern Illinois, USA	[256]
548	Bacteria	Proteobacteria	Gammaproteobacteria	Pseudomonadales	Moraxellaceae	*Alkanindiges*	*Alkanindiges illinoisensis*	Crude oil-contaminated soil from an oil field in southern Illinois	[256]

(*continued*)

Table 10.1 (Continued)

Index	Kingdom	Phylum	Class	Order	Family	Genus	Species	Isolation Source	Reference
549	Bacteria	Proteobacteria	Gammaproteobacteria	Pseudomonadales	Moraxellaceae	*Psychrobacter*	*Psychrobacter nivimaris*	Germany	[163]
550	Bacteria	Proteobacteria	Gammaproteobacteria	Pseudomonadales	Moraxellaceae	*Psychrobacter*	*Psychrobacter okhotskensis*	Germany	[163]
551	Bacteria	Proteobacteria	Gammaproteobacteria	Pseudomonadales	Pseudomonadaceae		Uncultured bacterium	Unknown	[145]
552	Bacteria	Proteobacteria	Gammaproteobacteria	Pseudomonadales	Pseudomonadaceae		Uncultured bacterium	A high-temperature petroleum reservoir at an offshore oil field, China	[143]
553	Bacteria	Proteobacteria	Gammaproteobacteria	Pseudomonadales	Pseudomonadaceae	*Pseudomonas*	Bacterium enrichment culture clone DT3-12	Daqing Oil Field, China	[169]
554	Bacteria	Proteobacteria	Gammaproteobacteria	Pseudomonadales	Pseudomonadaceae	*Pseudomonas*	Bacterium enrichment culture clone DT3-61	Daqing Oil Field, China	[169]
555	Bacteria	Proteobacteria	Gammaproteobacteria	Pseudomonadales	Pseudomonadaceae	*Pseudomonas*	*Pseudomonas aeruginosa*	Production water (an oil/water mixture) of the oil fields TPS "Thyna Petroleum Services," Tunisia	[178]
556	Bacteria	Proteobacteria	Gammaproteobacteria	Pseudomonadales	Pseudomonadaceae	*Pseudomonas*	*Pseudomonas pelagia*	Germany	[163]
557	Bacteria	Proteobacteria	Gammaproteobacteria	Pseudomonadales	Pseudomonadaceae	*Pseudomonas*	*Pseudomonas* sp. C2SS10	Production water (an oil/water mixture) of the oil fields TPS "Thyna Petroleum Services," Tunisia	[178]
558	Bacteria	Proteobacteria	Gammaproteobacteria	Pseudomonadales	Pseudomonadaceae	*Pseudomonas*	*Pseudomonas* sp. Da2	A low-temperature and low-salinity petroleum reservoir in Canada	[147]

559	Bacteria	Proteobacteria	Gammaproteobacteria	Pseudomonadales	Pseudomonadaceae	*Pseudomonas*	*Pseudomonas* sp. MIXRH13	The petroleum oil-contaminated soil (17.2 g total hydrocarbon kg^{-1} soil) from a landfill used for deposition of crude oil-contaminated soil from oil pumping sites in Zistersdorf, Austria	[151]
560	Bacteria	Proteobacteria	Gammaproteobacteria	Pseudomonadales	Pseudomonadaceae	*Pseudomonas*	*Pseudomonas* sp. SG-2	Sagara oil reservoir, Shizuoka Prefecture, Japan	[156]
561	Bacteria	Proteobacteria	Gammaproteobacteria	Pseudomonadales	Pseudomonadaceae	*Pseudomonas*	*Pseudomonas stutzeri*	Sagara oil reservoir, Shizuoka Prefecture, Japan	[156]
562	Bacteria	Proteobacteria	Gammaproteobacteria	Pseudomonadales	Pseudomonadaceae	*Pseudomonas*	Uncultured bacterium	Daqing Oil Field, China	[169]
563	Bacteria	Proteobacteria	Gammaproteobacteria	Pseudomonadales	Pseudomonadaceae	*Pseudomonas*	Uncultured bacterium	A continental high-temperature, waterflooded petroleum reservoir in the J-12 Unit at Huabei Oil field, Hebei Province, China	[161]
564	Bacteria	Proteobacteria	Gammaproteobacteria	Pseudomonadales	Pseudomonadaceae	*Pseudomonas*	Unidentified	Unknown	[170]
565	Bacteria	Proteobacteria	Gammaproteobacteria	PYR10d3			Uncultured bacterium	Daqing Oil Field, China	[169]
566	Bacteria	Proteobacteria	Gammaproteobacteria	Salinisphaerales	Solimonadaceae	*Polycyclovorans*	*Polycyclovorans algicola*	Nonaxenic laboratory culture of the marine diatom Skeletonema costatum CCAP1077/1C (origin, North Sea)	[175]

(continued)

Table 10.1 (Continued)

Index	Kingdom	Phylum	Class	Order	Family	Genus	Species	Isolation Source	Reference
567	Bacteria	Proteobacteria	Gammaproteobacteria	Thiotrichales	H_2-104-2		Uncultured bacterium	The Ban 876 Gas and Oil Field within the DaGang Area (39°32′N, 117°38′E), TianJin, China	[153]
568	Bacteria	Proteobacteria	Gammaproteobacteria	Uncultured			Uncultured bacterium	The Ban 876 Gas and Oil Field within the DaGang Area (39°32′N, 117°38′E), TianJin, China	[153]
569	Bacteria	Proteobacteria	Gammaproteobacteria	Uncultured			Uncultured gamma proteobacterium	The Statfjord Weld located in the Tampen Spur area in the Norwegian sector of the North Sea	[195]
570	Bacteria	Proteobacteria	Gammaproteobacteria	Vibrionales	Vibrionaceae	*Vibrio*	*Vibrio cyclitrophicus*	PAH contaminated marine sediment Eagle Harbor Puget Sound, Washington, USA	[247]
571	Bacteria	Proteobacteria	Gammaproteobacteria	Xanthomonadales	Nevskiaceae	*Hydrocarboniphaga*	Uncultured bacterium	A high-temperature petroleum reservoir at an offshore oil field, China	[143]
572	Bacteria	Proteobacteria	Gammaproteobacteria	Xanthomonadales	Nevskiaceae	*Hydrocarboniphaga*	Uncultured bacterium	A high-temperature petroleum reservoir at an offshore oil field, China	[143]
573	Bacteria	Proteobacteria	Gammaproteobacteria	Xanthomonadales	Rhodanobacteraceae	*Rhodanobacter*	*Rhodanobacter lindaniclasticus*	Hawaiian soils, Hawaii, USA	[157]

574	Bacteria	Proteobacteria	Gammaproteobacteria	Xanthomonadales	Xanthomonadaceae	*Pseudoxanthomonas*	*Pseudoxanthomonas* sp. ITRH31	The petroleum oil-contaminated soil (17.2 g total hydrocarbon kg^{-1} soil) from a landfill used for deposition of crude oil-contaminated soil from oil pumping sites in Zistersdorf, Austria	[151]
575	Bacteria	Proteobacteria	Gammaproteobacteria	Xanthomonadales	Xanthomonadaceae	*Pseudoxanthomonas*	*Stenotrophomonas* sp. MIXRI12	The petroleum oil-contaminated soil (17.2 g total hydrocarbon kg^{-1} soil) from a landfill used for deposition of crude oil-contaminated soil from oil pumping sites in Zistersdorf, Austria	[151]
576	Bacteria	Proteobacteria	Gammaproteobacteria	Xanthomonadales	Xanthomonadaceae	*Pseudoxanthomonas*	*Stenotrophomonas* sp. MIXRI12	The petroleum oil-contaminated soil (17.2 g total hydrocarbon kg^{-1} soil) from a landfill used for deposition of crude oil-contaminated soil from oil pumping sites in Zistersdorf, Austria	[151]

(*continued*)

Table 10.1 (Continued)

Index	Kingdom	Phylum	Class	Order	Family	Genus	Species	Isolation Source	Reference
577	Bacteria	Proteobacteria	Gammaproteobacteria	Xanthomonadales	Xanthomonadaceae	*Stenotrophomonas*	*Stenotrophomonas maltophilia*	Hawaiian soils, Hawaii, USA	[157]
578	Bacteria	Proteobacteria	Gammaproteobacteria	Xanthomonadales	Xanthomonadaceae	*Xylella*	*Pseudomonas* sp. ITRI24	The petroleum oil-contaminated soil (17.2 g total hydrocarbon kg^{-1} soil) from a landfill used for deposition of crude oil-contaminated soil from oil pumping sites in Zistersdorf, Austria	[151]
579	Bacteria	RBG-1 (Zixibacteria)					Uncultured bacterium	The Ban 876 Gas and Oil Field within the DaGang Area (39°32′N, 117°38′E), TianJin, China	[153]
580	Bacteria	Saccharibacteria					Uncultured bacterium	The Ban 876 Gas and Oil Field within the DaGang Area (39°32′N, 117°38′E), TianJin, China	[153]
581	Bacteria	Spirochaetae	Spirochaetes	Spirochaetales	Spirochaetaceae	*Sphaerochaeta*	Uncultured *Spirochaeta* sp.	Operating oil well NR-6 in the Niibori oil field, which is located in the northeast part of the main island of Japan (39° 43′ N, 139° 53′ E)	[137]
582	Bacteria	Spirochaetae	Spirochaetes	Spirochaetales	Spirochaetaceae	*Sphaerochaeta*	Unidentified	Unknown	[170]
583	Bacteria	Spirochaetae	Spirochaetes	Spirochaetales	Spirochaetaceae	Uncultured	Bacterium enrichment culture clone B312154	A disposal field that treats mixtures of crude oil-contaminated soil and oily sludge in the Shengli oil field, China	[152]

584	Bacteria	Spirochaetae	Spirochaetes	Spirochaetales	Spirochaetaceae	Uncultured	Spirochaetaceae bacterium enrichment culture clone B31131	A disposal field that treats mixtures of crude oil-contaminated soil and oily sludge in the Shengli oil field, China	[152]
585	Bacteria	Spirochaetae	Spirochaetes	Spirochaetales	Spirochaetaceae	Uncultured	Spirochaetaceae bacterium enrichment culture clone B31135	A disposal field that treats mixtures of crude oil-contaminated soil and oily sludge in the Shengli oil field, China	[152]
586	Bacteria	Spirochaetae	Spirochaetes	Spirochaetales	Spirochaetaceae	Uncultured	Spirochaetaceae bacterium enrichment culture clone B31155	A disposal field that treats mixtures of crude oil-contaminated soil and oily sludge in the Shengli oil field, China	[152]
587	Bacteria	Spirochaetae	Spirochaetes	Spirochaetales	Spirochaetaceae	Uncultured	Spirochaetaceae bacterium enrichment culture clone B31159	A disposal field that treats mixtures of crude oil-contaminated soil and oily sludge in the Shengli oil field, China	[152]
588	Bacteria	Spirochaetae	Spirochaetes	Spirochaetales	Spirochaetaceae	Uncultured	Spirochaetaceae bacterium enrichment culture clone B3116	A disposal field that treats mixtures of crude oil-contaminated soil and oily sludge in the Shengli oil field, China	[152]
589	Bacteria	Spirochaetae	Spirochaetes	Spirochaetales	Spirochaetaceae	Uncultured	Spirochaetaceae bacterium enrichment culture clone B31169	A disposal field that treats mixtures of crude oil-contaminated soil and oily sludge in the Shengli oil field, China	[152]

(*continued*)

Table 10.1 (Continued)

Index	Kingdom	Phylum	Class	Order	Family	Genus	Species	Isolation Source	Reference
590	Bacteria	Spirochaetae	Spirochaetes	Spirochaetales	Spirochaetaceae	Uncultured	Spirochaetaceae bacterium enrichment culture clone B312137	A disposal field that treats mixtures of crude oil-contaminated soil and oily sludge in the Shengli oil field, China	[152]
591	Bacteria	Spirochaetae	Spirochaetes	Spirochaetales	Spirochaetaceae	Uncultured	Spirochaetaceae bacterium enrichment culture clone B31296	A disposal field that treats mixtures of crude oil-contaminated soil and oily sludge in the Shengli oil field, China	[152]
592	Bacteria	Spirochaetae	Spirochaetes	Spirochaetales	Spirochaetaceae	Uncultured	Spirochaetaceae bacterium enrichment culture clone B31298	A disposal field that treats mixtures of crude oil-contaminated soil and oily sludge in the Shengli oil field, China	[152]
593	Bacteria	Spirochaetes	Spirochaetia	Spirochaetales	Spirochaetaceae	*Sediminispirochaeta*	*Sediminispirochaeta smaragdinae*	The production waters of the Emeraude oil fields, Congo	[257]
594	Bacteria	Synergistetes	Synergistia	Synergistales	Synergistaceae	*Acetomicrobium*	*Acetomicrobium thermoterrenum*	Production fluid from the Red Wash oil field in Utah, USA	[258]
595	Bacteria	Synergistetes	Synergistia	Synergistales	Synergistaceae	*Aminivibrio*	Uncultured bacterium	Unknown	[145]
596	Bacteria	Synergistetes	Synergistia	Synergistales	Synergistaceae	*Aminivibrio*	Uncultured *Synergistes* sp.	Production water sample from the mesothermic and highly degraded Schrader Bluff petroleum field in Alaska's North Slope region, USA	[146]

597	Bacteria	Synergistetes	Synergistia	Synergistales	Synergistaceae	*Anaerobaculum*	Bacterium enrichment culture clone EB32.1	Produced water samples of a high temperature and fractured chalk reservoir, the Ekofisk oil field, in block 2/4 of the Norwegian sector of the North Sea about 320 km southwest of Stavanger	[139]
598	Bacteria	Synergistetes	Synergistia	Synergistales	Synergistaceae	*Anaerobaculum*	*Uncultured Anaerobaculum* sp.	Two production wells (AR-80 and OR-79) in the Yabase oil field, a formation of tuffaceous sandstone of Miocene–Pliocene age, located around 1293–1436 m under the surface, with in situ temperature of 40–82°C and pressure of 5 MPa, Japan	[144]
599	Bacteria	Synergistetes	Synergistia	Synergistales	Synergistaceae	Syner-01	Uncultured bacterium	Unknown	[145]
600	Bacteria	Synergistetes	Synergistia	Synergistales	Synergistaceae	*Thermovirga*	Bacterium enrichment culture clone EB14.9	Produced water samples of a high temperature and fractured chalk reservoir, the Ekofisk oil field, in block 2/4 of the Norwegian sector of the North Sea about 320 km southwest of Stavanger	[139]

(continued)

Table 10.1 (Continued)

Index	Kingdom	Phylum	Class	Order	Family	Genus	Species	Isolation Source	Reference
601	Bacteria	Synergistetes	Synergistia	Synergistales	Synergistaceae	*Thermovirga*	Bacterium enrichment culture clone PW20.10B	Produced water samples of a high temperature and fractured chalk reservoir, the Ekofisk oil field, in block 2/4 of the Norwegian sector of the North Sea about 320 km southwest of Stavanger	[139]
602	Bacteria	Synergistetes	Synergistia	Synergistales	Synergistaceae	*Thermovirga*	*Thermovirga lienii*	An oil reservoir in the North Sea	[259]
603	Bacteria	Synergistetes	Synergistia	Synergistales	Synergistaceae	*Thermovirga*	Uncultured bacterium	Unknown	[145]
604	Bacteria	Synergistetes	Synergistia	Synergistales	Synergistaceae	*Thermovirga*	Uncultured *Synergistetes bacterium*	Operating oil well NR-6 in the Niibori oil field, which is located in the northeast part of the main island of Japan (39° 43′ N, 139° 53′ E)	[137]
605	Bacteria	Synergistetes	Synergistia	Synergistales	Synergistaceae	*Thermovirga*	Uncultured *Thermovirga* sp.	A high-temperature oil-bearing formation in the North Sea	[172]
606	Bacteria	Synergistetes	Synergistia	Synergistales	Synergistaceae	*Thermovirga*	Uncultured *Thermovirga* sp.	A high-temperature oil-bearing formation in the North Sea	[172]
607	Bacteria	Synergistetes	Synergistia	Synergistales	Synergistaceae	*Thermovirga*	Uncultured *Thermovirga* sp.	Operating oil well NR-6 in the Niibori oil field, which is located in the northeast part of the main island of Japan (39° 43′ N, 139° 53′ E)	[137]

608	Bacteria	Synergistetes	Synergistia	Synergistales	Synergistaceae	Uncultured	Synergistaceae bacterium enrichment culture clone B31171	A disposal field that treats mixtures of crude oil-contaminated soil and oily sludge in the Shengli oil field, China	[152]
609	Bacteria	Tenericutes	Mollicutes	NB1-n			Erysipelotrichaceae bacterium enrichment culture clone B312101	A disposal field that treats mixtures of crude oil-contaminated soil and oily sludge in the Shengli oil field, China	[152]
610	Bacteria	Thermodesulfo bacteria	Thermodesulfobacteria	Thermodesulfobacteriales	Thermodesulfobacteriaceae	*Thermodesulfobacterium*	Uncultured *Thermodesulfobacterium* sp.	Operating oil well NR-6 in the Niibori oil field, which is located in the northeast part of the main island of Japan (39° 43′ N, 139° 53′ E)	[137]
611	Bacteria	Thermotogae	Thermotogae				Uncultured *Thermotogae bacterium*	Operating oil well NR-6 in the Niibori oil field, which is located in the northeast part of the main island of Japan (39° 43′ N, 139° 53′ E)	[137]
612	Bacteria	Thermotogae	Thermotogae	EM3			Uncultured bacterium	A continental high-temperature, waterflooded petroleum reservoir in the J-12 Unit at Huabei Oil field, Hebei Province, China	[161]

(*continued*)

Table 10.1 (Continued)

Index	Kingdom	Phylum	Class	Order	Family	Genus	Species	Isolation Source	Reference
613	Bacteria	Thermotogae	Thermotogae	EM3			Uncultured *Thermotogae bacterium*	Operating oil well NR-6 in the Niibori oil field, which is located in the northeast part of the main island of Japan (39° 43′ N, 139° 53′ E)	[137]
614	Bacteria	Thermotogae	Thermotogae	Kosmotogales	Kosmotogaceae	*Kosmotoga*	Kosmotoga olearia TBF 19.5.1	The Troll B platform in the North Sea	[260]
615	Bacteria	Thermotogae	Thermotogae	Kosmotogales	Kosmotogaceae	*Kosmotoga*	*Kosmotoga shengliensis*	Oil-production fluid from Shengli oil field, China	[261]
616	Bacteria	Thermotogae	Thermotogae	Kosmotogales	Kosmotogaceae	*Mesotoga*	Thermotogaceae bacterium enrichment culture clone B31121	A disposal field that treats mixtures of crude oil-contaminated soil and oily sludge in the Shengli oil field, China	[152]
617	Bacteria	Thermotogae	Thermotogae	Kosmotogales	Kosmotogaceae	*Mesotoga*	Thermotogaceae bacterium enrichment culture clone B31176	A disposal field that treats mixtures of crude oil-contaminated soil and oily sludge in the Shengli oil field, China	[152]
618	Bacteria	Thermotogae	Thermotogae	Kosmotogales	Kosmotogaceae	*Mesotoga*	Thermotogaceae bacterium enrichment culture clone B312111	A disposal field that treats mixtures of crude oil-contaminated soil and oily sludge in the Shengli oil field, China	[152]

619	Bacteria	Thermotogae	Thermotogae	Kosmotogales	Kosmotogaceae	*Mesotoga*	Thermotogaceae bacterium enrichment culture clone B312114	A disposal field that treats mixtures of crude oil-contaminated soil and oily sludge in the Shengli oil field, China	[152]
620	Bacteria	Thermotogae	Thermotogae	Kosmotogales	Kosmotogaceae	*Mesotoga*	Thermotogaceae bacterium enrichment culture clone B312116	A disposal field that treats mixtures of crude oil-contaminated soil and oily sludge in the Shengli oil field, China	[152]
621	Bacteria	Thermotogae	Thermotogae	Kosmotogales	Kosmotogaceae	*Mesotoga*	Thermotogaceae bacterium enrichment culture clone B312127	A disposal field that treats mixtures of crude oil-contaminated soil and oily sludge in the Shengli oil field, China	[152]
622	Bacteria	Thermotogae	Thermotogae	Kosmotogales	Kosmotogaceae	*Mesotoga*	Thermotogaceae bacterium enrichment culture clone B312130	A disposal field that treats mixtures of crude oil-contaminated soil and oily sludge in the Shengli oil field, China	[152]
623	Bacteria	Thermotogae	Thermotogae	Petrotogales	Petrotogaceae	*Defluviitoga*	Uncultured *Thermotogaceae bacterium*	Two production wells (AR-80 and OR-79) in the Yabase oil field, a formation of tuffaceous sandstone of Miocene–Pliocene age, located around 1293–1436 m under the surface, with in situ temperature of 40–82°C and pressure of 5 MPa, Japan	[144]

(continued)

Table 10.1 (Continued)

Index	Kingdom	Phylum	Class	Order	Family	Genus	Species	Isolation Source	Reference
624	Bacteria	Thermotogae	Thermotogae	Petrotogales	Petrotogaceae	*Oceanotoga*	*Oceanotoga teriensis*		[262]
625	Bacteria	Thermotogae	Thermotogae	Petrotogales	Petrotogaceae	*Petrotoga*	Bacterium enrichment culture clone EB4.2	Produced water samples of a high temperature and fractured chalk reservoir, the Ekofisk oil field, in block 2/4 of the Norwegian sector of the North Sea about 320 km southwest of Stavanger	[139]
626	Bacteria	Thermotogae	Thermotogae	Petrotogales	Petrotogaceae	*Petrotoga*	Bacterium enrichment culture clone PW20.3	Produced water samples of a high temperature and fractured chalk reservoir, the Ekofisk oil field, in block 2/4 of the Norwegian sector of the North Sea about 320 km southwest of Stavanger	[139]
627	Bacteria	Thermotogae	Thermotogae	Petrotogales	Petrotogaceae	*Petrotoga*	*Petrotoga halophila*	Well TBM111 of the Tchibouella oil field, Congo	[263]
628	Bacteria	Thermotogae	Thermotogae	Petrotogales	Petrotogaceae	*Petrotoga*	*Petrotoga mexicana*	Oil/water mixtures taken from production wellheads (21-D) in Tabasco, Gulf of Mexico, Mexico	[264]
629	Bacteria	Thermotogae	Thermotogae	Petrotogales	Petrotogaceae	*Petrotoga*	*Petrotoga mobilis*	Production water taken from the water separator tanks on offshore oil platforms, North Sea	[265]

630	Bacteria	Thermotogae	Thermotogae	Petrotogales	Petrotogaceae	*Petrotoga*	*Petrotoga* sp. AR80	Yabase oil reservoir, Akita, Japan (39°43′ north latitude, 140°06′ east longitude)	[266]
631	Bacteria	Thermotogae	Thermotogae	Petrotogales	Petrotogaceae	*Petrotoga*	Uncultured *Petrotoga* sp.	Operating oil well NR-6 in the Niibori oil field, which is located in the northeast part of the main island of Japan (39° 43′ N, 139° 53′ E)	[137]
632	Bacteria	Thermotogae	Thermotogae	Petrotogales	Petrotogaceae	*Petrotoga*	Uncultured *Petrotoga* sp.	Two production wells (AR-80 and OR-79) in the Yabase oil field, a formation of tuffaceous sandstone of Miocene–Pliocene age, located around 1293–1436 m under the surface, with in situ temperature of 40–82°C and pressure of 5 MPa, Japan	[144]
633	Bacteria	Thermotogae	Thermotogae	Thermotogales	Fervidobacteriaceae	*Fervidobacterium*	Thermotogaceae bacterium enrichment culture clone B3112	A disposal field that treats mixtures of crude oil-contaminated soil and oily sludge in the Shengli oil field, China	[152]

(continued)

Table 10.1 (Continued)

Index	Kingdom	Phylum	Class	Order	Family	Genus	Species	Isolation Source	Reference
634	Bacteria	Thermotogae	Thermotogae	Thermotogales	Fervidobacteriaceae	*Fervidobacterium*	Uncultured bacterium	A high-temperature petroleum reservoir at an offshore oil field, China	[143]
635	Bacteria	Thermotogae	Thermotogae	Thermotogales	Fervidobacteriaceae	*Fervidobacterium*	Uncultured bacterium	A high-temperature petroleum reservoir at an offshore oil field, China	[143]
636	Bacteria	Thermotogae	Thermotogae	Thermotogales	Fervidobacteriaceae	*Fervidobacterium*	Uncultured bacterium	A continental high-temperature, waterflooded petroleum reservoir in the J-12 Unit at Huabei Oil field, Hebei Province, China	[161]
637	Bacteria	Thermotogae	Thermotogae	Thermotogales	Fervidobacteriaceae	*Thermosipho*	Uncultured *Thermosipho* sp.	Operating oil well NR-6 in the Niibori oil field, which is located in the northeast part of the main island of Japan (39° 43′ N, 139° 53′ E)	[137]
638	Bacteria	Thermotogae	Thermotogae	Thermotogales	Thermotogaceae	*Pseudothermotoga*	Uncultured bacterium	A high-temperature petroleum reservoir at an offshore oil field, China	[143]
639	Bacteria	Thermotogae	Thermotogae	Thermotogales	Thermotogaceae	*Pseudothermotoga*	Uncultured *Thermotoga* sp.	Operating oil well NR-6 in the Niibori oil field, which is located in the northeast part of the main island of Japan (39° 43′ N, 139° 53′ E)	[137]

640	Bacteria	Thermotogae	Thermotogae	Thermotogales	Thermotogaceae	*Pseudothermotoga*	Uncultured *Thermotoga* sp.	Two production wells (AR-80 and OR-79) in the Yabase oil field, a formation of tuffaceous sandstone of Miocene–Pliocene age, located around 1293–1436 m under the surface, with in situ temperature of 40–82°C and pressure of 5 MPa, Japan	[144]
641	Bacteria	Thermotogae	Thermotogae	Thermotogales	Thermotogaceae	*Thermotoga*	Bacterium enrichment culture clone EB40.9	Produced water samples of a high temperature and fractured chalk reservoir, the Ekofisk oil field, in block 2/4 of the Norwegian sector of the North Sea about 320 km southwest of Stavanger	[139]
642	Bacteria	Thermotogae	Thermotogae	Thermotogales	Thermotogaceae	*Thermotoga*	Bacterium enrichment culture clone PW30.2B	Produced water samples of a high temperature and fractured chalk reservoir, the Ekofisk oil field, in block 2/4 of the Norwegian sector of the North Sea about 320 km southwest of Stavanger	[139]

(continued)

Table 10.1 (Continued)

Index	Kingdom	Phylum	Class	Order	Family	Genus	Species	Isolation Source	Reference
643	Bacteria	Thermotogae	Thermotogae	Thermotogales	Thermotogaceae	*Thermotoga*	*Thermotoga naphthophila*	Production fluid of the Kubiki oil reservoir in Niigata, Japan	[267]
644	Bacteria	Thermotogae	Thermotogae	Thermotogales	Thermotogaceae	*Thermotoga*	*Thermotoga petrophila*	Kubiki oil reservoir located near the coast of the Sea of Japan in Niigata prefecture, Japan	[150]
645	Bacteria	Thermotogae	Thermotogae	Thermotogales	Thermotogaceae	*Thermotoga*	Uncultured bacterium	A high-temperature petroleum reservoir at an offshore oil field, China	[143]
646	Bacteria	Thermotogae	Thermotogae	Thermotogales	Thermotogaceae	*Thermotoga*	Uncultured *Thermotoga* sp.	A high-temperature oil-bearing formation in the North Sea	[172]
647	Bacteria	WS1					Uncultured candidate division WS1 bacterium	Operating oil well NR-6 in the Niibori oil field, which is located in the northeast part of the main island of Japan (39° 43′ N, 139° 53′ E)	[137]
648	Bacteria	WS2					Uncultured bacterium	Operating oil well NR-6 in the Niibori oil field, which is located in the northeast part of the main island of Japan (39° 43′ N, 139° 53′ E)	[137]
649	Bacteria	WS2					Uncultured bacterium	Daqing Oil Field, China	[169]

650	Bacteria	WS6	Uncultured candidate division WS6 bacterium	Operating oil well NR-6 in the Niibori oil field, which is located in the northeast part of the main island of Japan (39° 43′ N, 139° 53′ E)	[137]
651	Bacteria	WS6	Uncultured candidate division WS6 bacterium	Production water sample from the mesothermic and highly degraded Schrader Bluff petroleum field in Alaska's North Slope region, USA	[146]
652	Bacteria	WWE3	Bacterium enrichment culture clone B31295	A disposal field that treats mixtures of crude oil-contaminated soil and oily sludge in the Shengli oil field, China	[152]
653	Bacteria	WWE3	Uncultured bacterium	The Ban 876 Gas and Oil Field within the DaGang Area (39°32′N, 117°38′E), TianJin, China	[153]

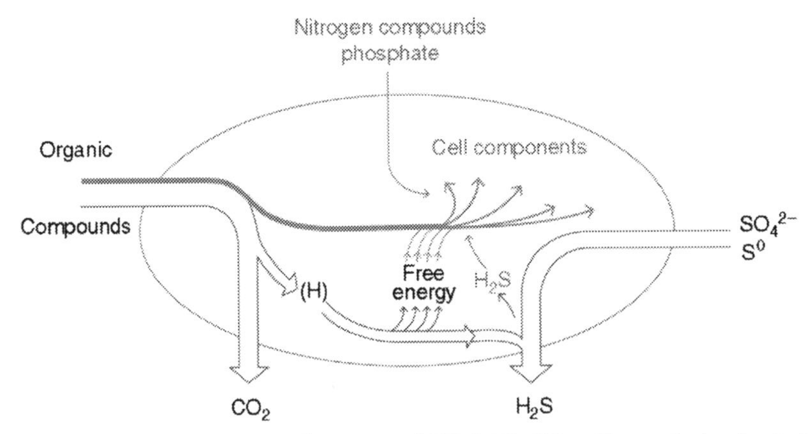

Figure 10.3 Schematic representation for action of SRB [319]. *SRB*, sulfate-reducing bacteria.

Further details about the methanogens from the oil fields can be found in literature [6,7]. Details about the methanogenesis is available elsewhere [301].

10.5.2.2 Sulfate-Reducing Bacteria (SRB)

SRB are the prokaryotic microorganisms, which gain energy via reducing sulfate (SO_4^{2-}) or partially oxidized compounds such as sulfite (SO_3^{2-}) and thiosulfate ($S_2O_3^{2-}$) in a non-assimilatory manner via anaerobic respiration [302] (see Fig. 10.3). In other words, these microorganisms (bacteria and archaea referred to as prokaryotes) utilize the sulfate as the terminal electron acceptor rather than oxygen for respiration [303,304]. However, it has been proved that SRB are able to tolerate the transient presence of oxygen [305–310]. Most of the prokaryotes capable to reduce the sulfate are bacteria [310], which supports the name "sulfate-reducing bacteria" rather than "sulfate-reducing prokaryotes." SRB reduce the majority of different terminal electron acceptors including inorganic sulfur compounds and various other inorganic and organic compounds [311–316]. It has been proved that SRB can utilize as wide range of low molecular organic compounds such as lactate, propionate, acetate, pyruvate, succinate, sugars, ethanol, etc. for growth through which SO_4^{2-} reduces to H_2S [304,317]. However, Barton and Fauque [310] stated that more than 100 compounds including sugars (e.g., fructose, glucose, etc.), monocarboxylic acids (e.g., acetate, propionate, butyrate, etc.), dicarboxylic acids (fumarate, succinate, malate, etc.), amino acids (glycine, serine, alanine, etc.), alcohols (e.g., methanol, ethanol, etc.) and aromatic compounds (benzoate, phenol, etc.) are potential electron donors for SRB [313,316]. Moreover, hydrocarbons in petroleum may also serve as electron donors for SRBs [304,318].

SRB are considered as one of the oldest bacterial life on the earth and their activity traces back to more than 3.5 billion years ago [310,320]. These prokaryotes have adopted themselves to almost all the possible ecosystems on the earth such as oil field environments [310]. The first microorganism isolated from the oil reservoirs were SRB and were recovered by Bastin et al. [67].

In 2009, Barton and Fauque [310] specified that 220 species of 60 genera of SRB have been described until now. They belong to four phyla within the bacteria and two phyla within the archaea [315,321–325]:

1. Proteobacteria;
2. Firmicutes;
3. Thermodesulfobacteria;
4. Nitrospirae; and

And two phyla of archaea:

1. Euryarchaeota;
2. Crenarchaeota;

10.5.2.3 Fermentative Microorganisms

Fermentation is defined as an anaerobic catabolism of a reduced carbon source to produce ATP within a strict internal redox balance [326]. A diversity of end-products can be generated by microorganisms through fermentation such as carbon dioxide, ethanol, lactate, butyrate, acetate, and propionate [326]. Two major forms of microbial catabolism are fermentation and respiration [301]. In the former, all the electrons from the organic substrates are put back onto the organic products, while in the latter, the removed electrons are finally transferred to an inorganic electron acceptor such as oxygen or nitrate [301]. Several respiratory microorganisms have been recovered from oil fields [177,226,228,286,327–334]. The main difference between the fermentation and respiration (the electron transport chain-dependent processes) is that the former is less energetically efficient due to retainment of lots of potential chemical energy in most of the end products [326]. Fermentation is associated with a negative ΔG due to breakdown of a large molecule to some more stable smaller products [301].

Youssef et al. [6] mentioned about isolation of some fermentative microorganisms from oil reservoirs. These authors specified dual metabolic abilities, i.e., fermentative and respiratory for many microorganisms in this group. The majority of the thermophilic fermentative microorganisms isolated from oil fields are belonging to phylum *Thermotogae* [263–265,267,335–338] and also the order *Thermoanaerobacteriales* [125,200,202,203,339,340] within the class *Clostridia*, phylum *Firmicutes* [6]. Acetate and hydrogen are the end products of the *Thermotogae* phylum. Moreover, Youssef et al. [6] mentioned other isolated species capable of fermenting organic acid [258] and amino acid [259]. In addition some other fermentative bacteria have been isolated from the oil fields [147,171,194,204,341–344]. More details are available in literature [6,345].

10.5.2.4 Nitrate-Reducing Bacteria

It is well established that nitrate or nitrite is effective in controlling souring [138,346–360]. This increased the interest in nitrate or nitrite-reducing bacteria (NRB) in oil fields [345,346,361,362]. NRB can be classified in two main classes of heterotrophic NRB (hNRB) and sulfide-oxidizing NRB (SO-NRB) [361]. Various NRB have been isolated from oil fields [186,333,334,350,363–367]. More details about the NRB and their action can be found in Section 10.12.14 and also literature [6,345].

10.5.2.5 Iron-Reducing Bacteria (IRB)

IRB have been isolated from several oil reservoirs [177,286,334,368,369]. Further details about IRB is available in literature [6,345].

10.5.3 Microorganisms Based on Metabolic Processes

Considering the metabolic processes, microorganisms are classified as aerobic and anaerobic. Through the metabolism, which is a chemical process, a living organisms gain energy and nutrient to continue living [63]. Aerobes employ oxygen (O_2) as the terminal electron acceptor while the anaerobes incorporate inorganic compounds such as sulfate, nitrate, or carbon dioxide [63]. The mentioned four compounds supply the metabolic needs for almost all reservoir microorganisms [63].

Volk et al. [22] mentioned the higher potential of anaerobic processes in MEOR considering the particular microbiology of the oil field waters, which is vastly biased toward the anaerobic microorganisms. For example, the studies on the oil fields in Western Canada revealed none of the oil field waters contain high numbers of aerobic microorganisms [370,371]. Aerobic condition is not common in the petroleum reservoirs. The aerobic microorganisms would have consumed the existed oxygen already. The other point is that iron and sulfur, which are commonplace in oil reservoirs, deplete the free oxygen too [22].

10.5.3.1 Aerobic Microorganisms

Aerobes metabolism is the most energy efficient of the mentioned four processes and can reproduce at a rapid rate [63]. This chemical process can be represented as

$$\text{oil} + O_2 + \text{bacteria} \rightarrow CO_2 + H_2O + \text{more bacteria} \tag{10.4}$$

In a few early filed trials, aerobic bacteria were utilized. Injection of oxygen as well as nutrients to stimulate the hydrocarbon metabolism was an early method for additional oil recovery [24,372,373]. In this respect, several researchers provided evidences linking microbial activity with oil recovery [18,275,276,374–386]. Most of the underground wells lack oxygen. Injection of oxygen to the well may result in metal corrosion and damages to the equipment and downhole piping [24]. Moreover, oxygen is an electron acceptor compound that may cause imbalances in the microbial environment [31]. The undesired effects of the air injection into the reservoir and doubts in their effectiveness dispirited their utilization [23]; however, still there are supporters for aerobic MEOR applications [18,387].

10.5.3.2 Anaerobic Microorganisms

As it was mentioned before, most of anaerobes available on oil reservoirs utilize carbon dioxide (CO_2), sulfate (SO_4^{2-}), or nitrate (NO_3^-) [63]. Methanogens incorporate CO_2 for their metabolic processes and produce methane (CH_4):

$$\text{oil} + CO_2 + \text{bacteria} \rightarrow CH_4 + \text{more bacteria} \tag{10.5}$$

This biodegradation process is the least energy efficient of the mentioned four metabolic processes [63].

SRB employ the sulfate (SO_4^{2-}) as the terminal electron acceptor. Through this metabolic process, sulfate is reduced to hydrogen sulfide, which is a notorious bioproduct due its deleterious corrosive effects on equipment and pipelines.

$$\text{oil} + SO_4^{2-} + \text{bacteria} \rightarrow CO_2 + H_2S + \text{more bacteria} \tag{10.6}$$

Nitrate-Reducing Bacteria (NRB) employ nitrate (NO_3^-) as the terminal electron acceptor for the metabolic process, which is called denitrification [63]. This process is the most energy efficient among the anaerobic processes and involves the stepwise reduction of nitrate (NO_3^-) to nitrite (NO_2^-), nitric oxide (NO), and nitrogen (N_2) [63]:

$$\text{oil} + NO_3^- + \text{bacteria} \rightarrow CO_2 + N_2 + \text{more bacteria} \tag{10.7}$$

In case of injecting adequate amount of nitrate, the microbial activity shifts to denitrification. As nitrate reduction is more energy efficient, NRB can outcompete the SRB so that H_2S production will reduce. More details are provided in Section 10.12.14.

Based on literature, the capacity of some bacteria to anaerobically metabolize hydrocarbons by microorganisms was not recognized until the late 1980 [388]. Far ahead, researchers reported that microorganisms can anaerobically degrade the oil in the subsurface within the reservoir [389,390]. More details are provided in Section 10.11.1. Most of the field trials with the high percentage of success have used anaerobic microorganisms [24]. There is an active area of research on anaerobic extremophiles such as the ones tolerant to extreme salinity (halophiles), pressure (piezophiles or barophiles), and temperature (thermophiles) to develop better adaptation to the reservoir harsh condition [391–394].

10.6 MICROBE SELECTION FOR MEOR

For a successful MEOR, it is mandatory to choose the right candidate of microorganisms to fulfill the desired objective by adequately production of target bioproducts. Biochemical production by different microorganisms will be debated in details in sections 10.1 to 10.5. Acquiring a thorough knowledge about the physicochemical condition of the reservoir is of great importance in choosing the right microorganisms. The right candidate not only should be capable to adapt itself to the reservoir condition, but also should produce the desired bioproducts in adequate volumes. Majority of the successful MEOR field trials have used anaerobic bacteria [23]. Lazar et al. [108] mentioned four different sources suitable to isolate microorganisms beneficial for MEOR processes:

1. Formation waters;
2. Sediments from formation water purification plants;
3. Sludge from biogas operations; and
4. Effluents from sugars.

Both pure and mixed cultures have been used in field trials. Youssef et al. [6] mentioned the improvement in the technology of MEOR as using mixture cultures adapted to the nutrients and reservoir condition such as temperature and pressure and also incorporating larger volumes of nutrients [105–108,110,111,113,114]. Adetunji [84] specified that mixed cultures have exhibited higher efficiency in enhancing oil recovery. There have been several different microorganisms used in field trials. Some of them are listed in the following:

1. *Bacillus*. This bacterium commonly produces biosurfactants, bioalcohols, and biogases [395,396].
2. *Clostridium*. This bacterium commonly produces bioacids and biogases. The process of methane production is referred to as methanogenesis [25,395,396]. This bacterium has been proved to be effective in both sandstone and carbonate reservoirs [25,395,397]. For the carbonate reservoirs, matrix acidizing to improve the permeability as well as gas generation for displacing oil have been successful mechanisms incorporating this bacterium. For the sandstone reservoirs, *clostridium* utilization has been successful to lower the oil viscosity.
3. *Pseudomonas*. This bacterium produces biosurfactants and biopolymers and is effective in permeability profile modification [396].
4. SRB: These bacteria result in biodegradation of oil large molecules to reduce the oil viscosity. In addition, they produce methane through methanogenesis [25].

Among the mentioned microorganisms, *Bacillus* and more frequently *Clostridium* are the most common ones. These microorganisms have exhibited higher percentage of success in field trials [24].

10.7 NUTRIENTS

Providing the nutrients is the main expense in MEOR trials. The nutrients are very important as the right combination and quantity has a key role in MEOR success. The main considerations to select the nutrients are the desired outcome and the involving organisms [6]. The need to the carbon source can be satisfied either by exogenous (usually sugar) or indigenous (crude oil itself) sources. Molasses (or black treacle, is a viscous product resulting from refining sugarcane or sugar beets into sugar) are the most commonly used carbon source as it is widely available, its injection to the well is an easy process, and contains essential minerals and vitamins [24,84]. Updegraff and Wren [398] were the first researchers, who proposed using molasses as the substrate. Although using ex situ carbon sources may induce more microbial activity, the carbon source might be expensive and it is of more interest in economical point of view to utilize microorganisms that mainly consumes residual oil as their carbon source [22]. Moreover, this would be excellent for heavy oil production as this process reduces the carbon chain of heavy oil and consequently increases the crude quality [84,399,400]. However, in case of utilizing the in situ oil as the carbon source, the generation of the by-products noticeably retards and the growth can be very slow [25,84]. The other necessary nutrients are inorganic nitrates and phosphorous salts. These compounds are usually provided by fertilizers such as ammonium phosphate ($(NH_4)_3PO_4$), superphosphate ($Ca(H_2PO_4)_2$), ammonium nitrate (NH_4NO_3), and sodium nitrate ($NaNO_3$) [25].

For selective plugging and modifying the permeability profile, biomass production is the main MEOR solution. Supplying nitrate, which is the electron acceptor, can maximize the biomass generation [6]. The other example to show the influence of nutrients on the microbial activity is *Leuconostoc* sp., which generates dextran (a type of biopolymer) only when sucrose is supplied [401,402]. In cases intended to produce biosurfactant, there should be a delicate balance between the carbon and nitrogen sources [6]. It is reported that limiting the nitrate source can promote the surfactin production by *Bacillus subtilis*, biosurfactant by *Candida tropicalis*, and rhamnolipid by *Pseudomonas* sp. [403–405]. As an example for nutrients, laboratory experiments have shown that corn steep liquor (carbon source) along with diammonium phosphate)nitrogen and phosphorus source) is an effective nutrient for microbial growth for efficient plugging scenarios [406]. Finding the optimum nutrient blends to stimulate the target bacteria is of great importance. For aerobic microorganisms, oxygen is the other essential nutrient. As it was debated before, injection of oxygen is associated with some adverse effects on the equipment and pipeline. The other issue is that oxygen solubility in water is limited [407].

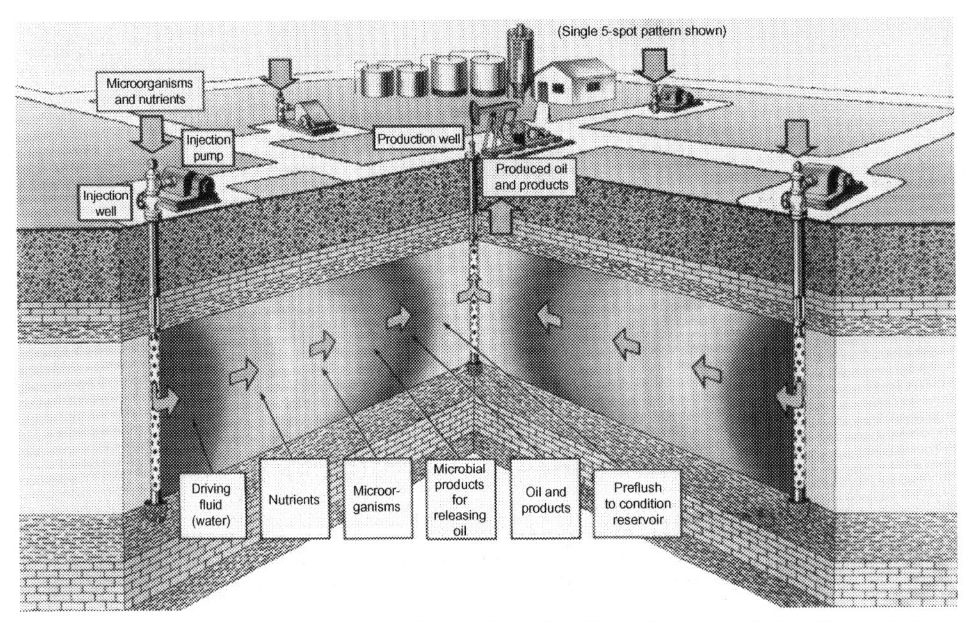

Figure 10.4 Schematic representation of microbial flooding. *Courtesy of the National Energy Technology Laboratory and the U.S. Department of Energy.*

10.8 MEOR APPLYING APPROACHES IN FIELD

10.8.1 Microbial Flooding

This method employs the effects of microbial metabolites for enhancing the oil recovery. In this approach, by adding the nutrients to the water flood the indigenous microorganisms are stimulated to produce desired bioproducts for mobilizing the residual oil or modifying the flow paths of the waterflood (see Fig. 10.4). In cases in which the target microorganisms are not present, they will be coinjected along with the nutrients via the injection wells. The mobilized oil will be produced in the producing wells. Several studies denotes the effectivity of this approach [11,76,108,408–410] and being economically feasible [108,409].

10.8.2 Cyclic Microbial Recovery

For cyclic microbial recovery, the solution containing the nutrients such as fermentable carbohydrates like molasses (and microorganisms in cases the indigenous microorganisms are not present) is injected down a well, which is close to its economic limit. Depending on the permeability and depth of the reservoir, injection time is different. After that, the well will be shut in for days or weeks before fluid is allowed to flow, which is referred to as the incubation period. Meanwhile, the microorganisms will produce the desired metabolites to facilitate the oil movement in the porous media and consequently increase the oil recovery. At the end of the incubation period, the well is backed to production and oil along with the produced biochemicals will be produced. The injection rate and also the kinetics of the microbial process will determine the area to be affected by bacteria [407]. This process might be repeated several times to maximize the oil recovery, hence, this process has been referred as microbial huff and puff too [411].

McInenery et al. [412] categorized the cyclic oil recovery in two main categories of (1) well stimulation and (2) fermentative microbially enhanced waterflooding processes. The difference is that in the latter, the injected nutrients (and microorganisms in cases the indigenous microorganisms are not present) are injected deep into the reservoir rather than to the well vicinity. Well stimulation approach was used in several early field trials due its simplicity [78,413,414]. Hitzman [78] specified this approach has been the most effective in carbonate well with a temperature range of 35–40°C, oil gravity of 875–965 kg/m^3, and salinities less than 100 g/L. It should be noted there have been inconsistencies in results acquired from several field trials and also little changes in the oil production from sandstone reservoirs has been observed [78,87] (Fig. 10.5).

10.9 MEOR METHODS

There are three general MEOR applying approaches, which are debated in the following.

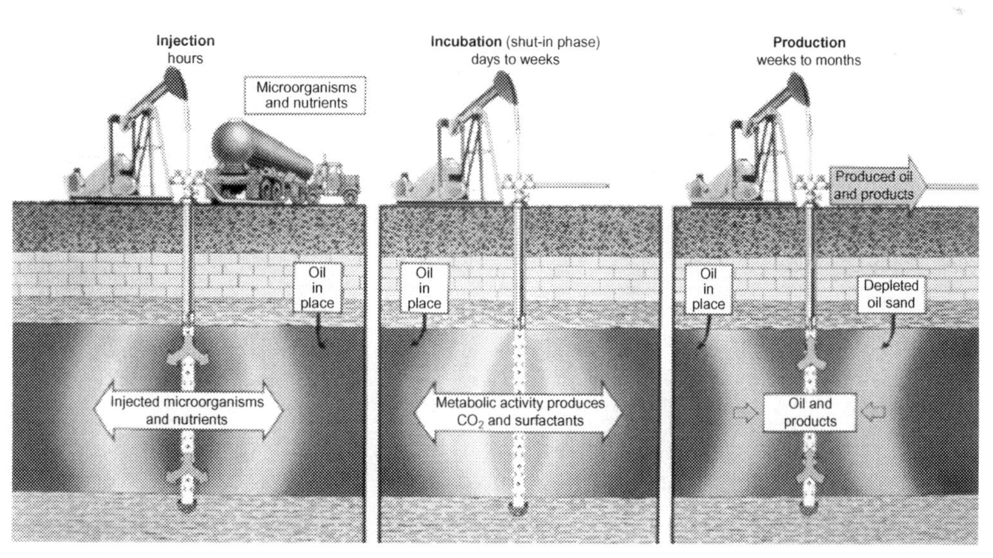

Figure 10.5 Schematic representation of cyclic microbial recovery. *Courtesy of the National Energy Technology Laboratory and the U.S. Department of Energy.*

10.9.1 Injection of Microbial Bioproducts

If there is no suitable indigenous microorganisms present in the target formation and it is not possible to inject exogenous microorganisms maybe due to the prevailing harsh condition, it is possible to directly inject the ex situ produced bioproducts. In this approach, the metabolic biochemicals are generated outside the well in laboratories and then the bioproducts maybe along with some synthetic chemicals will be injected. The advantage of this approach is that the operators are able to exert more direct control on the process. It is possible to directly select the desired produced compounds to be injected to the well. The bioproducts will be mixed with water before waterflooding. To be brief, in this approach, the metabolic bioproducts are injected to well rather than the microorganisms or nutrients. By this approach, there is a concern about the loss of injected biochemicals [6]. The other drawback of this approach is the high costs of laboratory equipment, maintenance of the bioreactors, facilities, and purification practices compared with lower yield of the induced production [24,31]. Because of this, direct injection of bioproducts may not be economically feasible.

10.9.2 Stimulation of Indigenous Microorganisms

In this approach, the microorganisms indigenous to the reservoir will be stimulated to generate the desired bioproducts. For this, it is necessary the microorganisms doing the preferred function (bioplugging or producing biochemicals) are present. If the proper microorganisms are already present, the next step is deciding how to stimulate them. Analysis of the produced fluids as well as core sample (if available) can be helpful in making the decision [6]. A standardized framework such as specialized sampling in coring techniques is essential for evaluation of microbial activity [415]. There are some procedures that minimize the contamination problems during the core material sampling [416]. Youssef et al. [6] mentioned both molecular and microbial techniques to confirm the presence of the suitable microorganisms. After this, the next step is to apply more tests to verify the production of desired biochemicals or activity [6]. Despite the exogenous microorganisms, which may be unable to adapt with the reservoir condition, the indigenous ones have more chance to thrive under the reservoir condition [23]. In economic point of view, the indigenous bacteria are more desired as it is only necessary to supply the nutrients. It is important to find a way to selectively stimulate only the target microorganisms. In order to stimulate the aerobic microorganisms, it is necessary to inject oxygen or chemical, which can be converted to oxygen (such as hydrogen peroxide (H_2O_2)) [6].

10.9.3 Injection and Stimulation of Exogenous Microorganisms

In cases in which the desired microorganisms are not present, an alternative is injecting and stimulating exogenous microorganisms. In this approach, the microorganisms along with the corresponding nutrients are injected to the well with the hope that the desired biochemicals will be generated within the reservoir [31]. The number of injected microorganisms should be in the range of 10^5-10^6/mL injected water [417]. An advantage corresponding with this method is the possibility to design a specific nutrient package to stimulate the growth of injected microorganisms in the reservoir [418]. One challenge associated with this method is the transport abilities of the exogenous microorganisms [6]. Mentioning conflicting recommendations on the utilization of nonstarved versus starved cells [419,420], Youssef et al. [6] specified microorganisms injection should be such that their adsorption to the reservoir rock material is minimal. As a matter of the fact, the starved microorganisms are relatively small and smaller cells are less likely for retention. Thus, utilizing starved cells provide more efficient transport [421]. The other point about the starved cells is more effective penetrating capacity to the porous material as it is proved by laboratorial studies [422,423]. It is also possible to utilize microbial spores in this regard [424,425] (spores are thick-walled and highly resistant to survive under undesired conditions. When the condition becomes suitable the microorganisms will rise again) [6,426]. In applying this approach, there is the presumption that the injected microorganisms will dominate the microorganisms already present in the reservoir and has adapted themselves with the reservoir condition and became fitted. Unfortunately, this is not always the case [31]. Another constraint with this approach is about the co-injection of nutrients and microorganisms. In reservoirs characterized with small pore throat sizes, the nutrient, which is coinjected in solution, has faster propagation rate and may nourish and invigorate the preexisting microorganisms in the reservoir. This makes it difficult for the injected microorganisms to establish themselves over their indigenous counterparts [22].

10.10 PRODUCE BIOCHEMICALS AND THEIR ROLE IN MEOR

The main metabolically produced biochemicals by microorganisms beneficial for MEOR processes are:

1. Biosurfactants and bioemulsifiers;
2. Biopolymers;
3. Bioacids;
4. Biosolvents;
5. Biogases; and
6. Biomass.

Each of the mentioned items are discussed in the following.

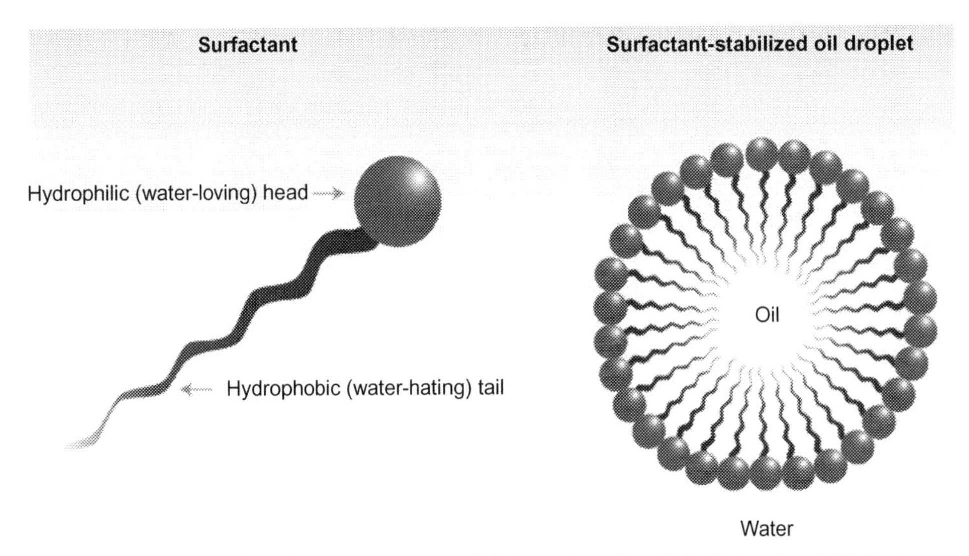

Figure 10.6 Structure of surfactants. *Courtesy of Science Learning Hub, University of Waikato, www.sciencelearn.org.nz [440].*

10.10.1 Biosurfactants and Bioemulsifiers

Biosurfactants or bioemulsifiers (high molecular weight biosurfactants) are a heterogeneous group of amphipathic molecules, i.e., they have both hydrophilic and hydrophobic groups in their structure [6,22,427,428]. The oil droplets would be trapped in tiny pores within the rock matrix by strong capillary pressure at about end of the secondary oil recovery. To release this entrapped oil it is necessary to largely decrease the IFT between the oil and water phase [17] (see Section 10.3). In addition to altering the surface and interfacial tensions, these biochemicals can generate micro-emulsions in which hydrocarbons are solubilized in water or vice versa [22] (Fig. 10.6). McInerney et al. [17] mentioned that biosurfactant production has been traditionally considered as a mechanism to enhance the biodegradation of hydrocarbons via promoting their apparent aqueous solubility [429–438] or through promoting the interaction of microbial cell with hydrocarbons [435,439].

Biosurfactants would be appealing alternatives for classic chemically synthesized surfactants conventionally used in petroleum industry. The main advantages of biosurfactants over the chemically sensitized ones are being temperature tolerant, biodegradable, general nontoxicity to humans, pH-resistant, possibility to be produced in-situ in oil reservoirs, and being less expensive and more environmental friendly [22,31]. Moreover, biosurfactants can tolerate salt concentration as high as 10%, while only 2% NaCl is enough to deactivate the synthetic conventional surfactants [441,442]. The other appealing characteristic of biosurfactant making them more fruitful than

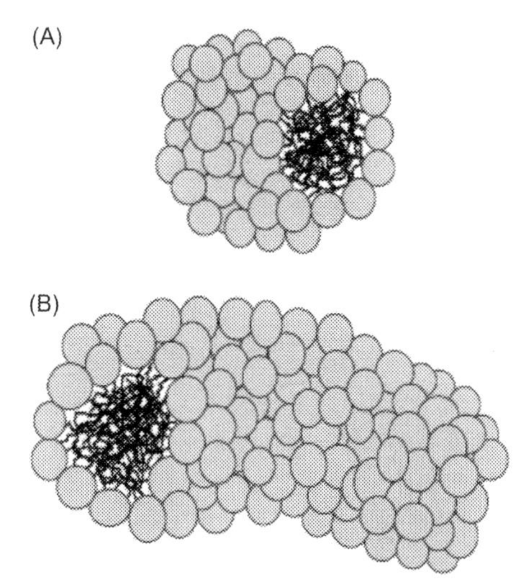

Figure 10.7 Different micelle structures of (A) spherical and (B) cylindrical [60].

traditional synthesized surfactant is that lower concentrations of which can provide the same results [15,443]. In other words, introducing a relatively low concentration of biosurfactant can induce the desired effects [31]. It can be said that biosurfactant often shows better performance than their synthetic counterparts [444].

Aggregates of surfactant monomers known as micelles form when sufficient amount of surfactant is added to the solution. In low concentrations, micelles have a spherical shape containing several hundred surfactant's monomers. The charged head of the molecule is oriented toward the aqueous phase while the hydrophobic ends develop a hydrophobic environment in the core of the micelle [445]. Cylindrical micelle can be formed at higher surfactant concentrations. Fig. 10.7 shows both spherical and cylindrical micelles.

The threshold concentration at which the micelles starts to form is referred as the critical micelle concentration (CMC). At concentrations above CMC, addition of surfactant to the aqueous solution will not increase the number of monomers but results in development of additional micelles [48].

The required surfactant concentration to form micelles is typically small and in range of 10–2000 mg/L [445]. This depends on the parameters such as temperature, water hardness, and also surfactant type. Moreover, as the hydrophobicity of the surfactant increases its CMC generally decreases [60]. Comparatively, biosurfactants exhibits much lower CMC than their synthetic counterparts [51,52,446–449], which makes it possible to use relatively lower concentrations of them compared with synthetic ones [54]. This is in range of 20 − 50 mg/L [17]. As an example, 65% of the residual kerosene from the sand-packed columns were recovered using a very low concentration of lipopeptides purified from *Bacillus subtilis* MTCC 1427 cultures [450]. This indicates effectiveness of biosurfactants at much lower concentrations.

As the main function of surfactants, these biochemicals reduce the interfacial and surface tensions, which results in flooding recovery [444]. The roles of biosurfactants are chiefly (1) altering the surface and IFTs, (2) adsorbing on the immiscible interfaces, (3) emulsifying the crude oil, and (4) increasing the mobility of bacterial cells [15,31,451]. Moreover, they can alter the wettability, clean the contaminated soil, and promote the long alkyl chains degradation [15,443,444,451]. A sign of biosurfactant generation is alteration of some characteristics of the crude oil such as viscosity, cloud point, and pour point [6,452–455]. In MEOR trials, sometimes further compounds such as different metal ions are mixed with biosurfactants. Because of polar interactions between the biosurfactants and metal ions the action will be improved [456].

Beside several advantageous of biosurfactants, there are some drawbacks associated with them in MEOR projects. There is a fact that both quantity and quality of biosurfactant are affected by nutrient supply and also environmental condition [457]. Biological cytotoxicity of metabolites such as the case for surfactin, which is reported to exhibits antimicrobial activities [458], may make some constraints in microbial growth in oil reservoirs [22]. Several microorganisms (including aerobic bacteria, which are not proper for in situ MEOR purposes [412]) are reported to generate biosurfactants; however, only a few of them are proper candidates for MEOR purposes to be able to sufficiently reduce the oil – water IFT [22]. It is likely that biosurfactants may undergo biodegradation by other microorganism in mixed cultures under fermentative, aerobic, nitrate-reducing, and sulfate-reducing conditions [459,460]. This makes the surviving of the produced biosurfactants a challenging issue. Comparatively, the soluble biosurfactant is a much more accessible nutrient for the microorganisms than the crude oil [66]. Biosurfactant activity and solubility can be influenced by salinity and pH [461,462]. The effective pH range is 4–10. It is found that at pHs less than 4, many biosurfactants starts to precipitate [461,462]. This can be attributed to the fact that their isoelectric point is near pH = 4 [121]. It is reported that surfactants' activity increases in NaCl concentration from 0% to 8% [462,463] while some authors have reported no considerable impact on activity [464,465].

There are several types of biosurfactants with particular characteristics, which can be produced by microorganisms [4]. Neu [466] categorized the biological surface active compounds as

1. Biosurfactants;
2. Amphiphilic polymers (e.g., lipopolysaccharides and lipoteichoic acids); and
3. Polyphilic polymers (e.g., hydrophobic polysaccharides and emulsan).

The major categories of biosurfactants include [441,467–469] (see Table 10.2):

1. Glycolipids;
2. Lipopeptides and lipoproteins;

Table 10.2 Different Types of Biosurfactants Along with the Typical Producing Microorganisms

Surfactant type	Producing Microorganisms	References
Glycolipids		
Rhamnolipids	*Pseudomonas aeruginosa, Acinetobacter calcoaceticus, Enterobacter asburiae, Pseudomonas fluorescens,*	[473–477]
Sophorolipids	*Candida bombicola, Candida apicola, Candida bogoriensis, Wickerhamiella domericqiae, Candida riodocensis, Candida stellata, Candida batistae, Candida lipolytica*	[478–486]
Trehalolipids	*Rhodococcus erythropolis, Mycobacterium fortuitum, Mycobacterium tuberculosis, Mycobacterium smegmatis, Corynebacterium matruchotii, Pseudomonas fluorescence, Arthrobacter parafineus, Brevibacteria sp., Arthrobacter sp., Corynebacterium spp., Nocardia spp., Rhodococcus fascians, Rhodococcus sp. H13-A, Rhodococcus opacus, Rhodococcus wratislaviensis, Micrococcus luteus.*	[487–507]
Glucose lipids	*Alcanyvorax borkumensis, Alcaligenes sp*	[508,509]
Lipopeptides and lipoproteins		
Peptide–lipid	*Bacillus licheniformis*	[441,510]
Viscosin	*Pseudomonas fluorescens*	[441,510]
Serrawettin	*Serratia marcenscens*	[441,510]
Surfactin	*Bacillus subtilis*	[441,510,526–529]
Subtilisin	*Bacillus subtilis*	[441,510]
Gramicidin	*Bacillus brevis*	[441,510]
Polymyxin	*Bacillus polymyxia*	[441,510]
Lichenysin	*Bacillus licheniformis, Bacillus mojavensis*	[527,530–532]
Fatty acids, neutral lipids, and phospholipids		
Fatty acid	*Corynebacterium lepus, Capnoytophaga sp., Penicillium spiculisporum, Arthrobacter paraffineus, Talaramyces trachyspermus, Norcadia erythropolis*	[441,510]
Neutral lipids	*Nocardia erythropolis*	[441,510]
Phospholipids	*Thiobacillus thiooxidans, Acinetocbacter sp., Corynebacterium lepus*	[486,511]
Polymeric surfactants		
Emulsan	*Acinetobacter calcoaceticus*	[514,533,534]
Alasan	*Acinetobacter radioresistens*	[452,486,512,513]

(continued)

Table 10.2 (Continued)

Surfactant type	Producing Microorganisms	References
Biodispersan	*Acinetobacter calcoaceticus*	[464,514,515]
Liposan	*Candida lipolytica*	[516–518]
Carbohydrate–lipid–protein	*Pseudomonas fluorescens, Yarrowia lipolytica, Pseudomonas nautica*	[441,519,520]
Mannan–lipid–protein	*Candida tropicalis*	[521–524]
Particulate surfactant		
Vesicles	*Acinetobacter calcoaceticus*	[441,468,525]
Cells	*Various bacteria*	[441,510]

Source: Data are gathered from different sources [441,464,468,470–525].

3. Fatty acids, neutral lipids and phospholipids;
4. Polymeric surfactants; and
5. Particulate surfactant.

Several different microorganisms produce different types of biosurfactants, a list of which is presented in Table 10.2 [441,464,468,470–525].

Regarding the molecular weight, surfactants can be categorized as the low and high molecular weight categories [19,524]. The former, are generally glycolipids or lipopeptides, while the latter include amphipathic polysaccharides, proteins, lipopolysaccharides, lipoproteins or complex mixtures of these biopolymers. The main functions of low and high molecular weight biosurfactant in MEOR applications are lowering of surface and IFTs and stabilizing oil-in-water emulsions, respectively, therefore called emulsifiers [404,457,524,535–537]. Many microorganisms including yeasts, bacteria, and some filamentous fungi are capable of producing biosurfactants with different surface activities and molecular structures [538].

Van Hamme et al. [19] listed the major biosurfactants produced by microorganisms classified as low and high molecular weight microorganisms.

Youssef et al. [6] mentioned the most common low molecular weight biosurfactants used in MEOR as [427,457,541–546]

1. Lipopeptides produced by *Bacillus* and some *Pseudomonas* spp.;
2. Glycolipids (rhamnolipids) produced by *Pseudomonas* sp.; and
3. Trehalose lipids produced by *Rhodococcus* sp.

And high molecular weight ones as the bioemulsifiers as [182,524,547–549]:

1. Emulsan produced by genus *Acinetobacter;*
2. Heteropolysaccharides produced by *Halomonas eurihalina and Pseudomonas tralucida*;
3. Protein complexes produced by *Methanobacterium thermoautotrophicus*;
4. Protein–polysaccharide–lipid complex produced by *Bacillus stearothermophilus*;
5. Carbohydrate–protein complex as liposan produced by *Candida lipolytica*; and
6. Mannan protein produce by *Saccharomyces cerevisiae.*

The IFT between the mineral oils and water is typically in the range of the 35–60 mN/m [60]. Rhamnolipid and lipopeptides biosurfactants can reduce the IFT between the hydrocarbon and water phases as low as 0.1 mN/m or even lower [51–53,433,524,536,539,550,551]. It is reported that in comparison with synthetic surfactants, rhamnolipid was proved to be 20 times more effective in solubilizing hexadecane [436] and could mobilize 75% of the residual hexadecane from the sand-packed columns [430,552] but a large number of pore volume of 40–70 was required. A glycolipid biosurfactant showed great resistance to harsh condition and was stable at temperatures as high as 120°C, pH range of 2–12, and salinity up to 10%. [553]. Comparing lipopeptide and rhamnolipid, the effective concentration of the former is much lower [17,554]. Lipopeptides are reported to be able to reduce the surface tension of water from 72 to 27.9 mN/m at a CMC of 0.017 g/L [555]. A report by Banat [556] indicates up to 95% oil recovery from sand-packed columns. McInerney et al. [17] recounted about the effectiveness of lipopeptide produced by *Bacillus mojavensis* strain JF-2 in mobilizing large amounts of oil from sand-packed columns in low concentrations and in less than one pore volume injection containing about 900 mg of biosurfactant/liter. Lipopeptides can function in reservoir harsh condition of temperatures as high as 100°C, pH range of 6–10 and salt concentrations as high as 8% [404,450,557,558]. Many laboratory experiments have been done to evaluate the effect of biosurfactants on recovering residual oil [124,387,393,412,444,450,451,457,462,467,531,540,543–545,550,553,556–584]. Moreover, several patents have been issued, a list of which is presented in Table 10.3.

As it was mentioned before, bioemulsifiers are high molecular weight amphiphilic compounds produced by different microorganisms [182,606], the role of which is establishing stable emulsion with hydrocarbons usually as the form of oil-in-water and less commonly vice versa [6,182]. Despite the low molecular biosurfactant, the high molecular weight ones, i.e., bioemulsifiers are not likely to reduce the IFT [182,607]. Bioemulsifiers are adequately effective in low concentrations just as biosurfactants [182]. The most common bioemulsifier is emulsan, which is an anionic heteropolysaccharide and protein complex [6]. It is reported that emulsan just emulsifies the hydrocarbon mixtures rather than the pure hydrocarbons [547,548].

Biosurfactant can be used for several other purposes [441,608], including separating the oil from the bottom of the tanks [609,610] and bioremediation (such as removing the crude oil from the oil-contaminated soil [541,611,612]). Bioemulsifier may have many different roles in petroleum industry other than MEOR such as emulsion-based fuels, emulsion-facilitated petroleum transport, oil tank clean-up, preventing paraffin deposition, and environmental protection and remediation [17,182,457,540,607,613], which are none of this study's concerns.

Table 10.3 Some Issued Patents About Applications of Biosurfactants in Oil Recovery

Year	Inventor	Assignee	Patent Title	US Patent No.	Reference
1985	Michael J. McInerney, Gary E. Jenneman, Roy M. Knapp, Donald E. Menzie	The Board of Regents for the University of Oklahoma	Biosurfactant and enhanced oil recovery	4,522,261	[531]
1989	David L. Gutnick, Eirik Nestaas, Eugene Rosenberg, Nechemia Sar	Petroleum Fermentations N.V.	Bioemulsifier production by *Acinetobacter calcoaceticus* strains	4,883,757	[585]
1990	Alan Sheehy	B.W.N. Live-Oil Pty. Ltd.	Recovery of oil from oil reservoirs	4,971,151	[586]
1990	Rebecca S. Bryant	IIT Research Institute	Microbial enhanced oil recovery and compositions therefore	4,905,761	[77]
1991	Catherine N. Mulligan, Terry Y. Chow	Her Majesty the Queen in right of Canada, as represented by the National	Enhanced p roduction of biosurfactant through the use of a mutated B. subtilis strain	5,037,758	[587]
1992	James B. Clark, Gary E. Jenneman	Phillips Petroleum Company	Nutrient injection method for subterranean microbial processes	5,083,611	[588]
1992	Alan Sheehy	B. W. N. Live-Oil Pty. Ltd.	Recovery of oil from oil reservoirs	5,083,610	[571]
1993	Paolo Carrera, Paola Cosmina, Guido Grandi	Eniricerche S.p.A.	Method of producing surfactin with the use of mutant of Bacillus subtilis	5,227,294	[589]
1994	Tadayuki Imanaka, Shoji Sakurai	Nikko Bio Technica Co., Ltd.	Biosurfactant cyclopeptide compound produced by culturing a specific Arthrobacter microorganism	5,344,913	[590]
1998	Eugene Rosenberg, Eliora Z. Ron	RAMOT University Authority for Applied Research & Indfustrial	Bioemulsifiers	5,840,547	[591]
1999	Willem P. C. Duyvesteyn, Julia Rose Budden, Merijn Amilcare Picavet	BHP Minerals International Inc.	Extraction of bitumen from bitumen froth and biotreatment of bitumen froth tailings generated from tar sands	5,968,349	[592]
1999	Carlos Ali Rocha, Dosinda Gonzalez, Maria Lourdes Iturralde, Ulises Leonardo Lacoa, Fernando Antonio Morales	Universidad Simon Bolivar	Production of oily emulsions mediated by a microbial tenso-active agent	5,866,376	[593]
2000	Willem P. C. Duyvesteyn, Julia Rose Budden, Bernardus Josephus Huls	BHP Minerals International Inc.	Biochemical treatment of bitumen froth tailings	6,074,558	[594]
2000	Giulio Prosperi, Marcello Camilli, Francesco Crescenzi, Eugenio Fascetti, Filippo Porcelli, Pasquale Sacceddu	EniTecnologie S.P.A.	Lipopolysaccharide biosurfactant	6,063,602	[595]

(*continued*)

Table 10.3 (Continued)

Year	Inventor	Assignee	Patent Title	US Patent No.	Reference
2000	Carlos Ali Rocha, Dosinda Gonzalez, Maria Lourdes Iturralde, Ulises Leonardo Lacoa, Fernando Antonio Morales	Universidad Simon Bolivar	Production of oily emulsions mediated by a microbial tenso-active agent	6,060,287	[596]
2003	David R. Converse, Stephen M. Hinton, Glenn B. Hieshima, Robert S. Barnum, Mohankumar R. Sowlay	ExxonMobil Upstream Research Company	Process for stimulating microbial activity in a hydrocarbon-bearing, subterranean formation	6,543,535	[597]
2006	James B. Crews	Baker Hughes Incorporated	Bacteria-based and enzyme-based mechanisms and products for viscosity reduction breaking of viscoelastic fluids	7,052,901	[598]
2009	Robin L. Brigmon, Christopher J. Berry	Savannah River Nuclear Solutions, LLC	Biological enhancement of hydrocarbon extraction	7,472,747	[599]
2009	Banwari Lal, Mula Ramajaneya Varaprasada Reddy, Anil Agnihotri, Ashok Kumar, Munish Prasad Sarbhai, Nimmi Singh, Raj Karan Khurana, Shinben Kishen Khazanchi, Tilak Ram Misra	The Energy And Resource Institute, Institute Of Reservoir Studies	Process for enhanced recovery of crude oil from oil wells using novel microbial consortium	7,484,560	[562]
2011	Fallon; Robert D.	E. I. du Pont de Nemours and Company (Wilmington, DE)	Methods for improved hydrocarbon and water compatibility	7,992,639	[600]
2011	Frederick D. Busche, John B. Rollins, Harold J. Noyes, James G. Bush	International Business Machines Corporation	System and method for preparing near-surface heavy oil for extraction using microbial degradation	7,922,893	[601]
2013	Sharon Jo Keeler, Robert D. Fallon, Edwin R. Hendrickson, Linda L. Hnatow, Scott Christopher Jackson, Michael P. Perry	E.I. Du Pont De Nemours and Company	Identification, characterization, and application of Pseudomonas stutzeri (LH4:15), useful in microbially enhanced oil release	8,357,526	[602]
2014	Michael Raymond Pavia, Thomas Ishoey, Stuart Mark Page, Egil Sunde	Glori Energy Inc.	Systems and methods of microbial enhanced oil recovery	8,826,975	[603]
2016	Edwin R Hendrickson, Abigail K Luckring, Michael P Perry	E I Du Pont De Nemours and Company	Altering the interface of hydrocarbon-coated surfaces	9,499,842	[604]
2016	William J. Kohr, Zhaoduo Zhang, David J. Galgoczy	Geo Fossil Fuels, Llc	Alkaline microbial enhanced oil recovery	9,290,688	[605]

10.10.2 Biopolymers

Bacteria within the reservoir tend to produce surface molecules in the form of biopolymers [31]. The majority of these metabolic biopolymers are exopolysaccharides, which can promote the cell adhesion to the surface and prevent desiccation and predation of the bacteria [17,43,391,409,614–618].

Biopolymers can act as the bioplugging agent as the biofilms. Generally, biofilms are composed of exopolysaccharides bound clusters of cells, which can induce clogging effects [22]. The metabolically generated biopolymers can be favorably used for selective plugging and modifying the permeability profile within the reservoir and also reducing the viscous fingering of waterflood [17,19,21,42,124,391,422,616,617, 619–635]. In addition to the laboratory experiments, application of biopolymers on diverting the waterflood from the high-permeable to the low-permeable zones has been tested in fields too [617,619,620,636–638]. Microbial plugging is usually associated with supplying nutrients to microorganisms either injected or exogenous [15,21,616,628,639]. The injected fluid including the microorganisms and/or nutrients tends to flow through the more permeable pathways. The provided nutrients stimulate the microorganisms to grow and this would decrease the permeability of region that once were highly permeable and consequently modify the permeability profile. It is likely that in situ production of biopolymers plugs the useful pore and consequently results in formation damage. Just as the case for biosurfactants, a sign of bioemulsifier generation is alteration of some characteristics of the crude oil such as viscosity, cloud point, and pour point [6,452–455].

Table 10.4 lists the major biobiopolymers used in MEOR along with the producing microorganisms. Two important biopolymers are xanthan gum [624,640–642] and relatively less effective, curdlan [643] (a high-molecular-weight polymer of glucose) [4]. The other reported biopolymers are levan [621], pullulan [622], dextran [617], scleroglucan [24,617,619,634,644]. Polysaccharides such as xanthan gum have thickening effect in waterflooding (increase the drive water viscosity). This thermally stable heteropolysaccharide can be produced by several *Xanthomonas* species via fermenting carbohydrates [15,24]. The efficiency of this biopolymer has been validated in laboratorial studies [626,637,645,646]. The properties making the xanthan gum an ideal polymer for EOR applications include its suitable viscosity, temperature, and salt tolerance, and also being shear resistant [15,623,624,647]. The disadvantages are being relatively expensive and its susceptibility for bacterial degradation [15,619,644]. Curdlan develop an insoluble gel in lower pH values as it is soluble in alkaline pH [620,632,648–650]. Khachatoorian et al. [651] tested the effect of a number of biopolymers in reducing the permeability of the flow system in a sand-packed column and they concluded that the poly-β-hydroxybutyrate has been the most effective than the others (xanthan gum, guar gum, polyglutamic acid, and chitosan). More details

Table 10.4 Important Biopolymers With the Producing Microorganism

Biopolymer	Producing Microorganisms	References
Xanthan gum	*Xanthomonas comprestris* sp.,	[652–655]
Curdlan	*Agrobacterium* sp., *Paenibacillus* sp., *Pseudomonas* sp. *QL212, Alcaligenes faecalis*	[656–663]
Levan	*Lactobacillus reuteri, Zymomonas mobilis., streptococcus salivarius, Serratia* sp., *Bacillus subtilis*	[664–669]
Pullulan	*Aureobasidium pullulans, Pullularia pullulans*	[670–672]
Dextran	*Leuconostoc mesenteroides, cariogenic streptococcus, Pediococcus pentosaceus, Weissella cibaria*	[673–677]
Scleroglucan	*Sclerotium rolfsii*	[678–680]
Poly-β-hydroxybutyrate (PHB)	*Azotobacter vinelandii, Bacillus* spp., *Alcaligenes eutrophus*	[651,681,682]
Polyglutamic acid (PGA)	*Bacillus licheniformis., Bacillus subtilis*	[651,683,684]

PHB, poly-3-hydroxybutyrate; *PGA*, polyglutamic acid.
This table is limited to the biopolymers can be used in MEOR processes.

about the using biopolymers in MEOR can be found in literature [16–19,21,24,617,619–622,624,634,640,643,644].

In addition to the microorganisms listed in Table 10.4, distinguished bacteria to produce the biopolymers include *Xanthomonous*, *Aureobasidium*, *Bacillus*, *Alcaligeness*, *Leuconostoc*, *Sclerotium*, *Brevibacterium*, and *Enterobacter* [15,24].

10.10.3 Bioacids

The produced acids by microbial metabolisms dissolve the carbonate rocks (and also sandstones cemented by carbonate minerals [66]) and subsequently increase the porosity and permeability, which leads to improve the oil migration [566,685]. Dissolution of iron scales in media containing magnetite and goethite is also reported due to acid production by bacteria [686]. In addition, acids can generate gases, which improve the oil displacement by reducing the viscosity along with several other fruitful effects such as oil swelling. Moreover, acids can reduce the permeability due to the clay movement [24]. The other role of bioacids is aiding emulsification [24]. In case of bioacid production in large volumes via in-situ microbial fermentation, it would have the potential to be an alternative for the conventional acid treatments [686]. Members of the genera *Bacillus* and *Clostridium* are the most commonly used microorganisms to produce bioacids, biogases, and biosolvents [6,22,24,88,114,397,408,567,685,687]. *Clostridium* spp. can produce acetate and butyrate acids, while *Bacillus* spp. produce acids of acetate, formate, lactate, etc. [6]. Moreover, lactic acid bacteria (LAB) can produce lactate [6]. Microorganisms including *Clostridium*, mixed *Acidogens*, *Dsulfovibrio*, and *Bacillus* generate bioacids such as carboxylic acids of various molecular weights, low molecular weight fatty acids, formic acid, propionic acid, (iso)butyric acid, etc. [22,24]. Bacteria such as *Clostridium* produce acetate and butyrate, which will produce corresponding bioacids [688]. *Lactobacillus* sp. and *Pediococcus* sp. are the other bacteria reported to be acid-producing [686]. Moreover, CO_2 and H_2S biogases are the other microbial products dissociation of which into the water makes acids [22]. Different studies suggest that aerobic hydrocarbon degraders can produce a mixture of bioacids and bioalcohols, which furtherly would be converted to methane by the methanogenic consortia [275,276,375,689,690].

10.10.4 Biosolvents

A possible microbial action is partial oxidization of hydrocarbons to biosolvents such as alcohols, aldehydes, and fatty acids [691]. As it was mentioned before, biosolvents can partially dissolve the carbonate rocks and consequently improve the reservoir permeability and porosity. The other action by the biosolvents, which improve the permeability, is dissolving heavy components from pore throats. Biosolvents can reduce the oil viscosity through dissolution of asphaltene and heavy components existing in the oil. In addition, dissolution of biosolvents in oil could reduce the viscosity too [17]. The other advantage of biosolvents is they have cosurfactant effects and are capable to reduce the IFT between oil/water and oil-rock [22,24]. Solvent can alter the rock wettability at the rock − oil interface [17]. Good solvent candidates for MEOR application include mainly lower alcohols (methanol, ethanol, 1-propanol, propan-2-diol, 1-butanol, which are water soluble), volatile fatty acids, and ketones such as propanone (acetone) and butanone [15,22,692]. As it was mentioned before, members of the genera *Bacillus* and *Clostridium* are the most commonly used microorganisms to produce bioacids, biogases, and biosolvents [6,88,114,397,408,567,685,687]. *Clostridium* spp. can produce alcohols of ethanol and butanol and also acetone as biosolvents. *Bacillus* spp. produce alcohols of ethanol and 2,3-buanediol [6]. In addition, LAB can produce CO_2 too [6]. Some important reported biosolvent generating bacteria are *Zymomonas mobilis*, *Clostridium acetobutylicum*, *Klebsiella*, *Arthrobacter*, and *Clostridium pasteurianum* [24,77].

Patel et al. [31] mentioned that biosolvents are unlikely to be produced in large enough volumes for direct injection into the wellbore. Considering the current technology, the best practice seems to be stimulation of the indigenous or injected microorganisms to generate the biosolvents rather than direct injection of generated biosolvents in laboratory.

10.10.5 Biogases

Microorganisms can produce biogases of H_2 (which is then rapidly consumed by further microbial activity), H_2S, N_2, CH_4, and CO_2 by fermenting the carbohydrates [15,22,24]. H_2 is produced in large volumes in anaerobic environments; however, it is then quickly consumed by methanogens (reducing CO_2 to CH_4), SRB (reducing sulfate to sulfide), homoacetogenic bacteria (reducing CO_2 to acetic acid), and NRB (reducing nitrate to N_2) [6,17]. As it was mentioned before, the other source for CO_2 can be the bioacids reaction with rock minerals. The biogenic gases can enhance oil recovery via different mechanisms of:

1. Reducing the oil viscosity;
2. Repressurization of reservoir [144,407];
3. Swelling the oil [693];
4. Altering pH of the formation water [694];
5. Changing IFT [22];
6. Increasing permeability by dissolving the carbonate rocks [24];
7. Bioplugging via precipitation of inorganic minerals such as calcium carbonate ($CaCO_3$) due to the metabolite CO_2 [22,695,696]; and
8. Changing pour point of oil [22].

Biogas production is reported to be an important mechanisms for oil recovery [697]. Several laboratory experiments using different microorganisms such as *Enterobacter* sp. [697], *Clostridium* strains [698], *Vibrio* sp. and *Bacillus polymyxa* [124], *Streptococcus* sp. [699], *Staphylococcus* sp. [699], *Clostridium acetobutylicum* [114,408], have shown the effect of biogas production on enhanced oil recovery. The produced gas can be absorbed by the oil and consequently decrease its viscosity. Lower viscosity facilitates the oil displacement. The other function of generated biogases can be the repressurizing the reservoir [144,407]. The induced pressure will exert extra forces to expel oil out of the pores [23,700]. The other beneficial effects of biogases are swelling the oil volume, displacing the immobile oil, and increasing the permeability by dissolving the carbonate rocks [24]. Induction of biogenic CO_2 into the formation water and consequent pH reduction might promote oil producibility, especially in fractured carbonate rocks [694]. It is unlikely that the generated biogases enhance oil recovery through miscible displacement due to limited generated volumes [24,407]. Production of biogases may be stopped at some points when the gas is swollen enough. This let production of more valuable components [10]. As it was mentioned before, members of the genera *Bacillus* and *Clostridium* are the most commonly used microorganisms to produce bioacids, biogases, and biosolvents [6,88,114,397,408,567, 685,687]. *Clostridium* spp. produces gases of CO_2 and H_2 and *Bacillus* spp. produces CO_2. In addition heterofermentative LAB produce CO_2 too [6]. Several different methanogens generate methane. Formation of mineral precipitates is another mechanism through which produced biogases may improve oil recovery [17]. CO_2 can result in oversaturation of the fluid with calcite ($CaCO_3$) and consequently its precipitation, which may reduce the formation porosity and permeability and hence modify the fluid flow path [701]. Observations of flow alterations in a hypersaline petroleum reservoir after addition of nutrient [702] can be explained by this mechanism.

As it was mentioned earlier, the process of methane production is referred to as methanogenesis [25,395,396], which is strictly anaerobic [22]. Details of the methanogenesis process by studies on laboratory microcosms can be found in literature [703]. Methanogenesis is the easiest way to recover the trapped oil; however, the produced methane is relatively cheap [10]. Because of this, when it was assured that the oil has swollen enough, this process may be stopped to recover more valuable components. To inhibit methanogenesis, it is possible to use inhibitors such as 2-bromoethane-sulfonic acid and methylfluoride for general and acetoclastic methanogenesis, respectively [704]. On the other hand, it should be mentioned that to generate 1 kWh energy, methane produces 0.569 kg CO_2 which is less than those for using oil (0.881 kg) and coal (0.963 kg) [705]. This can reduce the CO_2 emission by more than 35% in case of energy generation by the biogenic methane. Briefly speaking, harvesting energy from heavy oils and tar sand in form of gas rather than heavier hydrocarbons is more economic and environmentally friendly [22].

10.10.6 Biomass

Generally, microorganisms prefer to develop clusters and colonies in form of biomass, which can be beneficial considering their plugging effect. The biomass is comprised of bacteria, their produced exopolysaccharides, and water channels [15]. To be more specific, they are composed of 27% microbial bodies and the rest (73%−98%) are extracellular products (such as exopolysaccharides) and void space [618,636,637]. Bacteria multiply their mass at exponential rates [17]. Microbial plugging is usually associated with supplying nutrients for microorganisms either injected or exogenous [15,21,616,628,639]. The injected fluid including the microorganisms and/or nutrients tends to flow through the more permeable pathways [706,707]. The provided nutrients stimulate the microorganisms to grow and this would decrease the permeability of region that once were highly permeable and consequently modify the permeability profile. In another words, most of the injected nutrients will stimulate the bacteria available in the regions with high permeability to growth. Biomass can accumulate in the highly permeable zones, referred to as the thief zones, and then divert the waterflood to the oil-bearing zone [636]. The biomass, first, start to accumulate in wellbore along high-permeable channels, which diverts the fluid flow into the lower permeable zones. For this, cautions should be considered as the fast growth of biomass may adversely affect the well injectivity [24]. In the porous media, biomass will occur in a web-like structure, hence, they would be called bioweb at a certain stage of growth too [708]. There is a wealth of laboratorial evidences [423,425,639,709–713] and also some reported field applications [646,708,714] that in situ growth of bacteria and consequent biomass development can significantly reduce the permeability.

When the microorganisms grow within the reservoir, the molecules on their surfaces keep them attached to the substrates near where they feed. As the result, a biofilm is created that prevent the oil to be drawn into the porous zones [715]. Biofilms can be introduced as the populations of bacterial cells attached to a surface enclosed in an organic polymeric matrix [715–718]. As it was mentioned before, exopolysaccharides-bound clusters of cells commonly comprises the biofilms [22].

The biomass displace the oil by its growth. Considering the microscopic sweep efficiency, the biomass growth in the large pores diverts the fluid flow to the smaller pores because the large pores receive most of the nutrients [6]. This has been proved by pore size distribution analysis of fused-glass columns [719] and sandstone cores [720]. Moreover, biomass can act as the plugging agents and contribute to permeability profile modification. It should be mentioned that biopolymer production is necessary in addition to biomass development for significant permeability reduction in fused glass and sand-packed columns [6,401,719,721–729]. The other favorable effect is its viscosity and pour point reduction [24]. It is reported that biomass can play a role in oil emulsification and desulfurization [24]. There is a fact that bacteria tend to develop attached to a surface [715,730]. Biofilms are capable to alter the physical and chemical characteristics of the surfaces [731,732], thus it is possible they favorably amend the wettability of the reservoir rock. Some attempts have been done regarding starving the bacteria to reduce their size and increasing their penetration depth into the formation. After that the nutrient are provided for growing the bacteria colonies to act as the biomass [420]. In comparison with the bacteria suspended in the liquid phase, biofilms exhibits more resistance to the biocides [733–735].

10.11 MEOR MECHANISMS

At first, it should be mentioned MEOR takes place through a combination of several mechanisms and multiple biochemical processes rather than a single one [21,43,392,736]; however, it is possible that one mechanism do the most contribution. Youssef et al. [6] specified that utilizing a consortium of microorganisms with different properties and employing multiple mechanisms is an effective strategy for oil recovery [77,88,693,737]. Microorganisms can produce a broad range of biochemicals including biosurfactants and bioemulsifiers, biopolymers, biomass, bioacids, biosolvents, and biogases. As it was mentioned before, several parameters affect the subsurface microbial life including [119–121]:

1. Chemical factors such as nutrient composition, electrolyte composition, redox potential (activity of electrons (Eh)), and activity of hydrogen ions (pH);
2. Physical factors such as pressure, temperature, salinity, pore size, and pore geometry, porosity, permeability, dissolved solids; and
3. Biological factors such as cytotoxity of the microbial metabolites and also specific type of microorganisms.

10.11.1 Hydrocarbon Metabolisms and Biodegradation

Owing to their structural properties, hydrocarbons exhibits low chemical reactivity and their degradation demands special biochemical reactions [738]. Oxygen plays the main role in aerobic hydrocarbon degradation, which is the terminal electron acceptor and provide highly reactive oxygen species. Microbial degradation of alkanes and other aliphatic oil components has been investigated since the beginning of the 20th century [739–742]. In the 1990s, the anaerobic alkane degradation was discovered [743]. As a matter of fact, anoxic or anaerobic condition is dominant in many natural environments including the oil reservoirs [443,744]. Hydrocarbon biodegradation in anoxic condition is probably an evolutionary metabolic trait of microorganisms [738]. It seems that microbial degradation of long-chain *n*-alkanes under anaerobic conditions is more realistic and relevant for MEOR [121]. Anaerobic hydrocarbon biodegradation can affect the geochemistry of the petroleum reservoirs and also is applicable for bioremediation of contaminated sites [738].

Paraffin available in the crude oil may induce formation damages. Different methods including chemical (such as using solvents, dispersants, surfactants, and wax crystal inhibitors) and physical (such as physical scarping, downhole electrical heating, using hot hydrocarbons or water) methods are alternatives to remediate this problem to keep the wells producing [453,745–751], which are usually expensive [6]. As it was mentioned before, the most common microbial approach to deal with the paraffin deposits is stimulation of in situ hydrocarbon metabolism [6]. Based on literature, the wells treated with hydrocarbon degrading microorganisms exhibit less paraffin depositions on the production equipment such as sucker rods and do not require as frequent treatment with hot oil [615,691,752,753]. This would significantly decrease the operational costs and increase the economic lifetime of the producing wells [412]. However, some studies have reported the mentioned method is ineffective or inconclusive results were acquired [748,754].

It is well proofed that many microorganisms are capable to degrade hydrocarbons both aerobically and anaerobically [19,388,691,755−757]. Some commercial processes use proprietary mixtures of hydrocarbon biodegraders to prevent paraffin and asphaltene deposition [81,615,691,748,753,758,759]. Some microorganisms can attach to the long-chain hydrocarbons and break them to smaller-chain ones (the proportion of low-carbon number alkanes to high-carbon number alkanes increases), which owns generally lower viscosity and consequently better mobility [6,81,759−766]. Youssef et al. [6] mentioned that the conversion process of long-chain alkanes to the short-chain ones is unclear and there is no microorganisms known to catalyze such a reaction. The other assumed mechanism for hydrocarbon biodegradation is the partial hydrocarbon transformation to aldehydes, alcohols, and fatty acids, which could serve as biosolvents or biosurfactants [691]. Temperatures higher than 80°C, which is the common condition for many oil reservoirs, inhibit the oil anaerobic degradation in oil reservoirs [10,119,129].

An excellent review on biodegradation in petroleum reservoir by Head et al. can be found in literature [119]. Further details about aerobic [743] and anaerobic [738] hydrocarbon biodegradation can be found elsewhere.

10.11.2 Lowering the Entrapped Oil Viscosity

Two main microbial activities can reduce the oil viscosity [121,767]:

1. Microbial production of metabolites such as biogases, which alter the physical properties of the oil [119,351,443,768], and
2. Microbial biodegradation of heavy oil components to the lighter ones [24,338,443,757,760,769,770].

Microbial metabolisms produce carbon dioxide, hydrogen, nitrogen, and methane gases [24]. In addition, carbon dioxide may be produced through the reaction of metabolic acids with carbonate rocks. Provided that sufficient amount of gas is produced by the microorganisms, it can be absorbed in oil and resulted in a viscosity decrease. As it was mentioned before, some microorganisms can attach the long chain hydrocarbons and break them to smaller chain ones, which owns generally lower viscosity and consequently better mobility [6,81,759−766]. Biosolvents can reduce the oil viscosity through dissolution of asphaltene and heavy components existing in the oil [24]. Briefly speaking, biosolvents such as alcohols, ketones, and short-chain hydrocarbons as well as bioacids and biogases can reduce the oil viscosity [771].

10.11.3 Increasing the Water Viscosity

As it was mentioned before, an increase in the water viscosity promote the sweep efficiency. Another favorable effect is that it may contribute to miscibilization of entrapped oil and surrounding water. Metabolically generated bioproducts capable to increase the water viscosity are biofilms, biopolymers, long-chain alcohols, and fatty acids [771]. Several studies have proved the ability of bioproducts generated by microorganisms to increase the water viscosity [15,24,768,772].

10.11.4 Selective Plugging To Modify the Permeability Profile

In some reservoirs, there are highly permeable streaks, which reduce the overall water flooding sweep efficiency. As it would be explained in further details in the following sections, there is a fact that biomass preferentially plugs the large pores [719,720], which conduct the majority of the fluid flow. This would divert the flow to the much smaller pores, which results in permeability reduction. Based on the selective plugging concept, microorganisms can block the highly permeable channels (thief zones) and direct the floodwater to the less permeable zone with immobilized water to enhance oil recovery. The permeability profile modification will increase the sweep efficiency and subsequently more oil will be produced. Laboratory experiments have proved that permeability reduction is higher in higher permeable channels compared with lower permeable ones [773,774]. Permeability modification depends only on the microbial growth and it is independent of the bacteria type. It is mentioned that generation of biopolymers is more effective than biomass accumulation [15]. Moreover, emulsions can become adequately thick to promote the channeling for oil recovery [775]. In fact, permeability reduction occurs as the result of combination of pore throat plugging due to occlusion of bacterial cells and biofilm formation, which retards the fluid flow [17]. Considering the microbial activities, plugging may occur through three different mechanisms of

1. Physically via flourish of highly viscus organic matter such as biopolymers and biofilms [22,31,776] and emulsions [31,775];
2. Biologically via formation of biomass [22,777]; and
3. Chemically via precipitation of inorganic minerals such as calcium carbonate due to the metabolite CO_2 [22,695,696].

Myriad studies have reported about the microbial induced selective plugging [629,639,695,696,712,714,778–783]. Based on Fink [783] experiments, after injecting a mixture including biopolymer and bioacid producing bacteria into the Berea Sandstone core, the permeability was decreased from 850 to 2.99 mD and 904 to 4.86 mD. Four criteria are proposed in literature to conduct and effective selective plugging by biomass [777]. These criteria suggest

1. The cells must be transported through the rock matrix;
2. Necessary nutrient for growth should be supplied;
3. The bacteria community should adequately grow and/or generate bioproducts for selective plugging; and
4. The growth rate must be under control and should not be so rapid that it chokes the wellbore.

It has been reported that the *Bacillus licheniformis* BNP29 is a suitable strain for selective plugging and fulfill the mentioned four criteria [545]. In contrast, SRB nonselectively plugs the porous media and adversely influence the oil recovery. This shows the importance of stimulating the proper bacteria type. As it was mentioned, the other probable plugging agent is the inorganic biomass. Literatures indicate the capability of microorganisms to enhance precipitation processes generating solid sulfides [784,785] and carbonates [786–789]. The mentioned processes can be influenced by water chemistry, surface charge, nutrients, pH, fluid flow, and microbial physiology [17,618,619].

10.11.5 Dissolution of Some Parts of Reservoir Rocks

Bioacids and biosolvents such as acetone and ethanol can dissolve some parts of the carbonate rock to increase the porosity and improve the permeability and make it easier to access the hidden oil during the flooding [31,391]. Carbonate dissolution has been regarded as the reason for higher porosity [76,685]. It is possible to metabolically produce the bioacids and biosolvents using microorganisms. Acetic and propionic bioacids are particularly used for this aim [15]. Siegert et al. [10] mentioned bioacids produced by the microorganisms hydrolyzes the carbonates. Core flood experiments has proved that carbonate dissolution may mobilize the entrapped oil from carbonate rocks [8]. This method is also effective in sandstones cemented by carbonate minerals [66].

10.11.6 Wettability Alteration

Wettability has a key role in controlling the location, flow, and distribution of fluids in a reservoir as well as in multiphase flow problems such as oil migration from the source rocks and EOR [790,791]. Favorably changing the wetness state can improve the spontaneous imbibition of water. This would promote the waterflooding performance and consequently the oil recovery. Microbial activities can induce the desired alteration by different mechanisms in which produced biosurfactants, biopolymers, and biomass and even enzymes are involved such as [66]:

1. Direct attachment of the microorganism to the matrix surface. In such a condition, the contact angle is governed by a heterogeneous mixture of mineral and bacterial surface properties. This mechanism is applicable only in large pores that can be accessible by the bacteria. The pore size should be at least 1 μm;
2. Absorption of produced biochemicals on the mineral surface. Microbial metabolites such as biosurfactants can be absorbed on the mineral surface and potentially change the wettability by altering the surface to be more or less hydrophobic. There is no pore-size limitation for this mechanism; and
3. Coating the minerals by biopolymers or exopolysaccharides. This mechanism is somehow analogous to the direct attachment of the microorganisms to the mineral surface. As the polymers are associated with cells, this mechanisms is corresponded

with the cell attachment. Even in case of release of dying of cells, the biopolymers will be left behind [792].

Fractured carbonate reservoir contains most of the world's oil [793]. In this type of reservoirs, the matrix blocks are mixed to oil-wet. This characteristic leads to difficult water absorption onto the matrix blocks, which in turn diminishes the sweep efficiency. In fact, the oil trapped in this region will be unaffected by the floodwater [15]. It is proved that the oil recovery can be improved by turning the matrix surface to a more water-wet condition [16]. As it was mentioned before, this can be done by introducing surfactants, enzymes, or development of a biofilm on the reservoir rock surface by microorganisms. Wettability alteration plays a crucial role in oil and surfactant adsorption in carbonates and clays [10]. In spite of carbonates, wettability of the mineral surfaces plays a secondary role in oil displacement in sandstone formations. Thus, wettability reversal by microbial activity will not provide a promising mechanism for enhancing oil recovery in sandstone formations [66]. However, when the hydrocarbon degrading microorganisms coat the oil droplets with a biofilm, the role of wettability may become trivial [771,794]. Karimi et al. [715] through their experiments concluded that the effect of biofilm formation was much more significant compared with many other microbial products including biosurfactants in enhancement of wettability. It is reported that brine composition affects the wettability changes [795–797] (Fig. 10.8).

10.11.7 Emulsification

Microorganisms are capable to produce bioemulsifiers (high molecular weight biosurfactants). These biochemicals create micro-emulsion in which the water is solubilized in hydrocarbon or hydrocarbon is solubilized in water. Further details are provided in Section 10.10.1.

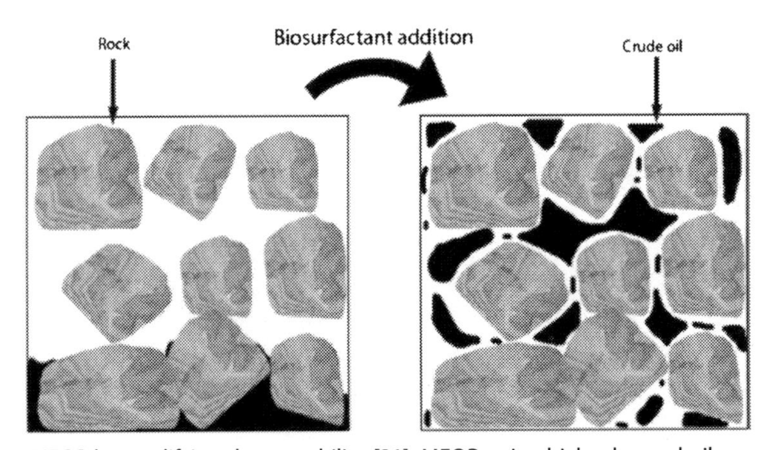

Figure 10.8 MEOR by modifying the wettability [31]. MEOR, microbial enhanced oil recovery.

10.11.8 Surface and Interfacial Tension Alteration

Reducing the surface tension between the oil and water increase the yield of flood recovery [31]. Surfactants are the chemicals incorporated to do this. The features determining the most optimum biosurfactant are strong facial activity, a low CMC, good solubility, adequate tolerance to pH and temperature, and high emulsion capacities [798]. Further details about surface and IFT alteration are provided in Section 10.10.1.

10.11.9 Repressurizing the Reservoir

Re-pressurization of the reservoir is a traditional enhanced oil recovery method [799]. Applying more pressure can force the remained oil out of the pores [23,700]. Microorganisms can produce gases in situ to repressurize the reservoir and subsequently promote gas drive. Bacteria can produce gases such as methane, hydrogen, and carbon dioxide through fermenting the carbohydrates [15]. N_2, which is produced by NRB, can increase the reservoir pressure too [18]. These metabolic gases can enhance the oil recovery by increasing the reservoir pressure. Further details can be found in Section 10.10.5.

10.11.10 Oil Swelling

Biogenic gases can be dissolved in the oil to increase it volume and therefore decrease its density. This process referred to as oil swelling ease the oil displacement [66]. Further details can be found in Section 10.10.5.

10.11.11 Well Stimulation via Removing the Wellbore Damages

During this process, the barriers such as paraffin and asphaltene deposits will be removed from the well bore. However, this process only increases the productivity index and accelerates the recovery. In other words, it makes no increase in the amount of produced oil [23]. In addition, the biologically produced acids, solvents, and gases can remove the debris and scales in the injection well so that the injectivity increases [6]. Although this is not an EOR process, such processes extend the life of a field by reducing the operating costs or enhancing the daily revenue for several cases [11,800]. There are several studies reporting successful microbial paraffin removal processes from individual wells [81,691,801,802]. Further details can be found in Section 10.11.1.

10.12 MEOR CONSTRAINTS AND SCREENING CRITERIA

One of the most important aspects of MEOR is the screening of the reservoir physical properties, which are greatly effective on the capability of the microorganisms

Table 10.5 MEOR Screening Criteria Suggested by Sheng [24] Based on the Literature Information [392,768]

Parameter	Suitable Range
Formation temperature	$<98°C$, preferably $< 80°C$
Pressure	10.5–20 MPa
Formation depth	<2400–3500 m
Porosity	>0.15
Permeability	>50 mD
Formation water TDS	NaCl < 10%–15%
pH	4–9
Oil density	$<0.966\ g/cm^3$
Oil viscosity	5–50 cP
Residual oil saturation	>0.25
Elements	Arsenic, mercury < 15 mg/L
Well spacing	40 ac

TDS, total dissolved solids.

to take the desired action. Sheehy [803] specified that a main reason for failure of MEOR field trials is insufficient consideration of the reservoir properties and the physiology of the microorganisms, which thrive in that condition. The reservoir physical and chemical parameters such as temperature, pressure, salinity, pH, redox potential, etc. affect the microbial activities such as microorganisms' growth, proliferation, survival, metabolism, and the ability to produce the desired volume of the biochemicals [84]. Proper prior planning is mandatory for success of an MEOR field trial. It should be mentioned that criteria proposed by different researchers might be slightly different.

Scrutinizing 407 field trials reported in literature, Maudgalya et al. [23] stated that a minimum reservoir permeability of 75 mD and temperatures less than 93°C have been the most suitable conditions for MEOR. More recently, Sheng [24] proposed the MEOR screening criteria as it is listed in Table 10.5. More details about the MEOR and screening criteria are debated in the following.

10.12.1 Reservoir Engineering Considerations

The first step in planning an MEOR project is acquiring a clear understanding about the factor limiting the oil production and is aimed to be solved [17,22]. After the problem was clearly identified, it can be decided which microbial processes or microorganism would be the best solution. This makes it essential to consider the problem in engineering point of view by the reservoir engineers [66,407]. For example, in case of mobilizing the trapped oil from the porous media, microorganisms such as biosurfactant-producing species should be utilized. On the other hand, to remediate the water channeling by the thief zones, in situ biopolymer generation may be the

Table 10.6 Production Problems and Suitable MEOR Processes Modified After Bryant and Rhonda [392] and Volk and Hendry [22]

Production Problems	MEOR Process	Helpful Microorganisms
Formation damage, low oil relative permeability	Well stimulation	Generally surfactant, gas, acid, and alcohol producers
Trapped oil due to capillary forces	Waterflooding	Generally surfactant, gas, acid, and alcohol producers
Poor seep efficiency, channeling	Permeability modification	Microorganisms that produce polymer and/or copious amounts of biomass
Paraffin problems, scaling	Wellbore cleanup	Microorganisms that produce emulsifiers, surfactants, and acids. Microorganisms that degrade hydrocarbons
Unfavorable mobility ratio low sweep efficiency	Polymer flooding	Microorganisms that produce polymer
Water or gas coning	Mitigation of coning and/or copious	Microorganisms that produce polymer and/or copious amounts of biomass

best MEOR solution. Table 10.6 shows the common production problems and their probable solution using MEOR applications. After clear determination of the problem constraining the oil production, the best MEOR practice can be formulated. For this, Volk and Hendry proposed [22] it is necessary to quantify parameters including the microbial growth rate, mass of the target product per unit mass substrate consumed and added, and concentration and fate of the bioproducts.

10.12.2 Considering Microbiological Principles

The other important step to have a beneficial MEOR project is to gain a thorough knowledge about the existing microbial biological community in the target reservoir. It is essential to now about the diversity of the microbial life and biochemical cycles. As it was mentioned before, it is unlikely that the pristine microbial diversity is preserved after production rate declines. Thus, understanding the status of microbial consortia before commencement of MEOR treatment is much more important than the status of the pristine microbial biology before the time production started [22]. To exploit the microbial activities for enhancing oil recovery, at least it should be clear that which microorganisms are present in the reservoir and what are the parameters affecting their growth and activity [17]. For employing in situ approaches, it is essential to ascertain that a diversity of microbial life capable for metabolisms even in extremely hypersaline and thermophilic oil reservoirs exists [17,125,804,805]. It is likely that the microorganisms already present in the reservoir take the desired actions.

If the microorganisms needed to perform specific metabolisms are absent, injection of microorganisms along with the nutrients is the alternative. It would be difficult for the exogenous microorganisms to establish themselves over the indigenous ones. As the supplied nutrients may stimulate the detrimental microbial activities such as souring and corrosion, extreme caution should be considered [17].

The oil fields worldwide exhibit complicated biological systems, the exact replication of which in laboratory is very challenging [15]. Maudgalya et al. [23], after reviewing several field trials, stated that in most of the cases the laboratory results cannot be replicated by the field trials. As an example, the performance of the lipopeptide-producing *Bacillus* strain JF-2 [806] was tested in both laboratory core flood and field [88,687] and the results were inconsistent. Emerge of new advanced technology will provide acceptable explanations by analyzing the data. Generally, it is observed that microbial behavior is inconsistent. Employment of a specific bacteria type may be successful at times but unsuccessful at other times [23,24]. It is very difficult to duplicate the prevailing dynamic environment of an underground reservoir in laboratory [84]. Moreover, in the laboratory studies, the cores' length is in the range of inches to a few feet, thus they cannot truly represent the characteristics of the petroliferous formations [736]. The other point is that interaction of the multiplying microorganisms with the porous media matrix leads to some chemical and physical changes within the reservoir duplication of which is not possible in the laboratory [84]. It goes without saying that the studied cores must be acquired from petroliferous formations. It should be mentioned that many patents on MEOR are based on laboratory studies [736].

10.12.3 Temperature

As it was mentioned before, temperature is the most controlling parameter of the microbial life in deep biosphere [118]. The most important criteria to apply MEOR is the reservoir temperature [24]. There is a direct relationship between depth and temperature, as temperature increases with an increase in depth [25,807]. For example, the temperature gradient in the North Seas is about 2.5°C/100 m [808]. This indicates a temperature as high as 90°C for depth of 3000 m. Temperatures greater than 80°C are common in oil reservoirs [119]. Bachman et al. [121] mentioned the natural temperature of oil reservoirs varies from 10°C in the Canadian Athabasca oil sands [688] to 124°C [8] with the majority in range of 40–80°C [561,809–811]. As the extreme temperatures, Donaldson et al. [25] mentioned maximum and minimum reservoir temperatures of 404.4°C and −5.6°C in Wyoming and Mississippi, respectively. Temperatures greater than 80°C may prevent anaerobic oil degradation [129]. High temperatures hinder the biochemical production by the microorganisms [25]. In addition, temperature highly influences the enzymes functionality [24]. Enzymes are the proteins acting as the catalyst in chemical reactions. Based on the temperature ranges the microorganisms can survive, they can classified as psychrophiles (<25°C), mesophiles (20–45°C), thermophiles (45–80°C), and hyperthermophiles (>80°C). Further details about the microbial thermophily can be found in literature [812–814]. For example, the upper limit temperature tolerance level of *Leuconostoc mesenteroides* is 40°C [4]. Some microorganisms are reported to be able to survive only up to 80°C [24], some at 115°C [24] and even higher at 121°C [122]. A study by Maudgalya et al. [23] reported the best temperature for MEOR projects is less than 93°C. In addition, Zahner et al. [39] reported the best temperature for MEOR projects is less than 93°C, which supports the Maudgalya et al. [23] statement. Certain hyperthermophiles may be present at reservoir extreme temperature condition [125,150,339,815]; however their indigenous nature is questionable [6,339,815].

10.12.4 Pressure

Depending on the geographical area, the pressure gradient varies between 0.43 and 1.0 psig/ft (0.973 and 2.262 MPa/100 m), however, is some areas the rate of pressure changes increases with increasing the depth [816]. The considerable hydrostatic pressure, which is in range of several tens of MPa will not prevent the microorganisms' life but adversely affect the growth of microorganisms, which are adapted to the atmospheric pressure condition [817–819]. Donaldson et al. [25] specified that extreme pressures impose considerable effects on growth and metabolism of microorganisms. Based on Schwarz et al. [820], increasing the pressure at ambient temperature will result in a considerable decrease in the rate of hydrocarbon metabolisms compared with that in atmospheric pressure. The most applicable pressure for EOR in producing wells is in range of 20–30 MPa [84]. Pressures lower tan 10–20 MPa generally do not make a tremendous impact on the microbial metabolism [25]. For many mesophilic microorganisms, a hydrostatic pressure in range of several tens of MPa will hinder the cell growth and it will be completely inhibited at about 50 MPa [821]. It is reported that some bacteria can only survive at pressure up to 20 mPa [24]. It is worthwhile to mention that pressure effects may depend on other physicochemical factors such as pH, temperature, composition of culture media, and oxygen supply [818,821,822]. In addition, microbial pressure maxima can be affected by the utilized nutrients by tens of MPa [823]. Not only the pressure value but also the duration of being exposed to the pressure will affect the microorganisms [821]. Moreover, an increase in the overburden pressure will reduce the permeability in sandstones [824]. The authors observed that an overburden pressure of 3000 psig decreased the permeability to the 59%–89% of the permeability measured under no overburden pressure. Hover, above a certain compaction pressure, permeability will not change by further pressure increase [25]. Donaldson et al. [25] mentioned that for most MEOR processes barotolerant microorganism will be necessary rather than the barophilic ones. Barotolerant microorganism can grow at severe pressures but their optimal growth does not depend on the high pressures. The pressure tolerance is dependent on the prevailing biophysical condition [825]. Microbial growth at high pressures depends on energy source present, inorganic present, Eh (redox potential), pH, and temperature [25]. Salts such as *NaCl* and also divalent cations such as Mg, Mg^{2+}, and Ca^{2+}, which are common in in petroleum reservoirs, can confer a greater pressure tolerance to some marine organisms [25,826–828].

10.12.5 Salinity

The importance of aqueous phase on microbial growth and metabolisms was debated in Section 10.5. The origin of the water found in mostly marine. As it was mentioned before, water is injected to the reservoir during EOR processes.

Reservoir water may contains high dissolved salt contents as high as 0.1 % to saturation [829]. Sodium chloride (NaCl) accounts for up to 90% of the total dissolved solids found in the reservoir brine [84]. Capability to tolerate the NaCl is an essential key factor for microorganisms to be used in MEOR [25]. Donaldson et al. [25] specified more likely candidate halophiles for MEOR processes would be those capable to grow over a wide range of salinities, often referred to as moderate halophiles. A study by Zahner et al. [39] testified successful MEOR at formation salinities as high as 140,000 ppm TDS. However, as it was mentioned earlier, most of the microorganisms utilized for MEOR cannot function in salinities greater than 100,000 ppm [23]. Grula et al. [698] reported about a significant reduction in solvents and gases production by *Clostridia* sp. at NaCl concentrations as high as 5% w/v. However, there are reports regarding isolation of halophilic methanogens among which one microorganism can grow optimally at NaCl concentration of 15% w/v [272,830,831]. It is found that moderate homophiles can dominate the extreme halophiles at high salinities and limiting nutrient concentrations, which is commonly the condition for MEOR processes [25,832]. Fujiwara et al. [833] stated that salinity along with pH affects the enzymatic activity and change the membrane thickness and cellular surface. Salinity of the brines acquired from the oil fields spans in range of a few thousand to 463,000 ppm TDS [834]. In addition, it is likely that the salinity gradient be different in the range of the same formation [84]. Donaldson et al. [25] mentioned that there is positive relationship between the microorganisms' growth at high salinities and their ability to grow at high temperatures, the condition, which is common in oil reservoir [835–838]. Extensive waterflooding of the high saline reservoir with low saline water may reduce the overall salinity and make the reservoir suitable for MEOR.

10.12.6 pH

A broad range of pH can be observed in reservoirs. This factor is considered one of the main environmental factors that affect the microbial growth [84]. Microorganisms grow superbly under slightly alkaline condition [10] and at low pH values the microbial activities will be adversely affected [84]. In general, the optimal pH range for microbial growth is 4.0−9.0 [84,839]. However, there are evidences of withstanding extreme pH values less than 2.0 at high temperatures by *Sulfolobus* [840]. Moreover, Donaldson et al. [25] mentioned about isolation of microorganisms capable of growth at pH values as low as 1.0 and as high as 12.0. Before injecting microorganisms to the reservoir, the growth capability under the reservoir condition such as pH should be confirmed as it is done in several laboratory studies [559,560,841]. The pH induces some effects on enzymatic activities and many enzymes are sensitive to pH [84,833]. Jenneman and Clark [842] stated that the prevalent pH range in the oil reservoirs may not hinder the microorganisms' growth; however, the pH gradients can influence the specific metabolic processes required for some certain MEOR processes. pH can indirectly affect the microorganisms' growth and metabolism by influencing the solubility of toxic materials [25].

10.12.7 Lithology

In MEOR activities, it is necessary to inject the nutrients (and in some cases microorganisms) in to the reservoir. The effects of rocks and clays on retentions of microbial cells and nutrients are of importance through the transport process [25,629,843,844]. Several different minerals build the oil reservoir rocks. Sedimentary rocks are the most common in oil reservoirs; however, hydrocarbons might be found in ingenious and metamorphic rocks too [25,816]. Sandstones and carbonates including limestones and dolomites are the main categories of sedimentary rocks in which hydrocarbons can be found [25]. Fractured carbonate reservoir contains most of the world's oil [793]. In this type of reservoirs, the matrix blocks are mixed to oil-wet. Carbonates and silicates do not significantly retard the microbial activity but adsorptive capacity of clays and some other minerals present in the reservoir rocks may interfere with microbial processes [25]. There are charges on the rocks surfaces, which can adsorb the microorganisms and prevent their transport. Amongst the clays, montmorillonite and kaolinites are the greatest and the least ion exchange capacities, respectively, and illites exhibit intermediate capacity [25]. Via swelling, clays absorb water, which impose some limitations for microorganisms' migration [25].

10.12.8 Porous Media and Microorganisms' Size

Scrutinizing several field trials, Maudgalya et al. [23] stated that a minimum reservoir permeability of 75 mD has been the most suitable for MEOR projects. The lower limit for effective microbial transport is reported to be 75–100 mD [842]; however, some studies indicate microbial transportation in permeabilities lower than 75 mD [646,845]. Moreover, the movement of microorganisms in a formation with permeability of 30 mD is proved too [417]. There have been the problem of plugging in some early MEOR experiments [72,846]. In a study, Davis and Updegraff [847] stated that to avoid plugging problems, the diameter of the pore entry should be at least twice the diameter of the injected microbial cells. In his patent, Hitzman [848] suggested to use spores instead of vegetative cells as their size is smaller. Later, it was discussed that spores could induce plugging and it was suggested to incorporate UMB due to their further smaller size [422]. Some years later, Jack et al. [849] concluded that the microorganisms to be injected should be small and the ideal one is with a size less than one-fifth of the pore throat size of the target formation. Bacteria can be found in different morphologies such as rods, curved rods, cocci, tetrads, chains, etc, and have typical dimensions of 0.5–10.0 μm in length and 0.5–2.0 μm in width [25]. This indicates that pore dimensions less than 0.5 μm pose severe restriction on microbial activity. It is suggested that the pore considerable bacterial activity was reported in media in which the interconnections of pores have at least 0.2 μm diameters [850]. In accordance to the permeability calculations by Stiles [851], Gray et al. [66] suggested that reservoirs with porosities less than 6% would be suitable cases for microbial plugging. Comparing the shales with sandstones, the former has much smaller pore-throat (less than 0.2 μm compared with up to 13 μm for sandstone) [852]. In such systems, the microorganisms may not be able to easily transport within the matrix and also the rate of nutrients diffusion will be slow too. Sheng [24] stated that pore geometry and size can influence the chemotaxis although it has not been proven in reservoir condition. In addition to the transfer of microorganisms, porous media characteristics such as porosity, permeably, and pore size can affect the microbial growth and metabolism and also size and number of the bacterial cells [25,853–855].

10.12.9 Oil Gravity

Successful MEOR trials have been reported for the oil gravity range of 0.82–0.96 g/cm^3 [39]. Pautz and Thomas [856] reported an API oil gravity range of 34–40 for several MEOR project performed all over the world.

10.12.10 Depth

Deep reservoirs are usually associated with high temperature, pressure, and salinity and also poor permeability, which adversely affect the MEOR efficiency [23,24]. Deep reservoirs are not favorable candidates for MEOR. The depth itself does not impose limitation on microbial growth; actually, its effect on the temperature and pressure influence the microbial growth and metabolism [25]. For the several MEOR projects all around the world the data of which have been collected by National Institute for Petroleum Energy Research (NIPER), the average and maximum depth have been about 550 m and 800 m, respectively [856].

10.12.11 Well Spacing

It is proved that the microorganisms can grow and travel in porous media as they are observed in wells nearby the injector well [857]. Thus, the injection well spacing is of great importance in MEOR success. This is important because the microorganisms consume all the nutrients as they move forward [23,407]. The time the microorganisms spend within the microbial incubation zone, referred to as the residence time, should be longer than the time needed by a bioproduct to reach to the desired concentration [407]. In cases with relatively small spacing, in order to reach the desired concentration of bioproducts, the metabolic rate should be higher or higher concentrations of microorganisms and nutrients should be employed. Sheng [24] proposed 40 ac for well spacing for MEOR.

10.12.12 Residual Oil Saturation

After extensive waterflooding as the secondary oil recovery, still a significant volume of oil remains in the reservoir, which is called residual oil [6]. For applying MEOR, the residual oil saturation should be high enough to justify the project economy. Sheng [24] suggested residual oil saturation greater than 0.25 is suitable for MEOR processes.

10.12.13 Metals

Heavy metals can act as very toxic materials to the microorganisms in levels highly more than what is required for nutrition, which is commonly in range of 10^{-3}–10^{-4} M [25]. Generally, heavy metals' concentration more than 10^{-3} M can become toxic to many microorganisms, while high concentration of light metals cations may induce inhibition or stimulation [25,858]. Parameters such as pH, temperature, pressure, and salinity can influence the solubilization of metals; thus determining of metal toxicity in suit is complicated [25]. For example, Bubela [859] reported that increasing temperature from 53 to 63°C makes the copper more toxic to *Bacillus stearothermophilus*, which is a thermophile. Heavy metals such as copper, zinc, ferric iron, etc. may be present at concentrations higher than the required level. Hitzman [860] mentioned the detrimental effects of heavy metals such as arsenic and lead on the microorganisms' growth and the fact that their concentration should not be in excessive quantities in the oil or the formation to be treated. It should be mentioned that organisms are influenced in different ways and there are some organisms that can tolerate very high concentrations of almost any heavy metal [861]. Sheng [24] mentioned the suitable concentration of arsenic and mercury less than 15 mg/L.

10.12.14 Souring Due to the Presence of Sulfate-Reducing Bacteria (SRB)

SRB plays a very negative role in MEOR processes [15,21,129]; however, recently some positive roles are reported [304,862] as they are capable to reduce the oil viscosity, replenish the declining pressure of reservoir, transform the heavy oil to light oil via the effect of produced acids, and gas (H_2S) as well as their wide availability in global oil reservoirs [304]. In addition, it is reported that SRB nonselectively plug the porous media [31], which adversely affect the oil recovery.

Reservoir flooding by the seawater or brine containing a high level of sulfate can be the starting point for souring due to sulfide production [6,37,352]. It is worthwhile to mention that SRB can survive extended starvation in sea water at both reservoir and surface temperatures [15]. Suitable condition for SRB to produce hydrogen sulfide can be provided by [6,37,863]

1. Supplying nitrogen, sulfate, and phosphorous sources by the injected water;
2. Reducing the reservoir temperature by the injected water, which is cooler; and
3. Presence of the electron donors (organic acids and hydrocarbons) in the reservoir.

The deleterious effects of presence of hydrogen sulfide or in other words souring can be summarized as [318,354,362,370,864–866]

1. Corrosion of pipelines and equipment;
2. Inducing additional costs to refine oil and gas;
3. Increasing the health risks as H_2S is highly toxic; and
4. Plugging the reservoir due to accumulation of sulfides minerals.

The microbially-influenced corrosion (MIC) may be the most important detrimental effect of the SRB. The cost related to MIC is hundreds of millions of dollars per year [867,868]. The role of SRB on corrosion of ferrous metals has been reviewed by several researchers [310,315,554,867,869–871].

Biocides such as bronopol, formaldehyde, glutaraldehyde, benzalkonium chloride, cocodiamine, and tetrakishydroxymethyl phosphonium sulfate are generally used to control the H_2S concentration [6,872]. Employing biocides is associated with some problems such as the need for high concentrations to achieve the desired results [863,873,874] and health concerns for the operators [6] as well as hazardous to the environment. However, there have been some efforts to evolve green and biodegradable biocides to remediate the SRB problem [875]. Addition of nitrate, nitrite, molybdate, and inorganic nutrients to the oil formation is suggested as an alternative to inhibit sulfate reduction and also stimulate the indigenous microorganisms to produce CO_2 [876]. As it was mentioned before, it is well established that nitrate or nitrite is effective in controlling souring [138,346–360]. This increased the interest in NRB in oil fields [345,346,361,362]. In case of using as the bacterial electron acceptor, nitrate provides more energy than sulfate, thus, the growth of NRB is enhanced, which out-compete the SRB [37,354,877]. The mechanisms through which nitrate addition can control souring are [6,863,878–882]

1. Competition for electron donors between the SRB and NRB;
2. Promoting the redox potential and consequently inhibition of SRB;
3. H_2S oxidation by NRB; and
4. Production of incompletely reduced nitrogen compounds such as NO_2, which inhibit the sulfate reduction pathway.

In different cases, it is probable that one mechanism dominate the others or multiple mechanisms do simultaneously [863]. As it was mentioned before, NRB are categorized as hNRB and SO-NRB. The former outcompete the SRB for common electron donors due to the fact that nitrate or nitrite reduction is more favorable than sulfate reduction in energetical point of view. In other words, nitrate reduction to nitrogen or ammonia provides more free energy than sulfate reduction [688]. This dictates greater molar growth yields compared with SRB [37]. The mechanism of SO-NRB is different. With nitrate or nitrite as the electron acceptor, SO-NRB oxidize the hydrogen sulfide to sulfate or sulfur and do not effect on the SRB growth [6,874,881]. Youssef et al. [6] specified the importance of SO-NRB in decreasing the sulfide concentration reported by several laboratory experiments [318,865,881,883,884].

Sulfurospirilum spp. is reported to be capable for both hNRB and SO-NRB metabolisms [885]. *Thiomicrospira* sp. strain CVO and *Arcobacter* sp. strain FWKO both are reported to be SO-NRB [363]. Other microorganisms capable to reducing nitrate are *Denitrovibrio acetiphilus*, *Proteobacteria*, *Campylobacter* sp. strains NO3A, NO2B, and KW, *Garciella nitratireducens*, belonging to cluster XII of the *Clostridiales*, moderately thermophilic members of the genus *Geobacillus* [333,361,364,878,886]. Gittel [887] mentioned that although the activity of hNRB and SO-NRB was not specifically assessed, the recovery of sequences affiliated with representatives of both types of nitrate reducers, including members of the *Epsilonproteobacteria* (*Sulfurospirillum* spp.,*Arcobacter* spp.) and the *Deferribacterales* (*Deferribacter* spp.), is promising for future studies.

10.13 FIELD TRIALS

The Socony Mobil Research laboratory performed the first MEOR field trial in the Lisbon field, Union County, Arkansas in 1954 and reported marginal success due to the increase in the wells oil follow rates [16,31,76–88] and the analyses mentioned the complexity of using microorganisms. However, Volk and Liu [32] mentioned that the pioneering field studies were performed in the United States in the 1930s and 1940s by Claude ZoBell et al. at the Scripps Institute of Oceanography in

La Jolla, California. A comprehensive review of the early works is available in literature by Premuzic and Woodhead [86].

The starting point of MEOR field trials in former Soviet Union backs to the 1960s. Encouraged by this, four Eastern European countries of former Czechoslovakia, Poland, Hungary, and Romania performed some field trials [31]. For example, in Romania, several MEOR field tests were performed between 1971 and 1982, reporting successful results [31,104]. Youssef et al. [6] mentioned the improvement in the technology of MEOR as using mixture cultures adapted to the nutrients and reservoir condition such as temperature and pressure and also incorporating larger volumes of nutrients [105−114].

In 2003, Van Hamme et al. [19] stated that more than 400 MEOR field projects have been done so far just in the United States. Based on Khire and Khan [43,115], MEOR projects has been applied on over 400 wells in the same country. In addition, this recovery method has been tested on more than 1000 wells in numerous oil fields in China [32]. Based on Thomas [116], an estimated amount of 2.5 million oil barrels per day were produced in 2007 using EOR method, of which the role of MEOR was negligible. On the other hand, based on a report by Chinese Ministry of Land and Resources (www.mlr.gov.cn), nearly 50 billion oil barrels in onshore Chinese oil fields have the potential to be treated by MEOR [32]. Youssef et al. [6] specified that the residual oil saturation in many MEOR field trials increased by 10%−340% for 2−8 years [78,92,99,101,106,108,117]. In recently performed reviews, Maudgalya [23] in 2007 mentioned 407 field trials reported in literature [11,76,78,87,108−110, 114,351,392,408−410,702,763,766,780,800,803,849,888−904]. In the investigated trials, among the different recovery mechanisms, the permeability profile modification has been the most successful. A list of some MEOR field trials is presented in Table 10.7.

10.14 ENZYME ENHANCED OIL RECOVERY

Incorporating the enzymes in petroleum industry was first suggested by Harris and Mckay in 1998 [946] for:

1. Enzyme pretreatment of biopolymers to improve them;
2. Handling characteristics and gel breaking in drilling to disrupt filter cake formation;
3. Desulfurization of hydrocarbons; and
4. Enzyme-based acid production for different purposes like formation damage treatment and matrix acidizing of carbonate, etc.

Table 10.7 Some MEOR Field Trials

Country	Technology	Cases	Cases Characteristics	Microbial system	Nutrient	Effects	Ref.
Argentina	*Batch, Squeeze*	Six wells in Piedras Coloradas oil field	Two separate reservoirs The area produces 430 M^3/D of very paraffinic oil from 85 active wells Average production per well is 5.8 M^3/D	Hydrocarbon degrading anaerobic facultative microorganisms	Nourishment from linear hydrocarbon	Oil production enhancement between 25.8% and 110 %; water cut reduction by 39.1%, 59.5%, 55.6%, 72.8%, 58.7% and 40% in different wells; oil viscosity reduction	[905]
Argentina	*MSPR*	Vizcacheras oil field (Papagayos formation)	Two main reservoirs (Papagayos and Barrancas) 60% of total production is from papagayos with: Temperature: 198°C Average permeability: 1000 mD Effective porosity: 25% Residual oil saturation: 25%	Hydrocarbon degrading anaerobic facultative microorganisms	Inorganic nutrient (N,P,K and oligoelements)	An increase in oil recovery; water retraction; fractional flow alteration	[410]
Argentina	APAC-Flow	Diadema field	Absolute permeability: 500 mD Depth: 900 m Temperature: 52°c Oil density: 21°API Irreducible water Saturation: 37% Original formation pressure: 70 kg/cm^2 Saturation Pressure: 59 kg/cm^2 Original oil viscosity: 55 cP Volume factor: 1.068 m^3/m^3	Typically 160 L of microbial formulation and 28 m^3 of nutrient solution	Reservoir brine	Microbial action was proved by the presence of CO_2, decreased emulsion and increased carbonate and bicarbonate concentration, in addition to the increase in oil production	[902]
Argentina	MIT[a]	Piedras Coloradas field	The area produces 430 M^3/D of very paraffinic oil becoming from 85 active wells Average production per well is 508M^3/D	Hydrocarbon degrading anaerobic facultative microorganisms	Nourishment from linear hydrocarbon	The technology is cost effective, easy to implement and complies very well with local environmental regulations and biosafety issues	[905]

(continued)

Table 10.7 (Continued)

Country	Technology	Cases	Cases Characteristics	Microbial system	Nutrient	Effects	Ref.
Argentina	MEOR	Vizcacheras field	60% of total production comes from Papagayos formation Temperature: 198°F Average permeability: 1000 mD Effective porosity: 25% Residual oil saturation: 25%	Hydrocarbon degrading anaerobic facultative microorganisms	Salts containing N, P, K, oligoelements	Oil increased and water decreased in neighbor producers of treated injector reduction of residual oil saturation by in situ biosurfactatnts with low interfacial tension improved oil mobility from short-chain solvent	[410]
Argentina	MEOR	La Ventana field	La Ventana block has a total of 230 producers. Current oil production is 11,950 BOPD (1900 M^3/D) and 214 MBWPD (34,000 M^3/D) of coproduced water	Hydrocarbon degrading anaerobic-facultative microorganisms	Injection water	MEOR Incremental Reserves (IR) totaled a range of 7893 bbls to 21,000 bbls of oil during the first year and 22,358 bbls to 40,371 bbls during the second year. Values ranging from 138 Mbbls to 256 Mbbls of oil for the next five years of MEOR was assumed as a conservative projection	[894]
Australia	BOS[b] system	Alton oil field in Queensland		Ultra microbacteria with surface active properties	Formulate suitable base media	Positive effect	[42,571]
Australia	MCF	Alton field	Temperature: 76°C Permeability: 11 to 884 md Porosity: 15.4 to 19.8%	Interactions between biometabo1ite producing species and the resident microbiota of the reservoir	Formation water	An approximate 40% increase in net oil production	[803]
Brazil	*CMR*	An onshore oil field located in the northeast of Brazil	The salinity of formation water is 2% NaCl Temperature: 45°C The salinity of injection water is 0.01% NaCl	Reservoir native bacterium	Adaptable nutrients	Plugging of the high permeability zone; Vertical sweep efficiency improvement; Biopolymer production	[906]
Bulgaria	CMR, ASMR[c]	All cases		Indigenous oil-oxidizing bacteria from water injection and water formation	Water containing air, ammonium and phosphate ions; Molasses 2%	Positive effect	[689]
Canada	MSPR	A heavy oil reservoir with high permeability zones	Unconsolidated Sand Depth: 650 m Temperature: 21°C Permeability: 1500 mD Porosity:30% Oil viscosity: heavy Density: 15°API	Biopolymer producing bacteria (Leuconostoc mesenteroides)	Dry sucrose; Sugar beet; Molasses; Fresh water	Surface tension reduction from 66.5 to 59.6; Restoring good fluid flow; PH reduction from 6.4 to 6; Bioproducts generation as: acetic acid, lactic acids, ethanol and propanol	[729, 910–910]

Canada	MEOR	Trial field in Saskatchewan	Average Porosity: 15.2% – 21.5% Average Permeability: 53–567 mD Reservoir Temp: 47°C Depth: 1200 m Average Recovery: 29% Oil Gravity: 22–24°API	Chemical nutrients solution mixed with injection water	Unknown	The Well produced at an average of 200% more oil with 10% decrease in ware cut	[911]
China	Huff & Puff	Fuyu oil field	Temperature: 28°C Permeability: 240 md Porosity: 0.27 Specific Gravity: 0.87 Oil Viscosity @Reservoir: 40 mPa.s	Microorganisms CJF-002 producing insoluble polymer	Molasses	Water cut decreased from 99% to 75% Oil production rate increased from 0.25 to 2.0 ton/d	[779,896]
China	Huff & Puff	Daqing oil field	Five-spot well patterns are applied in 6 injection wells and 11 production wells Original oil in place: 175.04 × 104 t Temperature: 45°C	*Pseudomonas aeruginosa*, *Xanthomonas campestris*, *Bacillus licheniformis*, and 5GA, which is like Bacteroides	K H_2PO_4, Na H_2PO_4 CaC12.7 H_2O, FeSO4.7 H_2O $(NH_4)2SO_4$, $MgSO_4$. 7 H_2O Molasses 5% Residue of sugar 4% Crude oil 5%	Oil recovery increased by 34.3%, and the residual oil recovery was 69.8%	[897,898]
China	MEOR	Xinjiang oil field	Water Content: 80% Temperature ~42°C Average Porosity: 29.9% Average Permeability: 522* 10^3 μm^2	Hydrocarbon-Degrading Bacteria (HDB) Nitrate-Reducing Bacteria (NRB) Sulfate-Reducing Bacteria (SRB) Methanogens	Hydrocarbon as the carbon source	Abundant microbial populations, including HDB, NRB, SRB, and methanogens, are ubiquitous in water-flooding reservoir. The reservoir has potential for MEOR and biological control of SRB propagation by stimulating NRB	[912,913]
China	MEOR	Fuyu oil field	Sandstone Reservoir Depth is from 300 to 450 m Temperature: 30°C Average Permeability: 240 mD Average Porosity: 27% Water Cut: 90%	Strain CJF-002 producing insoluble polymer	Molasses	MEOR can effectively increase oil recovery MEOR is expected to be an economically feasible technique	[833]

(continued)

Table 10.7 (Continued)

Country	Technology	Cases	Cases Characteristics	Microbial system	Nutrient	Effects	Ref.
China	MEOR	Daqing's Low permeability areas	Six oil layers Depth: 500 to 2200 m Permeability:$0.1 \star 10^{-3}$ to $200 \star 10^{-3}$ μm^2 Viscosity: 8 to 100 mPa.s Wax Content: 20 to 30%	*Brevibacillus brevis* and *Bacillus cereus*	Hydrocarbons as carbon source	MEOR was able to enhance oil recovery by 6.5% over that obtained by water flooding Viscosity of crude oil declined by 40% and wax content and gum content dropped by various degrees, improving the oil rheology The combination of MEOR fracturing deserves consideration for development of low permeability reservoirs IFT decreased by 50% after microbial treatment Water cut decreased by 45.2% to 38.6%	[762,914]
China	MEOR	Dagang oil field	Sandstone formation with two layers Temperature: 70 & 73°C Porosity: 27.6 & 24.9% Permeability: $468 \star 10^{-3}$ &$259 \star 10^{-3}$ μm^2 Viscosity: 75.8 & 42.5 mPa. s Density: 0.8787 & 0.8841 g/cm3	*Arthrobacter* sp, *Pseudomonas* sp. and *Bacillus* sp.	Crude oil (20 g/L), $Na_2HPO_4.12\ H_2O$ (0.8 g/L), K H_2PO4 (0.45 g/L), Yeast extract (0.25 g/L), Peptone (0.1 g/L), NH4Cl (2 g/L), Na2EDTA (0.25 g/L)	Microbial water-flooding techniques have a potential of enhancing oil recover y in high temperature oil reservoirs Microorganisms can thrive, proliferate and move in the high temperature reservoir matrix, The positive effect of the biotreatment first and mainly occurs in those production wells which have good connectivity with injection wells	[915]

China	MF	Wenmingzhai oil field	Porosity: 20%–30% Permeability: 60–726*10^{-3} μm^2 Oil saturation: 72.2% Average effective thickness: 35.7 m Density: 0.88–0.92 g/cm^3 Viscosity: 36–200 MPa.s Saturation pressure: 7–9.6 MPa	Fermentative Bacteria, Sulfate-Reducing Bacteria, Iron-Reducing Bacteria, Methane Bacteria	Formation Water	Microbial flooding resulted in stable wellbore pressure and better water injection profile Water cut was reduced and oil production rate was increased in corresponding oil wells, total oil increasment was 9536 tons Microbial flooding process is of great economic benefit and good application prospect	[916]
China	MEOR	Daqing oil field	Average depth: 1400 ft Average porosity: 16% Average effective thickness: 9.2 m. Wax content: 20% Temperature: 45°C Viscosity: 6.7 MPa	*Pseudomonas aeruginosa* (P-1) (isolated from the water contaminated by the crude oil)	Glucose, 20 g/L; Peptone, 2 g/L; Na_2HPO_4, 2 g/L; $(NH4)_2SO_4$, 2 g/L; K H_2PO_4, 3 g/L; MgSO4, 2 g/L; $CaCl_2$, 0.05 g/L	*Pseudomonas aeruginosa* (P-1) and its metabolic products (PIMP) of 10% could enhance the oil recovery in the model reservoir by 11.2% and also decrease injection pressure by 40.1%. PIMP (10%) could reduce the crude oil viscosity by 38.5%	[169,561, 626]
China	MEOR	Kongdian oil field	Sandstone oilbearing Depths: 1206–1435 m Temperature: 59°C Average porosity: 33% Stratum permeability: 1.878 μm^2 Density: 0.900 g/cm^3 53% saturated hydrocarbons 20% aromatic compounds 21.15% resins and asphaltenes	*Anaerobic thermophilic* microorganisms, including fermentative (102–105 cells/ml), sulfate-reducing (0–102 cells/mL), and methanogenic (0–103 cells/mL) microorganisms	Bacto tryptone (0.5 g/L), yeast extract (2.5 g/L), and glucose (1.0 g/L)	The method for enhancement of oil recovery based on the stratal microflora activation is appropriate for use at the Kongdian oil field	[917,918]

(continued)

Table 10.7 (Continued)

Country	Technology	Cases	Cases Characteristics	Microbial system	Nutrient	Effects	Ref.
China	MEOR	Liaohe oil field	Formation temperature: 40–90°C Water content: <80% Formation permeability: $>50 \star 10^{-3}\ \mu m^2$ Paraffin content: 15–25% Pour point: 25–35°C	*Bacillus* sp. (LWH 1), *Bacillus* sp. (LWH 2), and *Pseudomonas* sp. (LWH 3)	Solid wax as the sole carbon source	Tested wells obtained good effects after mixed bacteria treatment Oil production increased by 561 tones 16 cycles of thermal-washing treatment and 44 cycles of additives were eliminated from the four wells during 4 months of testing Considerable economic profit was achieved	[765]
China	CMR[d]; MFR[e]; MSPR[f]	Shengli oil field (pilot tests)		Microorganisms in all cases: Slime-forming bacteria: *Xanthomonas*, *Campestris*, *Brevibacterium viscogenes*, *Corynebacterium gumiform*; Mixed enriched bacterial cultures of *Bacillus*, *Pseudomonas*, *Eurobacterium*, *Fusobacterium*;	Bacteroides; Bacillus cereus; Brevibacillus brevis; Hydrocarbon-degrading strains Nutrients in all cases: Molasses 4%–6%; Molasses 5 %; Residue sugar 4 %; Crude oil 5 %; Xanthan 3 % in waterflooding.	Oil production increased in a range of 2001–122800 t in different cases; Water cut reduction; Natural decline rate alteration.	[16,762, 915, 921–930]
China Denmark	CMR[g]; MFR[h]; MSPR[i] MEOR	Dagang Kongdian oil field Xinjiang oil field Jilin oil field Huabe Baolige oil field Changing Jing'an Y9 oil field Daqing oil field (post polymer flooding)		Microorganisms in all cases: Slime-forming bacteria: *Xanthomonas*, *Campestris*, *Brevibacterium viscogenes*, *Corynebacterium gumiform*; Mixed enriched bacterial cultures of *Bacillus*, *Pseudomonas*,	Bacteroides; Bacillus cereus; Brevibacillus brevis; Hydrocarbon-degrading strains Nutrients in all cases: Molasses 4%–6 %; Molasses 5 %; Residue sugar 4 %; Crude oil 5 %;	Oil viscosity reduction by 7.7%; production improvement; Inter-well permeability profile modification; Surface tension reduction. Oil emulsification; Emulsion stability improvement; Water cut reduction;	[16,762, 915, 918–927] [84]

		Eight articles mostly dedicated to combination of different chemical and physical laboratory methods for experimentation, analysis and interpretation	*Eurobacterium*, *Fusobacterium*; Adapted strain of *Clostridium tyrobutyricum*	Xanthan 3 % in waterflooding. Molasses	Oil production improvement. Promotion of water injection profile. Rate of kinematic viscosity reduction increased. High permeability zone plugging; Improvement of oil displacement coefficient; Oil production improvement. Polymer plugging removal; Oil production increased by 165 %; Oil viscosity reduction; Alkane profile alteration. Treatment for microbial improved oil recovery from sand and carbonate packed columns by application of molasses and inoculums of adapted strain at a salt concentration of 90 g/L showed 38% improved recovery in sandstone and 25 % in carbonate rock after a shut in period	
England	MHAF[j], MSPR	Case 1	Naturally occurring anaerobic strain (high acid generator); Special starved bacteria (good producers of exopolymers)	Soluble carbohydrate sources; Suitable growth media	Positive/negative effect	[901]
Former Czecho-slovakia	CMR, MFR, ASMR	Preliminary trials in an oil deposit	Mixed cultures of sulfate-reducing bacteria and *Pseudomonas* SD.	Molasses	Microbial growth detection; Oil viscosity reduction; Stimulation of sulfate-Reducing Bacteria	[109,928]
Former East Germany	MFR, ASMR	A carbonate reservoir case	Mixed cultures of thermophilic: *Bacillus* and *Clostridium*; Indigenous brine microflora	Molasses 2%—4% with addition of nitrogen and phosphorous sources	Oil production enhancement; Water/oil ratio reduction from 88% to 34%	[929]

(*continued*)

Table 10.7 (Continued)

Country	Technology	Cases	Cases Characteristics	Microbial system	Nutrient	Effects	Ref.
Hungary	MFR	Demjen field	Sandstone and Limestone Depth: 650–8061 ft Temperature: 207°F Permeability: 10–700 mD	Mixed sewage-sludge cultures; Anaerobic thermophilic mixed cultures (*predominants: Clostridium, Desulfovibric* and *Pseudomonas*)	Molasses; Sucrose; KNO_3; Na_3PO_4	Increase of oil production by 12%–60 % for a few weeks up to 18 months; Gas production (CO_2); Decrease in PH, oil viscosity and oil/water ratio	[101]
India	*CMR, MSPR*	Indian oil fields		Multi-bacterial consortium: clostridium type thermo anaero bacterium sp. and *Thermococcus* sp.	Molasses 3%	An increase in oil production	[31]
Indonesia	MCF[k] Huff & Puff	Ledok field	Total depth: 186 m Porosity: 26.60% Permeability: 300 mD Temperature: 29.0°C	Indigenous microorganisms enriched with *Bacillus licheniformis*	Molasses	Increase in oil production rate, from 8.18 BOPD before the injection to 12.27BOPD, hopefully, after the injection	[899]
Malaysia	*CMR*	Three wells in Bokor offshore field	Porosity: 15–32% Permeability: 50–4000 mD Oil gravities range from 19O to 22O API in the shallower reservoirs (1500 - 3000 Ft. ss) to 37O API in the deep reservoirs (6300 Ft. ss)	Adaptive microorganisms	Adaptable nutrients	Water cut reduction; An increase in oil production by 15%, 36% and 120% in different wells; Permeability reduction; Skin reduction in two wells	[904,930, 931]
Malaysia	MCP[l]	Oil field		Isolation and combination of microorganisms in novel mixtures	Organic nutrients	Reduction of interfacial tension, decrease in oil viscosity, and improving the microscopic sweep efficiency of the water flood A 20%–50% increase in oil production rates	[932]
Mongolia		27 wells in Changqing oil field	Sandstone reservoirs Temp.: 43 – 54°C Porosity: 11.0%–17.8%	Mixed cultures of facultative anaerobes	They are motile and capable of using normal alkanes as their sole carbon food source	Increased production 17.5 t/ day or 18% 0%–48% increase in oil production; 2950 m^3 of incremental oil in 3–6 months. Treated another 20 wells; 18% increase in oil production for 15–30 days; wax content altered	[6,766]

Mongolia	MEOR	Changqing oil field	Sandstone reservoirs Effective thickness: 2.6 – 29.6 m Porosity: 11.0 – 17.8% Permeability: $1.66 - 149.2 \times 10^{-3}$ μm Temperature: 43 – 54°C Viscosity (at 50°C): 4.29 – 6.58 mPa·s Paraffin Content: 4.40 – 12.98 % Water Cut: 4% – 76 % Salinity:17.13 – 104.4 g/L	Mixed cultures of facultative anaerobes	Normal alkanes as sole carbon food source	Increased production 17.5 t/ day or 18% Reduced interfacial tension Improved crude oil mobility and the relationship between oil and water	[766]
Norway	–	North Sea MEOR field projects	Deep subsea oil reservoirs Temperatures:60–100°C Pressures: 200–400 bars Average recovery: 30%–40%	Nitrate-reducing bacteria naturally occurring in North Sea water	Nitrate and 1% carbohydrates addition to injected sea water	Negative effect	[415,933]
Peru		Talara oil field	32–36° API			36% and 46% increase in oil production; 3080 and 2200 m^3 incremental oil	[760]
Poland	MFR	16 tests in Carpathian crude oil reservoir	Depth: 1325–3753 ft Porosity:13%–25%	Mixed aerobic and anaerobic bacteria belonging to genera: *Arthrobacter*, *Clostridium*, *Mycobacterium*, *Pseudomonas* and *Peptococcus*	Molasses 4%	Significant increase of oil production up to 300%–360 % for 2–8 years; Fluids characteristics alteration	[102]
Romania	*CMR*	Romanian oil fields	Depth: 336–1559 ft Temperature: 27–55°C Permeability: 100–1500 mD Oil viscosity: 6–53 cP Density: 0.85–0.91 Kg/dm^3	Adapted mixed enrichment *Clostridium*, *Bacillus* and gram-negative rods	Molasses 4%	Oil production enhancement by 100%–200% up to 5 months; Reduction of water flooding injection pressure	[16,87, 106, 937–937]

(*continued*)

Table 10.7 (Continued)

Country	Technology	Cases	Cases Characteristics	Microbial system	Nutrient	Effects	Ref.
Romania	*CMR*	Romanian oil fields	Depth: 336–1559 ft Temperature: 27–55°C Permeability: 100–1500 mD Oil viscosity: 6–53 cp Density: 0.85–0.91 Kg/dm^3	Adapted mixed enrichment cultures predominated by *Clostridium*, *Bacillua* and gram-negative rods	Molasses 2%–4%	Oil production enhancement up to 200% in two wells for 1–4 years	[16,87, 935, 936]
Romania		Bragadiru oil field	Depth: 780 m Permeability: 150–300 mD Salinity: 0.06%–0.3% Oil viscosity: 9 cP	*Bacillus*, *Clostridium*, *Arthrobacter*, *Pseudomonas*, *Micrococcus*	Molasses	Cyclic microbial recovery, well-bore clean up	[15]
Romania	MWS[m], MEW[n], MWC[o]	Romanian's oil fields	Temperature: up to 55°C Salinity:100–150 g/L Deepness: 1000–1500 m Viscosity: up to 50 cP	Adapted mixed enrichment culture	Molasses	Positive effect	[110]
Russia	MSPR	Bashkiria reservoir		Aerobic and anaerobic activated sludge bacteria	Waste waters with addition of some biostimulators and chemical additives	Additional oil recovery of 1000–2000 t/year for each of 600 producing treated wells	[938]
Russia	NF[p]	Vyngapour oil field in west Siberia		Indigenous bacteria; *Lactobacteria*	Local industry wastes; Sources of nitrogen, phosphorus and potassium	Production of 2268.6 extra tons of oil; water extraction reduction	[939]
Russia	MFR	Three pilot tests in Romashkino field	Sandstones and silty sandstones Depth: 1500–1700 m Average porosity: 21.8% Average permeability: 500 mD Relative density: 0.871–0.876 Temperature: 30 – 40°C	Stratal microflora (aerobic and anaerobic) of flooded oil fields	Aerated fresh water with added mineral salts	32.9% additional oil recovery; Production of organic acids, surfactants, polysaccharides, methane and carbonic acid	[377]
Russia	MSPR	Case 4		Anaerobic and aerobic bacteria as sulfate-reducing dentrifying, putrefactive and acid butyric fermenting, cellulose digesting	Peat biomass and silt reach in hydrolysable substrates	Increase in oil production from 180 to 200–300 t oil/day	[940]
Russia	MFR	Case 5		Mixed aerobic and anaerobic bacteria	Molasses 4%	Oil production increased by 8 % for 4 months	[103]

Saudi Arabia	CMF, CMR, MFR, MSPR	Oil fields of seven Arab countries	Original oil in place in Saudi Arabia is about 700 billion barrels 35% of the total oil in place, can be produced by conventional methods	Adequate bacterial inoculum according to requirements of each technology	Adequate nutrients for each technology	Negative effect	[31,83, 941]
The Netherlands	MSPR	Case 1		*Betacocus dextranicus* (Slime-forming bacteria)	Sucrose; molasses 10%	Significant increase of oil production; The oil/water ratio changed from 1/20 to 1/50	[87,942]
Trinidad-Tobago	CMF	Trinidadian oil wells	Wells producing oils of gravities greater than 250 and oil production ranging from 5–12 bopd and some water; this represented approximately 10% of the active wells	Fac. anaerobic bacteria high producers of gases	Molasses 2%–4%	Negative effect	[943]
UAE	MEOR	UAE Oil Reservoirs	Limestone Cores Porosity: 6%–26% Permeability: 0.5–64 mD Temperature: 22°C	*Bacillus* and *Clostridium*	Inorganic powder nutrient	Increases in oil recovery by tertiary injection of thermophilic bacteria The mechanism for enhanced recovery seems to be through production of biosurfactant, biogas, and biomass by the bacteria Production of the biosurfactant reduced the IFT A successful bacteria EOR flood was obtained even in a tight limestone core	[543]
UK	MHAF[q]	Lidsey Field	Water-wet shallow marine limestone Seven cores taken from depths of 3346 – 3377 ft Porosity: 16.5%–19.8% Permeability: 0.62 – 4.3 8 mD	A naturally-occurring anaerobic bacterial strain capable of generating organic acid	Suitable carbohydrate sources	No well damage resulted Increased fracture length coupled with the employment of noncorrosive, nonhazardous and environmentally friendly feed stocks	[779,901]

(continued)

Table 10.7 (Continued)

Country	Technology	Cases	Cases Characteristics	Microbial system	Nutrient	Effects	Ref.
UK	MPPM[r]	North Blowhorn Creek Unit oil field	Sandstone Formation Depth: 2300 ft Initial Oil In-place: 16 million barrels Oil Production:3000 BOPD 20 injection wells and 32 producing wells	Nitrogen and phosphorus containing microbial nutrients added the injection water	Potassium, nitrate, sodium dihydrogen phosphate, and molasses	Recovery of 69,000 bbl of incremental oil during the first 42 months with a projected recovery of 400,000–600,000 bbl and an extension of the economic life of the field by 60–137 months	[11]
USA	*CMR*	Single well stimulation water-flooding case held in Tulsa, Oklahoma		Mixed cultures of: *Clostridium* sp., *Bacillus* sp., *Bacillus licheniformis*; and a gram-negative rod	Molasses 4%	Oil production improved up to 79%	[889]
USA	*CMR*	Scale up of microorganisms in a single well stimulation case		Anaerobic and facultative anaerobic bacteria high fermenting sucrose-molasses medium	Sucrose; Molasses; Phosphate salts; Nitrate salts; Yeast extract	Significant increase of oil production	[944]
USA	*CMR*	Univ.field (Oklahoma state)		*Clostridium*	Molasses (4%–8%); Dry milk solids (0.09%)	Oil production increased by 100% for 30 days; PH reduction; Production of gases, acids and solvents	[413]
USA	*CMR*	Single well stimulation case		Mixed anaerobic microbial cultures	Molasses	Oil production increased by 230% for 7 months	[108]
USA	*CMR*	Case 5		Mixed cultures of *Bacillus* and *Clostridium*	Molasses 4% with compatible mineral nutrients as $(NH4)_3PE_4$	Oil production increased up to 350%	[945]
USA	*CMR*	Case 6		Culture of *Clostridium* type	Molasses (4%–10%); Salts: Urea, ammonium nitrate, glycolate acetate	Considerable increase of oil production for 5 months	[16]
USA	MFR	Alpha environmental field test in Texas		Mixed cultures of hydrocarbon degrading bacteria	Inorganic nitrogen; and phosphate nutrients; Biocatalyst	Oil recovery improvement due to surfactant and CO_2 production; PH and paraffin reduction; Increase of API gravity	[900]

USA	MFR	Pilot test in the Loco filed (a heavy oil reservoir with API of 21)		Special adapted strain of *Clostridium*	Water; Free corn syrup; Some mineral salts	Oil viscosity Reduction caused by CO_2 production; Butanol and surfactant production; Improved mobility control and sweep efficiency	[890]
USA	MFR	Single well stimulation water-flooding case held in Tulsa, Oklahoma		Mixed cultures of *Clostridiun*, *Bacillus licheniformis* and a gram-negative rod	Molasses 2%–4%	Oil production enhancement by 13% and water/oil ratio reduction by 30% After 10 months	[889]
USA	MFR	Cretaceous Nacatoch formation in Arkansas	Depth: 1920 ft Temperature: 90–105°C Permeability: 5770 mD Porosity: 30.5% Oil viscosity: 4.48 cP Density: 36°API	*Clostridium acetobutvlicum*	Molasses 2 %	Oil production Increased by 250%	[76]
USA	MFR	Case 11			Molasses; Mineral salts	An average increase of oil production by 42% for all wells	[108]
USA	MSPR	Case 12		Surfactant and cosurfactant gas producing cultures; Polymer-polysaccharide gas producing cultures	Injection medium (ingredients not mentioned)	Oil production enhancement for a limited time	[108]
USA	MFFR[s]	Case 13		*Desulfovibrio hydrocarbon-lastus*	Ca or Na lactate; Ascorbic acid; Yeast extract; K_2HPO_4; NaCl; Agar gel agent	Oil production rate increased by 66 %	[108]
USA	PRT[t]	Case 14		A liquid culture of mixed marine source microorganisms	Saline solution of nutrients to control paraffin deposition	Oil production Increased by 166% for 3 months	[108]
USA		Chelsea-Alluwe oilfield, Bartlesville sandstone formation	Depth: 122 m Permeability: 16 md Salinity: 2.9% Oil viscosity: 6 cP	*Bacillus*, *Clostridium*	Cane molasses	20% increment in oil production	[15]

(continued)

Table 10.7 (Continued)

Country	Technology	Cases	Cases Characteristics	Microbial system	Nutrient	Effects	Ref.
USA		SE Vasser Vertz sand unit oilfield, Vertz sandstone formation	Depth: 550 m Permeability: 60–181 md Salinity: 11–19% Oil viscosity: 2.9 cp	Indigenous microflora	Molasses, NH_4NO_3	Decreased permeability	[15,702]
USA		72 producing oil wells in the Permian Basin.	39.4° API 25 cP viscosity	naturally occurring, non-pathogenic and no genetically engineered microorganism mixtures of live facultative anaerobes	unknown	Gravity increase of 2.5° API, viscosity reduction of 10 cpo at 100°F, pour point reduction of 17 OF and 12% increase in solvent composition.	[615]
USA	CMR, MFR, MSPR, ASMR, MCSC, MSDR, MPR			Pure or mixed cultures of *Bacillus*, *Clostridium*, *Pseudomonas*, gram-negative rods Mixed cultures of hydrocarbon degrading bacteria Mixed cultures of marine source bacteria Spore suspension of *Clostridium* Indigenous stratal microflora Slime-forming bacteria Ultra microbacteria	Molasses 2%–4% Molasses and ammonium nitrate addition Free corn syrup C mineral salts Maltodextrine and OPE Salt solution Sucrose 10% C Peptone 1% C NaCl 0.5%–30% Brine supplemented with nitrogen and phosphorous sources and nitrate Biodegradable paraffinic fractions + mineral salts Naturally contain inorganic and organic materials C N, P sources	Increment of oil production	[16]
USA		More than 2000 producing oil wells	70% sandstone 30% carbonates			For 14–44 months; incremental oil recovery ranged from 340 to 4110 m^3	[6,800]

USA		Prudhoe Bay oil field in Alaska	All the treated wells are completed in the Zulu formation. Top of structure averages 8975′ true vertical depth (TVD) with an average sand thickness of 80 feet. Reservoir temperature is 196°F while the reservoir pressure is estimated at 3400 psi				[754]
USA	MEW[u]	Delaware-Childers field	Sandstone formation 21 injection wells and 15 producing wells Well completions are open hole Average oil saturation of 30% Depth: 600 f t Temperature: 80°F Permeability: 52 mD Oil viscosity: 7 cP Density: 35°API	*Clostridium* sp. *Bacillus licheniformis* *Bacillus* sp. Gram-negative rod	Molasses	The rate of oil production improved about 13% The WOR at all monitored producing wells decreased as much as 35% No effects on injectivity were caused by the microbial treatment	[409]
USA	RMA[v]	Seven pilot areas of oil fields	Net pay thickness: 2.27 m Porosity : 0.187 Initial oil Saturation: 0.784 Absolute permeability: 0.328 m^2	Microbiological degradation of hydrocarbons formed in porous media	Injected water with mineral salts of nitrogen and phosphorus added	Recovering 70 thousand tonnes of additional oil	[892]
USA	MEOR	Case studies of successful projects	Reservoir depths: 4450 to 6900 ft Net thickness: 18 − 60 ft Porosity: 0.079 − 0.232 Effective permeability: 1.7 to 300 mD Temperatures: 110 to 18°F Water salinities: 8000 − 180,000 ppm chlorides	Facultative anaerobic bacteria capable of deriving nourishment	Salt water	Reducing oil viscosity and residual oil result in improved oil recovery Production rate increases range from 10% to 500%, averaging 39% Oil recovery increases thus far average 32%	[895]
USA	MMEOR[w]	Microbial field pilot	The pilot has one injector well and 3 production wells	Populations of anaerobic, facultative and aerobic halophilic bacteria	15% & 19% sal Molasses	Significant permeability reduction occurred 13.1 M^3 tertiary oil have been recovered CO_2 content has been increased	[888]

(continued)

Table 10.7 (Continued)

Country	Technology	Cases	Cases Characteristics	Microbial system	Nutrient	Effects	Ref.
USA	MEOR	NBU field	Sandstone formation Depth: 3000 ft Temperature: 40–45°C BHP: 800 Psi API gravity: 40° Viscosity: 3.0 cP	In situ growth of indigenous microorganisms	Maltodextrin (carbon source) and Ethyl-Acid phosphate (phosphate source)	33% drop in the effective permeability to the injection fluid and a negative skin factor	[780]
Venezuela	MFR	Venezuelan oil wells	Temperature: 65–70°C Pressure: 1400 psi	Adapted mixed enriched cultures	Molasses	Positive/negative effect	[16,31]
Venezuela		25 Lake Maracaibo wells operated in the Petroleos de Venezuela SA Lagunillas distric	10–19° API heavy crude	Para-Bac/S for controlling paraffin Ben-Bac for preventing asphaltene deposition and improving crude oil flow properties Corroso-Bac for protecting downhole and surface equipment from corrosion by sequestration, filming, and removing solids	Unknown	The microorganisms reduce paraffin accumulation, asphaltenes agglomerates, and other problems in the well bore area as well as the reservoir. During the stimulation treatment, bio produced surfactants and solvents decrease oil/water interfacial tension, altering effective permeability of oil by changing wettability characteristics, and lowering fluid surface tension	[764]

IR, incremental reserves; *SRB*, sulfate-reducing bacteria; *HDB*, hydrocarbon-degrading bacteria; *NRB*, nitrate-reducing bacteria; *PIMP*, pseudomonas aeruginosa (P-1) and its metabolic products; *OPE*, organic phosphate esters; *MEOR*, microbial enhanced oil recovery; *IFT*, interfacial tension.

[a]Microbial improvement technology.
[b]Biological stimulation of oil production.
[c]Activation of stratal microflora recovery.
[d]Cyclic microbial recovery.
[e]Microbial flooding recovery.
[f]Microbial selective plugging recovery.
[g]Cyclic microbial recovery.
[h]Microbial flooding recovery.
[i]Microbial selective plugging recovery.
[j]Microbial hydrocarbon anaerobic fermentation.
[k]Microbial core flooding.
[l]Microbial culture products.
[m]Microbial well stimulation.
[n]Microbial enhanced water flooding.
[o]Microbial wellbore cleanup.
[p]Nutritional flooding.
[q]Microbial Hydraulic Acid Fracturing.
[r]*Microbial permeability profile modification.*
[s]Microbial fracturing fluids recovery.
[t]Paraffin removal treatment.
[u]*Microbial enhanced water flooding.*
[v]*Reservoir microflora* activation.
[w]Multi-well microbial enhanced oil recovery.

A new concept in MEOR is to employ the enzymes to enhance oil recovery. Generally speaking, addition of enzymes to the waterflood can enhance oil recovery from both sandstone and carbonate cores [60]. Enzymes are a specific group of proteins synthesized by living cells capable to catalyze several biochemical reactions [947]. As the catalysts, enzymes reduce the activation energy of a reaction and consequently significantly enhance the rate of the reaction [948,949]. Enzymes can either degrade the unwanted chemicals or generate the desired ones [946]. Some enzymes having potential for enhancing oil recovery are [60]:

1. Greenzyme, which is a commercial EOR enzyme and consists of enzymes and stabilizers (surfactants),
2. The Zonase group consists of two types of pure enzymes, Zonase1 and Zonase2, which are protease enzymes and whose catalytic functions are to hydrolyze (breakdown) peptide bonds,
3. The Novozyme group consists of three types of pure enzymes, NZ2, NZ3 and NZ6, which are esterase enzymes and whose catalytic functions are to hydrolyze ester bonds, and
4. Alpha-lactalbumin, which is an important whey protein.

The main processes, aiding the enzyme enhanced oil recovery are the adsorption ability of enzymes – proteins as well the accompanying increase in the water-wetness [950]. Enzymes – proteins can improve the waterflooding efficiency by converting the wettability to a more water-wet condition, especially in oil-wet reservoirs [60,951,952]. Enzymes can amend the fluids – rock interfacial dynamics and influence the wettability and capillary action and also make emulsions, which can promote oil recovery [60,953]. Moreover, enzymes can improve the plugging methods by degrading the insoluble bacterial cells resulted from the fermentation process and also molecular aggregates or micro-gels for improving the injection efficiency of biopolymers [954]. The generated products during the catalyzed reactions can promote the plugging efficiency, thus, enzymes can be used to modify the reservoir permeability profile too [955]. Several authors have proposed the application of enzymes to modify the permeability profile of the porous media in conventional and fractures reservoirs [955,956]. Nemati and Voordouw [955] successfully utilized enzymes to catalyze a $CaCO_3$ formation to reduce the porous media permeability. It is proved that several hydrolases (the enzymes, which catalyze the hydrolysis of a chemical bond) can break down the crude oil components [60]. For example, hydrolase enzymes catalyze the bond cleavage via introduction of water, which may break down the crude oil components into either smaller molecules with increased water solubility and reduced interfacial activity or more polar molecules such as hydrolyzing ester to form acid and alcohol [60]. The mentioned process may affect the wettability and IFT (see Fig. 10.9). It is worthwhile to mention that Khusainova [957] specified that crude oil may contain esters in the form of either free compound such as dioctylphthalate [958]

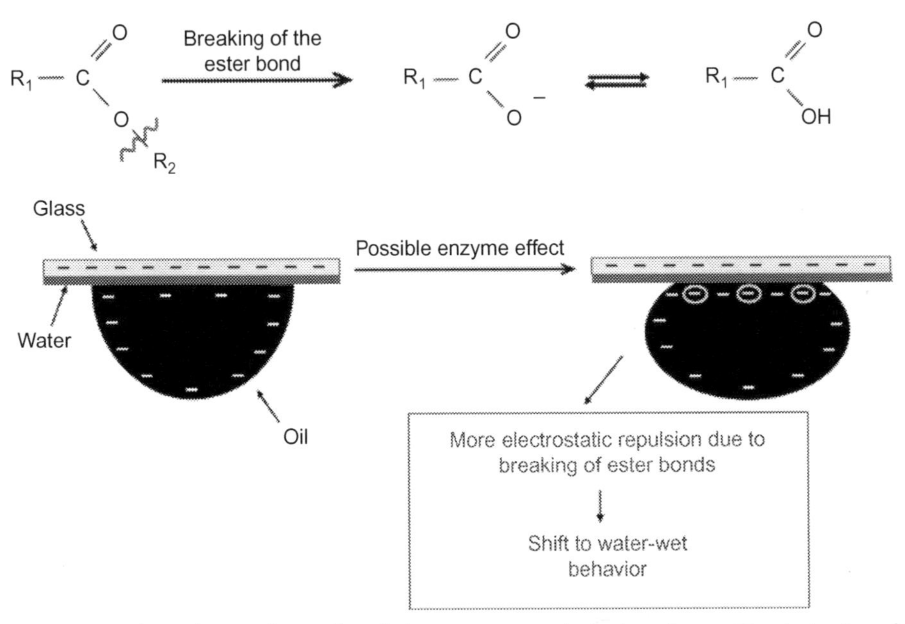

Figure 10.9 Breaking down of ester bonds by enzymes and altering the wetting behavior of the solid [60].

or as binding elements within high-molecular compounds [959]. Several authors have reported application of enzymes in removing formation damages resulted from drilling operations [960–964]. Nasiri [60] specified that effect of enzymes on the oil–water properties is trivial compared to the effect on the oil–water–solid properties.

The enzyme functionality is limited by temperature [24]. Water salinity and pH are the other parameters that influence the enzymatic activity and change the membrane thickness as well as cellular thickness [833]. However, some types of enzymes can resist the extreme condition of pH, salinity, and pressure [965].

10.15 GENETICALLY-ENGINEERED MICROBIAL ENHANCED OIL RECOVERY

The conventional MEOR methods incorporate the available native microorganisms with a limited number of applicable trait combinations. Each bacteria type owns their particular traits, which are associated with certain limitations [4]. As an example, a certain bacteria type, which adequately generates the desired bioacids, may not survive in the reservoir harsh pH condition. These limitations and constraints encouraged a new biotechnology in petroleum industry referred to as Genetically-Engineered

MEOR (GEMEOR). GEMEOR utilizes genetic engineering methods such as mutagenesis, recombineering, and protoplast fusion to combine the favorable traits from different microorganisms to make more efficient strains in enhancement of the oil recovery [6,966,967]. Using this method, it would be possible to acquire the biochemical with favorable properties using the engineered bacterial strains [31]. GEMEOR can make the enhanced oil recovery more economically feasible. The main advantages of the engineered strains, which they are designed for, are [31]

1. Tolerating harsh environmental condition;
2. Selectively producing biochemicals in substantial volumes; and
3. Ability to grow on cheaper substrates.

By the genetic engineering approaches, it is possible to insert the DNA sequence of an organism into a host through protoplast fusion or incorporation of recombinant plasmid DNA into the competent cells [968]. Hybrid strains can be developed using protoplast fusion [31].

REFERENCES

[1] World Population, in: International Programs, Census Bureau, International Data Base, US, 2017.

[2] C. Hall, et al., Hydrocarbons and the evolution of human culture, Nature 426 (2003) 318–322.

[3] International Energy Outlook DOE/EIA-0484, Energy Information Administration, U.S. Department of Energy, Washington, D.C., 2010. 2010.

[4] International Energy Outlook DOE/EIA-0484, Energy Information Administration, U.S. Department of Energy, Washington, D.C., 2017. 2017.

[5] Annual Energy Outlook DOE/EIA-0383, Energy Information Administration, U.S. Department of Energy, Washington, D.C., 2007. 2007.

[6] N. Youssef, et al., Microbial processes in oil fields: culprits, problems, and opportunities, Adv. Appl. Microbiol. 66 (2009) 141–251.

[7] B. Ollivier, M. Magot, Petroleum Microbiology, ASM Press, Washington, D.C., 2005.

[8] L.R. Brown, Microbial enhanced oil recovery (MEOR), Curr. Opin. Microbiol. 13 (2010) 316–320.

[9] A. Lundquist, et al., Energy for a new century: increasing domestic energy supplies, National Energy Policy, Report of the National Energy Policy Development Group, 2001, pp. 69–90.

[10] M. Siegert, et al., Starting up microbial enhanced oil recovery, in: A. Schippers, F. Glombitza, W. Sand (Eds.), Geobiotechnology II. Advances in Biochemical Engineering/Biotechnology, Springer, Berlin, Heidelberg, 2013, pp. 1–94.

[11] L. Brown, et al., Slowing production decline and extending the economic life of an oil field: new MEOR technology, SPE/DOE Improved Oil Recovery Symposium, Society of Petroleum Engineers, Tulsa, Oklahoma, 2000.

[12] R.A. Kerr, USGS optimistic on world oil prospects, Science 289 (2000). 237-237.

[13] M.D. Mehta, J.J. Gair, Social, political, legal and ethical areas of inquiry in biotechnology and genetic engineering, Technol. Soc. 23 (2001) 241–264.

[14] J. Giles, Oil exploration: every last drop, Nature 429 (2004) 694–695.

[15] R. Sen, Biotechnology in petroleum recovery: the microbial EOR, Prog. Energy Combust. Sci. 34 (2008) 714–724.

[16] I. Lazar, et al., Microbial enhanced oil recovery (MEOR), Pet. Sci. Technol. 25 (2007) 1353–1366.

[17] M.J. McInerney, et al., Microbially enhanced oil recovery: past, present, and future, in: B. Ollivier, M. Magot (Eds.), Petroleum Microbiology, ASM Press, Washington, DC, 2005, pp. 215–238.

[18] S. Belyaev, et al., Use of microorganisms in the biotechnology for the enhancement of oil recovery, Microbiology 73 (2004) 590–598.
[19] J.D. Van Hamme, et al., Recent advances in petroleum microbiology, Microbiol. Mol. Biol. Rev. 67 (2003) 503–549.
[20] B. Govreau, et al., Field applications of organic oil recovery-a new MEOR method-Chapter 21, in: J. Sheng (Ed.), Chapter 21 - Enhanced Oil Recovery Field Case Studies, Gulf Professional Publishing, Bostan, Massachusetts, 2013, pp. 572–605.
[21] C. Bass, H. Lappin-Scott, The bad guys and the good guys in petroleum microbiology, Oilfield Rev. 9 (1997) 17–25.
[22] H. Volk, P. Hendry, 3° Oil recovery: fundamental approaches and principles of microbially enhanced oil recovery, in: K.N. Timmis (Ed.), Handbook of Hydrocarbon and Lipid Microbiology, Springer, Berlin, Heidelberg, 2010, pp. 2727–2738.
[23] S. Maudgalya, et al., Microbially enhanced oil recovery technologies: a review of the past, present and future, Production and Operations Symposium, Society of Petroleum Engineers, Oklahoma City, Oklahoma, 2007.
[24] J.J. Sheng, Introduction to MEOR and its field applications in China, in: J.J. Sheng (Ed.), Enhanced Oil Recovery Field Case Studies, Gulf Professional Publishing, Bostan, Massachusetts, 2013, pp. 543–559.
[25] E.C. Donaldson, et al., Microbial Enhanced Oil Recovery, Elsevier, Amsterdam, Netherlands, 1989.
[26] P. Simandoux, et al., Managing the cost of enhanced oil recovery, Revue de l'Institut Français du Pétrole 45 (1990) 131–139.
[27] B. Shibulal, et al., Microbial enhanced heavy oil recovery by the aid of inhabitant spore-forming bacteria: an insight review, Sci. World J. 2014 (2014) 1–12.
[28] R.S. Bryant, et al., Chapter 14 Microbial enhanced oil recovery, in: E.C. Donaldson, G.V. Chilingarian, T.F. Yen (Eds.), Developments in Petroleum Science, Elsevier, Amsterdam, Netherlands, 1989, pp. 423–450.
[29] J. Monod, La technique de culture continue: theorie et applications, Ann. l'Inst. Pasteur 79 (1950) 390–410.
[30] D.W. Green, G.P. Willhite, Enhanced Oil Recovery, Society of Petroleum Engineers, Richardson, Texas, 1998.
[31] J. Patel, et al., Recent developments in microbial enhanced oil recovery, Renewable Sustainable Energy Rev. 52 (2015) 1539–1558.
[32] H. Volk, K. Liu, 3° Oil recovery: experiences and economics of microbially enhanced oil recovery (MEOR), in: K.N. Timmis (Ed.), Handbook of Hydrocarbon and Lipid Microbiology, Springer, Berlin, Heidelberg, 2010, pp. 2739–2751.
[33] M. Bao, et al., Laboratory study on activating indigenous microorganisms to enhance oil recovery in Shengli Oilfield, J. Pet. Sci. Eng. 66 (2009) 42–46.
[34] O.J. Hao, et al., Sulfate-reducing bacteria, Crit. Rev. Environ. Sci. Technol. 26 (1996) 155–187.
[35] R. Cord-Ruwisch, et al., Sulfate-reducing bacteria and their activities in oil production, J. Pet. Technol. 39 (1987) 97–106.
[36] W. Lee, et al., Role of sulfate-reducing bacteria in corrosion of mild steel: a review, Biofouling 8 (1995) 165–194.
[37] E. Sunde, T. Torsvik, Microbial control of hydrogen sulfide production in oil reservoirs, in: B. Ollivier, M. Magot (Eds.), Petroleum Microbiology, ASM Press, Washington, DC, 2005, pp. 201–214.
[38] J. Odom, Industrial and environmental activities of sulfate-reducing bacteria, in: J.M. Odom, R. Singleton (Eds.), The Sulfate-Reducing Bacteria: Contemporary Perspectives, Springer, New York, NY, 1993, pp. 189–210.
[39] R.L. Zahner, et al., What has been learned from a hundred MEOR applications, SPE Enhanced Oil Recovery Conference, Society of Petroleum Engineers, Kuala Lumpur, Malaysia, 2011.
[40] F.F. Craig, The Reservoir Engineering Aspects of Waterflooding, HL Doherty Memorial Fund of AIME, New York, NY, 1971.

[41] T. Jack, Microbially enhanced oil recovery, Biorecovery 1 (1988) 59–73.
[42] E.C. Donaldson, Microbial Enhancement of Oil Recovery-Recent Advances, Elsevier, Amsterdam, Netherlands, 1991.
[43] J. Khire, M. Khan, Microbially enhanced oil recovery (MEOR). Part 1. Importance and mechanism of MEOR, Enzyme Microb. Technol. 16 (1994) 170–172.
[44] A. Tarek, Reservoir Engineering Handbook, Butterworth-Heinemann, US, 2000.
[45] S.C. Ayirala, Surfactant-induced relative permeability modifications for oil recovery enhancement, in: The Department of Petroleum Engineering, Louisiana State University, Baton Rouge, Louisiana, 2002.
[46] W. Xu, Experimental investigation of dynamic interfacial interactions at reservoir conditions, in: The Craft and Hawkins Department of Petroleum Engineering Louisiana State University, Baton Rouge, Louisiana, 2005.
[47] J. Taber, Dynamic and static forces required to remove a discontinuous oil phase from porous media containing both oil and water, Soc. Pet. Eng. J. 9 (1969) 3–12.
[48] L. Lake, Enhanced Oil Recovery, Prentice Hall, Englewood Cliffs, New Jersey, 1989.
[49] Schlumberger, Capillary number, in: Oilfield Glossary, Schlumberger, 2017.
[50] R.L. Reed, R.N. Healy, Some physicochemical aspects of microemulsion flooding: a review, in: D. O. Shah, R.S. Schecheter (Eds.), Improved Oil Recovery by Surfactant and Polymer Flooding, Academic Press, New York, NY, 1977, pp. 383–437.
[51] S.-C. Lin, et al., Structural and immunological characterization of a biosurfactant produced by *Bacillus licheniformis* JF-2, Appl. Environ. Microbiol. 60 (1994) 31–38.
[52] M.J. McInerney, et al., Properties of the biosurfactant produced by*Bacillus licheniformis* strain JF-2, J. Ind. Microbiol. 5 (1990) 95–101.
[53] T.T. Nguyen, et al., Rhamnolipid biosurfactant mixtures for environmental remediation, Water Res. 42 (2008) 1735–1743.
[54] N.H. Youssef, et al., Basis for formulating biosurfactant mixtures to achieve ultra low interfacial tension values against hydrocarbons, J. Ind. Microbiol. Biotechnol. 34 (2007) 497–507.
[55] J.O. Amaefule, L.L. Handy, The effect of interfacial tensions on relative oil/water permeabilities of consolidated porous media, Soc. Pet. Eng. J. 22 (1982) 371–381.
[56] S. Kumar, et al., Relative permeability functions for high-and low-tension systems at elevated temperatures, SPE California Regional Meeting, Society of Petroleum Engineers, Bakersfield, California, 1985.
[57] T. Maldal, et al., Correlation of capillary number curves and remaining oil saturations for reservoir and model sandstones, In Situ 21 (1997) 239–269.
[58] P. Shen, et al., The influence of interfacial tension on water-oil two-phase relative permeability, SPE/DOE Symposium on Improved Oil Recovery, Society of Petroleum Engineers, Tulsa, Oklahoma, 2006.
[59] A.A. Hamouda, O. Karoussi, Effect of temperature, wettability and relative permeability on oil recovery from oil-wet chalk, Energies 1 (2008) 19–34.
[60] H. Nasiri, Enzymes for enhanced oil recovery (EOR), in: Centre for Integrated Petroleum Research, Department of Chemistry, University of Bergen, Bergen, Norway, 2011.
[61] S.M. Skjæveland, J. Kleppe, SPOR Monograph, Recent advances in improved oil recovery methods for north sea sandstone reservoirs, Norwegian Petroleum Directorate, Norway, 1992.
[62] P. Berger, C. Lee, Ultra-low concentration surfactants for sandstone and limestone floods, SPE/DOE Improved Oil Recovery Symposium, Society of Petroleum Engineers, Tulsa, Oklahoma, 2002.
[63] A. Cui, Experimental study of microbial enhanced oil recovery and its impact on residual oil in sandstones, in: Petroleum and Geosystems Engineering, The University of Texas at Austin, Austin, Texas 2016.
[64] R.D. Sydansk, L. Romero-Zerón, Reservoir Conformance Improvement, Society of Petroleum Engineers, Richardson, Texas, 2011.
[65] C. Hutchinson, Reservoir inhomogeneity assessment and control, Pet. Eng. 31 (1959) 19–26.
[66] M. Gray, et al., Potential microbial enhanced oil recovery processes: a critical analysis, SPE Annual Technical Conference and Exhibition, Society of Petroleum Engineers, Denver, Colorado, 2008.

[67] E.S. Bastin, The problem of the natural reduction of sulphates, Am. Assoc. Pet. Geol. Bull. 10 (1926) 1270–1299.
[68] J. Beckman, Action of bacteria on mineral oil, J. Ind. Eng. Chem. 4 (1926) 21–22.
[69] C.E. ZoBell, Bacterial release of oil from oil-bearing materials, Part I, World Oil 126 (1947) 1–11.
[70] C.E. ZoBell, Bacterial release of oil from oil-bearing materials, Part II, World Oil 127 (1947) 1–11.
[71] C.E. Zobell, Bacteriological process for treatment of fluid-bearing earth formations, in American Petroleum Inst US, 1946.
[72] J.V. Beck, Use of bacteria for releasing oil from sands, Prod. Mon. 11 (1947) 13–19.
[73] C.E. Zobell, Recovery of hydrocarbons, in Texaco Development Corp US, 1953.
[74] D.M. Updegraff, G.B. Wren, Secondary recovery of petroleum oil by Desulfovibrio, in ExxonMobil Oil Corp US, 1953.
[75] D.M. Updegraff, Recovery of petroleum oil, in ExxonMobil Oil Corp US, 1957.
[76] H. Yarbrough, V. Coty, Microbially enhanced oil recovery from the upper cretaceous nacatoch formation, Union County, Arkansas, in: E.C. Donaldson, J.B. Clark (Eds.), International Conference on Microbial Enhancement of Oil Recovery, NTIS, Afton, Oklahoma, 1982, pp. 149–153.
[77] R.S. Bryant, Microbial enhanced oil recovery and compositions therefor, in US Department of Energy ITT Research Institute US, 1990.
[78] D. Hitzman, Petroleum microbiology and the history of its role in enhanced oil recovery, in: E.C. Donaldson, J.B. Clark (Eds.), Proceedings of the International Conference on Microbial Enhancement of Oil Recovery, Afton, Oklahoma, 1982.
[79] M.M. Grula, H. Russell, Isolation and screening of anaerobic Clostridia for characteristics useful in enhanced oil recovery. Final report, October 1983 – February 1985, in Oklahoma State University, Department of Botany and Microbiology, Stillwater, Oklahoma, 1985.
[80] J. Zajic, S. Smith, Oil separation relating to hydrophobicity and microbes, in: N. Kosaric, W.L. Cairns, N.C.C. Gray (Eds.), Surfactant Science Series: Biosurfactants and Biotechnology, Marcel Dekker, New York, NY, 1987, p. 133.
[81] L. Nelson, D.R. Schneider, Six years of paraffin control and enhanced oil recovery with the microbial product, Para-Bac™, in: E.T. Premuzic, A. Woodhead (Eds.), Developments in Petroleum Science, Elsevier, Amsterdam, Netherlands, 1993, pp. 355–362.
[82] G.E. Jenneman, et al., A nutrient control process for microbially enhanced oil recovery applications, in: E.T. Premuzic, A. Woodhead (Eds.), Developments in Petroleum Science, Elsevier, Amsterdam, Netherlands, 1993, pp. 319–333.
[83] H. Al-Sulaimani, et al., Microbial biotechnology for enhancing oil recovery: current developments and future prospects, Biotechnol. Bioinf. Bioeng. 1 (2011) 147–158.
[84] J.I. Adetunji, Microbial enhanced oil recovery, Department of Chemistry, Biotechnology and Environment, Aalborg University, Esbjerg, Denmark, 2012.
[85] M. Safdel, et al., Microbial enhanced oil recovery, a critical review on worldwide implemented field trials in different countries, Renewable Sustainable Energy Rev. 74 (2017) 159–172.
[86] E.T. Premuzic, A. Woodhead, Microbial Enhancement of Oil Recovery-Recent Advances, Elsevier, Amsterdam, Netherlands, 1993.
[87] D. Hitzman, Review of microbial enhanced oil recovery field tests, in: T.E. Burchfield, R.S. Bryant (Eds.), Proceedings of the Applications of Microorganisms to Petroleum Technology, Bartlesville Project Office, US Department of Energy, Bartlesville, Oklahoma, 1988.
[88] R.S. Bryant, J. Douglas, Evaluation of microbial systems in porous media for EOR, SPE Reservoir Eng. 3 (1988) 489–495.
[89] S. Kuznetsov, et al., Introduction to Geological Microbiology, McGraw-Hill, New York, NY, 1963.
[90] M. Spurny, et al., A method of quantitative determination of sulphate reducing bacteria, Folia Biol. 3 (1957) 202–211.
[91] M. Dostálek, et al., The action of microorganisms on petroleum hydro-carbons, Českoslov. Mikrobiol. 2 (1957) 43–47.

[92] M. Dostalek, M. Spurny, Release of oil through the action of microorganisms. II: Effect of physical and physical-chemical conditions in oil-bearing rock, Cechoslov. Mikrobiol. 2 (1957) 307–317.
[93] M. Dostalek, Bacterial release of Oil. 3. A real distribution of effect of nutrient injection into deposit, Folia. Microbiol. 6 (1961) 10–16.
[94] M. Dostalek, M. Spurný, Geomicrobiological oil prospection, Folia. Microbiol. 7 (1962) 141–150.
[95] M. Dostalek, et al., Action of bacteria on petroleum loosening in collectors, Pr. Ustavu Naft. Výzk. 9 (1958) 29.
[96] M. Dostalek, Hydrocarbon bacteria in soils of oil-bearing regions, Chekhoslov. Biol. 2 (1953) 347.
[97] M. Spurny, M. Dostalek, Mikrobiologie naftovych poli, Pr. Ustavu Naft. Výzk. 59 (1956) 26–30.
[98] M. Dostalek, Characteristics of growth of hydrocarbons assimilating bacteria, Chekhoslov. Biol. 3 (1954) 99–107.
[99] I. Jaranyi, Beszamolo a nagylengyel terzegeben elvegzett Koolaj mikrobiologiai Kiserletkrol. M. All. Foldtani Intezet Evi Jelentese A, Evrol, (1968) 423–426.
[100] M. Dienes, I. Jaranyi, Increasing recovery of oil in Demjen field by populating formation with anaerobic bacteria, Int. Chem. Eng. 15 (1975) 240–244.
[101] M. Dienes, I. Yaranyi, Increase of oil recovery by introducing anaerobic bacteria into the formation Demjen field, kőolaj és földgáz, 106 (1973) 205–208.
[102] I. Karaskiewicz, Application des méthodes microbiologiques pour l'intensification de l'exploitation des gisements pétrolifères de la region des Carpathes, Śląsk, Kraków, Poland, 1974.
[103] V. Senyukov, et al., Microbial method of treating a petroleum deposit containing highly mineralized stratal waters, Mikrobiologiya 39 (1970) 705–710.
[104] I. Lazar, International MEOR applications for marginal wells, Pak. J. Hydrocarbon Res. 10 (1998) 11–30.
[105] I. Lazar, P. Constantinescu, Field trial results of microbial enhanced oil recovery, in: J.E. Zajic, E. C. Donaldson (Eds.), Microbes and Oil Recovery, Bioresearch Publications, El Paso, Texas, 1985, pp. 122–143.
[106] I. Lazar, Research on the microbiology of MEOR in Romania, in: J. King, D. Stevens (Eds.), Proceedings of the First International MEOR Workshop, Bartlesville, Oklahoma, 1987, pp. 124–153.
[107] I. Lazar, et al., Some considerations concerning nutrient support injected into reservoirs subjected to microbiological treatment, in: T.E. Burchfield, R.S. Bryant (Eds.), Proceedings of the Symposium on the Application of Microorganisms to Petroleum Technology, National Technical Information Service, Bartlesville, Oklahoma, 1987, pp. XIV 1–XIV 6.
[108] I. Lazar, Ch. A-1 MEOR field trials carried out over the world during the last 35 years, in: E.C. Donaldson (Ed.), Developments in Petroleum Science, Elsevier, Amsterdam, Netherlands, 1991, pp. 485–530.
[109] I. Lazar, et al., Ch. F-2 Preliminary results of some recent MEOR field trials in Romania, in: E.C. Donaldson (Ed.), Developments in Petroleum Science, Elsevier, Amsterdam, Netherlands, 1991, pp. 365–385.
[110] I. Lazar, et al., MEOR, the suitable bacterial inoculum according to the kind of technology used: results from Romania's last 20 years' experience, SPE/DOE Enhanced Oil Recovery Symposium, Society of Petroleum Engineers, Tulsa, Oklahoma, 1992.
[111] I. Lazar, The microbiology of MEOR, practical experience in Europe, International Biohydrometallurgy Symposium, Vol 2; Fossil Energy Materials Bioremediation, Microbial Physiology, The Minerals, Metals & Materials Society, Jackson Hole, Wyoming, 1993, pp. 329–338.
[112] I. Lazar, et al., MEOR, recent field trials in Romania: reservoir selection, type of inoculum, protocol for well treatment and line monitoring, in: E.T. Premuzic, A. Woodhead (Eds.), Developments in Petroleum Science, Elsevier, Amsterdam, Netherlands, 1993, pp. 265–287.

[113] I. Lazar, Microbial systems for enhancement of oil recovery used in Romanian oil fields, Miner. Process. Extr. Metall. Rev. 19 (1998) 379–393.
[114] M. Wagner, Microbial enhancement of oil recovery from carbonate reservoir with complex formation characteristics, in: E.C. Donaldson (Ed.), Micobial Enhancement of Oil Recovery-Recent Advances, Elsevier, Amsterdam, Netherlands, 1991, pp. 387–398.
[115] J. Khire, M. Khan, Microbially enhanced oil recovery (MEOR). Part 2. Microbes and the subsurface environment for MEOR, Enzyme Microb. Technol. 16 (1994) 258–259.
[116] S. Thomas, Enhanced oil recovery-an overview, Oil Gas Sci. Technol.-Rev. l'IFP 63 (2008) 9–19.
[117] J. Karaskiewicz, Studies on increasing petroleum oil recovery from Carpathian deposits using bacteria, Nafta (Pet.) 21 (1975) 144–149.
[118] B.B. Jørgensen, A. Boetius, Feast and famine—microbial life in the deep-sea bed, Nat. Rev. Microbiol. 5 (2007) 770–781.
[119] I.M. Head, et al., Biological activity in the deep subsurface and the origin of heavy oil, Nature 426 (2003) 344–352.
[120] J. Wang, et al., Monitoring exogenous and indigenous bacteria by PCR-DGGE technology during the process of microbial enhanced oil recovery, J. Ind. Microbiol. Biotechnol. 35 (2008) 619–628.
[121] R.T. Bachmann, et al., Biotechnology in the petroleum industry: an overview, Int. Biodeterior. Biodegrad. 86 (2014) 225–237.
[122] K. Kashefi, D.R. Lovley, Extending the upper temperature limit for life, Science 301 (2003). 934-934.
[123] J.B. Fisher, Distribution and occurrence of aliphatic acid anions in deep subsurface waters, Geochim. Cosmochim. Acta 51 (1987) 2459–2468.
[124] P.Fd Almeida, et al., Selection and application of microorganisms to improve oil recovery, Eng. Life Sci. 4 (2004) 319–325.
[125] G.S. Grassia, et al., A systematic survey for thermophilic fermentative bacteria and archaea in high temperature petroleum reservoirs, FEMS Microbiol. Ecol. 21 (1996) 47–58.
[126] W. Griffin, et al., Methods for obtaining deep subsurface microbiological samples by drilling, in: P. S. Amy, D.L. Haldeman (Eds.), The microbiology of the terrestrial deep subsurface, CRC Press, Boca Raton, Florida, 1997, pp. 23–44.
[127] L.R. Krumholz, et al., Confined subsurface microbial communities in Cretaceous rock, Nature 386 (1997) 64–66.
[128] L.-H. Lin, et al., Long-term sustainability of a high-energy, low-diversity crustal biome, Science 314 (2006) 479–482.
[129] M. Magot, et al., Microbiology of petroleum reservoirs, Antonie van Leeuwenhoek 77 (2000) 103–116.
[130] M. Magot, Indigenous microbial communities in oil fields, in: B. Ollivier, M. Magot (Eds.), Petroleum Microbiology, ASM Press, Washington, D.C., 2005, pp. 21–34.
[131] C. Vetriani, et al., Thermovibrio ammonificans sp. nov., a thermophilic, chemolithotrophic, nitrate-ammonifying bacterium from deep-sea hydrothermal vents, Int. J. Syst. Evol. Microbiol. 54 (2004) 175–181.
[132] K. Takai, et al., Thiomicrospira thermophila sp. nov., a novel microaerobic, thermotolerant, sulfur-oxidizing chemolithomixotroph isolated from a deep-sea hydrothermal fumarole in the TOTO caldera, Mariana Arc, Western Pacific, Int. J. Syst. Evol. Microbiol. 54 (2004) 2325–2333.
[133] M.R. Mormile, et al., Isolation of *Halobacterium salinarum* retrieved directly from halite brine inclusions, Environ. Microbiol. 5 (2003) 1094–1102.
[134] R.H. Vreeland, et al., Halosimplex carlsbadense gen. nov., sp. nov., a unique halophilic archaeon, with three 16S rRNA genes, that grows only in defined medium with glycerol and acetate or pyruvate, Extremophiles 6 (2002) 445–452.
[135] R. Vreeland, et al., Isolation of live Cretaceous (121–112 million years old) halophilic Archaea from primary salt crystals, Geomicrobiol. J. 24 (2007) 275–282.
[136] P. Yilmaz, et al., The SILVA and "All-species Living Tree Project (LTP)" taxonomic frameworks, Nucleic Acids Res. 42 (2014) D643–D648.

[137] H. Kobayashi, et al., Phylogenetic diversity of microbial communities associated with the crude-oil, large-insoluble-particle and formation-water components of the reservoir fluid from a non-flooded high-temperature petroleum reservoir, J. Biosci. Bioeng. 113 (2012) 204–210.
[138] K.M. Kaster, et al., Effect of nitrate and nitrite on sulfide production by two thermophilic, sulfate-reducing enrichments from an oil field in the North Sea, Appl. Microbiol. Biotechnol. 75 (2007) 195–203.
[139] K.M. Kaster, et al., Characterisation of culture-independent and-dependent microbial communities in a high-temperature offshore chalk petroleum reservoir, Antonie van Leeuwenhoek 96 (2009) 423–439.
[140] M.R. Bonfá, et al., Biodegradation of aromatic hydrocarbons by Haloarchaea and their use for the reduction of the chemical oxygen demand of hypersaline petroleum produced water, Chemosphere 84 (2011) 1671–1676.
[141] H.-Y. Ren, et al., Comparison of microbial community compositions of injection and production well samples in a long-term water-flooded petroleum reservoir, PLoS One 6 (2011) e23258.
[142] T. Nazina, et al., Phylogenetic diversity and activity of anaerobic microorganisms of high-temperature horizons of the Dagang Oilfield (China), Mikrobiologiia 75 (2006) 70–81.
[143] H. Li, et al., Molecular phylogenetic diversity of the microbial community associated with a high-temperature petroleum reservoir at an offshore oilfield, FEMS Microbiol. Ecol. 60 (2007) 74–84.
[144] H. Kobayashi, et al., Analysis of methane production by microorganisms indigenous to a depleted oil reservoir for application in microbial enhanced oil recovery, J. Biosci. Bioeng. 113 (2012) 84–87.
[145] C.M. Callbeck, et al., Microbial community succession in a bioreactor modeling a souring low-temperature oil reservoir subjected to nitrate injection, Appl. Microbiol. Biotechnol. 91 (2011) 799–810.
[146] V.D. Pham, et al., Characterizing microbial diversity in production water from an Alaskan mesothermic petroleum reservoir with two independent molecular methods, Environ. Microbiol. 11 (2009) 176–187.
[147] A. Grabowski, et al., Microbial diversity in production waters of a low-temperature biodegraded oil reservoir, FEMS Microbiol. Ecol. 54 (2005) 427–443.
[148] L. Cheng, et al., Methermicoccus shengliensis gen. nov., sp. nov., a thermophilic, methylotrophic methanogen isolated from oil-production water, and proposal of Methermicoccaceae fam. nov, Int. J. Syst. Evol. Microbiol. 57 (2007) 2964–2969.
[149] M.L. Miroshnichenko, et al., Isolation and characterization of *Thermococcus sibiricus* sp. nov. from a Western Siberia high-temperature oil reservoir, Extremophiles 5 (2001) 85–91.
[150] Y. Takahata, et al., Distribution and physiological characteristics of hyperthermophiles in the Kubiki oil reservoir in Niigata, Japan, Appl. Environ. Microbiol. 66 (2000) 73–79.
[151] S. Yousaf, et al., Phylogenetic and functional diversity of alkane degrading bacteria associated with Italian ryegrass (*Lolium multiflorum*) and Birdsfoot trefoil (*Lotus corniculatus*) in a petroleum oil-contaminated environment, J. Hazard. Mater. 184 (2010) 523–532.
[152] L. Cheng, et al., Enrichment and dynamics of novel syntrophs in a methanogenic hexadecane-degrading culture from a Chinese oilfield, FEMS Microbiol. Ecol. 83 (2013) 757–766.
[153] F. Zhang, et al., Molecular biologic techniques applied to the microbial prospecting of oil and gas in the Ban 876 gas and oil field in China, Appl. Microbiol. Biotechnol. 86 (2010) 1183–1194.
[154] I. Yumoto, et al., Dietzia psychralcaliphila sp. nov., a novel, facultatively psychrophilic alkaliphile that grows on hydrocarbons, Int. J. Syst. Evol. Microbiol. 52 (2002) 85–90.
[155] X.-B. Wang, et al., Degradation of petroleum hydrocarbons (C6–C40) and crude oil by a novel Dietzia strain, Bioresour. Technol. 102 (2011) 7755–7761.
[156] T. Nunoura, et al., Vertical distribution of the subsurface microorganisms in Sagara oil reservoir, in: AGU 2002 Fall Meeting, American Geophysical Union, San Francisco, California, 2002.
[157] C.T. Hennessee, et al., Polycyclic aromatic hydrocarbon-degrading species isolated from Hawaiian soils: *Mycobacterium crocinum* sp. nov., *Mycobacterium pallens* sp. nov., *Mycobacterium rutilum* sp. nov., *Mycobacterium rufum* sp. nov. and *Mycobacterium aromaticivorans* sp. nov, Int. J. Syst. Evol. Microbiol. 59 (2009) 378–387.

[158] P. Willumsen, et al., *Mycobacterium frederiksbergense* sp. nov., a novel polycyclic aromatic hydrocarbon-degrading Mycobacterium species, Int. J. Syst. Evol. Microbiol. 51 (2001) 1715–1722.

[159] M. Kästner, et al., Enumeration and characterization of the soil microflora from hydrocarbon-contaminated soil sites able to mineralize polycyclic aromatic hydrocarbons (PAH), Appl. Microbiol. Biotechnol. 41 (1994) 267–273.

[160] K. Derz, et al., *Mycobacterium pyrenivorans* sp. nov., a novel polycyclic-aromatic-hydrocarbon-degrading species, Int. J. Syst. Evol. Microbiol. 54 (2004) 2313–2317.

[161] H. Li, et al., Molecular analysis of the bacterial community in a continental high-temperature and water-flooded petroleum reservoir, FEMS Microbiol. Lett. 257 (2006) 92–98.

[162] Y. Xue, et al., *Gordonia paraffinivorans* sp. nov., a hydrocarbon-degrading actinomycete isolated from an oil-producing well, Int. J. Syst. Evol. Microbiol. 53 (2003) 1643–1646.

[163] A. Schippers, et al., *Microbacterium oleivorans* sp. nov. and *Microbacterium hydrocarbonoxydans* sp. nov., novel crude-oil-degrading Gram-positive bacteria, Int. J. Syst. Evol. Microbiol. 55 (2005) 655–660.

[164] S. Dore, et al., Naphthalene-utilizing and mercury-resistant bacteria isolated from an acidic environment, Appl. Microbiol. Biotechnol. 63 (2003) 194–199.

[165] M. Kubota, et al., *Nocardioides aromaticivorans* sp. nov., a dibenzofuran-degrading bacterium isolated from dioxin-polluted environments, Syst. Appl. Microbiol. 28 (2005) 165–174.

[166] A. Schippers, et al., *Nocardioides oleivorans* sp. nov., a novel crude-oil-degrading bacterium, Int. J. Syst. Evol. Microbiol. 55 (2005) 1501–1504.

[167] M. Cai, et al., *Salinarimonas ramus* sp. nov. and *Tessaracoccus oleiagri* sp. nov., isolated from a crude oil-contaminated saline soil, Int. J. Syst. Evol. Microbiol. 61 (2011) 1767–1775.

[168] C. Balachandran, et al., Petroleum and polycyclic aromatic hydrocarbons (PAHs) degradation and naphthalene metabolism in Streptomyces sp.(ERI-CPDA-1) isolated from oil contaminated soil, Bioresour. Technol. 112 (2012) 83–90.

[169] F. Zhang, et al., Response of microbial community structure to microbial plugging in a mesothermic petroleum reservoir in China, Appl. Microbiol. Biotechnol. 88 (2010) 1413–1422.

[170] E.R. Hendrickson, et al., Method of improving oil recovery from an oil reservoir using an enriched anaerobic steady state microbial consortium, in E I du Pont de Nemours and Co US, 2013.

[171] A. Grabowski, et al., *Petrimonas sulfuriphila* gen. nov., sp. nov., a mesophilic fermentative bacterium isolated from a biodegraded oil reservoir, Int. J. Syst. Evol. Microbiol. 55 (2005) 1113–1121.

[172] H. Dahle, et al., Microbial community structure analysis of produced water from a high-temperature North Sea oil-field, Antonie van Leeuwenhoek 93 (2008) 37–49.

[173] L.-Y. Wang, et al., Characterization of an alkane-degrading methanogenic enrichment culture from production water of an oil reservoir after 274 days of incubation, Int. Biodeterior. Biodegrad. 65 (2011) 444–450.

[174] J.-Y. Ying, et al., *Cyclobacterium lianum* sp. nov., a marine bacterium isolated from sediment of an oilfield in the South China Sea, and emended description of the genus Cyclobacterium, Int. J. Syst. Evol. Microbiol. 56 (2006) 2927–2930.

[175] T. Gutierrez, et al., *Polycyclovorans algicola* gen. nov., sp. nov., an aromatic-hydrocarbon-degrading marine bacterium found associated with laboratory cultures of marine phytoplankton, Appl. Environ. Microbiol. 79 (2013) 205–214.

[176] I. Szabó, et al., *Olivibacter oleidegradans* sp. nov., a hydrocarbon-degrading bacterium isolated from a biofilter clean-up facility on a hydrocarbon-contaminated site, Int. J. Syst. Evol. Microbiol. 61 (2011) 2861–2865.

[177] A.C. Greene, et al., *Deferribacter thermophilus* gen. nov., sp. nov., a novel thermophilic manganese- and iron-reducing bacterium isolated from a petroleum reservoir, Int. J. Syst. Evol. Microbiol. 47 (1997) 505–509.

[178] S. Mnif, et al., Simultaneous hydrocarbon biodegradation and biosurfactant production by oilfield-selected bacteria, J. Appl. Microbiol. 111 (2011) 525–536.

[179] C. Zheng, et al., Hydrocarbon degradation and bioemulsifier production by thermophilic *Geobacillus pallidus* strains, Bioresour. Technol. 102 (2011) 9155–9161.

[180] C.D. Da Cunha, et al., Oil biodegradation by Bacillus strains isolated from the rock of an oil reservoir located in a deep-water production basin in Brazil, Appl. Microbiol. Biotechnol. 73 (2006) 949–959.

[181] Y.-H. She, et al., Investigation of biosurfactant-producing indigenous microorganisms that enhance residue oil recovery in an oil reservoir after polymer flooding, Appl. Biochem. Biotechnol. 163 (2011) 223–234.

[182] S. Dastgheib, et al., Bioemulsifier production by a halothermophilic Bacillus strain with potential applications in microbially enhanced oil recovery, Biotechnol. Lett. 30 (2008) 263–270.

[183] T.N. Nazina, et al., *Geobacillus jurassicus* sp. nov., a new thermophilic bacterium isolated from a high-temperature petroleum reservoir, and the validation of the Geobacillus species, Syst. Appl. Microbiol. 28 (2005) 43–53.

[184] N. Kuisiene, et al., *Geobacillus lituanicus* sp. nov, Int. J. Syst. Evol. Microbiol. 54 (2004) 1991–1995.

[185] J. Zhang, et al., Isolation of a thermophilic bacterium, Geobacillus sp. SH-1, capable of degrading aliphatic hydrocarbons and naphthalene simultaneously, and identification of its naphthalene degrading pathway, Bioresour. Technol. 124 (2012) 83–89.

[186] T. Nazina, et al., Taxonomic study of aerobic thermophilic bacilli: descriptions of *Geobacillus subterraneus* gen. nov., sp. nov. and *Geobacillus uzenensis* sp. nov. from petroleum reservoirs and transfer of *Bacillus stearothermophilus*, *Bacillus thermocatenulatus*, *Bacillus thermoleovorans*, *Bacillus kaustophilus*, *Bacillus thermodenitrificans* to Geobacillus as the new combinations *G. stearothermophilus*, G. th, Int. J. Syst. Evol. Microbiol. 51 (2001) 433–446.

[187] L. Feng, et al., Genome and proteome of long-chain alkane degrading Geobacillus thermodenitrificans NG80-2 isolated from a deep-subsurface oil reservoir, Proc. Natl. Acad. Sci. USA 104 (2007) 5602–5607.

[188] T. Kato, et al., Isolation and characterization of long-chain-alkane degrading Bacillus thermoleovorans from deep subterranean petroleum reservoirs, J. Biosci. Bioeng. 91 (2001) 64–70.

[189] M.L. Gana, et al., Antagonistic activity of Bacillus sp. obtained from an Algerian oilfield and chemical biocide THPS against sulfate-reducing bacteria consortium inducing corrosion in the oil industry, J. Ind. Microbiol. Biotechnol. 38 (2011) 391–404.

[190] H. Li, et al., Biodegradation of benzene and its derivatives by a psychrotolerant and moderately haloalkaliphilic Planococcus sp. strain ZD22, Res. Microbiol. 157 (2006) 629–636.

[191] M. Engelhardt, et al., Isolation and characterization of a novel hydrocarbon-degrading, Gram-positive bacterium, isolated from intertidal beach sediment, and description of *Planococcus alkanoclasticus* sp. nov, J. Appl. Microbiol. 90 (2001) 237–247.

[192] R. Das, B.N. Tiwary, Isolation of a novel strain of Planomicrobium chinense from diesel contaminated soil of tropical environment, J. Basic Microbiol. 53 (2013) 723–732.

[193] M. Lavania, et al., Biodegradation of asphalt by Garciaella petrolearia TERIG02 for viscosity reduction of heavy oil, Biodegradation 23 (2012) 15–24.

[194] G. Ravot, et al., *Fusibacter paucivorans* gen. nov., sp. nov., an anaerobic, thiosulfate-reducing bacterium from an oil-producing well, Int. J. Syst. Evol. Microbiol. 49 (1999) 1141–1147.

[195] G. Bødtker, et al., Microbial analysis of backflowed injection water from a nitrate-treated North Sea oil reservoir, J. Ind Microbiol. Biotechnol. 36 (2009) 439–450.

[196] U. Kunapuli, et al., *Desulfitobacterium aromaticivorans* sp. nov. and *Geobacter toluenoxydans* sp. nov., iron-reducing bacteria capable of anaerobic degradation of monoaromatic hydrocarbons, Int. J. Syst. Evol. Microbiol. 60 (2010) 686–695.

[197] Y.-J. Lee, et al., *Desulfosporosinus youngiae* sp. nov., a spore-forming, sulfate-reducing bacterium isolated from a constructed wetland treating acid mine drainage, Int. J. Syst. Evol. Microbiol. 59 (2009) 2743–2746.

[198] R.K. Nilsen, et al., *Desulfotomaculum thermocisternum* sp. nov., a sulfate reducer isolated from a hot North Sea oil reservoir, Int. J. Syst. Evol. Microbiol. 46 (1996) 397–402.

[199] B. Morasch, et al., Degradation of o-xylene and m-xylene by a novel sulfate-reducer belonging to the genus Desulfotomaculum, Arch. Microbiol. 181 (2004) 407–417.

[200] M.B. Salinas, et al., *Mahella australiensis* gen. nov., sp. nov., a moderately thermophilic anaerobic bacterium isolated from an Australian oil well, Int. J. Syst. Evol. Microbiol. 54 (2004) 2169–2173.

[201] K. Lysnes, et al., Microbial response to reinjection of produced water in an oil reservoir, Appl. Microbiol. Biotechnol. 83 (2009) 1143–1157.

[202] M.-L. Fardeau, et al., *Thermoanaerobacter subterraneus* sp. nov., a novel thermophile isolated from oil-field water, Int. J. Syst. Evol. Microbiol. 50 (2000) 2141–2149.

[203] J.-L. Cayol, et al., Description of *Thermoanaerobacter brockii* subsp. lactiethylicus subsp. nov., isolated from a deep subsurface French oil well, a proposal to reclassify Thermoanaerobacter finnii as *Thermoanaerobacter brockii* subsp. finnii comb. nov., and an emended description of *Thermoanaerobacter brockii*, Int. J. Syst. Evol. Microbiol. 45 (1995) 783–789.

[204] G. Ravot, et al., *Haloanaerobium congolense* sp. nov., an anaerobic, moderately halophilic, thiosulfate-and sulfur-reducing bacterium from an African oil field, FEMS Microbiol. Lett. 147 (1997) 81–88.

[205] S.L.C. Shartau, et al., Ammonium concentrations in produced waters from a mesothermic oil field subjected to nitrate injection decrease through formation of denitrifying biomass and anammox activity, Appl. Environ. Microbiol. 76 (2010) 4977–4987.

[206] X.-C. Pan, et al., *Nitratireductor shengliensis* sp. nov., isolated from an oil-polluted saline soil, Curr. Microbiol. 69 (2014) 561–566.

[207] M. Cai, et al., *Rubrimonas shengliensis* sp. nov. and *Polymorphum gilvum* gen. nov., sp. nov., novel members of Alphaproteobacteria from crude oil contaminated saline soil, Syst. Appl. Microbiol. 34 (2011) 321–327.

[208] T.U. Harwati, et al., *Tropicibacter naphthalenivorans* gen. nov., sp. nov., a polycyclic aromatic hydrocarbon-degrading bacterium isolated from Semarang Port in Indonesia, Int. J. Syst. Evol. Microbiol. 59 (2009) 392–396.

[209] T.U. Harwati, et al., *Tropicimonas isoalkanivorans* gen. nov., sp. nov., a branched-alkane-degrading bacterium isolated from Semarang Port in Indonesia, Int. J. Syst. Evol. Microbiol. 59 (2009) 388–391.

[210] J.-Y. Ying, et al., *Wenxinia marina* gen. nov., sp. nov., a novel member of the Roseobacter clade isolated from oilfield sediments of the South China Sea, Int. J. Syst. Evol. Microbiol. 57 (2007) 1711–1716.

[211] Y. Kodama, et al., *Thalassospira tepidiphila* sp. nov., a polycyclic aromatic hydrocarbon-degrading bacterium isolated from seawater, Int. J. Syst. Evol. Microbiol. 58 (2008) 711–715.

[212] B. Zhao, et al., *Thalassospira xianhensis* sp. nov., a polycyclic aromatic hydrocarbon-degrading marine bacterium, Int. J. Syst. Evol. Microbiol. 60 (2010) 1125–1129.

[213] D.L. Balkwill, et al., Taxonomic study of aromatic-degrading bacteria from deep-terrestrial-subsurface sediments and description of *Sphingomonas aromaticivorans* sp. nov., *Sphingomonas subterranea* sp. nov., and *Sphingomonas stygia* sp. nov, Int. J. Syst. Evol. Microbiol. 47 (1997) 191–201.

[214] W. Chung, G. King, Isolation, characterization, and polyaromatic hydrocarbon degradation potential of aerobic bacteria from marine macrofaunal burrow sediments and description of *Lutibacterium anuloederans* gen. nov., sp. nov., and *Cycloclasticus spirillensus* sp. nov, Appl. Environ. Microbiol. 67 (2001) 5585–5592.

[215] J. Yuan, et al., *Novosphingobium indicum* sp. nov., a polycyclic aromatic hydrocarbon-degrading bacterium isolated from a deep-sea environment, Int. J. Syst. Evol. Microbiol. 59 (2009) 2084–2088.

[216] S. Suzuki, A. Hiraishi, *Novosphingobium naphthalenivorans* sp. nov., a naphthalene-degrading bacterium isolated from polychlorinated-dioxin-contaminated environments, J. Gen. Appl. Microbiol. 53 (2007) 221–228.

[217] J.H. Sohn, et al., *Novosphingobium pentaromativorans* sp. nov., a high-molecular-mass polycyclic aromatic hydrocarbon-degrading bacterium isolated from estuarine sediment, Int. J. Syst. Evol. Microbiol. 54 (2004) 1483–1487.

[218] G. Harms, et al., Anaerobic oxidation of o-xylene, m-xylene, and homologous alkylbenzenes by new types of sulfate-reducing bacteria, Appl. Environ. Microbiol. 65 (1999) 999–1004.

[219] C. Cravo-Laureau, et al., *Desulfatibacillum aliphaticivorans* gen. nov., sp. nov., an n-alkane-and n-alkene-degrading, sulfate-reducing bacterium, Int. J. Syst. Evol. Microbiol. 54 (2004) 77–83.

[220] C. Cravo-Laureau, et al., *Desulfatibacillum alkenivorans* sp. nov., a novel n-alkene-degrading, sulfate-reducing bacterium, and emended description of the genus Desulfatibacillum, Int. J. Syst. Evol. Microbiol. 54 (2004) 1639–1642.

[221] C. Cravo-Laureau, et al., *Desulfatiferula olefinivorans* gen. nov., sp. nov., a long-chain n-alkene-degrading, sulfate-reducing bacterium, Int. J. Syst. Evol. Microbiol. 57 (2007) 2699–2702.

[222] T. Lien, J. Beeder, *Desulfobacter vibrioformis* sp. nov., a sulfate reducer from a water-oil separation system, Int. J. Syst. Evol. Microbiol. 47 (1997) 1124–1128.

[223] R. Rabus, et al., Complete oxidation of toluene under strictly anoxic conditions by a new sulfate-reducing bacterium, Appl. Environ. Microbiol. 59 (1993) 1444–1451.

[224] H. Ommedal, T. Torsvik, *Desulfotignum toluenicum* sp. nov., a novel toluene-degrading, sulphate-reducing bacterium isolated from an oil-reservoir model column, Int. J. Syst. Evol. Microbiol. 57 (2007) 2865–2869.

[225] T. Lien, et al., *Desulfobulbus rhabdoformis* sp. nov., a sulfate reducer from a water-oil separation system, Int. J. Syst. Evol. Microbiol. 48 (1998) 469–474.

[226] C. Tardy-Jacquenod, et al., *Desulfovibrio gabonensis* sp. nov., a new moderately halophilic sulfate-reducing bacterium isolated from an oil pipeline, Int. J. Syst. Evol. Microbiol. 46 (1996) 710–715.

[227] P. Rueter, et al., Anaerobic oxidation of hydrocarbons in crude oil by new types of sulphate-reducing bacteria, Nature 372 (1994) 455–458.

[228] M. Magot, et al., *Desulfovibrio bastinii* sp. nov. and *Desulfovibrio gracilis* sp. nov., moderately halophilic, sulfate-reducing bacteria isolated from deep subsurface oilfield water, Int. J. Syst. Evol. Microbiol. 54 (2004) 1693–1697.

[229] A.C. Greene, et al., *Geoalkalibacter subterraneus* sp. nov., an anaerobic Fe (III)-and Mn (IV)-reducing bacterium from a petroleum reservoir, and emended descriptions of the family Desulfuromonadaceae and the genus Geoalkalibacter, Int. J. Syst. Evol. Microbiol. 59 (2009) 781–785.

[230] J.D. Coates, et al., *Geobacter hydrogenophilus*, *Geobacter chapellei* and *Geobacter grbiciae*, three new, strictly anaerobic, dissimilatory Fe (III)-reducers, Int. J. Syst. Evol. Microbiol. 51 (2001) 581–588.

[231] D.R. Lovley, et al., *Geobacter metallireducens* gen. nov. sp. nov., a microorganism capable of coupling the complete oxidation of organic compounds to the reduction of iron and other metals, Arch. Microbiol. 159 (1993) 336–344.

[232] I.A. Davidova, et al., *Desulfoglaeba alkanexedens* gen. nov., sp. nov., an n-alkane-degrading, sulfate-reducing bacterium, Int. J. Syst. Evol. Microbiol. 56 (2006) 2737–2742.

[233] J. Beeder, et al., *Thermodesulforhabdus norvegicus* gen. nov., sp. nov., a novel thermophilic sulfate-reducing bacterium from oil field water, Arch. Microbiol. 164 (1995) 331–336.

[234] A. Rodríguez-Blanco, et al., *Gallaecimonas pentaromativorans* gen. nov., sp. nov., a bacterium carrying 16S rRNA gene heterogeneity and able to degrade high-molecular-mass polycyclic aromatic hydrocarbons, Int. J. Syst. Evol. Microbiol. 60 (2010) 504–509.

[235] M.J. Gauthier, et al., *Marinobacter hydrocarbonoclasticus* gen. nov., sp. nov., a new, extremely halotolerant, hydrocarbon-degrading marine bacterium, Int. J. Syst. Evol. Microbiol. 42 (1992) 568–576.

[236] M.B. Salinas, et al., *Petrobacter succinatimandens* gen. nov., sp. nov., a moderately thermophilic, nitrate-reducing bacterium isolated from an Australian oil well, Int. J. Syst. Evol. Microbiol. 54 (2004) 645–649.

[237] R. Rabus, F. Widdel, Anaerobic degradation of ethylbenzene and other aromatic hydrocarbons by new denitrifying bacteria, Arch. Microbiol. 163 (1995) 96–103.

[238] B. Song, et al., Taxonomic characterization of denitrifying bacteria that degrade aromatic compounds and description of *Azoarcus toluvorans* sp. nov. and *Azoarcus toluclasticus* sp. nov, Int. J. Syst. Evol. Microbiol. 49 (1999) 1129–1140.

[239] J.D. Coates, et al., Anaerobic benzene oxidation coupled to nitrate reduction in pure culture by two strains of Dechloromonas, Nature 411 (2001) 1039–1043.

[240] S.A. Weelink, et al., A strictly anaerobic betaproteobacterium *Georgfuchsia toluolica* gen. nov., sp. nov. degrades aromatic compounds with Fe (III), Mn (IV) or nitrate as an electron acceptor, FEMS Microbiol. Ecol. 70 (2009) 575–585.
[241] B. Song, et al., Identification of denitrifier strain T1 as *Thauera aromatica* and proposal for emendation of the genus Thauera definition, Int. J. Syst. Evol. Microbiol. 48 (1998) 889–894.
[242] Y. Shinoda, et al., Aerobic and anaerobic toluene degradation by a newly isolated denitrifying bacterium, Thauera sp. strain DNT-1, Appl. Environ. Microbiol. 70 (2004) 1385–1392.
[243] T. Gutierrez, et al., *Porticoccus hydrocarbonoclasticus* sp. nov., an aromatic hydrocarbon-degrading bacterium identified in laboratory cultures of marine phytoplankton, Appl. Environ. Microbiol. 78 (2012) 628–637.
[244] P.M. Sarma, et al., Degradation of polycyclic aromatic hydrocarbons by a newly discovered enteric bacterium, *Leclercia adecarboxylata*, Appl. Environ. Microbiol. 70 (2004) 3163–3166.
[245] M. Chamkha, et al., Isolation and characterization of *Klebsiella oxytoca* strain degrading crude oil from a Tunisian off-shore oil field, J. Basic Microbiol. 51 (2011) 580–589.
[246] S.E. Dyksterhouse, et al., *Cycloclasticus pugetii* gen. nov., sp. nov., an aromatic hydrocarbon-degrading bacterium from marine sediments, Int. J. Syst. Evol. Microbiol. 45 (1995) 116–123.
[247] A.D. Geiselbrecht, et al., Enumeration and phylogenetic analysis of polycyclic aromatic hydrocarbon-degrading marine bacteria from Puget sound sediments, Appl. Environ. Microbiol. 62 (1996) 3344–3349.
[248] C. Liu, Z. Shao, *Alcanivorax dieselolei* sp. nov., a novel alkane-degrading bacterium isolated from sea water and deep-sea sediment, Int. J. Syst. Evol. Microbiol. 55 (2005) 1181–1186.
[249] A. Bruns, L. Berthe-Corti, *Fundibacter jadensis* gen. nov., sp. nov., a new slightly halophilic bacterium, isolated from intertidal sediment, Int. J. Syst. Evol. Microbiol. 49 (1999) 441–448.
[250] G. Wu, et al., *Halomonas daqingensis* sp. nov., a moderately halophilic bacterium isolated from an oilfield soil, Int. J. Syst. Evol. Microbiol. 58 (2008) 2859–2865.
[251] Z.B.A. Gam, et al., *Modicisalibacter tunisiensis* gen. nov., sp. nov., an aerobic, moderately halophilic bacterium isolated from an oilfield-water injection sample, and emended description of the family Halomonadaceae Franzmann et al. 1989 emend Dobson and Franzmann 1996 emend. Ntougias et al. 2007, Int. J. Syst. Evol. Microbiol. 57 (2007) 2307–2313.
[252] B.P. Hedlund, et al., Polycyclic aromatic hydrocarbon degradation by a new marine bacterium, *Neptunomonas naphthovorans* gen. nov., sp. nov, Appl. Environ. Microbiol. 65 (1999) 251–259.
[253] M. Teramoto, et al., *Oleibacter marinus* gen. nov., sp. nov., a bacterium that degrades petroleum aliphatic hydrocarbons in a tropical marine environment, Int. J. Syst. Evol. Microbiol. 61 (2011) 375–380.
[254] M.M. Yakimov, et al., *Oleispira antarctica* gen. nov., sp. nov., a novel hydrocarbonoclastic marine bacterium isolated from Antarctic coastal sea water, Int. J. Syst. Evol. Microbiol. 53 (2003) 779–785.
[255] M.M. Yakimov, et al., *Thalassolituus oleivorans* gen. nov., sp. nov., a novel marine bacterium that obligately utilizes hydrocarbons, Int. J. Syst. Evol. Microbiol. 54 (2004) 141–148.
[256] B.W. Bogan, et al., *Alkanindiges illinoisensis* gen. nov., sp. nov., an obligately hydrocarbonoclastic, aerobic squalane-degrading bacterium isolated from oilfield soils, Int. J. Syst. Evol. Microbiol. 53 (2003) 1389–1395.
[257] M. Magot, et al., *Spirochaeta smaragdinae* sp. nov., a new mesophilic strictly anaerobic spirochete from an oil field, FEMS Microbiol. Lett. 155 (1997) 185–191.
[258] G.N. Rees, et al., *Anaerobaculum thermoterrenum* gen. nov., sp. nov., a novel, thermophilic bacterium which ferments citrate, Int. J. Syst. Evol. Microbiol. 47 (1997) 150–154.
[259] H. Dahle, N.-K. Birkeland, *Thermovirga lienii* gen. nov., sp. nov., a novel moderately thermophilic, anaerobic, amino-acid-degrading bacterium isolated from a North Sea oil well, Int. J. Syst. Evol. Microbiol. 56 (2006) 1539–1545.
[260] J.L. DiPippo, et al., *Kosmotoga olearia* gen. nov., sp. nov., a thermophilic, anaerobic heterotroph isolated from an oil production fluid, Int. J. Syst. Evol. Microbiol. 59 (2009) 2991–3000.

[261] Y. Feng, et al., *Thermococcoides shengliensis* gen. nov., sp. nov., a new member of the order Thermotogales isolated from oil-production fluid, Int. J. Syst. Evol. Microbiol. 60 (2010) 932–937.
[262] H.S. Jayasinghearachchi, B. Lal, *Oceanotoga teriensis* gen. nov., sp. nov., a thermophilic bacterium isolated from offshore oil-producing wells, Int. J. Syst. Evol. Microbiol. 61 (2011) 554–560.
[263] E. Miranda-Tello, et al., *Petrotoga halophila* sp. nov., a thermophilic, moderately halophilic, fermentative bacterium isolated from an offshore oil well in Congo, Int. J. Syst. Evol. Microbiol. 57 (2007) 40–44.
[264] E. Miranda-Tello, et al., *Petrotoga mexicana* sp. nov., a novel thermophilic, anaerobic and xylanolytic bacterium isolated from an oil-producing well in the Gulf of Mexico, Int. J. Syst. Evol. Microbiol. 54 (2004) 169–174.
[265] T. Lien, et al., Petrotoga mobilis sp. nov., from a North Sea oil-production well, Int. J. Syst. Evol. Microbiol. 48 (1998) 1007–1013.
[266] I.A. Purwasena, et al., *Petrotoga japonica* sp. nov., a thermophilic, fermentative bacterium isolated from Yabase Oilfield in Japan, Arch. Microbiol. 196 (2014) 313–321.
[267] Y. Takahata, et al., *Thermotoga petrophila* sp. nov. and *Thermotoga naphthophila* sp. nov., two hyperthermophilic bacteria from the Kubiki oil reservoir in Niigata, Japan, Int. J. Syst. Evol. Microbiol. 51 (2001) 1901–1909.
[268] V. Ekzertsev, S. Kuznetsov, Examination of microflora of oil fields of the Second Baku, Mikrobiologiya 23 (1954) 3–14.
[269] S. Kuznetsov, Examination of the possibility of contemporary methanogenesis in gas-and petroleum-bearing facies of the Saratov and Buguruslan province, Mikrobiologiya 19 (1950) 193–202.
[270] T. Nazina, E. Rozanova, Ecologic conditions for the spread of methane-forming bacteria in the petroleum strata of Apsheron, Mikrobiologiia 49 (1980) 123–129.
[271] S. Belyaev, M. Ivanov, Bacterial methanogenesis in underground waters, Ecol. Bull. (1983) 273–280.
[272] S. Belyaev, et al., Methanogenic bacteria from the Bondyuzhskoe oil field: general characterization and analysis of stable-carbon isotopic fractionation, Appl. Environ. Microbiol. 45 (1983) 691–697.
[273] M. Ivanov, et al., Microbiological formation of methane in the oil-field development, Geokhimiya 11 (1983) 1647–1654.
[274] M. Ivanov, et al., Development dynamic of microbiological processes after oxidation of oil field aquifers, Mikrobiologiia 54 (1985) 293–300.
[275] T.N. Nazina, et al., Microbial oil transformation processes accompanied by methane and hydrogen-sulfide formation, Geomicrobiol. J. 4 (1985) 103–130.
[276] T.N. Nazina, et al., Occurrence and geochemical activity of microorganisms in high-temperature, water-flooded oil fields of Kazakhstan and Western Siberia, Geomicrobiol. J. 13 (1995) 181–192.
[277] S. Belyaev, et al., Characteristics of rod-shaped methane-producing bacteria from an oil pool and description of methanobacterium-ivanovii sp-nov, Microbiology 55 (1986) 821–826.
[278] J.L. Sanz, Methanogens, in: M. Gargaud, R. Amils, J.C. Quintanilla, H.J. Cleaves, W.M. Irvine, D.L. Pinti, M. Viso (Eds.), Encyclopedia of Astrobiology, Springer, Berlin, Heidelberg, 2011, pp. 1037–1038.
[279] I. Davydova-Charakhch'yan, et al., Methanogenic rod-shaped bacteria from the oil fields of Tataria and western Siberia, Microbiologiya 61 (1992) 299–305.
[280] B. Ollivier, et al., *Methanoplanus petrolearius* sp. nov., a novel methanogenic bacterium from an oil-producing well, FEMS Microbiol. Lett. 147 (1997) 51–56.
[281] C. Jeanthon, et al., Hyperthermophilic and methanogenic archaea in oil fields, in: B. Ollivier, M. Magot (Eds.), Petroleum Microbiology, ASM Press, Washington, DC, 2005, pp. 55–69.
[282] T.K. Ng, et al., Possible nonanthropogenic origin of two methanogenic isolates from oil-producing wells in the san miguelito field, ventura county, California, Geomicrobiol. J. 7 (1989) 185–192.
[283] V. Orphan, et al., Culture-dependent and culture-independent characterization of microbial assemblages associated with high-temperature petroleum reservoirs, Appl. Environ. Microbiol. 66 (2000) 700–711.

[284] R.K. Nilsen, T. Torsvik, Methanococcus thermolithotrophicus isolated from North Sea oil field reservoir water, Appl. Environ. Microbiol. 62 (1996) 728–731.
[285] B. Ollivier, et al., *Methanocalculus halotolerans* gen. nov., sp. nov., isolated from an oil-producing well, Int. J. Syst. Evol. Microbiol. 48 (1998) 821–828.
[286] A.I. Slobodkin, et al., Dissimilatory reduction of Fe (III) by thermophilic bacteria and archaea in deep subsurface petroleum reservoirs of Western Siberia, Curr. Microbiol. 39 (1999) 99–102.
[287] G. Ravot, Nouvelles approches microbiologiques de la thiosulfato-réduction en milieu pétrolier, in Aix-Marseille 1, Marseille, France, 1996.
[288] M.-L. Fardeau, et al., H_2 oxidation in the presence of thiosulfate, by a Thermoanaerobacter strain isolated from an oil-producing well, FEMS Microbiol. Lett. 113 (1993) 327–332.
[289] S. Ni, D.R. Boone, Isolation and characterization of a dimethyl sulfide-degrading methanogen, *Methanolobus siciliae* HI350, from an oil well, characterization of *M. siciliae* T4/MT, and emendation of M. siciliae, Int. J. Syst. Evol. Microbiol. 41 (1991) 410–416.
[290] A.Y. Obraztsova, et al., Properties of the coccoid methylotrophic methanogen, *Methanococcoides euhalobius* sp. nov., Microbiology 56 (1987) 523–527.
[291] A. Obraztsova, et al., Biological properties of halophilic methanogen isolated from oil deposits, Dokl. Akad. Nauk SSSR 278 (1984) 227–230.
[292] S.N. Doerfert, et al., *Methanolobus zinderi* sp. nov., a methylotrophic methanogen isolated from a deep subsurface coal seam, Int. J. Syst. Evol. Microbiol. 59 (2009) 1064–1069.
[293] H. König, K.O. Stetter, Isolation and characterization of *Methanolobus tindarius*, sp. nov., a coccoid methanogen growing only on methanol and methylamines, ZBL. Bakt. Mik. Hyg. I. C. 3 (1982) 478–490.
[294] A. Obraztsova, et al., Biological properties of methanosarcina not utilizing carbonic-acid and hydrogen, Microbiology 56 (1987) 807–812.
[295] E.A. Bonch-Osmolovskaya, et al., Radioisotopic, culture-based, and oligonucleotide microchip analyses of thermophilic microbial communities in a continental high-temperature petroleum reservoir, Appl. Environ. Microbiol. 69 (2003) 6143–6151.
[296] V. Orphan, et al., Geochemical influence on diversity and microbial processes in high temperature oil reservoirs, Geomicrobiol. J. 20 (2003) 295–311.
[297] K. Revesz, et al., Methane production and consumption monitored by stable H and C isotope ratios at a crude oil spill site, Bemidji, Minnesota, Appl. Geochem. 10 (1995) 505–516.
[298] M.A. Dojka, et al., Microbial diversity in a hydrocarbon-and chlorinated-solvent-contaminated aquifer undergoing intrinsic bioremediation, Appl. Environ. Microbiol. 64 (1998) 3869–3877.
[299] C. Bolliger, et al., Characterizing intrinsic bioremediation in a petroleum hydrocarbon-contaminated aquifer by combined chemical, isotopic, and biological analyses, Bioremediat. J. 4 (2000) 359–371.
[300] C.G. Struchtemeyer, et al., Evidence for aceticlastic methanogenesis in the presence of sulfate in a gas condensate-contaminated aquifer, Appl. Environ. Microbiol. 71 (2005) 5348–5353.
[301] J.W. Foster, J.L. Slonczewski, Microbiology: An Evolving Science, fourth ed., W. W. Norton & Company Incorporated, New York, NY, 2017.
[302] F. Gomez, Sulfate reducers, in: M. Gargaud, W.M. Irvine, R. Amils, H.J. Cleaves, D.L. Pinti, J.C. Quintanilla, D. Rouan, T. Spohn, S. Tirard, M. Viso (Eds.), Encyclopedia of Astrobiology, Springer, Berlin, Heidelberg, 2015, pp. 2409-2409.
[303] S. Al Zuhair, et al., Sulfate inhibition effect on sulfate reducing bacteria, J Biochem. Technol. 1 (2008) 39–44.
[304] W. Song, et al., The role of sulphate-reducing bacteria in oil recovery, Int. J. Curr. Microbiol. Appl. Sci. 7 (2014) 385–398.
[305] H. Cypionka, Oxygen respiration by Desulfovibrio species, Annu. Rev. Microbiol. 54 (2000) 827–848.
[306] W. Dilling, H. Cypionka, Aerobic respiration in sulfate-reducing bacteria, FEMS Microbiol. Lett. 71 (1990) 123–127.
[307] A. Dolla, et al., Oxygen defense in sulfate-reducing bacteria, J. Biotechnol. 126 (2006) 87–100.

[308] A. Dolla, et al., Biochemical, proteomic and genetic characterization of oxygen survival mechanisms in sulphate reducing bacteria of the genus Desulfovibrio, in: L. Barton, W. Hamilton (Eds.), Sulphate-Reducing Bacteria Environmental and Engineered Systems, Cambridge University Press, New York, NY, 2007, pp. 185–214.
[309] M. Santana, Presence and expression of terminal oxygen reductases in strictly anaerobic sulfate-reducing bacteria isolated from salt-marsh sediments, Anaerobe 14 (2008) 145–156.
[310] L.L. Barton, G.D. Fauque, Chapter 2 Biochemistry, physiology and biotechnology of sulfate-reducing bacteria, Advances in Applied Microbiology, Academic Press, 2009, pp. 41–98.
[311] G.D. Fauque, Ecology of sulfate-reducing bacteria, in: L.L. Barton (Ed.), Sulfate-Reducing Bacteria, Springer, Bostan, Massachusetts, 1995, pp. 217–241.
[312] G. Fauque, B. Ollivier, Anaerobes: the sulfate-reducing bacteria as an example of metabolic diversity, in: A. Bull (Ed.), Microbial Diversity and Bioprospecting, ASM Press, Washington, D.C., 2004, pp. 169–176.
[313] G. Fauque, et al., Sulfate-reducing and sulfur-reducing bacteria, in: J.M. Shively, L.L. Barton (Eds.), Variations in Autotrophic Life, Academic Press, New York, NY, 1991, pp. 271–337.
[314] J. LeGall, G. Fauque, Dissimilatory reduction of sulfur compounds, Biol. Anaerobic Microorg. (1988) 587–639.
[315] G. Muyzer, A.J. Stams, The ecology and biotechnology of sulphate-reducing bacteria, Nat. Rev. Microbiol. 6 (2008) 441–454.
[316] R. Rabus, et al., Dissimilatory sulfate-and sulfur-reducing prokaryotes, in: E. Rosenberg, E.F. DeLong, S. Lory, E. Stackebrandt, F. Thompson (Eds.), The Prokaryotes, Springer, Berlin, Heidelberg, 2006, pp. 659–768.
[317] A. Sherry, et al., Anaerobic biodegradation of crude oil under sulphate-reducing conditions leads to only modest enrichment of recognized sulphate-reducing taxa, Int. Biodeterior. Biodegrad. 81 (2013) 105–113.
[318] M. Nemati, et al., Mechanistic study of microbial control of hydrogen sulfide production in oil reservoirs, Biotechnol. Bioeng. 74 (2001) 424–434.
[319] R. Rabus, et al., Dissimilatory sulfate-and sulfur-reducing prokaryotes, in: E. Rosenberg, E.F. DeLong, S. Lory, E. Stackebrandt, F. Thompson (Eds.), The Prokaryotes, Springer, Berlin, Heidelberg, 2013, pp. 309–404.
[320] E.D. Schulze, H.A. Mooney, Biodiversity and Ecosystem Function, Springer-Verlag, Berlin, Heidelberg, 1994.
[321] H.F. Castro, et al., Phylogeny of sulfate-reducing bacteria, FEMS Microbiol. Ecol. 31 (2000) 1–9.
[322] T. Itoh, et al., *Thermocladium modestius* gen. nov., sp. nov., a new genus of rod-shaped, extremely thermophilic crenarchaeote, Int. J. Syst. Evol. Microbiol. 48 (1998) 879–887.
[323] T. Itoh, et al., *Caldivirga maquilingensis* gen. nov., sp. nov., a new genus of rod-shaped crenarchaeote isolated from a hot spring in the Philippines, Int. J. Syst. Evol. Microbiol. 49 (1999) 1157–1163.
[324] K. Mori, et al., A novel lineage of sulfate-reducing microorganisms: Thermodesulfobiaceae fam. nov., *Thermodesulfobium narugense*, gen. nov., sp. nov., a new thermophilic isolate from a hot spring, Extremophiles 7 (2003) 283–290.
[325] B. Ollivier, et al., Sulphate-reducing bacteria from oil field environments and deep-sea hydrothermal vents, in: L.L. Barton, W.A. Hamilton (Eds.), Sulphate-Reducing Bacteria: Environmental and Engineered Systems, Cambridge University Press, Cambridge, England, 2007, pp. 305–328.
[326] J. Peretó, Fermentation, in: M. Gargaud, W.M. Irvine, R. Amils, H.J. Cleaves, D.L. Pinti, J.C. Quintanilla, D. Rouan, T. Spohn, S. Tirard, M. Viso (Eds.), Encyclopedia of Astrobiology, Springer, Berlin, Heidelberg, 2015, pp. 848–849.
[327] N.-K. Birkeland, Sulfate-reducing bacteria and archaea, in: B. Ollivier, M. Magot (Eds.), Petroleum Microbiology, ASM Press, ASM Press, 2005, pp. 35–54.
[328] J.Y. Leu, et al., The same species of sulphate-reducing Desulfomicrobium occur in different oil field environments in the North Sea, Lett. Appl. Microbiol. 29 (1999) 246–252.
[329] M. Magot, et al., *Desulfovibrio longus* sp. nov., a sulfate-reducing bacterium isolated from an oil-producing well, Int. J. Syst. Evol. Microbiol. 42 (1992) 398–402.

[330] E. Miranda-Tello, et al., *Desulfovibrio capillatus* sp. nov., a novel sulfate-reducing bacterium isolated from an oil field separator located in the Gulf of Mexico, Anaerobe 9 (2003) 97–103.
[331] P.N. Dang, et al., *Desulfovibrio vietnamensis* sp. nov., a halophilic sulfate-reducing bacterium from Vietnamese oil fields, Anaerobe 2 (1996) 385–392.
[332] E. Rozanova, et al., Isolation of a new genus of sulfate-reducing bacteria and description of a new species of this genus, *Desulfomicrobium apsheronum* gen. nov., sp. nov, Microbiology (Mikrobiologiya) 57 (1988) 514–520.
[333] S. Myhr, T. Torsvik, *Denitrovibrio acetiphilus*, a novel genus and species of dissimilatory nitrate-reducing bacterium isolated from an oil reservoir model column, Int. J. Syst. Evol. Microbiol. 50 (2000) 1611–1619.
[334] T. Nazina, et al., Occurrence of sulfate-and iron-reducing bacteria in stratal waters of the Romashkinskoe oil field, Microbiology 64 (1995) 203–208.
[335] M.E. Davey, et al., Isolation of three species of Geotoga and Petrotoga: two new genera, representing a new lineage in the bacterial line of descent distantly related to the "Thermotogales", Syst. Appl. Microbiol. 16 (1993) 191–200.
[336] S. Haridon, et al., *Thermosipho geolei* sp. nov., a thermophilic bacterium isolated from a continental petroleum reservoir in Western Siberia, Int. J. Syst. Evol. Microbiol. 51 (2001) 1327–1334.
[337] S. L'Haridon, et al., *Petrotoga olearia* sp. nov. and *Petrotoga sibirica* sp. nov., two thermophilic bacteria isolated from a continental petroleum reservoir in Western Siberia, Int. J. Syst. Evol. Microbiol. 52 (2002) 1715–1722.
[338] I.A. Purwwasena, et al., Estimation of the Potential of an Oil-Viscosity-Reducing Bacteria, Petrotoga Isolated from an Oilfield for MEOR, in: International Petroleum Technology Conference, International Petroleum Technology Conference, Doha, Qatar, 2009.
[339] S. L'haridon, et al., Hot subterranean biosphere in a continental oil reservoir, Nature 377 (1995) 223–224.
[340] M.-L. Fardeau, et al., Isolation from oil reservoirs of novel thermophilic anaerobes phylogenetically related to Thermoanaerobacter subterraneus: reassignment of *T. subterraneus*, *Thermoanaerobacter yonseiensis*, *Thermoanaerobacter tengcongensis* and *Carboxydibrachium pacificum* to *Caldanaerobacter subterraneus* gen. nov., sp. nov., comb. nov. as four novel subspecies, Int. J. Syst. Evol. Microbiol. 54 (2004) 467–474.
[341] V. Bhupathiraju, et al., *Haloanaerobium salsugo* sp. nov., a moderately halophilic, anaerobic bacterium from a subterranean brine, Int. J. Syst. Evol. Microbiol. 44 (1994) 565–572.
[342] V.K. Bhupathiraju, et al., *Haloanaerobium kushneri* sp. nov., an obligately halophilic, anaerobic bacterium from an oil brine, Int. J. Syst. Evol. Microbiol. 49 (1999) 953–960.
[343] M. Magot, et al., *Dethiosulfovibrio peptidovorans* gen. nov., sp. nov., a new anaerobic, slightly halophilic, thiosulfate-reducing bacterium from corroding offshore oil wells, Int. J. Syst. Evol. Microbiol. 47 (1997) 818–824.
[344] I. Davydova-Charakhch'yan, et al., Acetogenic bacteria from oil fields of Tataria and wester Siberia, Microbiology-AIBS-C 61 (1992). 208-208.
[345] B. Ollivier, J.-L. Cayol, Fermentative, iron-reducing, and nitrate-reducing microorganisms, in: B. Ollivier, M. Magot (Eds.), Petroleum Microbiology, ASM Press, Washington, D.C., 2005, pp. 71–88.
[346] G. Jenneman, et al., Sulfide removal in reservoir brine by indigenous bacteria, SPE Prod. Facil. 14 (1999) 219–225.
[347] J. Larsen, et al., Prevention of Reservoir Souring in the Halfdan Field by Nitrate Injection, in: CORROSION 2004, NACE International, New Orleans, Louisiana, 2004.
[348] E. Sunde, et al., H_2S inhibition by nitrate injection on the Gullfaks field, in: CORROSION 2004, NACE International, New Orleans, Louisiana, 2004.
[349] T. Thorstenson, et al., Biocide replacement by nitrate in sea water injection systems, in: CORROSION 2002, NACE International, Denver, Colorado, 2002.
[350] A.J. Telang, et al., Effect of nitrate injection on the microbial community in an oil field as monitored by reverse sample genome probing, Appl. Environ. Microbiol. 63 (1997) 1785–1793.

[351] D.O. Hitzman, et al., Recent successes: MEOR using synergistic H_2S prevention and increased oil recovery systems, SPE/DOE Symposium on Improved Oil Recovery, Society of Petroleum Engineers, Tulsa, Oklahoma, 2004.
[352] M. McInerney, et al., Ch. F-7 Microbial control of the production of sulfide, in: E.C. Donaldson (Ed.), Developments in Petroleum Science, Elsevier, Amsterdam, Netherlands, 1991, pp. 441–449.
[353] C. Hubert, et al., Corrosion risk associated with microbial souring control using nitrate or nitrite, Appl. Microbiol. Biotechnol. 68 (2005) 272–282.
[354] M. Nemati, et al., Control of biogenic H_2S production with nitrite and molybdate, J. Ind. Microbiol. Biotechnol. 26 (2001) 350–355.
[355] M. Nemati, et al., Impact of nitrate-mediated microbial control of souring in oil reservoirs on the extent of corrosion, Biotechnol. Prog. 17 (2001) 852–859.
[356] E.A. Greene, et al., Synergistic inhibition of microbial sulfide production by combinations of the metabolic inhibitor nitrite and biocides, Appl. Environ. Microbiol. 72 (2006) 7897–7901.
[357] A. Gittel, et al., Prokaryotic community structure and sulfate reducer activity in water from high-temperature oil reservoirs with and without nitrate treatment, Appl. Environ. Microbiol. 75 (2009) 7086–7096.
[358] K. Londry, J. Suflita, Use of nitrate to control sulfide generation by sulfate-reducing bacteria associated with oily waste, J. Ind. Microbiol. Biotechnol. 22 (1999) 582–589.
[359] J.J. Arensdorf, et al., Mitigation of reservoir souring by nitrate in a produced-water reinjection system in Alberta, SPE International Symposium on Oilfield Chemistry, Society of Petroleum Engineers, The Woodlands, Texas, 2009.
[360] C. Kuijvenhoven, et al., 1 year experience with the injection of nitrate to control souring in Bonga Deepwater Development Offshore Nigeria, International Symposium on Oilfield Chemistry, Society of Petroleum Engineers, Houston, Texas, 2007.
[361] M.J. Mcinerney, et al., Oil field microbiology, in: C. Hurst, R. Crawford, J. Garland, D. Lipson, A. Mills, L. Stetzenbach (Eds.), Manual of Environmental Microbiology, third ed., ASM Press, Washington, DC, 2007, pp. 898–911.
[362] I. Davidova, et al., The influence of nitrate on microbial processes in oil industry production waters, J. Ind. Microbiol. Biotechnol. 27 (2001) 80–86.
[363] D. Gevertz, et al., Isolation and characterization of strains CVO and FWKO B, two novel nitrate-reducing, sulfide-oxidizing bacteria isolated from oil field brine, Appl. Environ. Microbiol. 66 (2000) 2491–2501.
[364] E. Miranda-Tello, et al., *Garciella nitratireducens* gen. nov., sp. nov., an anaerobic, thermophilic, nitrate-and thiosulfate-reducing bacterium isolated from an oilfield separator in the Gulf of Mexico, Int. J. Syst. Evol. Microbiol. 53 (2003) 1509–1514.
[365] Y. Kodama, K. Watanabe, Isolation and characterization of a sulfur-oxidizing chemolithotroph growing on crude oil under anaerobic conditions, Appl. Environ. Microbiol. 69 (2003) 107–112.
[366] G. Voordouw, et al., Characterization of 16S rRNA genes from oil field microbial communities indicates the presence of a variety of sulfate-reducing, fermentative, and sulfide-oxidizing bacteria, Appl. Environ. Microbiol. 62 (1996) 1623–1629.
[367] N.B. Huu, et al., *Marinobacter aquaeolei* sp. nov., a halophilic bacterium isolated from a Vietnamese oil-producing well, Int. J. Syst. Evol. Microbiol. 49 (1999) 367–375.
[368] C. Pickard, et al., Oil field and freshwater isolates of *Shewanella putrefaciens* have lipopolysaccharide polyacrylamide gel profiles characteristic of marine bacteria, Can. J. Microbiol. 39 (1993) 715–717.
[369] K. Semple, D. Westlake, Characterization of iron-reducing *Alteromonas putrefaciens* strains from oil field fluids, Can. J. Microbiol. 33 (1987) 366–371.
[370] R. Eckford, P. Fedorak, Chemical and microbiological changes in laboratory incubations of nitrate amendment "sour" produced waters from three western Canadian oil fields, J. Ind. Microbiol. Biotechnol. 29 (2002) 243–254.
[371] R. Eckford, P. Fedorak, Planktonic nitrate-reducing bacteria and sulfate-reducing bacteria in some western Canadian oil field waters, J. Ind. Microbiol. Biotechnol. 29 (2002) 83–92.

[372] I. Andreevskii, The influence of the microflora of the third stratum of the Yaregskoe oil field on the composition and properties of oil, Trudy Inst. Mikrobiol 9 (1961) 75–81.
[373] I. Andreevskii, Application of oil microbiology to the oil-extracting industry, Trudy Vses. Nauch.-Issled. Geol.-Razved. Inst 131 (1959) 403–413.
[374] S. Belyaev, et al., Activation of the geochemical activity of stratal microflora as the basis of a biotechnology for enhancement of oil recovery, Microbiology 67 (1998) 708–714.
[375] S. Belyaev, et al., Microbiological processes in the critical zone of injection wells in oilfields, Microbiology 51 (1982) 793–797.
[376] M. Ivanov, S. Belyaev, Microbial activity in waterflooded oilfields and its possible regulation, in: E. C. Donaldson, J.B. Clark (Eds.), Proceedings,1982 International Conference on Microbial Enhancement of Oil Recovery, NTIS, Springfield, Virginia, 1982, pp. 48–57.
[377] M. Ivanov, et al., Additional oil production during field trials in Russia, in: E.T. Premuzic, A. Woodhead (Eds.), Developments in Petroleum Science, Elsevier, Amsterdam, Netherlands, 1993, pp. 373–381.
[378] T. Nazina, et al., Microbiological investigation of the stratal waters, Microbiology 68 (1999) 214–221.
[379] T.N. Nazina, et al., Microorganisms of the high-temperature Liaohe oil field, Resour. Environ. Biotechnol. 3 (2000) 149–160.
[380] T.N. Nazina, et al., Diversity and activity of microorganisms in the daqing oil, Resour. Environ. Biotechnol. 3 (2000) 161–172.
[381] E. Rozanova, et al., Microbiological processes in a high-temperature oil field, Microbiology 70 (2001) 102–110.
[382] E. Yulbarisov, Evaluation of the effectiveness of the biological method for enhancing oil recovery of a reservoir, Neftyanoe Khozyaistvo 11 (1976) 27–30.
[383] E. Yulbarisov, On the enhancement of oil recovery of flooded oil strata, Neftyanoe Khozyaistvo 3 (1981) 36–40.
[384] E. Yulbarisov, Microbiological method for EOR, Revue de l'Institut Français du Pétrole 45 (1990) 115–121.
[385] E. Yulbarisov, N. Zhdanova, On the microbial enhancement of oil recovery of flooded oil strata, Neftyanoe khozyaistvo 3 (1984) 28–32.
[386] E. Kulik, et al., Hexadecane oxidation in a porous system with the formation of fatty acids, Mikrobiologiya 54 (1985) 381–385.
[387] E. Kowalewski, et al., Microbial improved oil recovery—bacterial induced wettability and interfacial tension effects on oil production, J. Pet. Sci. Eng. 52 (2006) 275–286.
[388] J. Heider, et al., Anaerobic bacterial metabolism of hydrocarbons, FEMS Microbiol. Rev. 22 (1998) 459–473.
[389] C.M. Aitken, et al., Anaerobic hydrocarbon biodegradation in deep subsurface oil reservoirs, Nature 431 (2004) 291–294.
[390] K.G. Kropp, et al., Anaerobic oxidation of n-dodecane by an addition reaction in a sulfate-reducing bacterial enrichment culture, Appl. Environ. Microbiol. 66 (2000) 5393–5398.
[391] R.S. Bryant, Potential uses of microorganisms in petroleum recovery technology, Proc. Oklahoma Acad. Sci. 67 (1987) 97–104.
[392] R.S. Bryant, R.P. Lindsey, World-wide applications of microbial technology for improving oil recovery, SPE/DOE Improved Oil Recovery Symposium, Society of Petroleum Engineers, Tulsa, Oklahoma, 1996.
[393] R. Illias, et al., Production of biosurfactant and biopolymer from Malaysian oil fields isolated microorganisms, SPE Asia Pacific Improved Oil Recovery Conference, Society of Petroleum Engineers, Kuala Lumpur, Malaysia, 1999.
[394] M. Tango, M. Islam, Potential of extremophiles for biotechnological and petroleum applications, Energy Sources 24 (2002) 543–559.
[395] T. Marsh, et al., Mechanisms of microbial oil recovery by *Clostridium acetobutylicum* and Bacillus strain JF-2, in: R.S. Bryant (Ed.), The Fifth International Conference on Microbial Enhanced Oil Recovery and Related Biotechnology for Solving Environmental Problems, BDM Oklahoma, Inc., Bartlesville, OK (United States), Dallas, Texas, 1995, pp. 593–610.

[396] S. Kianipey, E. Donaldson, Mechanisms of oil displacement by microorganisms, SPE Annual Technical Conference and Exhibition, Society of Petroleum Engineers, New Orleans, Louisiana, 1986.

[397] R. Tanner, et al., The potential for MEOR from carbonate reservoirs: literature review and recent research, in: E.T. Premuzic, A. Woodhead (Eds.), Developments in Petroleum Science, Elsevier, Amsterdam, Netherlands, 1993, pp. 391−396.

[398] D. Updegraff, G.B. Wren, The release of oil from petroleum-bearing materials by sulfate-reducing bacteria, Appl. Microbiol. 2 (1954) 309.

[399] D. Cooper, et al., Isolation and identification of biosurfactants produced during anaerobic growth of Clostridium pasteurianum, J. Ferment. Technol. 58 (1980) 83−86.

[400] V. Moses, Ch. I-3 MEOR in the field: why so little? in: E.C. Donaldson (Ed.), Developments in Petroleum Science, Elsevier, Amsterdam, Netherlands, 1991, pp. 21−28.

[401] R.E. Lappan, H.S. Fogler, Leuconostoc mesenteroides growth kinetics with application to bacterial profile modification, Biotechnol. Bioeng. 43 (1994) 865−873.

[402] G. Jenneman, et al., Bacterial profile modification with bulk dextran gels produced by the in-situ growth and metabolism of leuconostoc species, SPE/DOE Improved Oil Recovery Symposium, Society of Petroleum Engineers, Tulsa, Oklahoma, 2000.

[403] D. Davis, et al., The production of surfactin in batch culture by *Bacillus subtilis* ATCC 21332 is strongly influenced by the conditions of nitrogen metabolism, Enzyme Microb. Technol. 25 (1999) 322−329.

[404] S.S. Cameotra, R. Makkar, Synthesis of biosurfactants in extreme conditions, Appl. Microbiol. Biotechnol. 50 (1998) 520−529.

[405] K. Gautam, V. Tyagi, Microbial surfactants: a review, J. Oleo. Sci. 55 (2006) 155−166.

[406] C. Yao, et al., Laboratory experiment, modeling and field application of indigenous microbial flooding, J. Pet. Sci. Eng. 90-91 (2012) 39−47.

[407] S.L. Bryant, T.P. Lockhart, Reservoir engineering analysis of microbial enhanced oil recovery, SPE Reservoir Eval. Eng. 5 (2002) 365−374.

[408] M. Wagner, et al., Development and application of a new biotechnology of the molasses in-situ method; detailed evaluation for selected wells in the Romashkino carbonate reservoir, in: R.S. Bryant (Ed.), The Fifth International Conference on Microbial Enhanced Oil Recovery and Related Biotechnology for Solving Environmental Problems, BDM Oklahoma, Inc., Dallas, Texas, 1995, pp. 153−173.

[409] R. Bryant, et al., Ch. F-4 Microbial enhanced waterflooding: a pilot study, in: E.C. Donaldson (Ed.), Developments in Petroleum Science, Elsevier, Amsterdam, Netherlands, 1991, pp. 399−419.

[410] L. Strappa, et al., A novel and successful MEOR pilot project in a strong water-drive reservoir Vizcacheras Field, Argentina, SPE/DOE Symposium on Improved Oil Recovery, Society of Petroleum Engineers, Tulsa, Oklahoma, 2004.

[411] C.H. Gao, A. Zekri, Applications of microbial-enhanced oil recovery technology in the past decade, Energy Sources, Part A 33 (2011) 972−989.

[412] M. McInerney, et al., Development of microorganisms with improved transport and biosurfactant activity for enhanced oil recovery, in University of Oklahoma, US, 2005.

[413] E. Grula, et al., Field trials in central Oklahoma using Clostridium strains for microbially enhanced oil recovery, Microbes Oil Recovery 1 (1985) 144−150.

[414] G. Petzet, B. Williams, Operators trim basic EOR research, Oil Gas J. 84 (1986) 41−45.

[415] A.R. Awan, et al., A survey of North Sea enhanced-oil-recovery projects initiated during the years 1975 to 2005, SPE Reservoir Eval. Eng. 11 (2008) 497−512.

[416] T.L. Kieft, et al., Drilling, coring, and sampling subsurface environments, in: C. Hurst, R. Crawford, J. Garland, D. Lipson, A. Mills, L. Stetzenbach (Eds.), Manual of Environmental Microbiology, third ed., ASM Press, Washington, DC, 2007, pp. 799−817.

[417] W.-D. Wang, MEOR studies and pilot tests in the Shengli oilfield, in: C.-Z. Yan, Y. Li (Eds.), Tertiary Oil Recovery Symposium, Petroleum Industry Press, Beijing, China, 2005, pp. 123−128.

[418] N. Youssef, et al., In situ biosurfactant production by Bacillus strains injected into a limestone petroleum reservoir, Appl. Environ. Microbiol. 73 (2007) 1239–1247.
[419] A.K. Camper, et al., Effects of motility and adsorption rate coefficient on transport of bacteria through saturated porous media, Appl. Environ. Microbiol. 59 (1993) 3455–3462.
[420] A.B. Cunningham, et al., Effects of starvation on bacterial transport through porous media, Adv. Water Resour. 30 (2007) 1583–1592.
[421] D.E. Fontes, et al., Physical and chemical factors influencing transport of microorganisms through porous media, Appl. Environ. Microbiol. 57 (1991) 2473–2481.
[422] H. Lappin-Scott, et al., Nutrient resuscitation and growth of starved cells in sandstone cores: a novel approach to enhanced oil recovery, Appl. Environ. Microbiol. 54 (1988) 1373–1382.
[423] F. MacLeod, et al., Plugging of a model rock system by using starved bacteria, Appl. Environ. Microbiol. 54 (1988) 1365–1372.
[424] L.-K. Jang, et al., Selection of bacteria with favorable transport properties through porous rock for the application of microbial-enhanced oil recovery, Appl. Environ. Microbiol. 46 (1983) 1066–1072.
[425] J. Bae, et al., Microbial profile modification with spores, SPE Reservoir Eng. 11 (1996) 163–167.
[426] I.L. Gullapalli, et al., Laboratory design and field implementation of microbial profile modification process, SPE Reservoir Eval. Eng. 3 (2000) 42–49.
[427] N.H. Youssef, et al., Comparison of methods to detect biosurfactant production by diverse microorganisms, J. Microbiol. Methods 56 (2004) 339–347.
[428] R. Marchant, I.M. Banat, Microbial biosurfactants: challenges and opportunities for future exploitation, Trends Biotechnol. 30 (2012) 558–565.
[429] M. García-Junco, et al., Bioavailability of solid and non-aqueous phase liquid (NAPL)-dissolved phenanthrene to the biosurfactant-producing bacterium *Pseudomonas aeruginosa* 19SJ, Environ. Microbiol. 3 (2001) 561–569.
[430] D.C. Herman, et al., Rhamnolipid (biosurfactant) effects on cell aggregation and biodegradation of residual hexadecane under saturated flow conditions, Appl. Environ. Microbiol. 63 (1997) 3622–3627.
[431] I. Ivshina, et al., Oil desorption from mineral and organic materials using biosurfactant complexes produced by Rhodococcus species, World J. Microbiol. Biotechnol. 14 (1998) 711–717.
[432] W.R. Jones, Biosurfactants, bioavailability and bioremediation, Stud. Env. Sci. 66 (1997) 379–391.
[433] R. Maier, G. Soberon-Chavez, *Pseudomonas aeruginosa* rhamnolipids: biosynthesis and potential applications, Appl. Microbiol. Biotechnol. 54 (2000) 625–633.
[434] A.C. Morán, et al., Enhancement of hydrocarbon wastebiodegradation by addition of a biosurfactantfrom *Bacillus subtilis* O9, Biodegradation 11 (2000) 65–71.
[435] E.Z. Ron, E. Rosenberg, Natural roles of biosurfactants, Environ. Microbiol. 3 (2001) 229–236.
[436] S. Thangamani, G.S. Shreve, Effect of anionic biosurfactant on hexadecane partitioning in multiphase systems, Environ. Sci. Technol. 28 (1994) 1993–2000.
[437] A. Toren, et al., Solubilization of polyaromatic hydrocarbons by recombinant bioemulsifier AlnA, Appl. Microbiol. Biotechnol. 59 (2002) 580–584.
[438] Y. Zhang, R.M. Miller, Enhanced octadecane dispersion and biodegradation by a *Pseudomonas rhamnolipid* surfactant (biosurfactant), Appl. Environ. Microbiol. 58 (1992) 3276–3282.
[439] R.A. Al-Tahhan, et al., Rhamnolipid-induced removal of lipopolysaccharide from *Pseudomonas aeruginosa*: effect on cell surface properties and interaction with hydrophobic substrates, Appl. Environ. Microbiol. 66 (2000) 3262–3268.
[440] Science Learning Hub. (2012). Cleaning up the oil spill. Retrieved from www.sciencelearn.org.nz/resources/1140-cleaning-up-the-oil-spill.
[441] H. Sobrinho, et al., Biosurfactants: classification, properties and environmental applications, Recent Dev. Biotechnol. 11 (2013) 1–29.
[442] M. Nitschke, S. Costa, Biosurfactants in food industry, Trends Food Sci. Technol. 18 (2007) 252–259.
[443] E.J. Gudiña, et al., Isolation and study of microorganisms from oil samples for application in microbial enhanced oil recovery, Int. Biodeterior. Biodegradation. 68 (2012) 56–64.

[444] O. Pornsunthorntawee, et al., Isolation and comparison of biosurfactants produced by *Bacillus subtilis* PT2 and *Pseudomonas aeruginosa* SP4 for microbial surfactant-enhanced oil recovery, Biochem. Eng. J. 42 (2008) 172–179.
[445] P.C. Hiemenz, R. Rajagopalan, Principles of Colloid and Surface Chemistry, revised and expanded, CRC press, New York, NY, 1997.
[446] G. Georgiou, et al., Surface–active compounds from microorganisms, Nat. Biotechnol. 10 (1992) 60–65.
[447] R.S. Makkar, K.J. Rockne, Comparison of synthetic surfactants and biosurfactants in enhancing biodegradation of polycyclic aromatic hydrocarbons, Environ. Toxicol. Chem. 22 (2003) 2280–2292.
[448] M. Morikawa, et al., A study on the structure–function relationship of lipopeptide biosurfactants, Biochim. Biophys. Acta (BBA)-Mol. Cell Biol. Lipids 1488 (2000) 211–218.
[449] C. Schippers, et al., Microbial degradation of phenanthrene by addition of a sophorolipid mixture, J. Biotechnol. 83 (2000) 189–198.
[450] R. Makkar, S.S. Cameotra, Production of biosurfactant at mesophilic and thermophilic conditions by a strain of *Bacillus subtilis*, J. Ind. Microbiol. Biotechnol. 20 (1998) 48–52.
[451] P. Yan, et al., Oil recovery from refinery oily sludge using a rhamnolipid biosurfactant-producing Pseudomonas, Bioresour. Technol. 116 (2012) 24–28.
[452] T. Barkay, et al., Enhancement of solubilization and biodegradation of polyaromatic hydrocarbons by the bioemulsifier alasan, Appl. Environ. Microbiol. 65 (1999) 2697–2702.
[453] A. Etoumi, Microbial treatment of waxy crude oils for mitigation of wax precipitation, J. Pet. Sci. Eng. 55 (2007) 111–121.
[454] M. Rosenberg, et al., Adherence of Bacteria to Hydrocarbons, Penn Well Publications Company, Tulsa, Oklahoma, 1983.
[455] G.T. de Acevedo, M.J. McInerney, Emulsifying activity in thermophilic and extremely thermophilic microorganisms, J. Ind. Microbiol. 16 (1996) 1–7.
[456] L. Thimon, et al., Interactions of surfactin, a biosurfactant from *Bacillus subtilis*, with inorganic cations, Biotechnol. Lett. 14 (1992) 713–718.
[457] I.M. Banat, Biosurfactants production and possible uses in microbial enhanced oil recovery and oil pollution remediation: a review, Bioresour. Technol. 51 (1995) 1–12.
[458] P. Das, et al., Antimicrobial potential of a lipopeptide biosurfactant derived from a marine *Bacillus circulans*, J. Appl. Microbiol. 104 (2008) 1675–1684.
[459] P.K. Mohan, et al., Biokinetics of biodegradation of surfactants under aerobic, anoxic and anaerobic conditions, Water Res. 40 (2006) 533–540.
[460] P.K. Mohan, et al., Biodegradability of surfactants under aerobic, anoxic, and anaerobic conditions, J. Environ. Eng. 132 (2006) 279–283.
[461] H. Amani, et al., Comparative study of biosurfactant producing bacteria in MEOR applications, J. Pet. Sci. Eng. 75 (2010) 209–214.
[462] S. Al-Bahry, et al., Biosurfactant production by *Bacillus subtilis* B20 using date molasses and its possible application in enhanced oil recovery, Int. Biodeterior. Biodegrad. 81 (2013) 141–146.
[463] P. Darvishi, et al., Biosurfactant production under extreme environmental conditions by an efficient microbial consortium, ERCPPI-2, Colloids Surf. B: Biointerfaces 84 (2011) 292–300.
[464] F. Freitas, et al., Emulsifying behaviour and rheological properties of the extracellular polysaccharide produced by *Pseudomonas oleovorans* grown on glycerol byproduct, Carbohydr. Polym. 78 (2009) 549–556.
[465] M. Shavandi, et al., Emulsification potential of a newly isolated biosurfactant-producing bacterium, Rhodococcus sp. strain TA6, Colloids Surf. B: Biointerfaces 82 (2011) 477–482.
[466] T.R. Neu, Significance of bacterial surface-active compounds in interaction of bacteria with interfaces, Microbiol. Rev. 60 (1996) 151–166.
[467] I.M. Banat, et al., Microbial biosurfactants production, applications and future potential, Appl. Microbiol. Biotechnol. 87 (2010) 427–444.
[468] S. Vijayakumar, V. Saravanan, Biosurfactants-types, sources and applications, Res. J. Microbiol. 10 (2015) 181–192.

[469] P.K. Rahman, E. Gakpe, Production, characterisation and applications of biosurfactants-review, Biotechnology 7 (2008) 360–370.
[470] K. Muthusamy, et al., Biosurfactants: properties, commercial production and application, Curr. Sci. 94 (2008) 736–747.
[471] M. Pacwa-Płociniczak, et al., Environmental applications of biosurfactants: recent advances, Int. J. Mol. Sci. 12 (2011) 633–654.
[472] A. Toren, et al., The active component of the bioemulsifier alasan from *Acinetobacter radioresistens* KA53 is an OmpA-like protein, J. Bacteriol. 184 (2002) 165–170.
[473] H. Chong, Q. Li, Microbial production of rhamnolipids: opportunities, challenges and strategies, Microb. Cell. Fact. 16 (2017) 137.
[474] A.P. Rooney, et al., Isolation and characterization of rhamnolipid-producing bacterial strains from a biodiesel facility, FEMS Microbiol. Lett. 295 (2009) 82–87.
[475] M. Hošková, et al., Structural and physiochemical characterization of rhamnolipids produced by *Acinetobacter calcoaceticus*, *Enterobacter asburiae* and *Pseudomonas aeruginosa* in single strain and mixed cultures, J. Biotechnol. 193 (2015) 45–51.
[476] M. Hošková, et al., Characterization of rhamnolipids produced by non-pathogenic Acinetobacter and Enterobacter bacteria, Bioresour. Technol. 130 (2013) 510–516.
[477] M. Healy, et al., Microbial production of biosurfactants, Resour., Conserv. Recycl. 18 (1996) 41–57.
[478] M. Konishi, et al., Production of new types of sophorolipids by *Candida batistae*, J. Oleo. Sci. 57 (2008) 359–369.
[479] J. Chen, et al., Production, structure elucidation and anticancer properties of sophorolipid from *Wickerhamiella domercqiae*, Enzyme Microb. Technol. 39 (2006) 501–506.
[480] X. Song, Wickerhamiella domercqiae Y2A for producing sophorose lipid and its uses, in 山东大学 China, 2006.
[481] G. Soberón-Chávez, Biosurfactants: From Genes to Applications, Springer-Verlag, Berlin Heidelberg, 2010.
[482] N.P. Price, et al., Structural characterization of novel sophorolipid biosurfactants from a newly identified species of Candida yeast, Carbohydr. Res. 348 (2012) 33–41.
[483] C.P. Kurtzman, et al., Production of sophorolipid biosurfactants by multiple species of the *Starmerella* (Candida) *bombicola* yeast clade, FEMS Microbiol. Lett. 311 (2010) 140–146.
[484] A. Tulloch, et al., A new hydroxy fatty acid sophoroside from *Candida bogoriensis*, Can. J. Chem. 46 (1968) 345–348.
[485] A.E. Elshafie, et al., Sophorolipids production by *Candida bombicola* ATCC 22214 and its potential application in microbial enhanced oil recovery, Front. Microbiol. 6 (2015) 1–11.
[486] A. Hatha, et al., Microbial biosurfactants–review, J. Mar. Atmos. Res. 3 (2007) 1–17.
[487] A. Marqués, et al., The physicochemical properties and chemical composition of trehalose lipids produced by *Rhodococcus erythropolis* 51T7, Chem. Phys. Lipids 158 (2009) 110–117.
[488] B. Tuleva, et al., Isolation and characterization of trehalose tetraester biosurfactants from a soil strain *Micrococcus luteus* BN56, Process Biochem. 44 (2009) 135–141.
[489] B. Tuleva, et al., Production and structural elucidation of trehalose tetraesters (biosurfactants) from a novel alkanothrophic *Rhodococcus wratislaviensis* strain, J. Appl. Microbiol. 104 (2008) 1703–1710.
[490] Y. Tokumoto, et al., Structural characterization and surface-active properties of a succinoyl trehalose lipid produced by Rhodococcus sp. SD-74, J. Oleo. Sci. 58 (2009) 97–102.
[491] Y. Uchida, et al., Extracellular accumulation of mono-and di-succinoyl trehalose lipids by a strain of *Rhodococcus erythropolis* grown on n-alkanes, Agric. Biol. Chem. 53 (1989) 757–763.
[492] F. Peng, et al., An oil-degrading bacterium: *Rhodococcus erythropolis* strain 3C-9 and its biosurfactants, J. Appl. Microbiol. 102 (2007) 1603–1611.
[493] S. Niescher, et al., Identification and structural characterisation of novel trehalose dinocardiomycolates from n-alkane-grown *Rhodococcus opacus* 1CP, Appl. Microbiol. Biotechnol. 70 (2006) 605–611.
[494] M.V. Singer, et al., Physical and chemical properties of a biosurfactant synthesized by Rhodococcus species H13-A, Can. J. Microbiol. 36 (1990) 746–750.

[495] M.V. Singer, W. Finnerty, Physiology of biosurfactant synthesis by Rhodococcus species H13-A, Can. J. Microbiol. 36 (1990) 741–745.
[496] J.-S. Kim, et al., Microbial glycolipid production under nitrogen limitation and resting cell conditions, J. Biotechnol. 13 (1990) 257–266.
[497] A. Kretschmer, F. Wagner, Characterization of biosynthetic intermediates of trehalose dicorynomycolates from *Rhodococcus erythropolis* grown on n-alkanes, Biochim. Biophys. Acta (BBA)-Lipids Lipid Metab. 753 (1983) 306–313.
[498] A. Kretschmer, et al., Chemical and physical characterization of interfacial-active lipids from *Rhodococcus erythropolis* grown on n-alkanes, Appl. Environ. Microbiol. 44 (1982) 864–870.
[499] M. Yakimov, et al., Characterization of antarctic hydrocarbon-degrading bacteria capable of producing bioemulsifiers, New Microbiol. 22 (1999) 249–256.
[500] S.W. Esch, et al., A novel trisaccharide glycolipid biosurfactant containing trehalose bears ester-linked hexanoate, succinate, and acyloxyacyl moieties: NMR and MS characterization of the underivatized structure, Carbohydr. Res. 319 (1999) 112–123.
[501] D. Schulz, et al., Marine biosurfactants, I. Screening for biosurfactants among crude oil degrading marine microorganisms from the North Sea, Z. Naturforsch. C 46 (1991) 197–203.
[502] A. Passeri, et al., Marine biosurfactants, II. Production and characterization of an anionic trehalose tetraester from the marine bacterium Arthrobacter sp. EK 1, Z. Naturforsch. C 46 (1991) 204–209.
[503] A. Desai, et al., Emulsifier production by *Pseudomonas fluorescens* during the growth on hydrocarbons, Curr. Sci. 57 (1988) 500–501.
[504] A.K. Datta, K. Takayama, Isolation and purification of trehalose 6-mono-and 6, 6′-di-corynomycolates from *Corynebacterium matruchotii*. Structural characterization by 1H NMR, Carbohydr. Res. 245 (1993) 151–158.
[505] B. Mompon, et al., Isolation and structural determination of a "cord-factor"(trehalose 6, 6′ dimycolate) from *Mycobacterium smegmatis*, Chem. Phys. Lipids 21 (1978) 97–101.
[506] G.S. Besra, et al., Structural elucidation of a novel family of acyltrehaloses from *Mycobacterium tuberculosis*, Biochemistry 31 (1992) 9832–9837.
[507] N. Gautier, et al., Structure of mycoside F, a family of trehalose-containing glycolipids of *Mycobacterium fortuitum*, FEMS Microbiol. Lett. 98 (1992) 81–87.
[508] K. Poremba, et al., Marine biosurfactants, III. Toxicity testing with marine microorganisms and comparison with synthetic surfactants, Z. Naturforsch. C 46 (1991) 210–216.
[509] W.-R. Abraham, et al., Novel glycine containing glucolipids from the alkane using bacterium *Alcanivorax borkumensis*, Biochim. Biophys. Acta (BBA)-Lipids Lipid Metab. 1393 (1998) 57–62.
[510] S. Mehta, et al., Biomimetic amphiphiles: properties and potential use, in: R. Sen (Ed.), Biosurfactants. Advances in Experimental Medicine and Biology, Springer, New York, NY, 2010, pp. 102–120.
[511] N. Karanth, et al., Microbial production of biosurfactants and their importance, Curr. Sci. 77 (1999) 116–126.
[512] A. Toren, et al., Emulsifying activities of purified alasan proteins from *Acinetobacter radioresistens* KA53, Appl. Environ. Microbiol. 67 (2001) 1102–1106.
[513] S. Navon-Venezia, et al., Alasan, a new bioemulsifier from *Acinetobacter radioresistens*, Appl. Environ. Microbiol. 61 (1995) 3240–3244.
[514] E. Rosenberg, E.Z. Ron, Bioemulsans: microbial polymeric emulsifiers, Curr. Opin. Biotechnol. 8 (1997) 313–316.
[515] N. Kosaric, F.V. Sukan, Biosurfactants: Production and Utilization—Processes, Technologies, and Economics, CRC Press, Boca Raton, Florida, 2014.
[516] M.C. Cirigliano, G.M. Carman, Isolation of a bioemulsifier from *Candida lipolytica*, Appl. Environ. Microbiol. 48 (1984) 747–750.
[517] R. Vazquez-Duhalt, R. Quintero-Ramirez, Petroleum Biotechnology: Developments and Perspectives, Elsevier, Amsterdam, Netherlands, 2004.
[518] R. Diniz Rufino, et al., Characterization and properties of the biosurfactant produced by *Candida lipolytica* UCP 0988, Electron. J. Biotechnol. 17 (2014). 6-6.

[519] D. Husain, et al., The effect of temperature on eicosane substrate uptake modes by a marine bacterium *Pseudomonas nautica* strain 617: relationship with the biochemical content of cells and supernatants, World J. Microbiol. Biotechnol. 13 (1997) 587–590.
[520] S.S. Zinjarde, A. Pant, Emulsifier from a tropical marine yeast, *Yarrowia lipolytica* NCIM 3589, J. Basic Microbiol. 42 (2002) 67–73.
[521] O. Käppeli, et al., Chemical and structural alterations at the cell surface of *Candida tropicalis*, induced by hydrocarbon substrate, J. Bacteriol. 133 (1978) 952–958.
[522] R. Shepherd, et al., Novel bioemulsifiers from microorganisms for use in foods, J. Biotechnol. 40 (1995) 207–217.
[523] M. Singh, J. Desai, Hydrocarbon emulsification by *Candida tropicalis* and *Debaryomyces polymorphus*, Indian J. Exp. Biol. 27 (1989) 224–226.
[524] E. Rosenberg, E. Ron, High-and low-molecular-mass microbial surfactants, Appl. Microbiol. Biotechnol. 52 (1999) 154–162.
[525] O. Käppeli, W. Finnerty, Partition of alkane by an extracellular vesicle derived from hexadecane-grown Acinetobacter, J. Bacteriol. 140 (1979) 707–712.
[526] S.-C. Lin, et al., Continuous production of the lipopeptide biosurfactant of *Bacillus licheniformis* JF-2, Appl. Microbiol. Biotechnol. 41 (1994) 281–285.
[527] M. Folmsbee, et al., Re-identification of the halotolerant, biosurfactant-producing *Bacillus licheniformis* strain JF-2 as *Bacillus mojavensis* strain JF-2, Syst. Appl. Microbiol. 29 (2006) 645–649.
[528] K. Arima, et al., Surfactin, a crystalline peptidelipid surfactant produced by *Bacillus subtilis*: Isolation, characterization and its inhibition of fibrin clot formation, Biochem. Biophys. Res. Commun. 31 (1968) 488–494.
[529] D. Cooper, et al., Enhanced production of surfactin from *Bacillus subtilis* by continuous product removal and metal cation additions, Appl. Environ. Microbiol. 42 (1981) 408–412.
[530] A.S. Nerurkar, Structural and molecular characteristics of lichenysin and its relationship with surface activity, in: R. Sen (Ed.), Biosurfactants. Advances in Experimental Medicine and Biology, Springer, New York, NY, 2010, pp. 304–315.
[531] M.J. McInerney, et al., Biosurfactant and Enhanced Oil Recovery, University of Oklahoma, US, 1985.
[532] M.M. Yakimov, et al., A putative lichenysin A synthetase operon in *Bacillus licheniformis*: initial characterization, Biochim. Biophys. Acta (BBA)-Gene Struct. Expression 1399 (1998) 141–153.
[533] C. Rubinovitz, et al., Emulsan production by *Acinetobacter calcoaceticus* in the presence of chloramphenicol, J. Bacteriol. 152 (1982) 126–132.
[534] R. Rautela, S.S. Cameotra, Role of biopolymers in industries: their prospective future applications, in: M. Fulekar, B. Pathak, R. Kale (Eds.), Environment and Sustainable Development, Springer, New Delhi, 2014, pp. 133–142.
[535] J.D. Desai, I.M. Banat, Microbial production of surfactants and their commercial potential, Microbiol. Mol. Biol. Rev. 61 (1997) 47–64.
[536] S.C. Lin, Biosurfactants: recent advances, J. Chem. Technol. Biotechnol. 66 (1996) 109–120.
[537] A. Franzetti, et al., Production and applications of trehalose lipid biosurfactants, Eur. J. Lipid Sci. Technol. 112 (2010) 617–627.
[538] J.M. Campos, et al., Microbial biosurfactants as additives for food industries, Biotechnol. Prog. 29 (2013) 1097–1108.
[539] R. Makkar, S. Cameotra, An update on the use of unconventional substrates for biosurfactant production and their new applications, Appl. Microbiol. Biotechnol. 58 (2002) 428–434.
[540] I.M. Banat, et al., Potential commercial applications of microbial surfactants, Appl. Microbiol. Biotechnol. 53 (2000) 495–508.
[541] I. Banat, Characterization of biosurfactants and their use in pollution removal—state of the art, Eng. Life Sci. 15 (1995) 251–267.
[542] A.A. Bodour, R.M. Maier, Biosurfactants: types, screening methods, and applications, in: G. Bitton (Ed.), Encyclopedia of Environmental Microbiology, Wiley, New York, NY, 2003, pp. 750–770.
[543] A.Y. Zekri, et al., Project of increasing oil recovery from UAE reservoirs using bacteria flooding, SPE Annual Technical Conference, Society of Petroleum Engineers, Houston, Texas, 1999.

[544] E.H. Sugihardjo, S.W. Pratomo, Microbial core flooding experiments using indigenous microbes, SPE Asia Pacific Improved Oil Recovery Conference, Society of Petroleum Engineers, Kuala Lumpur, Malaysia, 1999.
[545] M.M. Yakimov, et al., The potential of *Bacillus licheniformis* strains for in situ enhanced oil recovery, J. Pet. Sci. Eng. 18 (1997) 147–160.
[546] H. Yonebayashi, et al., Fundamental studies on MEOR with anaerobes. Flooding experiments using sandpack for estimation of capabilities of microbes, Sekiyu Gijutsu Kyokaishi 62 (1997) 195–202.
[547] E. Rosenberg, et al., Emulsifier of Arthrobacter RAG-1: specificity of hydrocarbon substrate, Appl. Environ. Microbiol. 37 (1979) 409–413.
[548] E. Rosenberg, et al., Emulsifier of Arthrobacter RAG-1: isolation and emulsifying properties, Appl. Environ. Microbiol. 37 (1979) 402–408.
[549] F. Martínez-Checa, et al., Characteristics of bioemulsifier V2-7 synthesized in culture media added of hydrocarbons: chemical composition, emulsifying activity and rheological properties, Bioresour. Technol. 98 (2007) 3130–3135.
[550] Q. Wang, et al., Engineering bacteria for production of rhamnolipid as an agent for enhanced oil recovery, Biotechnol. Bioeng. 98 (2007) 842–853.
[551] E. Acosta, et al., Linker-modified microemulsions for a variety of oils and surfactants, J. Surfactants Deterg. 6 (2003) 353–363.
[552] D.C. Herman, et al., Formation and removal of hydrocarbon residual in porous media: effects of attached bacteria and biosurfactants, Environ. Sci. Technol. 31 (1997) 1290–1294.
[553] A. Abu-Ruwaida, et al., Isolation of biosurfactant-producing bacteria, product characterization, and evaluation, Eng. Life Sci. 11 (1991) 315–324.
[554] W. Hua, C.-h Liang, Effect of sulfate reduced bacterium on corrosion behavior of 10CrMoAl steel, J. Iron Steel Res., Int. 14 (2007) 74–78.
[555] R. Sen, T. Swaminathan, Characterization of concentration and purification parameters and operating conditions for the small-scale recovery of surfactin, Process Biochem. 40 (2005) 2953–2958.
[556] I.M. Banat, The isolation of a thermophilic biosurfactant producing Bacillus sp, Biotechnol. Lett. 15 (1993) 591–594.
[557] R.S. Makkar, S.S. Cameotra, Utilization of molasses for biosurfactant production by two Bacillus strains at thermophilic conditions, J. Am. Oil Chem. Soc. 74 (1997) 887–889.
[558] S. Joshi, et al., Biosurfactant production using molasses and whey under thermophilic conditions, Bioresour. Technol. 99 (2008) 195–199.
[559] M. Sayyouh, Microbial enhanced oil recovery: research studies in the Arabic area during the last ten years, SPE/DOE Improved Oil Recovery Symposium, Society of Petroleum Engineers, Tulsa, Oklahoma, 2002.
[560] N. Abtahi, et al., Biosurfactant production in MEOR for improvement of Iran's oil reservoirs' production experimental approach, SPE International Improved Oil Recovery Conference in Asia Pacific, Society of Petroleum Engineers, Kuala Lumpur, Malaysia, 2003.
[561] Q. Li, et al., Application of microbial enhanced oil recovery technique to Daqing Oilfield, Biochem. Eng. J. 11 (2002) 197–199.
[562] B. Lal, et al., Process for enhanced recovery of crude oil from oil wells using novel microbial consortium, in Energy and Resources InstituteInstitute of Reservoir Studies US, 2009.
[563] S. Maudgalya, et al., Development of bio-surfactant based microbial enhanced oil recovery procedure, SPE/DOE Symposium on Improved Oil Recovery, Society of Petroleum Engineers, Tulsa, Oklahoma, 2004.
[564] S. Maudgalya, et al., Tertiary oil recovery with microbial biosurfactant treatment of low-permeability Berea sandstone cores, SPE Production Operations Symposium, Society of Petroleum Engineers, Oklahoma City, Oklahoma, 2005.
[565] J.P. Adkins, et al., Microbially enhanced oil recovery from unconsolidated limestone cores, Geomicrobiol. J. 10 (1992) 77–86.
[566] J.P. Adkins, et al., Microbial composition of carbonate petroleum reservoir fluids, Geomicrobiol. J. 10 (1992) 87–97.

[567] Y.I. Chang, Preliminary studies assessing sodium pyrophosphate effects on microbially mediated oil recovery, Ann. N.Y. Acad. Sci. 506 (1987) 296–307.
[568] K. Das, A.K. Mukherjee, Comparison of lipopeptide biosurfactants production by *Bacillus subtilis* strains in submerged and solid state fermentation systems using a cheap carbon source: some industrial applications of biosurfactants, Process Biochem. 42 (2007) 1191–1199.
[569] S. Joshi, et al., Production of biosurfactant and antifungal compound by fermented food isolate *Bacillus subtilis* 20B, Bioresour. Technol. 99 (2008) 4603–4608.
[570] S.J. Johnson, et al., Using biosurfactants produced from agriculture process waste streams to improve oil recovery in fractured carbonate reservoirs, International Symposium on Oilfield Chemistry, Society of Petroleum Engineers, Houston, Texas, 2007.
[571] A. Sheehy, Recovery of oil from oil reservoirs, in B W N Live-Oil Pty Ltd, US, 1992.
[572] G. Okpokwasili, A. Ibiene, Enhancement of recovery of residual oil using a biosurfactant slug, Afr. J. Biotechnol. 5 (2006) 453–456.
[573] K. Das, A.K. Mukherjee, Characterization of biochemical properties and biological activities of biosurfactants produced by *Pseudomonas aeruginosa* mucoid and non-mucoid strains isolated from hydrocarbon-contaminated soil samples, Appl. Microbiol. Biotechnol. 69 (2005) 192–199.
[574] N. Bordoloi, B. Konwar, Microbial surfactant-enhanced mineral oil recovery under laboratory conditions, Colloids and surfaces B: Biointerfaces 63 (2008) 73–82.
[575] R. Thavasi, et al., Effect of biosurfactant and fertilizer on biodegradation of crude oil by marine isolates of *Bacillus megaterium*, *Corynebacterium kutscheri* and *Pseudomonas aeruginosa*, Bioresour. Technol. 102 (2011) 772–778.
[576] E.J. Gudiña, et al., Biosurfactant-producing and oil-degrading *Bacillus subtilis* strains enhance oil recovery in laboratory sand-pack columns, J. Hazard. Mater. 261 (2013) 106–113.
[577] T.B. Lotfabad, et al., An efficient biosurfactant-producing bacterium *Pseudomonas aeruginosa* MR01, isolated from oil excavation areas in south of Iran, Colloids Surf. B: Biointerfaces 69 (2009) 183–193.
[578] A. Najafi, et al., Interactive optimization of biosurfactant production by *Paenibacillus alvei* ARN63 isolated from an Iranian oil well, Colloids Surf. B: Biointerfaces 82 (2011) 33–39.
[579] W. Zhang, et al., An experimental study on the bio-surfactant-assisted remediation of crude oil and salt contaminated soils, J. Environ. Sci. Health, Part A 46 (2011) 306–313.
[580] Q. Liu, et al., Production of surfactin isoforms by *Bacillus subtilis* BS-37 and its applicability to enhanced oil recovery under laboratory conditions, Biochem. Eng. J. 93 (2015) 31–37.
[581] V. Pruthi, S.S. Cameotra, Production of a biosurfactant exhibiting excellent emulsification and surface active properties by *Serratia marcescens*, World J. Microbiol. Biotechnol. 13 (1997) 133–135.
[582] V. Pruthi, S.S. Cameotra, Production and properties of a biosurfactant synthesized by *Arthrobacter protophormiae*—an antarctic strain, World J. Microbiol. Biotechnol. 13 (1997) 137–139.
[583] R. Makkar, S.S. Cameotra, Biosurfactant production by a thermophilic *Bacillus subtilis* strain, J. Ind. Microbiol. Biotechnol. 18 (1997) 37–42.
[584] G. Jenneman, et al., A halotolerant, biosurfactant producing Bacillus species potentially useful for enhanced oil recovery, Dev. Ind. Microbiol. 24 (1983) 485–492.
[585] D.L. Gutnick, et al., Bioemulsifier production by *Acinetobacter calcoaceticus* strains, in Emulsan Biotechnologies Inc, US, 1989.
[586] A. Sheehy, Recovery of oil from oil reservoirs, in B W N Live-Oil Pty Ltd, US, 1990.
[587] C.N. Mulligan, T.Y. Chow, Enhanced production of biosurfactant through the use of a mutated *B. subtilis* strain, in National Research Council of Canada, US, 1991.
[588] J.B. Clark, G.E. Jenneman, Nutrient injection method for subterranean microbial processes, in ConocoPhillips Co, US, 1992.
[589] P. Carrera, et al., Method of producing surfactin with the use of mutant of *Bacillus subtilis*, in EniTecnologie SpA, US, 1993.
[590] T. Imanaka, S. Sakurai, Biosurfactant cyclopeptide compound produced by culturing a specific Arthrobacter microorganism, in Nikko Bio Technica Co Ltd, US, 1994.
[591] E. Rosenberg, E.Z. Ron, Bioemulsifiers, in Ramot at Tel Aviv University Ltd US, 1998.

[592] W.P.C. Duyvesteyn, et al., Extraction of bitumen from bitumen froth and biotreatment of bitumen froth tailings generated from tar sands, in BHP Minerals International Inc, US, 1999.
[593] C.A. Rocha, et al., Production of oily emulsions mediated by a microbial tenso-active agent, in Universidad Simon Bolivar, US, 1999.
[594] W.P.C. Duyvesteyn, et al., Biochemical treatment of bitumen froth tailings, in BHP Minerals International Inc, US, 2000.
[595] G. Prosperi, et al., Lipopolysaccharide biosurfactant, in EniTecnologie SpA US, 2000.
[596] C.A. Rocha, et al., Production of oily emulsions mediated by a microbial tenso-active agent, in Universidad Simon Bolivar, US, 2000.
[597] D.R. Converse, et al., Process for stimulating microbial activity in a hydrocarbon-bearing, subterranean formation, in ExxonMobil Upstream Research Co US, 2003.
[598] J.B. Crews, Bacteria-based and enzyme-based mechanisms and products for viscosity reduction breaking of viscoelastic fluids, in Baker Hughes Inc, US, 2006.
[599] R.L. Brigmon, C.J. Berry, Biological enhancement of hydrocarbon extraction, in Savannah River Nuclear Solutions LLC US, 2009.
[600] R.D. Fallon, Methods for improved hydrocarbon and water compatibility, in E I du Pont de Nemours and Co US, 2011.
[601] F.D. Busche, et al., System and method for preparing near-surface heavy oil for extraction using microbial degradation, in System and method for preparing near-surface heavy oil for extraction using microbial degradation, US, 2011.
[602] S.J. Keeler, et al., Identification, characterization, and application of Pseudomonas stutzeri (LH4:15), useful in microbially enhanced oil release, in, E I du Pont de Nemours and Co, US, 2013.
[603] M.R. Pavia, et al., Systems and methods of microbial enhanced oil recovery, in New Aero Technology LLC, US, 2014.
[604] E.R. Hendrickson, et al., Altering the interface of hydrocarbon-coated surfaces, in E I du Pont de Nemours and Co US, 2016.
[605] W.J. Kohr, et al., Alkaline microbial enhanced oil recovery, in Geo Fossil Fuels LLC, US, 2016.
[606] G. Bognolo, Biosurfactants as emulsifying agents for hydrocarbons, Colloids Surf. A: Physicochem. Eng. Aspects 152 (1999) 41−52.
[607] H. Bach, D. Gutnick, Potential applications of bioemulsifiers in the oil industry, Stud. Surf. Sci. Catal. 151 (2004) 233−281.
[608] A.K. Mukherjee, K. Das, Microbial surfactants and their potential applications: an overview, in: R. Sen (Ed.), Biosurfactants. Advances in Experimental Medicine and Biology, Springer, New York, NY, 2010, pp. 54−64.
[609] A. Perfumo, et al., Rhamnolipid production by a novel thermophilic hydrocarbon-degrading *Pseudomonas aeruginosa* AP02-1, Appl. Microbiol. Biotechnol. 72 (2006) 132−138.
[610] A. Singh, et al., Surfactants in microbiology and biotechnology: Part 2. Application aspects, Biotechnol. Adv. 25 (2007) 99−121.
[611] K. Urum, et al., Optimum conditions for washing of crude oil-contaminated soil with biosurfactant solutions, Process Saf. Environ. Protect. 81 (2003) 203−209.
[612] M.S. Kuyukina, et al., Effect of biosurfactants on crude oil desorption and mobilization in a soil system, Environ. Int. 31 (2005) 155−161.
[613] A. Wentzel, et al., Bacterial metabolism of long-chain n-alkanes, Appl. Microbiol. Biotechnol. 76 (2007) 1209−1221.
[614] A. Poli, et al., Synthesis, production, and biotechnological applications of exopolysaccharides and polyhydroxyalkanoates by archaea, Archaea 2011 (2011) 1−13.
[615] F. Brown, Microbes: The practical and environmental safe solution to production problems, enhanced production, and enhanced oil recovery, Permian Basin Oil and Gas Recovery Conference, Society of Petroleum Engineers, Midland, Texas, 1992.
[616] C. Whitfield, Bacterial extracellular polysaccharides, Can. J. Microbiol. 34 (1988) 415−420.
[617] D.-S. Kim, H.S. Fogler, The effects of exopolymers on cell morphology and culturability of *Leuconostoc mesenteroides* during starvation, Appl. Microbiol. Biotechnol. 52 (1999) 839−844.

[618] S.E. Fratesi, Distribution and morphology of bacteria and their byproducts in microbial enhanced oil recovery operations, in Mississippi State University, Oktibbeha County, Mississippi, 2002.
[619] P.A. Sandford, Exocellular, microbial polysaccharides, Adv. Carbohydr. Chem. Biochem. 36 (1979) 265–313.
[620] C.S. Buller, S. Vossoughi, Subterranean permeability modification by using microbial polysaccharide polymers, in Kansas, A State Educational Institution of Kansas, University of University of Kansas, US, 1990.
[621] J.A. Ramsay, et al., Effects of oil reservoir conditions on the production of water-insoluble Levan by *Bacillus licheniformis*, Geomicrobiol. J. 7 (1989) 155–165.
[622] D.H. Cho, et al., Synthesis and characterization of a novel extracellular polysaccharide by *Rhodotorula glutinis*, Appl. Biochem. Biotechnol. 95 (2001) 183–193.
[623] M. Salome, Mutant strain of *Xanthomonas campestris*, process of obtaining xanthan, and non-viscous xanthan, in Google Patents, US, 1996.
[624] T.J. Pollock, L. Thorne, *Xanthomonas campestris* strain for production of xanthan gum, in Shin-Etsu Bio Inc, US, 1994.
[625] A. Lachke, Xanthan—a versatile gum, Resonance 9 (2004) 25–33.
[626] T. Nazina, et al., Production of oil-releasing compounds by microorganisms from the Daqing oil field, China, Microbiology 72 (2003) 173–178.
[627] M.J. McInerney, et al., Situ microbial plugging process for subterranean formations, in Board of Regents for University of OK, A Legal Entity of State of OK University of Oklahoma, US, 1985.
[628] M.J. McInerney, et al., Use of indigenous or injected microorganisms for enhanced oil recovery, in: C. Bell, M. Brylinsky, P. Johnson-Green (Eds.) Proceedings of the 8th International Symposium on Microbial Ecology, Atlantic Canada Society for Microbial Ecology, Halifax, Nova Scotia, 1999.
[629] G.E. Jenneman, et al., Experimental studies of in-situ microbial enhanced oil recovery, Soc. Pet. Eng. J. 24 (1984) 33–37.
[630] M.K. Dabbous, Displacement of polymers in waterflooded porous media and its effects on a subsequent micellar flood, Soc. Pet. Eng. J. 17 (1977) 358–368.
[631] M.K. Dabbous, L.E. Elkins, Preinjection of polymers to increase reservoir flooding efficiency, SPE Improved Oil Recovery Symposium, Society of Petroleum Engineers, Tulsa, Oklahoma, 1976.
[632] T. Harrah, et al., Microbial exopolysaccharides, in: E. Rosenberg, E.F. DeLong, S. Lory, E. Stackebrandt, F. Thompson (Eds.), The Prokaryotes, Springer, Berlin, Heidelberg, 2006, pp. 766–776.
[633] S.P. Trushenski, et al., Micellar flooding-fluid propagation, interaction, and mobility, Soc. Pet. Eng. J. 14 (1974) 633–645.
[634] S. Cao, et al., Engineering behavior and characteristics of water-soluble polymers: implication on soil remediation and enhanced oil recovery, Sustainability 8 (2016) 205.
[635] H.Y. Jang, et al., Enhanced oil recovery performance and viscosity characteristics of polysaccharide xanthan gum solution, J. Ind. Eng. Chem. 21 (2015) 741–745.
[636] L.R. Brown, Method for increasing oil recovery, in BP Corporation North America Inc, US, 1984.
[637] P. Vandevivere, P. Baveye, Effect of bacterial extracellular polymers on the saturated hydraulic conductivity of sand columns, Appl. Environ. Microbiol. 58 (1992) 1690–1698.
[638] R. Mitchell, Z. Nevo, Effect of bacterial polysaccharide accumulation on infiltration of water through sand, Appl. Microbiol. 12 (1964) 219–223.
[639] R.A. Raiders, et al., Microbial selective plugging and enhanced oil recovery, J. Ind. Microbiol. Biotechnol. 4 (1989) 215–229.
[640] A. Hove, et al., Visualization of xanthan flood behavior in core samples by means of X-ray tomography, SPE Reservoir Eng. 5 (1990) 475–480.
[641] G. Holzwarth, Xanthan and scleroglucan: structure and use in enhanced oil recovery, Dev. Ind. Microbiol. 26 (1984) 271–280.
[642] J.F. Kennedy, I. Bradshaw, Production, properties and applications of xanthan, Prog. Ind. Microbiol. 19 (1984) 319–371.

[643] W. Cannella, et al., Prediction of xanthan rheology in porous media, SPE annual Technical Conference and Exhibition, Society of Petroleum Engineers, Houston, Texas, 1988.
[644] D.G. Allison, I.W. Sutherland, The role of exopolysaccharides in adhesion of freshwater bacteria, Microbiology 133 (1987) 1319–1327.
[645] P. Vandevivere, P. Baveye, Saturated hydraulic conductivity reduction caused by aerobic bacteria in sand columns, Soil Sci. Soc. Am. J. 56 (1992) 1–13.
[646] P. Kalish, et al., The effect of bacteria on sandstone permeability, J. Pet. Technol. 16 (1964) 805–814.
[647] T.J. Pollock, et al., Production of non-native bacterial exopolysaccharide in a recombinant bacterial host, in Shin-Etsu Chemical Co Ltd Shin-Etsu Bio Inc US, 2000.
[648] S. Bailey, et al., Design of a novel alkaliphilic bacterial system for triggering biopolymer gels, J. Ind. Microbiol. Biotechnol. 24 (2000) 389–395.
[649] Y. Sun, et al., Preparation and characterization of novel curdlan/chitosan blending membranes for antibacterial applications, Carbohydr. Polym. 84 (2011) 952–959.
[650] T. Harada, et al., Curdlan: a bacterial gel-forming β-1, 3-glucan, Arch. Biochem. Biophys. 124 (1968) 292–298.
[651] R. Khachatoorian, et al., Biopolymer plugging effect: laboratory-pressurized pumping flow studies, J. Pet. Sci. Eng. 38 (2003) 13–21.
[652] S. Rosalam, R. England, Review of xanthan gum production from unmodified starches by *Xanthomonas comprestris* sp, Enzyme Microb. Technol. 39 (2006) 197–207.
[653] A. Palaniraj, V. Jayaraman, Production, recovery and applications of xanthan gum by *Xanthomonas campestris*, J. Food Eng. 106 (2011) 1–12.
[654] H. Funahashi, et al., Effect of glucose concentrations on xanthan gum production by *Xanthomonas campestris*, J. Ferment. Technol. 65 (1987) 603–606.
[655] I. Rottava, et al., Xanthan gum production and rheological behavior using different strains of Xanthomonas sp, Carbohydr. Polym. 77 (2009) 65–71.
[656] H. Zhang, et al., Component identification of electron transport chains in curdlan-producing Agrobacterium sp. ATCC 31749 and its genome-specific prediction using comparative genome and phylogenetic trees analysis, J. Ind. Microbiol. Biotechnol. 38 (2011) 667–677.
[657] A.M. Ruffing, et al., Genome sequence of the curdlan-producing Agrobacterium sp. strain ATCC 31749, J. Bacteriol. 193 (2011) 4294–4295.
[658] M.H. El-Sayed, et al., Optimization, purification and physicochemical characterization of curdlan produced by Paenibacillus sp. strain NBR-10, Biosci. Biotechnol. Res. Asia 13 (2016) 901–909.
[659] M. Yang, et al., Production and optimization of curdlan produced by Pseudomonas sp. QL212, Int. J. Biol. Macromol. 89 (2016) 25–34.
[660] I. Lee, et al., Production of curdlan using sucrose or sugar cane molasses by two-step fed-batch cultivation of Agrobacterium species, J. Ind. Microbiol. Biotechnol. 18 (1997) 255–259.
[661] J.-h Lee, I.Y. Lee, Optimization of uracil addition for curdlan (β-1 → 3-glucan) production by Agrobacterium sp, Biotechnol. Lett. 23 (2001) 1131–1134.
[662] K.R. Phillips, et al., Production of curdlan-type polysaccharide by *Alcaligenes faecalis* in batch and continuous culture, Can. J. Microbiol. 29 (1983) 1331–1338.
[663] M. Kim, et al., Residual phosphate concentration under nitrogen-limiting conditions regulates curdlan production in Agrobacterium species, J. Ind. Microbiol. Biotechnol. 25 (2000) 180–183.
[664] S.A. van Hijum, et al., Purification of a novel fructosyltransferase from *Lactobacillus reuteri* strain 121 and characterization of the levan produced, FEMS Microbiol. Lett. 205 (2001) 323–328.
[665] K.H. Kim, et al., Cosmeceutical properties of levan produced by *Zymomonas mobilis*, J. Cosmet. Sci. 56 (2005) 395–406.
[666] E. Newbrun, S. Baker, Physico-chemical characteristics of the levan produced by *Streptococcus salivarius*, Carbohydr. Res. 6 (1968) 165–170.
[667] P.J. Simms, et al., The structural analysis of a levan produced by *Streptococcus salivarius* SS2, Carbohydr. Res. 208 (1990) 193–198.
[668] I. Kojima, et al., Characterization of levan produced by Serratia sp, J. Ferment. Bioeng. 75 (1993) 9–12.

[669] L. Shih, Y.-T. Yu, Simultaneous and selective production of levan and poly (γ-glutamic acid) by *Bacillus subtilis*, Biotechnol. Lett. 27 (2005) 103–106.
[670] D. Wang, et al., The mechanism of improved pullulan production by nitrogen limitation in batch culture of *Aureobasidium pullulans*, Carbohydr. Polym. 127 (2015) 325–331.
[671] R. Taguchi, et al., Structural uniformity of pullulan produced by several strains of *Pullularia pullulans*, Agric. Biol. Chem. 37 (1973) 1583–1588.
[672] A. Lazaridou, et al., Characterization of pullulan produced from beet molasses by *Aureobasidium pullulans* in a stirred tank reactor under varying agitation, Enzyme Microb. Technol. 31 (2002) 122–132.
[673] R.K. Purama, et al., Structural analysis and properties of dextran produced by *Leuconostoc mesenteroides* NRRL B-640, Carbohydr. Polym. 76 (2009) 30–35.
[674] W.J. Lewick, et al., Determination of the structure of a broth dextran produced by a cariogenic streptococcus, Carbohydr. Res. 17 (1971) 175–182.
[675] S. Patel, et al., Structural analysis and biomedical applications of dextran produced by a new isolate *Pediococcus pentosaceus* screened from biodiversity hot spot Assam, Bioresour. Technol. 101 (2010) 6852–6855.
[676] A. Aman, et al., Characterization and potential applications of high molecular weight dextran produced by *Leuconostoc mesenteroides* AA1, Carbohydr. Polym. 87 (2012) 910–915.
[677] R.Z. Ahmed, et al., Characterization of high molecular weight dextran produced by *Weissella cibaria* CMGDEX3, Carbohydr. Polym. 90 (2012) 441–446.
[678] J. Farina, et al., High scleroglucan production by *Sclerotium rolfsii*: influence of medium composition, Biotechnol. Lett. 20 (1998) 825–831.
[679] J. Farina, et al., Isolation and physicochemical characterization of soluble scleroglucan from Sclerotium rolfsii. Rheological properties, molecular weight and conformational characteristics, Carbohydr. Polym. 44 (2001) 41–50.
[680] S.A. Survase, et al., Production of scleroglucan from Sclerotium rolfsii MTCC 2156, Bioresour. Technol. 97 (2006) 989–993.
[681] G.-Q. Chen, W.J. Page, Production of poly-b-hydroxybutyrate by *Azotobacter vinelandii* in a two-stage fermentation process, Biotechnol. Techn. 11 (1997) 347–350.
[682] M. Yilmaz, et al., Determination of poly-β-hydroxybutyrate (PHB) production by some Bacillus spp, World J. Microbiol. Biotechnol. 21 (2005) 565–566.
[683] Y. Ogawa, et al., Efficient production of γ-polyglutamic acid by *Bacillus subtilis* (natto) in jar fermenters, Biosci. Biotechnol. Biochem. 61 (1997) 1684–1687.
[684] M. Kambourova, et al., Regulation of polyglutamic acid synthesis by glutamate in *Bacillus licheniformis* and *Bacillus subtilis*, Appl. Environ. Microbiol. 67 (2001) 1004–1007.
[685] E. Udegbunam, et al., Assessing the effects of microbial metabolism and metabolites on reservoir pore structure, SPE Annual Technical Conference and Exhibition, Society of Petroleum Engineers, Dallas, Texas, 1991.
[686] J. Coleman, et al., Enhanced Oil Recovery, in, Archaeus Technology Group Ltd 1992.
[687] R.S. Bryant, et al., Microbial-enhanced waterflooding: mink unit project, SPE Reservoir Eng. 5 (1990) 9–13.
[688] N.K. Harner, et al., Microbial processes in the Athabasca oil sands and their potential applications in microbial enhanced oil recovery, J. Ind. Microbiol. Biotechnol. 38 (2011) 1761–1775.
[689] V. Groudeva, et al., Enhanced oil recovery by stimulating the activity of the indigenous microflora of oil reservoirs, Biohydrometall. Technol. 2 (1993) 349–356.
[690] E. Rozanova, T. Nazina, Hydrocarbon oxidizing bacteria and their activity in oil pools, Microbiology 51 (1982) 287–293.
[691] J.W. Pelger, Ch. F-8 Microbial enhanced oil recovery treatments and wellbore stimulation using microorganisms to control paraffin, emulsion, corrosion, and scale formation, Dev. Pet. Sci. 31 (1991) 451–466.
[692] L. Pinilla, et al., Bioethanol production in batch mode by a native strain of *Zymomonas mobilis*, World J. Microbiol. Biotechnol. 27 (2011) 2521–2528.

[693] R.S. Bryant, T.E. Burchfield, Review of microbial technology for improving oil recovery, SPE Reservoir Eng. 4 (1989) 151–154.
[694] R.S. Tanner, et al., Microbially enhanced oil recovery from carbonate reservoirs, Geomicrobiol. J. 9 (1991) 169–195.
[695] A. Kantzas, et al., A novel method of sand consolidation through bacteriogenic mineral plugging, in: Annual Technical Meeting, Petroleum Society of Canada, Calgary, Alberta, 1992.
[696] F. Ferris, et al., Bacteriogenic mineral plugging, in: Technical Meeting/Petroleum Conference of The South Saskatchewan Section, Petroleum Society of Canada, Regina, Saskatchewan, 1992.
[697] T. Jack, et al., The potential for use of microbes in the production of heavy oil, in: E.C. Donaldson, J.B. Clark (Eds.) Proceedings of the International Conference on Microbial Enhancement of Oil Recovery, SPE/DOE, Afton, Oklahoma, 1983, pp. 88–93.
[698] E. Grula, et al., Isolation and screening of clostridia for possible use in microbially enhanced oil recovery, in: E.C. Donaldson, J.B. Clark (Eds.), International Conference on Microbial Enhancement of Oil Recovery, NTIS, Springfield, Virginia, 1982, pp. 43–47.
[699] S. Desouky, et al., Modelling and laboratory investigation of microbial enhanced oil recovery, J. Pet. Sci. Eng. 15 (1996) 309–320.
[700] J. Chisholm, et al., Microbial enhanced oil recovery: interfacial tension and gas-induced relative permeability effects, SPE Annual Technical Conference and Exhibition, Society of Petroleum Engineers, New Orleans, Louisiana, 1990.
[701] F. Chapelle, et al., Alteration of aquifer geochemistry by microorganisms, in: C. Hurst, G. Knudsen, M. McInerney, L. Stetzenbach, M. Walter (Eds.), Manual of Environmental Microbiology, ASM press, Washington, DC, 1997, pp. 558–564.
[702] R. Knapp, et al., Design and implementation of a microbially enhanced oil recovery field pilot, Payne County, Oklahoma, SPE Annual Technical Conference and Exhibition, Society of Petroleum Engineers, Washington, DC, 1992.
[703] D. Jones, et al., Crude-oil biodegradation via methanogenesis in subsurface petroleum reservoirs, Nature 451 (2008) 176–180.
[704] R.P. Gunsalus, et al., Preparation of coenzyme M analogs and their activity in the methyl coenzyme M reductase system of *Methanobacterium thermoautotrophicum*, Biochemistry 17 (1978) 2374–2377.
[705] Carbon dioxide emissions from the generation of electric power in the United States, in U.S. Department of Energy, U.S. Environmental Protection Agency, Washington, D.C., 2000.
[706] P.B. Crawford, Possible reservoir damage from microbial enhanced oil recovery, in: E.C. Donaldson, J.B. Clark (Eds.), Proceedings of the 1982 International Conference on Microbial Enhancement of Oil Recovery, Afton, Oklahoma, 1982, pp. 76–79.
[707] P.B. Crawford, Water technology: continual changes observed in bacterial stratification rectification, Prod. Mon. 26 (1962) 12–13.
[708] J.E. Paulsen, et al., Biofilm morphology in porous media, a study with microscopic and image techniques, Water Sci. Technol. 36 (1997) 1–9.
[709] F. Cusack, et al., Advances in microbiology to enhance oil recovery, Appl. Biochem. Biotechnol. 24 (1990) 885–898.
[710] T.R. Jack, E. Diblasio, Selective plugging for heavy oil recovery, in: J.E. Zajic, E.C. Donaldson (Eds.), Microbial Enhanced Oil Recovery, Bioresearch Publications, El Paso, Texas, 1985.
[711] R. Raiders, et al., Selectivity and depth of microbial plugging in Berea sandstone cores, J. Ind. Microbiol. 1 (1986) 195–203.
[712] H.A. Khan, et al., Mechanistic models of microbe growth in heterogeneous porous media, SPE Symposium on Improved Oil Recovery, Society of Petroleum Engineers, Tulsa, Oklahoma, 2008.
[713] R. Knapp, et al., Mechanisms of microbial enhanced oil recovery in high salinity core environments, AIChE Symp. Ser. 87 (1991) 134–140.
[714] M.E. Davey, et al., Microbial selective plugging of sandstone through stimulation of indigenous bacteria in a hypersaline oil reservoir, Geomicrobiol. J. 15 (1998) 335–352.
[715] M. Karimi, et al., Investigating wettability alteration during MEOR process, a micro/macro scale analysis, Colloids Surf. B: Biointerfaces 95 (2012) 129–136.

[716] B. Carpentier, O. Cerf, Biofilms and their consequences, with particular reference to hygiene in the food industry, J. Appl. Microbiol. 75 (1993) 499–511.
[717] J.W. Costerton, et al., Bacterial biofilms: a common cause of persistent infections, Science 284 (1999) 1318–1322.
[718] G. Wolf, et al., Optical and spectroscopic methods for biofilm examination and monitoring, Rev. Environ. Sci. Biotechnol. 1 (2002) 227–251.
[719] T.L. Stewart, H. Scott Fogler, Pore-scale investigation of biomass plug development and propagation in porous media, Biotechnol. Bioeng. 77 (2002) 577–588.
[720] H. Torbati, et al., Effect of microbial growth on pore entrance size distribution in sandstone cores, J. Ind. Microbiol. Biotechnol. 1 (1986) 227–234.
[721] G.G. Geesey, et al., Evaluation of slime-producing bacteria in oil field core flood experiments, Appl. Environ. Microbiol. 53 (1987) 278–283.
[722] R. Lappan, H.S. Fogler, Effect of bacterial polysaccharide production on formation damage, SPE Prod. Eng. 7 (1992) 167–171.
[723] E. Robertson, The use of bacteria to reduce water influx in producing oil wells, SPE Prod. Facil. 13 (1998) 128–132.
[724] J.C. Shaw, et al., Bacterial fouling in a model core system, Appl. Environ. Microbiol. 49 (1985) 693–701.
[725] N. Ross, et al., Clogging of a limestone fracture by stimulating groundwater microbes, Water Res. 35 (2001) 2029–2037.
[726] R.E. Lappan, H.S. Fogler, Reduction of porous media permeability from in situ *Leuconostoc mesenteroides* growth and dextran production, Biotechnol. Bioeng. 50 (1996) 6–15.
[727] B.F. Wolf, H.S. Fogler, Alteration of the growth rate and lag time of *Leuconostoc mesenteroides* NRRL-B523, Biotechnol. Bioeng. 72 (2001) 603–610.
[728] R.S. Silver, et al., Bacteria and its use in a microbial profile modification process, in Chevron Research and Technology Co US, 1989.
[729] T. Jack, G. Stehmeier, Selective plugging in watered out oil reservoirs, in: T.E. Burchfield, R.S. Bryant (Eds.), Proceedings of the Symposium on the Application of Microorganisms to Petroleum Technology, National Technical Information Service, Bartlesville, Oklahoma, 1987, pp. VII 1–VII 14.
[730] J. Costerton, H. Lappin-Scott, Behavior of bacteria in biofilms, ASM News 55 (1989) 650–654.
[731] E.J. Polson, et al., An environmental-scanning-electron-microscope investigation into the effect of biofilm on the wettability of quartz, SPE J. 15 (2010) 223–227.
[732] P. Sharma, K.H. Rao, Analysis of different approaches for evaluation of surface energy of microbial cells by contact angle goniometry, Adv. Colloid. Interface. Sci. 98 (2002) 341–463.
[733] R.M. Donlan, J.W. Costerton, Biofilms: survival mechanisms of clinically relevant microorganisms, Clin. Microbiol. Rev. 15 (2002) 167–193.
[734] W.M. Dunne, Bacterial adhesion: seen any good biofilms lately? Clin. Microbiol. Rev. 15 (2002) 155–166.
[735] K. Lewis, Persister cells and the riddle of biofilm survival, Biochemistry (Moscow) 70 (2005) 267–274.
[736] L. Brown, 20 – The use of microorganisms to enhance oil recovery, in: J. Sheng (Ed.), Enhanced Oil Recovery Field Case Studies, Gulf Professional Publishing, Boston, Massachusetts, 2013, pp. 552–571.
[737] M. Rauf, et al., Enhanced oil recovery through microbial treatment, J. Trace Microprobe Techn. 21 (2003) 533–541.
[738] R. Rabus, Biodegradation of hydrocarbons under anoxic conditions, in: B. Ollivier, M. Magot (Eds.), Petroleum Microbiology, ASM Press, Washington, D.C., 2005, pp. 277–300.
[739] O. Rahn, Ein Paraffin zersetzender Schimmelpilz, Zentralbl. Bakt., Parasit. Infektionskr., Abteilung II 16 (1906) 382–384.
[740] N. Söhngen, Über bakterien, welche methan als kohlenstoffnahrung und energiequelle gebrauchen, Zentralbl. Bakt., Parasit. Infektionskr. 15 (1906) 513–517.
[741] H. Kaserer, Über die Oxydation des Wasserstoffes und des Methans durch Mikroorganismen, Zentralbl. Bakt., Parasit. Infektionskr., Abteilung II (1905) 573–576.

[742] N. Söhngen, Benzin, petroleum, paraffinöl und paraffin als kohlenstoff-und energiequelle für mikroben, Zentralbl. Bakt., Parasit. Infektionskr. 37 (1913) 595–609.
[743] J.B. van Beilen, B. Witholt, Diversity, function, and biocatalytic applications of alkane oxygenases, in: B. Ollivier, M. Magot (Eds.), Petroleum Microbiology, ASM Press, Washington, D.C., 2005, pp. 259–276.
[744] B.B. Skaare, et al., Alteration of crude oils from the Troll area by biodegradation: analysis of oil and water samples, Org. Geochem. 38 (2007) 1865–1883.
[745] D. Denney, Paraffin treatments: Hot oil/hot water vs. crystal modifiers, J. Pet. Technol. 53 (2001) 56–57.
[746] W. Ford, et al., Dispersant solves paraffin problems, Am. Oil Gas Rep. 43 (2000) 91–94.
[747] M. Sanjay, et al., Paraffin problems in crude oil production and transportation: a review, SPE Prod. Facil. 10 (1995) 50–54.
[748] I. Lazar, et al., The use of naturally occurring selectively isolated bacteria for inhibiting paraffin deposition, J. Pet. Sci. Eng. 22 (1999) 161–169.
[749] F.G. Bosch, et al., Evaluation of downhole electric impedance heating systems for paraffin control in oil wells, IEEE Trans. Ind. Appl. 28 (1992) 190–195.
[750] D. Shock, et al., Studies of the mechanism of paraffin deposition and its control, J. Pet. Technol. 7 (1955) 23–28.
[751] J.W. McManus, et al., Method and apparatus for removing oil well paraffin, in McManus James W Backus James US, 1985.
[752] M. Santamaria, R. George, Controlling paraffin-deposition-related problems by the use of bacteria treatments, SPE Annual Technical Conference and Exhibition, Society of Petroleum Engineers, Dallas, Texas, 1991.
[753] L. Streeb, F. Brown, MEOR-Altamont/Bluebell field project, SPE Rocky Mountain Regional Meeting, Society of Petroleum Engineers, Casper, Wyoming, 1992.
[754] K.R. Ferguson, et al., Microbial pilot test for the control of paraffin and asphaltenes at Prudhoe Bay, SPE Annual Technical Conference and Exhibition, Society of Petroleum Engineers, Denver, Colorado, 1996.
[755] A.M. Spormann, F. Widdel, Metabolism of alkylbenzenes, alkanes, and other hydrocarbons in anaerobic bacteria, Biodegradation 11 (2000) 85–105.
[756] A. Sadeghazad, N. Ghaemi, Microbial prevention of wax precipitation in crude oil by biodegradation mechanism, SPE Asia Pacific Oil and Gas Conference and Exhibition, Society of Petroleum Engineers, Jakarta, Indonesia, 2003.
[757] H.K. Kotlar, et al., Wax control by biocatalytic degradation in high-paraffinic crude oils, International Symposium on Oilfield Chemistry, Society of Petroleum Engineers, Houston, Texas, 2007.
[758] L. Giangiacomo, Paraffin Control Project, in: Rocky Mountain Oilfield Testing Center Project Test Reports, US Department of Energy, Virginia, 1997.
[759] T.L. Smith, G. Trebbau, MEOR treatments boost heavy oil recovery in Venezuela, Pet. Eng. Int. 71 (1998) 45–50.
[760] M.A. Maure, et al., Biotechnology applications to EOR in Talara offshore oil fields, Northwest Peru, SPE Latin American and Caribbean Petroleum Engineering Conference, Society of Petroleum Engineers, Rio de Janeiro, Brazil, 2005.
[761] F. Brown, et al., Microbial-induced controllable cracking of normal and branched alkanes in oils, in Microbes Inc, US, 2005.
[762] W. Guo, et al., Microbe-enhanced oil recovery technology obtains huge success in low-permeability reservoirs in Daqing oilfield, SPE Eastern Regional Meeting, Society of Petroleum Engineers, Canton, Ohio, 2006.
[763] G. Trebbau, et al., Microbial stimulation of lake maracaibo oil wells, SPE Annual Technical Conference, Society of Petroleum Engineers, Houston, Texas, 1999.
[764] C.J. Partidas, et al., Microbes aid heavy oil recovery in Venezuela, Oil Gas J. 96 (1998) 62–64.
[765] Z. He, et al., A pilot test using microbial paraffin-removal technology in Liaohe oilfield, Pet. Sci. Technol. 21 (2003) 201–210.
[766] D. Deng, et al., Systematic extensive laboratory studies of microbial EOR mechanisms and microbial EOR application results in Changqing Oilfield, SPE Asia Pacific Oil and Gas Conference and Exhibition, Society of Petroleum Engineers, Jakarta, Indonesia, 1999.

[767] A. Soudmand-asli, et al., The in situ microbial enhanced oil recovery in fractured porous media, J. Pet. Sci. Eng. 58 (2007) 161–172.

[768] K. Ohno, et al., Implementation and performance of a microbial enhanced oil recovery field pilot in Fuyu oilfield, China, SPE Asia Pacific Oil and Gas Conference and Exhibition, Society of Petroleum Engineers, Jakarta, Indonesia, 1999.

[769] I.A. Purwwasena, et al., Estimation of the potential of an oil-viscosity-reducing *Bacterium Petrotoga* sp. isolated from an oil field for MEOR, International Petroleum Technology Conference, Society of Petroleum Engineers, Doha, Qatar, 2009.

[770] A.Y. Halim, et al., Microbial enhanced oil recovery: an investigation of bacteria ability to live and alter crude oil physical characteristics in high pressure condition, Asia Pacific Oil and Gas Conference & Exhibition, Society of Petroleum Engineers, Jakarta, Indonesia, 2009.

[771] Z. Kang, et al., Hydrophobic bacteria at the hexadecane–water interface: examination of micrometre-scale interfacial properties, Colloids Surf. B: Biointerfaces 67 (2008) 59–66.

[772] R. Illias, et al., Isolation and characterization of thermophilic microorganisms from Malaysian oil fields, SPE Annual Technical Conference, Society of Petroleum Engineers, Houston, Texas, 1999.

[773] X.-C. Li, et al., Research and application results of microbial technologies in the Fuyu field, Jielin, in: C.-Z. Yan, Y. Li (Eds.), Tertiary Oil Recovery Symposium, Petroleum Industry Press, Beijing, China, 2005, pp. 279–286.

[774] L. Yu, Microbial technologies to enhance oil recovery, in: P.-P. Shen (Ed.), Technical Advances in Enhanced Oil Recovery, Petroleum Industry Press, Beijing, China, 2006, pp. 276–312.

[775] C. Zheng, et al., Investigation of a hydrocarbon-degrading strain, *Rhodococcus ruber* Z25, for the potential of microbial enhanced oil recovery, J. Pet. Sci. Eng. 81 (2012) 49–56.

[776] B. Thompson, T. Jack, Method of enhancing oil recovery by use of exopolymer producing micro-organisms, in Nova Husky Research Corp Ltd US, 1984.

[777] G.E. Jenneman, Chapter 3 The potential for in-situ microbial applications, in: E.C. Donaldson, G. V. Chilingarian, T.F. Yen (Eds.), Developments in Petroleum Science, Elsevier, Amsterdam, Netherlands, 1989, pp. 37–74.

[778] Y. Sugai, et al., Simulation studies on the mechanisms and performances of MEOR using a polymer producing microorganism Clostridium sp. TU-15A, Asia Pacific Oil and Gas Conference and Exhibition, Society of Petroleum Engineers, Jakarta, Indonesia, 2007.

[779] K. Nagase, et al., Improvement of sweep efficiency by microbial EOR process in Fuyu oilfield, China, SPE Asia Pacific Oil and Gas Conference and Exhibition, Society of Petroleum Engineers, Jakarta, Indonesia, 2001.

[780] G. Jenneman, et al., Application of a microbial selective-plugging process at the North Burbank Unit: prepilot tests, SPE Prod. Facil. 11 (1996) 11–17.

[781] H. Suthar, et al., Selective plugging strategy-based microbial-enhanced oil recovery using *Bacillus licheniformis* TT33, J. Microbiol. Biotechnol. 19 (2009) 1230–1237.

[782] J. Vilcáez, et al., Reactive transport modeling of induced selective plugging by *Leuconostoc mesenteroides* in carbonate formations, Geomicrobiol. J. 30 (2013) 813–828.

[783] J. Fink, Petroleum Engineer's Guide to Oil Field Chemicals and Fluids, Gulf Professional Publishing, Waltham, Massachusetts, 2011.

[784] E. Drobner, et al., Pyrite formation linked with hydrogen evolution under anaerobic conditions, Nature 346 (1990) 742–744.

[785] I. Spark, et al., The effects of indigenous and introduced microbes on deeply buried hydrocarbon reservoirs, North Sea, Clay. Miner. 35 (2000) 5–12.

[786] S. Stocks-Fischer, et al., Microbiological precipitation of $CaCO_3$, Soil Biol. Biochem. 31 (1999) 1563–1571.

[787] C. Dupraz, et al., Processes of carbonate precipitation in modern microbial mats, Earth-Sci. Rev. 96 (2009) 141–162.

[788] C. Glunk, et al., Microbially mediated carbonate precipitation in a hypersaline lake, Big Pond (Eleuthera, Bahamas), Sedimentology 58 (2011) 720–736.

[789] T. Bosak, D.K. Newman, Microbial nucleation of calcium carbonate in the Precambrian, Geology 31 (2003) 577–580.

[790] W.G. Anderson, Wettability literature survey-part 1: rock/oil/brine interactions and the effects of core handling on wettability, J. Pet. Technol. 38 (1986) 1125–1144.
[791] N.R. Morrow, Wettability and its effect on oil recovery, J. Pet. Technol. 42 (1990) 1476–1484.
[792] G. Wanger, et al., Structural and chemical characterization of a natural fracture surface from 2.8 kilometers below land surface: biofilms in the deep subsurface, Geomicrobiol. J. 23 (2006) 443–452.
[793] M. Salehi, et al., Mechanistic study of wettability alteration using surfactants with applications in naturally fractured reservoirs, Langmuir 24 (2008) 14099–14107.
[794] Z. Kang, et al., Mechanical properties of hexadecane–water interfaces with adsorbed hydrophobic bacteria, Colloids Surf. B: Biointerfaces 62 (2008) 273–279.
[795] N.R. Morrow, et al., Prospects of improved oil recovery related to wettability and brine composition, J. Pet. Sci. Eng. 20 (1998) 267–276.
[796] D.L. Zhang, et al., Wettability alteration and spontaneous imbibition in oil-wet carbonate formations, J. Pet. Sci. Eng. 52 (2006) 213–226.
[797] L. Yu, et al., Analysis of the wettability alteration process during seawater imbibition into preferentially oil-wet chalk cores, SPE Symposium on Improved Oil Recovery, Society of Petroleum Engineers, Tulsa, Oklahoma, 2008.
[798] V. Walter, et al., Screening concepts for the isolation of biosurfactant producing microorganisms, in: R. Sen (Ed.), Biosurfactants. Advances in Experimental Medicine and Biology, Springer, New York, NY, 2010, pp. 1–13.
[799] R. Moore, et al., A guide to high pressure air injection (HPAI) based oil recovery, SPE/DOE Improved Oil Recovery Symposium, Society of Petroleum Engineers, Tulsa, Oklahoma, 2002.
[800] J. Portwood, A commercial microbial enhanced oil recovery technology: evaluation of 322 projects, SPE Production Operations Symposium, Society of Petroleum Engineers, Oklahoma City, Oklahoma, 1995.
[801] C. Oppenheimer, F. Hiebert, Microbiological techniques for paraffin reduction in producing oil wells, in Alpha Environmental Corporation, Austin, Texas, 1989, pp. 1–67.
[802] J. Portwood, A commercial microbial enhanced oil recovery process: statistical evaluation of a multi-project database, in: R.S. Bryant (Ed.), The Fifth International Conference on Microbial Enhanced Oil Recovery and Related Biotechnology for Solving Environmental Problems, BDM Oklahoma, Inc., Bartlesville, OK (United States), Dallas, Texas, 1995, pp. 51–76.
[803] A.J. Sheehy, Field studies of microbial EOR, SPE/DOE Enhanced Oil Recovery Symposium, Society of Petroleum Engineers, Tulsa, Oklahoma, 1990.
[804] F. Bernard, et al., Indigenous microorganisms in connate water of many oil fields: a new tool in exploration and production techniques, SPE Annual Technical Conference and Exhibition, Society of Petroleum Engineers, Washington, D.C., 1992.
[805] V.K. Bhupathiraju, et al., Pretest studies for a microbially enhanced oil recovery field pilot in a hypersaline oil reservoir, Geomicrobiol. J. 11 (1993) 19–34.
[806] M. Javaheri, et al., Anaerobic production of a biosurfactant by *Bacillus licheniformis* JF-2, Appl. Environ. Microbiol. 50 (1985) 698–700.
[807] D.W. Hilchie, Applied Openhole Log Interpretation (for Geologists and Engineers), Douglas W. Hilchie, Inc., Golden, Colorado, 1978.
[808] J. Vermooten, et al., Quality control, correction and analysis of temperature borehole data in offshore Netherlands, Report A: Quality Control and Correction of Temperature Borehole Data; Part B: Analysis and Interpretation of Corrected Temperatures From Wells in Offshore NETHERLANDS 'Influence of Zechstein Salt Diapirs and Pillows on the Geothermal Gradient', Netherlands Institute of Applied Geoscience TNO, Utrecht, Netherlands, 2004, pp. 1–75.
[809] R. Hao, et al., Effect on crude oil by thermophilic bacterium, J. Pet. Sci. Eng. 43 (2004) 247–258.
[810] H. Ghojavand, et al., Enhanced oil recovery from low permeability dolomite cores using biosurfactant produced by a *Bacillus mojavensis* (PTCC 1696) isolated from Masjed-I Soleyman field, J. Pet. Sci. Eng. 81 (2012) 24–30.

[811] F. Zhang, et al., Impact of an indigenous microbial enhanced oil recovery field trial on microbial community structure in a high pour-point oil reservoir, Appl. Microbiol. Biotechnol. 95 (2012) 811–821.
[812] R. Amelunxen, A. Murdock, Microbial life at high temperatures: mechanisms and molecular aspects, in: D.J. Kushner (Ed.), Microbial Life in Extreme Environments, Academic Press, London, England, 1978, pp. 217–278.
[813] M. Tansey, T. Brock, Microbial life at high temperatures: ecological aspects, in: D.J. Kushner (Ed.), Microbial life in extreme environments, Academic Press, London, England, 1978, pp. 159–216.
[814] T.D. Brock, Thermophilic Microorganisms and Life at High Temperatures, Springer, New York, NY, 1978.
[815] K. Stetter, et al., Hyperthermophilic archaea are thriving in deep North Sea and Alaskan oil reservoirs, Nature 365 (1993) 743–745.
[816] J.W. Amyx, et al., Petroleum Reservoir Engineering: Physical Properties, McGraw-Hill College, New York, NY, 1960.
[817] F. Abe, et al., Pressure-regulated metabolism in microorganisms, Trends Microbiol. 7 (1999) 447–453.
[818] D. Bartlett, Pressure effects on in vivo microbial processes, Biochim. Biophys. Acta (BBA)-Protein Struct. Mol. Enzymol. 1595 (2002) 367–381.
[819] F. Abe, Piezophysiology of yeast: occurrence and significance, Cell. Mol. Biol. (Noisy-le-Grand, France) 50 (2004) 437–445.
[820] J. Schwarz, et al., Deep-sea bacteria: growth and utilization of n-hexadecane at in situ temperature and pressure, Can. J. Microbiol. 21 (1975) 682–687.
[821] F. Abe, Exploration of the effects of high hydrostatic pressure on microbial growth, physiology and survival: perspectives from piezophysiology, Biosci. Biotechnol. Biochem. 71 (2007) 2347–2357.
[822] R.E. Marquis, High pressure microbiology, in: P.B. Bennett, R.E. Marquis (Eds.), Basic and Applied High Pressure Biology, University of Rochester Press, Rochester, New York, 1994, pp. 1–14.
[823] R.E. Marquis, Microbial barobiology, Bioscience 32 (1982) 267–271.
[824] I. Fatt, D. Davis, Reduction in permeability with overburden pressure, J. Pet. Technol. 4 (1952). 16-16.
[825] R. Marquis, P. Matsumura, Microbial life under pressure, in: D.J. Kushner (Ed.), Microbial Life in Extreme Environments, Academic Press, London, England, 1978, pp. 105–158.
[826] J. Landau, D. Pope, Recent advances in the area of barotolerant protein synthesis in bacteria and implications concerning barotolerant and barophilic growth, Adv. Aquat. Microbiol. 2 (1980) 49–76.
[827] L.J. Albright, J.F. Henigman, Seawater salts—hydrostatic pressure effects upon cell division of several bacteria, Can. J. Microbiol. 17 (1971) 1246–1248.
[828] R.E. Marquis, C.E. ZoBell, Magnesium and calcium ions enhance barotolerance of streptococci, Arch. Microbiol. 79 (1971) 80–92.
[829] N.J. Hyne, Nontechnical Guide to Petroleum Geology, Exploration, Drilling, and Production, third ed., PennWell Books, Tulsa, Oklahoma, 2012.
[830] R. Paterek, P. Smith, Isolation of a halophilic methanogenic bacterium from the sediments of Great Salt Lake and a San Francisco Bay saltern, Abstracts of the 83rd Annual Meeting of the American Society for Microbiology, ASM Press, New Orleans, Louisiana, 1983.
[831] I. Yu, R. Hungate, Isolation and characterization of an obligately halophilic methanogenic bacterium, Annual Meeting of the American Society for Microbiology, Abstract 1, 1 (1983) p. 139.
[832] F. Rodriguez-Valera, et al., Behaviour of mixed populations of halophilic bacteria in continuous cultures, Can. J. Microbiol. 26 (1980) 1259–1263.
[833] K. Fujiwara, et al., Biotechnological approach for development of microbial enhanced oil recovery technique, Stud. Surf. Sci. Catal. 151 (2004) 405–445.
[834] J.K. Otton, T. Mercier, Produced water brine and stream salinity, in: Science for changing world, US Geological Survey, 2012.
[835] S.O. Stanley, R.Y. Morita, Salinity effect on the maximal growth temperature of some bacteria isolated from marine environments, J. Bacteriol. 95 (1968) 169–173.

[836] A.Z. Bilsky, J.B. Armstrong, Osmotic reversal of temperature sensitivity in *Escherichia coli*, J. Bacteriol. 113 (1973) 76–81.
[837] T. Novitsky, D. Kushner, Influence of temperature and salt concentration on the growth of a facultatively halophilic "Micrococcus" sp, Can. J. Microbiol. 21 (1975) 107–110.
[838] D. Keradjopoulos, A. Holldorf, Thermophilic character of enzymes from extreme halophilic bacteria, FEMS Microbiol. Lett. 1 (1977) 179–182.
[839] T.D. Brock, Microbial growth under extreme conditions, Symposium of the Society for General Microbiology, vol. 19, 1969, pp. 15–41.
[840] T.D. Brock, et al., Sulfolobus: a new genus of sulfur-oxidizing bacteria living at low pH and high temperature, Arch. Microbiol. 84 (1972) 54–68.
[841] M. Nourani, et al., Laboratory studies of MEOR in the micromodel as a fractured system, Eastern Regional Meeting, Society of Petroleum Engineers, Lexington, Kentucky, 2007.
[842] G. Jenneman, J. Clark, The effect of in-situ pore pressure on MEOR processes, SPE/DOE Enhanced Oil Recovery Symposium, Society of Petroleum Engineers, Tulsa, Oklahoma, 1992.
[843] L. Jang, et al., An investigation transport of bacteria through porous media, in: E.C. Donaldson, J. B. Clark (Eds.), International Conference on Microbial Enhancement of Oil Recovery, NTIS, Springfield, Virginia, 1982, pp. 60–70.
[844] G.E. Jenneman, et al., Transport phenomena and plugging in Berea sandstone using microorganisms, in: E.C. Donaldson, J.B. Clark (Eds.), International Conference on Microbial Enhancement Oil Recovery, NTIS, Springfield, Virginia, 1982, pp. 71–75.
[845] R. Hart, et al., The plugging effect of bacteria in sandstone systems, Can. Min. Metall. Bull. 53 (1960) 495–501.
[846] O. O'Bryan, T. Ling, The effect of the bacteria Vibrio desulfuricans on the permeability of limestone cores, Texas J. Sci. 1 (1949) 117–128.
[847] J.B. Davis, D.M. Updegraff, Microbiology in the petroleum industry, Bacteriol. Rev. 18 (1954) 215.
[848] D.O. Hitzman, Microbiological secondary recovery, in ConocoPhillips Co US, 1962.
[849] T. Jack, et al., Ch. F-6 Microbial selective plugging to control water channeling, in: E.C. Donaldson (Ed.), Developments in Petroleum Science, Elsevier, Amsterdam, Netherlands, 1991, pp. 433–440.
[850] J. Fredrickson, et al., Pore-size constraints on the activity and survival of subsurface bacteria in a late cretaceous shale-sandstone sequence, northwestern New Mexico, Geomicrobiol. J. 14 (1997) 183–202.
[851] W.E. Stiles, Use of permeability distribution in water flood calculations, J. Pet. Technol. 1 (1949) 9–13.
[852] L.R. Krumholz, Microbial communities in the deep subsurface, Hydrogeol. J. 8 (2000) 4–10.
[853] D. Zvyagintsev, Development of micro-organisms in fine capillaries and films, Mikrobiologiya 39 (1970) 161–165.
[854] D. Zvyagintsev, A. Pitryuk, Growth of microorganisms in capillaries of various sizes under continuous-flow and static conditions, Microbiology 42 (1973) 60–64.
[855] A. Nazarenko, et al., Development of methane-oxidizing bacteria in glass capillary tubes, Mikrobiologiia 43 (1974) 146–151.
[856] J. Pautz, R. Thomas, Applications of EOR technology in field projects-1990 update (NIPER-513), in IIT Research Institute, National Institute Petroleum Energy Research, Bartlesville, Oklahoma, 1991.
[857] A. Sarkar, et al., Ch. R-21 Compositional numerical simulation of MEOR processes, Dev. Pet. Sci. 31 (1991) 331–343.
[858] I.J. Kugelman, K.K. Chin, Toxicity, synergism, and antagonism in anaerobic waste treatment processes, in: F.G. Pohland (Ed.), Anaerobic Biological Treatment Processes, ACS Publications, Washington, DC, 1971, pp. 55–90.
[859] B. Bubela, Combined effects of temperature and other environmental stresses on microbiologically enhanced oil recovery, in: E.C. Donaldson, J.B. Clark (Eds.), International Conference on Microbial Enhancement of Oil Recovery, NTIS, Springfield, Virginia, 1982, pp. 118–123.

[860] D.O. Hitzman, Enhanced oil recovery using microorganisms, in ConocoPhillips Co US, 1984.
[861] S. Daniels, The adsorption of microorganisms onto solid surfaces: a review, Dev. Ind. Microbiol. Ser. 13 (1972) 211−253.
[862] C.M. Callbeck, et al., Acetate production from oil under sulfate-reducing conditions in bioreactors injected with sulfate and nitrate, Appl. Environ. Microbiol. 79 (2013) 5059−5068.
[863] I. Vance, D.R. Thrasher, Reservoir souring: mechanisms and prevention, in: B. Ollivier, M. Magot (Eds.), Petroleum Microbiology, ASM Press, Washington, D.C., 2005, pp. 123−142.
[864] C.I. Chen, et al., Kinetic investigation of microbial souring in porous media using microbial consortia from oil reservoirs, Biotechnol. Bioeng. 44 (1994) 263−269.
[865] S. Myhr, et al., Inhibition of microbial H_2S production in an oil reservoir model column by nitrate injection, Appl. Microbiol. Biotechnol. 58 (2002) 400−408.
[866] B. Ollivier, D. Alazard, The oil reservoir ecosystem, in: K.N. Timmis (Ed.), Handbook of Hydrocarbon and Lipid Microbiology, Springer, Berlin, Heidelberg, 2010, pp. 2259−2269.
[867] I.B. Beech, J. Sunner, Sulphate-reducing bacteria and their role in corrosion of ferrous materials, in: L.L. Barton, W.A. Hamilton (Eds.), Sulphate-Reducing Bacteria: Environmental and Engineered Systems, Cambridge University Press, Cambridge, England, 2007, pp. 459−482.
[868] H.-C. Flemming, Biofouling and microbiologically influenced corrosion (MIC)-an economical and technical overview, in: E. Heitz, W. Sand, H.-C. Flemming (Eds.), Microbial Deterioration of Materials, Springer, Berlin, Heidelberg, 1996, pp. 5−14.
[869] R. Cord-Ruwisch, Microbially influenced corrosion of steel, in: D. Lovley (Ed.), Environmental Microbe-Metal Interactions, ASM Press, Washington, D.C., 2000, pp. 159−173.
[870] W. Hamilton, Microbially influenced corrosion as a model system for the study of metal microbe interactions: a unifying electron transfer hypothesis, Biofouling 19 (2003) 65−76.
[871] F. Lopes, et al., Interaction of *Desulfovibrio desulfuricans* biofilms with stainless steel surface and its impact on bacterial metabolism, J. Appl. Microbiol. 101 (2006) 1087−1095.
[872] C.S. Cheung, I.B. Beech, The use of biocides to control sulphate-reducing bacteria in biofilms on mild steel surfaces, Biofouling 9 (1996) 231−249.
[873] E.D. Burger, Method for inhibiting microbially influenced corrosion, in Arkion Life Sciences US, 1998.
[874] B.V. Kjellerup, et al., Monitoring of microbial souring in chemically treated, produced-water biofilm systems using molecular techniques, J. Ind. Microbiol. Biotechnol. 32 (2005) 163−170.
[875] J. Wen, et al., A green biocide enhancer for the treatment of sulfate-reducing bacteria (SRB) biofilms on carbon steel surfaces using glutaraldehyde, Int. Biodeterior. Biodegradation. 63 (2009) 1102−1106.
[876] G.T. Sperl, et al., Use of natural microflora, electron acceptors and energy sources for enhanced oil recovery, in: E.T. Premuzic, A. Woodhead (Eds.), Developments in Petroleum Science, Elsevier, Amsterdam, Netherlands, 1993, pp. 17−25.
[877] H. Alkan, et al., An integrated German MEOR project, update: risk management and huff'n puff design, SPE Improved Oil Recovery Conference, Society of Petroleum Engineers, Tulsa, Oklahoma, 2016.
[878] C. Hubert, et al., Containment of biogenic sulfide production in continuous up-flow packed-bed bioreactors with nitrate or nitrite, Biotechnol. Prog. 19 (2003) 338−345.
[879] G. Voordouw, et al., Is souring and corrosion by sulfate-reducing bacteria in oil fields reduced more efficiently by nitrate or by nitrite? in: Corrosion 2004, NACE International, New Orleans, Louisiana, 2004.
[880] G.E. Jenneman, et al., Effect of nitrate on biogenic sulfide production, Appl. Environ. Microbiol. 51 (1986) 1205−1211.
[881] A.D. Montgomery, et al., Microbial control of the production of hydrogen sulfide by sulfate-reducing bacteria, Biotechnol. Bioeng. 35 (1990) 533−539.
[882] M. Reinsel, et al., Control of microbial souring by nitrate, nitrite or glutaraldehyde injection in a sandstone column, J. Ind. Microbiol. 17 (1996) 128−136.
[883] M.J. McInerney, et al., Microbial control of hydrogen sulfide production in a porous medium, Appl. Biochem. Biotechnol. 57 (1996) 933−944.

[884] M.J. Mclnerney, et al., Evaluation of a microbial method to reduce hydrogen sulfide levels in a porous rock biofilm, J. Ind. Microbiol. 11 (1992) 53–58.
[885] C. Hubert, G. Voordouw, Oil field souring control by nitrate-reducing Sulfurospirillum spp. that outcompete sulfate-reducing bacteria for organic electron donors, Appl. Environ. Microbiol. 73 (2007) 2644–2652.
[886] C.R. Hubert, Control of hydrogen sulfide production in oil fields by managing microbial communities through nitrate or nitrite addition, in University of Calgary, Calgary, Alberta, 2004.
[887] A. Gittel, Problems caused by microbes and treatment strategies monitoring and preventing reservoir souring using molecular microbiological methods (MMM), in: C. Whitby, T. Skovhus (Eds.), Applied Microbiology and Molecular Biology in Oilfield Systems, Springer, Dordrecht, Netherlands, 2010, pp. 103–107.
[888] R. Knapp, et al., Microbial field pilot study. Final report, in The University of Oklahoma, Norman, Oklahoma, 1993.
[889] R. Bryant, et al., Microbial enhanced waterflooding field tests, SPE/DOE Improved Oil Recovery Symposium, Society of Petroleum Engineers, Tulsa, Oklahoma, 1994.
[890] S. Davidson, H. Russell, A MEOR pilot test in the Loco field, in: T.E. Burchfield, R.S. Bryant (Eds.), Proceedings of the Symposium on the Application of Microorganisms to Petroleum Technology, National Technical Information Service, Bartlesville, Oklahoma, 1987, pp. VII 1–VII 12.
[891] I. Lazar, Microbial enhancement of oil recovery in Romania, in: E.C. Donaldson, J.B. Clark (Eds.), International Conference on Microbial Enhancement of Oil Recovery, NTIS, Springfield, Virginia, 1982, pp. 140–148.
[892] A.A. Matz, et al., Commercial (pilot) test of microbial enhanced oil recovery methods, SPE/DOE Enhanced Oil Recovery Symposium, Society of Petroleum Engineers, Tulsa, Oklahoma, 1992.
[893] M. Arinbasarov, et al., Chemical and biological monitoring of MIOR on the pilot area of Vyngapour oil field, West Sibera, Russia, in: R.S. Bryant (Ed.), The Fifth International Conference on Microbial Enhanced Oil Recovery and Related Biotechnology for Solving Environmental Problems, BDM Oklahoma, Inc., Dallas, Texas, 1995, pp. 365–374.
[894] A. Maure, et al., Waterflooding optimization using biotechnology: 2-year field test, La Ventana Field, Argentina, SPE Latin American and Caribbean Petroleum Engineering Conference, Society of Petroleum Engineers, Buenos Aires, Argentina, 2001.
[895] F.L. Dietrich, et al., Microbial EOR technology advancement: case studies of successful projects, SPE Annual Technical Conference and Exhibition, Society of Petroleum Engineers, Denver, Colorado, 1996.
[896] K. Nagase, et al., A successful field test of microbial EOR process in Fuyu Oilfield, China, SPE/DOE Improved Oil Recovery Symposium, Society of Petroleum Engineers, Tulsa, Oklahoma, 2002.
[897] Y. Zhang, et al., Microbial EOR laboratory studies and application results in Daqing oilfield, SPE Asia Pacific Oil and Gas Conference and Exhibition, Society of Petroleum Engineers, Jakarta, Indonesia, 1999.
[898] C.Y. Zhang, J.C. Zhang, A pilot test of EOR by in-situ microorganism fermentation in the Daqing oilfield, in: E.T. Premuzic, A. Woodhead (Eds.), Developments in Petroleum Science, Elsevier, Amsterdam, Netherlands, 1993, pp. 231–244.
[899] A. Yusuf, S. Kadarwati, Field test of the indigenous microbes for oil recovery, Ledok Field, Central Java, SPE Asia Pacific Improved Oil Recovery Conference, Society of Petroleum Engineers, Kuala Lumpur, Malaysia, 1999.
[900] H. Oppenheimer, K. Heibert, Microbial enhanced oil production field tests in Texas, in: T.E. Burchfield, R.S. Bryant (Eds.), Proceedings of the Symposium on the Application of Microorganisms to Petroleum Technology, National Technical Information Service, Bartlesville, Oklahoma, 1987, pp. XII 1–XII 15.
[901] V. Moses, et al., Microbial hydraulic acid fracturing, in: E.T. Premuzic, A. Woodhead (Eds.), Developments in Petroleum Science, Elsevier, Amsterdam, Netherlands, 1993, pp. 207–229.
[902] J. Buciak, et al., Enhanced oil recovery by means of microorganisms: pilot test, SPE Adv. Technol. Ser. 4 (1996) 144–149.

[903] J. Portwood, F. Hiebert, Mixed culture microbial enhanced waterflood: tertiary MEOR case study, SPE Annual Technical Conference and Exhibition, Society of Petroleum Engineers, Washington, D.C., 1992.
[904] M. Abd Karim, et al., Microbial enhanced oil recovery (MEOR) technology in Bokor Field, Sarawak, SPE Asia Pacific Improved Oil Recovery Conference, Society of Petroleum Engineers, Kuala Lumpur, Malaysia, 2001.
[905] M. Maure, et al., Microbial enhanced oil recovery pilot test in Piedras Coloradas field, Argentina, Latin American and Caribbean Petroleum Engineering Conference, Society of Petroleum Engineers, Caracas, Venezuela, 1999.
[906] R. Reksidler, et al., A microbial enhanced oil recovery field pilot in a Brazilian Onshore Oilfield, SPE Improved Oil Recovery Symposium, Society of Petroleum Engineers, Tulsa, Oklahoma, 2010.
[907] F. Cusack, et al., The use of ultramicrobacteria for selective plugging in oil recovery by waterflooding, International Meeting on Petroleum Engineering, Society of Petroleum Engineers, Beijing, China, 1992.
[908] T. Jack, MORE to MEOR: an overview of microbially enhanced oil recovery, in: E.T. Premuzic, A. Woodhead (Eds.), Developments in Petroleum Science, Elsevier, Amsterdam, Netherlands, 1993, pp. 7−16.
[909] T.R. Jack, Microbial enhancement of oil recovery, Curr. Opin. Biotechnol. 2 (1991) 444−449.
[910] L.G. Stehmeier, T.R. Jack, B.A. Blakely, J.M. Campbell Test of a microbial plugging system at standard hill, Saskatchewan to inhibit oilfield water encroachment, in: Proceedings of the International Symposium on Biobydrometaliurgy, Canmet, EMR, Jackson Hole, Wyoming, 1990.
[911] K. Town, et al., MEOR success in southern Saskatchewan, SPE Reservoir Eval. Eng. 13 (2010) 773−781.
[912] P. Gao, et al., Microbial diversity and abundance in the Xinjiang Luliang long-term water-flooding petroleum reservoir, Microbiologyopen 4 (2015) 332−342.
[913] F. Zhang, et al., Microbial diversity in long-term water-flooded oil reservoirs with different in situ temperatures in China, Sci. Rep. 2 (2012) 1−10.
[914] Z. Hou, et al., The application of hydrocarbon-degrading bacteria in Daqing's low permeability, high paraffin content oilfields, SPE Symposium on Improved Oil Recovery, Society of Petroleum Engineers, Tulsa, Oklahoma, 2008.
[915] L. Jinfeng, et al., The field pilot of microbial enhanced oil recovery in a high temperature petroleum reservoir, J. Pet. Sci. Eng. 48 (2005) 265−271.
[916] H. Zhao, et al., Field pilots of microbial flooding in high-temperature and high-salt reservoirs, SPE Annual Technical Conference and Exhibition, Society of Petroleum Engineers, Dallas, Texas, 2005.
[917] T. Nazina, et al., Microbiological investigations of high-temperature horizons of the Kongdian petroleum reservoir in connection with field trial of a biotechnology for enhancement of oil recovery, Microbiology 76 (2007) 287−296.
[918] T. Nazina, et al., Microbiological and production characteristics of the high-temperature Kongdian petroleum reservoir revealed during field trial of biotechnology for the enhancement of oil recovery, Microbiology 76 (2007) 297−309.
[919] J. Le, et al., A field test of activation indigenous microorganism for microbial enhanced oil recovery in reservoir after polymer flood, Acta Pet. Sin. 35 (2014) 99−106.
[920] W. Guo, et al., Study on Pilot test of microbial profile modification after polymer flooding in Daqing Oilfield, Acta Pet. Sin. 17 (2006) 86−90.
[921] S. Sun, Field practice and analysis of MEOR in Shengli oilfield, J. Oil Gas Technol. 36 (2014) 149−152.
[922] Q.-x Feng, et al., Pilot test of indigenous microorganism flooding in Kongdian Oilfield, Pet. Explor. Dev. 32 (2005) 125.
[923] Y.-F. Long, et al., The research on bio-chemical combination drive and its application in Y9 reservoir in block ZJ2 of Jing'an oilfield, J. Oil .Gas Technol. 35 (2013) 142−145.
[924] W.-k Guo, et al., The recovery mechanism and application of *Brevibacillus brevis* and *Bacillus cereus* in extra-low permeability reservoir of Daqing, Pet. Explor. Dev. 34 (2007) 73.

[925] Z.Y. Yang, et al., Study on authigenous microorganism community distribution and oil recovery mechanism in daqing oilfield, Acta Pet. Sin. 27 (2006) 95–100.
[926] X.-L. Wang, et al., Microbial flooding in Guan 69 Block, Dagang Oilfield, Pet. Explor. Dev. 32 (2005) 107–109.
[927] H. Guo, et al., Progress of microbial enhanced oil recovery in China, SPE Asia Pacific Enhanced Oil Recovery Conference, Society of Petroleum Engineers, Kuala Lumpur, Malaysia, 2015.
[928] D.M. Updegraff, Early research on microbial enhanced oil recovery, Dev. Ind. Microbiol. 31 (1990) 135–142.
[929] e.a. M. Wagner, Microbially improved oil recovery from carbonate, Biohydrometall. Technol. 2 (1993) 695–710.
[930] B. Sabut, et al., Further evaluation of microbial treatment technology for improved oil production in Bokor Field, Sarawak, SPE International Improved Oil Recovery Conference in Asia Pacific, Society of Petroleum Engineers, Kuala Lumpur, Malaysia, 2003.
[931] Z. Ibrahim, et al., Simulation analysis of microbial well treatment of Bokor field, Malaysia, SPE Asia Pacific Oil and Gas Conference and Exhibition, Society of Petroleum Engineers, Perth, Australia, 2004.
[932] S. Bailey, et al., Microbial enhanced oil recovery: diverse successful applications of biotechnology in the oil field, SPE Asia Pacific Improved Oil Recovery Conference, Society of Petroleum Engineers, Kuala Lumpur, Malaysia, 2001.
[933] S. Rassenfoss, From bacteria to barrels: microbiology having an impact on oil fields, J. Pet. Technol. 63 (2011) 32–38.
[934] I. Lazar, et al., Characteristics of bacterial inoculum applied on oil reservoir in field experiments, in: Proceedings of the 6th Industrial Microbiology and Biotechnology, Iasi, Romania, 1987, pp. 801–806.
[935] E.C. Donaldson, et al., Enhanced Oil Recovery, I: Fundamentals and Analyses, Elsevier, Amsterdam, Netherlands, 1985.
[936] I. Lazar, et al., Procedeu de injectare a sondelor in vederea cresterii recuperarii titeiului din zacaminte, Brevet de Inventie 98 (1989) 528–530.
[937] I. Lazar, et al., Characteristics to the bacterial inoculum used in some recent MEOR field trials in Romania, in: Proceedings of the MEOR International Conference, Norman, Oklahoma, 1990.
[938] E.M. Yulbarisov, Microbiological method for EOR, in: Proceedings of the Filth European Symposium on Improved Oil Recovery, Budapest. Hungary, 1989, pp. 695–704.
[939] V. Murygina, et al., Oil field experiments of microbial improved oil recovery in Vyngapour, West Siberia, Russia, in: R.S. Bryant (Ed.), The Fifth International Conference on Microbial Enhanced Oil Recovery and Related Biotechnology for Solving Environmental Problems, BDM Oklahoma, Inc., Dallas, Texas, 1995, pp. 87–94.
[940] E. Yulbarisov, Results from analysis of petroleum gas on introduction of biochemical processes in the oil formation, Gazovc Delo 4 (1972) 26–29.
[941] L. Sim, et al., Production and characterisation of a biosurfactant isolated from *Pseudomonas aeruginosa* UW-1, J. Ind. Microbiol. Biotechnol. 19 (1997) 232–238.
[942] J. Von Heningen, et al., Process for the recovery of petroleum from rocks, in: Netherlands Patent, Elsevier, Netherlands 1958.
[943] U. Maharaj, et al., The application of microbial enhanced oil recovery to Trinidadian Oil Wells, in: E.T. Premuzic, A. Woodhead (Eds.), Developments in Petroleum Science, Elsevier, Amsterdam, Netherlands, 1993, pp. 245–263.
[944] E. Zajic, Scale up of microbes for single well injection, in: J. King, D. Stevens (Eds.), Proceedings of the First International MEOR Workshop, U.S. Department of Energy, Bartlesville, Oklahoma, 1987, pp. 241–246.
[945] A. Johnson, Microbial oil release technique for enhanced oil recovery, in: Proceedings of the Conference on Microbiological Processes Useful in Enhanced Oil Recovery, San Diego, California, 1979, pp. 30–34.

[946] R. Harris, I. McKay, New applications for enzymes in oil and gas production, European Petroleum Conference, Society of Petroleum Engineers, The Hague, Netherlands, 1998, pp. 20–22.

[947] R.A. Copeland, Enzymes: a Practical Introduction to Structure, Mechanism, and Data Analysis, second ed., Wiley-VCH, New York, NY, 2000.

[948] J.M. Reiner, Behavior of enzyme systems, Q. Rev. Biol. 45 (1969) 66–67.

[949] G.F. Bickerstaff, Enzymes in Industry and Medicine, Edward Arnold, London, England, 1987.

[950] S. Sun, et al., Exopolysaccharide production by a genetically engineered *Enterobacter cloacae* strain for microbial enhanced oil recovery, Bioresour. Technol. 102 (2011) 6153–6158.

[951] Q.-x Feng, et al., EOR pilot tests with modified enzyme in China, EUROPEC/EAGE Conference and Exhibition, Society of Petroleum Engineers, London, U.K., 2007.

[952] H. Nasiri, et al., Use of enzymes to improve waterflood performance, in: International Symposium of the Society of Core Analysis, Noordwijk, Netherlands, 2009, pp. 27–30.

[953] L. Daoshan, et al., Research on pilot test of biologic-enzyme enhanced oil recovery in Gangxi oil reservoir of Dagang Oilfield, Pet. Geol. Recovery Effic. 4 (2009) 023.

[954] N. Kohler, et al., Injectivity improvement of xanthan gums by enzymes: process design and performance evaluation, J. Pet. Technol. 39 (1987) 835–843.

[955] M. Nemati, G. Voordouw, Modification of porous media permeability, using calcium carbonate produced enzymatically in situ, Enzyme Microb. Technol. 33 (2003) 635–642.

[956] J. Larsen, et al., Plugging of fractures in chalk reservoirs by enzyme-induced calcium carbonate precipitation, SPE Prod. Oper. 23 (2008) 478–483.

[957] A. Khusainova, Enhanced oil recovery with application of enzymes, in: Department of Chemical and Biochemical Engineering, Technical University of Denmark, Kongens Lyngby, Denmark, 2016.

[958] H.F. Phillips, I.A. Breger, Isolation and identification of an ester from a crude oil, Geochim. Cosmochim. Acta 15 (1958) 51–56.

[959] V. Kam'yanov, et al., Asphaltenes of Dzhafarly crude oil, Pet. Chem. USSR 30 (1990) 1–8.

[960] T.E. Suhy, R.P. Harris, Application of polymer specific enzymes to clean up drill-in fluids, SPE Eastern Regional Meeting, Society of Petroleum Engineers, Pittsburgh, Pennsylvania, 1998.

[961] J.E. Hanssen, et al., New enzyme process for downhole cleanup of reservoir drilling fluid filtercake, SPE International Symposium on Oilfield Chemistry, Society of Petroleum Engineers, Houston, Texas, 1999, pp. 79–91.

[962] E. Battistel, et al., Enzyme breakers for chemically modified starches, SPE European Formation Damage Conference, Society of Petroleum Engineers, Sheveningen, Netherlands, 2005.

[963] M.A.A. Siddiqui, H.A. Nasr-El-Din, Evaluation of special enzymes as a means to remove formation damage induced by drill-in fluids in horizontal gas wells in tight reservoirs, SPE Prod. Facil. 20 (2005) 177–184.

[964] M. Samuel, et al., A novel alpha-amylase enzyme stabilizer for applications at high temperatures, SPE Prod. Oper. 25 (2010) 398–408.

[965] W.K. Ott, et al., EEOR success in Mann field, Myanmar, SPE Enhanced Oil Recovery Conference, Society of Petroleum Engineers, Kuala Lumpur, Malaysia, 2011.

[966] Y. Dong, et al., Engineering of LadA for enhanced hexadecane oxidation using random- and site-directed mutagenesis, Appl. Microbiol. Biotechnol. 94 (2012) 1019–1029.

[967] J. Zhao, et al., Genome shuffling of *Bacillus amyloliquefaciens* for improving antimicrobial lipopeptide production and an analysis of relative gene expression using FQ RT-PCR, J. Ind. Microbiol. Biotechnol. 39 (2012) 889–896.

[968] R.W. Old, S.B. Primrose, Principles of Gene Manipulation: An Introduction to Genetic Engineering, University of California Press, 1981.

INDEX

Note: Page numbers followed by "*f*" and "*t*" refer to figures and tables, respectively.

D

E

F

G

H

K

L

M

N

O

P

R

S

T

U

V

Printed in the United States
By Bookmasters